Composites Science and Technology

Series Editor

Mohammad Jawaid, Laboratory of Biocomposite Technology, Universiti Putra Malaysia, INTROP, Serdang, Malaysia

This book series publishes cutting edge research monographs comprehensively covering topics in the field of composite science and technology. The books in this series are edited or authored by top researchers and professional across the globe. The series aims at publishing state-of-the-art research and development in areas including, but not limited to:

- Conventional Composites from natural and synthetic fibers
- Advanced Composites from natural and synthetic fibers
- Chemistry and biology of Composites and Biocomposites
- Fatigue damage modelling of Composites and Biocomposites
- Failure Analysis of Composites and Biocomposites
- Structural Health Monitoring of Composites and Biocomposites
- Durability of Composites and Biocomposites
- Biodegradability of Composites and Biocomposites
- Thermal properties of Composites and Biocomposites
- Flammability of Composites and Biocomposites
- Tribology of Composites and Biocomposites
- Applications of Composites and Biocomposites

Review Process

The proposal for each volume is reviewed by the main editor and/or the advisory board. The chapters in each volume are individually reviewed single blind by expert reviewers (at least two reviews per chapter) and the main editor.

Ethics Statement for this series can be found in the Springer standard guidelines here - https://www.springer.com/us/authors-editors/journal-author/journal-author-helpdesk/before-you-start/before-you-start/1330#c14214

Imran Uddin · Irfan Ahmad
Editors

Synthesis and Applications of Nanomaterials and Nanocomposites

Springer

Editors
Imran Uddin
University of Pannonia
Veszprém, Hungary

Irfan Ahmad
School of Engineering Sciences
and Technnology
Jamia Hamdard, New Delhi, India

ISSN 2662-1819 ISSN 2662-1827 (electronic)
Composites Science and Technology
ISBN 978-981-99-1349-7 ISBN 978-981-99-1350-3 (eBook)
https://doi.org/10.1007/978-981-99-1350-3

This Springer imprint is published by the registered company Springer Nature Singapore Pte Ltd.
The registered company address is: 152 Beach Road, #21-01/04 Gateway East, Singapore 189721, Singapore

Foreword

We live in a technological world that is ever more dependent on the special properties of engineered nanomaterials. All our modern electronic devices, including computers and smartphones, function by using semiconductor components; our energy-efficient lighting is based on light-emitting diodes; sensors and built-in controls in modern vehicles make travel safer and more efficient; energy from sunlight and wind is converted into electricity by solar panels and wind turbines, and stored in lithium-ion batteries; in all of these examples, the physical and chemical processes at the heart of the functioning of our devices are made possible by materials carefully engineered on the nanometer scale. While electronics and energy production are perhaps the most well-known areas where the roles of nanomaterials are obvious, their use is not restricted to non-biological systems. Whoever takes a pill or uses cosmetic products will probably be exposed to nanoparticle formulations. Nanoparticles play increasingly important roles in medicine, both in diagnostic and therapeutic applications. Magnetic and other particles are used as contrast agents in medical imaging and as drug carriers or agents of hyperthermic treatments. A striking example of the importance of nanoparticle engineering in medicine was provided by the ingenious and rapid development of vaccines for COVID-19: both the creation of the immunization agent (either RNA-, DNA-, or protein-based) and its packaging (into either lipid micelles or adenovirus vectors) are nanotechnological feats that benefit our society all over the globe. Modern food technology also increasingly relies on nanotechnology, producing both nanoparticulate food components and new packaging materials. Many more examples could be cited, the point being that nanotechnology is a buzzword of the twenty-first century.

Several factors, some rooted in the physical nature of nanomaterials, some in the scientific progress made over the last century, and others in the historical changes in the needs of society, contributed to the increasing use of nanomaterials. The properties of materials as known on a macroscopic scale can change drastically as the realm of the nanoworld is reached. Although definitions vary, particles are typically considered "nano" if they are smaller than 100 nm. Within this size range, the increasing surface-to-volume ratio becomes rather important. For crystalline substances, the surface of a particle can be considered a defective region, with different properties than those of its periodically ordered interior. Such highly reactive surfaces can be used for a number of applications, including catalysis, adsorption of specific molecules, or functionalization for biomedical purposes, to name a few. At the core of the operation of our electronic devices are layered nanostructures (such as solid-state transistors), in which the diffusion of electrons and holes is confined to nanometer-scale lengths. Materials that consist of a single atomic layer, such as graphene, can have electronic or even mechanical properties not seen in their macroscopic equivalents. Some nanomaterials, for example, viruses and antibodies used in biomedicine or clay minerals used as fillers or adsorbents, have nanometric sizes by nature. In fact, the use of clay in pottery is as old as our civilization; thus, nanotechnology was invented long before the twentieth century.

Nevertheless, while our ancestors may have used materials without knowing their nanoscale character, scientific progress resulted in the development of techniques for both material synthesis and characterization that allow us to intentionally tailor and measure the properties of substances in the nanometer range. The current book introduces the reader to a broad range of synthesis methods, from high-temperature layer deposition to various solution-based techniques, as well as intriguing novelties like synthesis in nanoscale confinements. Many of these approaches that were once cutting-edge scientific novelties found their way into industry and now form the basis of commercial material production, which is also aided by artificial intelligence, another new tool developed on the platform of nanotechnology. In addition to synthesis, spectacular progress has been made over the last hundred years in the field of material characterization. Giant strides in science, starting with the dawn of crystallography, followed by the invention of electron microscopy, coupled with various new analytical techniques using electromagnetic radiation, nuclear energy, and other signals, and then the revolution of molecular biology in the second part of the twentieth century, laid the foundation for the nanoscience of today. Many of the current standard techniques are on display in the present book and provide the reader with detailed and, in some cases, visually appealing information about nanomaterial characteristics.

While science and technology provided the means for the design of new nanomaterials, population growth and society's push for a more comfortable life drove the transition from pure science results to innovation and then to the production of nanomaterials. Within a generation, digital communication technology transformed the work, entertainment, travel, and study habits of people. More complex functions must be packed into handheld and desktop devices, forcing technology to invent electronic components with ever smaller dimensions. The global population is about to reach 8 billion people; feeding so many people and providing them with a desired high standard of living, while our Earth and its resources are limited, require the design of highly efficient systems for both agriculture and energy production. Such a viewpoint anticipates the growing importance of nanotechnological solutions. Ample examples are given in this book of research that strives to improve energy production (by developing metal halides and organic hybrids for photovoltaic applications and designing new ways of water splitting) and storage (by developing high-performance batteries from ionic liquids and storing hydrogen using graphene oxide). In electronics, the direction of progress is toward making electronic components smaller or more efficient (examples in this book include the development of components using graphene, for example) or suitable for new applications (by synthesizing metal-oxide nanorods for new types of electrodes, silver, and copper nanowires for transparent conductors; quantum dots for heavy metal sensing; magnetic ferrites; and ceria-based nanocomposites for catalysis, to name a few examples). Aims in the fields of environmental and medical sciences, with examples in the present book, include wastewater treatment using graphene oxide, the use of new biocompatible materials, such as graphene, inorganic nanocomposites, and coinage-metal nanoparticles for diagnostics and therapeutics, as well as reducing the material need and waste of food production (by designing biodegradable food packaging material).

The majority of nanomaterials used in technology or medicine today are composites, which are hybrids of two or more substances that benefit from the different properties of their constituents, as illustrated by the diverse range of such materials presented in this book. Nanocomposite materials are not exclusively human inventions, since nature developed them long before us. Organic layers and calcium carbonate platelets alternate in the shells of mollusks to provide both mechanical strength and flexibility; apatite nanocrystals interbedded with collagen fibers provide support and elasticity to our bones; and organic membrane-wrapped magnetic nanocrystals assist some bacteria in navigating concentration gradients. In many cases, technological inventions were based on lessons learned from nature. Current nanotechnology uses several synthesis techniques that mimic biological systems, as is also shown in some chapters of this book.

On the other hand, nature is also affected by human activities, and, as with any new technology, the benefits of nanomaterials come with downsides. Some industrial processes have been polluting the air, soil, rivers, and oceans with toxic nanoparticles since the start of the industrial revolution, and the scale of this pollution is just beginning to be appreciated. Currently, nanoparticles are produced in formerly unprecedented amounts, and their effects on human health and the environment in many cases are far from being understood. Because of their sizes, nanoparticles are easily inhaled, as is the case with natural biological nanoparticles, e.g., viruses that are transmitted through the air, and can reach the deepest parts of the lungs. There have been reports of the nanoparticle toxicity of materials that in larger sizes are considered harmless, and environmental magnetic nanoparticles have been found even in the human brain. Only science is able to assess the harmful effects of nanomaterials and provide solutions.

In contrast to posing a health threat, nanoparticles can be used for environmental remediation. A huge field of nanoparticle technology has emerged that is concerned with cleaning soils, waste, and underground water. Examples for such applications are also provided in this book. Moreover, nanotechnology will be crucial in the quest to preserve the habitability of our planet. As I am writing these words, several parts of the world are experiencing one of the most severe droughts in their history, while in other places, floods wreak havoc on communities, accentuating the effects of man-caused climate change. Our civilization's mere existence depends on whether we can improve the efficiency of clean energy production and distribution, preserve our water supplies, and be able to feed 8 billion people without expanding beyond the limits of Earth's resources. If solutions exist at all, they should include, along with changes in economic drivers and societal attitudes, rapid progress in technology, undoubtedly including new nanomaterials and advances in nanotechnology. Therefore, by providing readers with an overview of the state of the art in nanomaterials research, this book is both timely and important. I wish you good reading.

Prof. Mihaly Posfai
Nanolab, Research Institute for
Biomolecular and Chemical
Engineering
University of Pannonia
Veszprem, Hungary

Preface

We are immensely pleased to present our book, *Synthesis and Application of Nanotechnology and Nanomaterials*. As researchers and instructors of nanotechnology, we often struggled with one reference book in the field that can summarize crucial topics of the field with reasonable details to the early researchers in the field. A book that can elucidate topics in the field with sufficient details, without being too elementary in covering a lot of topics or being too specialized in a core topic. Furthermore, that covers essential applications and frontier research topics. The inception was a result of discussions during numerous tea breaks at the AMU campus, a place renowned for its tea enthusiasts.

This book brings together multiple essential topics of nanotechnology and its applications. Diverse aspects of the emerging research fields are strung together to make it a useful reference book, especially for early career research scholars in the field of nanotechnology and nanocomposites. A sufficient number of topics are included on the synthesis of nanoparticles, nanorods, graphene, graphene oxide-metal composites, characterization of these materials, and various aspects of emerging and existing applications, including biomedical, food packaging, heavy metal sensing, etc. In order to balance the theoretical details, chapters on DFT and Monte-Carlo analytical calculations have been included.

The book is divided into three parts. The first part contains five chapters and it describes the synthesis of various nanomaterials, analytical calculation of nanomaterials, exfoliation synthesis of graphene, graphene oxide, and its metal nanocomposites. The book commences with a report on recent advances in the synthesis of halide perovskite nanoparticles from a team of researchers from IIT-Bombay, India. The synthesis results of halide perovskite nanoparticles are complemented by analytical findings. The next chapter is a semi-review on graphene and its nanocomposites with an emphasis on the recently reported liquid exfoliation-based synthesis of graphene by our group. The subsequent chapter describes synthesis of graphene oxide and its nanocomposite with metals. The next chapter reports the synthesis of self-stabilized metal-hydroxide and metal-oxide nanorods by utilizing the electrochemical method. Optimization of the aspect ratio of nanorods with temperature and the applied voltage is discussed in the chapter. The last chapter in this part is contributed by a team of researchers from IIT-Delhi, India, and it adds a molecular dynamics perspective on nanoscale confinement.

The second part of the book is on biomedical/bioengineering applications with three chapters. The first chapter in this part describes biomedical applications of nanomaterials prepared from coinage metal. The subsequent chapter reports the role of nanocomposites in the drug delivery systems. The last chapter in this part describes the synthesis of bio-nanocomposite materials for food packaging applications.

The last part of the book contains five chapters on various applications of nanocomposites and nanomaterials. This part begins with an extensive report by a team of veteran researchers from the University of Rochester, USA, and CSU, Changsha, China, on the synthesis of metal nanowires and their applications in transparent conductors. The next chapter describes the synthesis of quantum dots and their photoluminescence properties. The application of quantum dots in sensing heavy metals is also demonstrated in this chapter. The next chapter reports on synthesis of TiO_2-based nanomaterials for water-splitting application compiled by a team of researchers from Egypt, Jordan, and Germany. The subsequent chapter is on the synthesis of metal-oxide nanoparticles and their photocatalytic activities. The next chapter is reporting advanced nanocomposite electrodes for high-performance supercapacitors. The last chapter is on ceria-based nanocomposites for versatile catalytic application.

The currently proposed content of the book makes it a comprehensive reference book for the researchers in the field, in particular, for the senior graduate, M.Tech., and MS students of nanotechnology. The book would also serve as a crucial reference book to the researchers working in the other areas of materials science, who are looking to contribute to the field of nanomaterials and nanocomposites.

We acknowledge the support of all the contributing authors, in particular, Prof. Yongli Gao, Prof. Aftab Alam, Prof. M. Aslam, Prof. Hemant Kashyap, and Prof. Biswajit Chowdhury. We would also like to appreciate researchers, who initially agreed to contribute, but due to their health conditions during the COVID-19 pandemic could not submit the chapters. We would like to thank all our students who, with their numerous thought-provoking questions, have been a constant motivation to contribute to the field. Last but not least, we would like to extend our sincere appreciation to the family members, who had to shoulder extra responsibilities during the preparation of this book.

Imran Uddin
University of Pannonia
Veszprem, Hungary

Irfan Ahmad
SBIR Coordinator
Alien Innovation Consultant
Hyderabad/Delhi-NCR, India

Contents

Applications: Miscellaneous

About the Editors

Imran Uddin completed his PhD in Biotechnology from CSIR-National Chemical Laboratory (NCL), Pune, India. He has served as a postdoctoral researcher at Rhodes University, South Africa; Korea University, Seoul, Republic of Korea; and Teri-Deakin Nanobiotechnology Centre, Delhi, India. In 2017, he joined as an assistant professor at Aligarh Muslim University (AMU). In 2022, he joined as a senior research scientist at Nanolab, Environmental Mineralogy Research Group, Research Institute of Biomolecular and Chemical Engineering, University of Pannonia, Hungary. His research interest lies in the fabrication of biocompatible nanostructures for nanobiosensors and flexible electronics. Moreover, his research work includes studying the interface of biology with inorganic nanomaterials, understanding the underlying biological process, developing new industrially relevant renewable nanomaterials as well as eco-toxicological aspects. He is the recipient of a number of awards and fellowships, including the Startup Research Grant (Young Scientists) Scheme funded by Life Sciences, Science & Engineering Research Board (SERB), Department of Science & Technology, Government of India; Senior Research Associateship (Scientists' Pool Scheme) and Senior Research Fellowship funded by CSIR (Council of Scientific and Industrial Research), Government of India; and the National Eligibility Test (NET) conducted by Agriculture Scientist Requirement Board (ASRB) for Assistant Professor/Lectureship in Indian Universities.

Irfan Ahmad is currently a coordinator at Alien Innovation Consulting (AlienTT group), managing a team of technical writers/PhDs and helping US/European technology startups interested in securing SBIR/EIC funds. He has extensive experience as a technical writer and a researcher. His expertise domains include materials science, renewable energy, and nanomaterials. He received his PhD in physics from the University of Rochester, USA. He is the leading author on multiple peer-reviewed articles that have been cited by a Nobel Laureate and top research institutions including MIT, Stanford and Princeton. He is a recipient of the Irish Ovshinsky Award from American Physical Society (APS) in 2011 for one of his PhD research works. He has worked as a scientist at Alameda Applied Science Corp. (AASC), a tech-startup in SF Bay Area, USA, for four years. At AASC, he has served as a key scientist/investigator on three awarded phase II DOE SBIR and several awarded phase I research grants, totaling >$4M research funds. He has worked as an assistant professor of physics for two years each at AMU, Aligarh and Jamia Hamdard, New Delhi. He enjoys learning about innovative technology ideas that would result in futuristic products, reshaping the frontiers of science and technology.

Synthesis

Advances in Synthesis and Defect Properties of Halide Perovskite Nanocrystals: Experimental and Theoretical Perspectives

Bhawna, Supriti Ghorui, Aftab Alam, and M. Aslam

Abstract Metal halide perovskites are gaining attention for their interesting optoelectronic properties which result in their promising use for commercial applications. The ease in the fabrication and processability of perovskite nanocrystals with high photoluminescence quantum yield has tremendously attracted the research community and since then various approaches for their synthesis have been developed. This chapter focuses on various size and shape-controlled solution-based synthesis methods for perovskite nanocrystals. Furthermore, synthesis of defect-free perovskite nanocrystals to attain long-term stability along with high efficiency is the primary focus of the growing research community. Therefore, deeper insight into defect properties in perovskite nanocrystals is crucial despite of their defect-tolerant nature. Thus, the purpose of this chapter is twofold (i) to give a complete understanding of various facile synthesis strategies for metal halide perovskite nanocrystals while discussing their advantages and limitations and (ii) to facilitate an in-depth insight about the formation of native defects both from the experimental and theoretical perspectives.

Keywords Hybrid perovskites · Photovoltaic · Synthesis Strategies · Defects · DFT simulation

1 Introduction

Metal halide perovskites (MHPs) have been known since a long time and the first work was reported by H. L. Wells in 1893 [1]. However, it was only the 1990s when they got traction from the scientific community due to their promising electronic and optical properties [2, 3]. The term perovskite was coined after the famous Russian

Bhawna · S. Ghorui · A. Alam · M. Aslam (✉)
Department of Physics, Indian Institute of Technology Bombay, Mumbai, India
e-mail: m.aslam@iitb.ac.in

A. Alam
e-mail: aftab@iitb.ac.in

I. Uddin and I. Ahmad (eds.), *Synthesis and Applications of Nanomaterials and Nanocomposites*, Composites Science and Technology,
https://doi.org/10.1007/978-981-99-1350-3_1

mineralogist Lev Perovski and belongs to the class of materials which have a general formula ABX_3. Here, in halide perovskites, A can be an inorganic cation like cesium (Cs) or organic cation like methylammonium (MA), formamidinium (FA) or their mixture, B typically refers to lead (Pb) or tin (Sn) and X is a halide anion namely chloride (Cl^-), bromide (Br^-) or iodide (I^-). Based on the kind of A cation, the halide-based perovskites can be classified in hybrid (A belongs to organic group) and all-inorganic (A is an inorganic cation) perovskites. Initially, lead halide perovskites (LHPs) were used as visible-light sensitizers in dye-sensitized solar cells in 2009 [4]. It took another 3 years to realize the impressive charge transport properties of these MHPs, in addition to the strong bandgap absorption, to achieve modest power-conversion efficiency (PCE) of 3–7% [5, 6]. Since then, a surge in research output in LHPs has been seen which resulted in a certified PCE of over 25% for single-junction perovskite-based solar cells in a very short duration [7]. The reason behind such a high PCE is high absorption coefficient (>1×10^4 cm^{-1}), low Urbach energies, long charge carrier diffusion lengths (>100 nm), high carrier mobility and long carrier lifetime [8, 9].

Surprisingly, LHPs not only provide a strong ground for the photovoltaic application but also pose as attractive candidates for the light emitting diodes (LEDs) due to strong direct band-to-band transitions, relatively low non-radiative recombination rates and high color purity [10–13]. However, the low photoluminescence quantum yield (PLQY) associated with the bulk perovskites limit their use in LEDs. This is mainly because of the small exciton binding energies (resulting in low radiative recombination) and the existence of mobile ionic defects (characterised by low formation energies) in the bulk perovskite structures. Halide perovskite films are easily grown from their precursor solution via facile solution processable method at low temperatures, but, the defects in the bulk as well as at the grain boundaries or interfaces are created and act as centres for fast non-radiative decays [14, 15]. This motivated the researchers to move to perovskite nanocrystals (NCs) in search for better PLQYs. Interestingly, optical and electronic properties of MHPs are size-dependent and can be tuned by changing the size of the crystals to the nano-regime. Further, one can play with the dimensionality (from 3D to *quasi*-2D to 2D, also known as layered perovskites) of MHPs by using long-chain alkylammonium cations as separators of the octahedron slabs. The number of the octahedral layers ($m = 1$ to $\propto$) provides flexibility in tuning emission wavelength and the exciton binding energies in layered perovskites [16–18]. The MHP NCs facilitate efficient light emission with near-unity PLQY as opposed to the low PLQYs seen in bulk perovskites making them a potential candidate for LEDs [19].

The research on MHP NCs took off after the first report on colloidal halide perovskite $MAPbBr_3$ with a PLQY of ~20% by Schmidt et al. in 2014 [20]. The selection and use of appropriate capping ligands in the preparation routine enable not only the control on the size of these crystals in the nanometer range but also help in passivating the surface defects. Gonzalez-Carrero et al. further improved the synthesis of highly luminescent $MAPbBr_3$ with an impressive PLQY of ~80%, however, the particles obtained were highly irregular in shape [21]. In early 2015, the hot-injection (HI) method was introduced by Protesescu and group in order to gain narrow particle size distribution of colloidal LHP NCs. This method resulted in the synthesis of monodisperse $CsPbX_3$ (X = halide or their mixed halides) NCs. Later, the room-temperature synthesis of colloidal $MAPbX_3$ NCs were proposed by Zhang et al. using ligand-assisted reprecipitation (LARP) method which resulted in the efficient emission (PLQY ~70%) with the added benefit of color-tunability [22]. Since then, several attempts have been made to regulate the shape and size of MHP NCs by changing reaction temperature, capping ligands etc. This resulted in attaining different morphologies ranging from nanowires (NWs) to nanosheets (NSs) to nanoplatelets (NPLs) to quantum dots (QDs) [23–27]. Extensive research on improving the synthesis methods of LHP NCs have been done in order to produce highly monodisperse with near-unity emission and high colloidal stability [28, 29]. Further, the composition and size of these NCs can be tuned using post-synthetic transformations such as through exfoliation or ion migration [30, 31]. However, due to the presence of surface defects/trap states and the loss of ligand binding to the NCs during the purification and isolation procedures, the PLQYs of MHPs get affected strongly [17, 32]. In this regard, the post-synthetic treatment with functional molecules/metal halides helps in improving their PLQYs [33, 34]. On one hand, ease in synthesis of MHPs has led the researchers to utilize these active materials for applications such as lasers, solar cells, LEDs, photodetectors etc., whereas, on the other hand, they possess substantial health hazards due to the presence of lead and also suffer from instability issues which includes degradation at interfaces [35, 36] and inter-diffusion of ions from different materials [37]. To address these concerns and to preserve the crystal structure of perovskite, replacement of Pb^{2+} with heterovalent substitution of a monovalent, B (Ag^+, Au^+, Cu^+, Na^+) and a trivalent cation, B' (Bi^{3+}, Au^{3+}, In^{3+}, Sb^{3+}) to form a quaternary double perovskite (elapsolite), $A_2B(I)B'(III)X_6$ (where A = Cs, and X = Cl, Br, I) have been proposed recently [38–46]. However, these newly developed structures are facing various other problems such as indirect bandgap and low PLQYs.

In MHPs, the presence of different kinds of chemical interactions results in their characteristic "soft" structure, which exposes the structure to a large number of defects. Synthesis of defect-free MHP NCs with a single radiative (excitonic) recombination pathway to achieve a near-unity PLQY is the main focus of the growing research community. Various studies show the creation of favourable halogen vacancy due to their low formation energies as opposed to Cs/Pb vacancies, antisites or interstitial sites in these NCs [47]. Moreover, the large surface area in addition to the increased number of the dangling bonds in case of NCs result in the increment in the density of trap states than in single-crystals or polycrystalline thin films [48]. In this light, it becomes imperative to understand the role of defects in the photophysics of MHPs [49, 50].

2 Introducing MHP NCs: Crystal Structure and Defect-Tolerance

Since the first observation of optical-related quantum effects in the nm-sized semiconductors in early 1980s [51, 52], tremendous attention has been drawn towards developing strategies to synthesize high-quality MHP NCs for their further utilisation in practical applications [53–55]. The fabrication of MHP NCs not only helps in achieving high PLQY (~90%) but also helps in attaining enhanced emission which is tunable across the whole visible range and in screening the surface traps [56]. The captivating results shifted the attention of research community to work on these interesting class of semiconducting materials.

3D perovskites belong to the class with general formula $A^{1+}B^{2+}X_3$, where, A is a monovalent cation (metal or organic cation; Cs^+, MA^+, FA^+ or mixed cation), B is a divalent cation (Pb^{2+}, Sn^{2+} etc.) and X is a halide anion (Cl^-, Br^- and I^-). The structure comprises of the corner-sharing BX_6 octahedra which repeat in all the directions to give 3D framework whereas, A lies in the cuboctahedral cavities (co-ordinating with 12 X anions) formed by this continuous network and puts a limit on the size of the enclosed cation [57]. To assess the structural stability and the size of A^+ cation that can effectively fit in the 3D perovskite structure, the Goldschmidt tolerance factor, $t = (R_A + R_X)/\sqrt{2}(R_B + R_X)$ and the octahedral factor, $\mu = R_B/R_X$ are utilised (Fig. 1). Here, ionic radii of A, B and X ions are represented by R_A, R_B and R_X. The perovskite structure is stable when $0.8 < t < 1$ and $0.442 < \mu < 0.895$ [58, 59]. The perovskite structures such as $CsPbI_3$ and $FAPbI_3$ have tolerance factor of ~0.8 and ~1 (at the edge of the limit), respectively and therefore, readily undergo transition to stable phases i.e., orthorhombic and tetragonal phases even at room temperature (RT), popularly known as "yellow phases" (Fig. 1) [60, 61].

The prime reason behind the easy fabrication of highly crystalline MHP thin films even at room temperature is the presence of ionic bonding in the perovskite structure [62]. Another significant feature regarding the success of MHPs in the opto-electronic applications is their defect-tolerance which helps a semiconductor to regain its properties despite the presence of a large number of crystallographic defects (see Fig. 2). Various factors for instance, preparation of precursor solution, heat treatment, crystallization etc. strongly influence the defect formation during perovskite film fabrication [63]. Density functional theory (DFT) calculations suggest the introduction of shallow trap states due to the formation of point defects such as halogen vacancies with low formation energies. In semiconductors, due to the antibonding and bonding character of conduction band (CB) and valence band (VB), respectively, deep defect states are generally formed after removal of an atom [64]. On the other hand, MHPs have CB and VB both with antibonding character, where, the CB edge is formed from the interaction between Pb–s and X-p orbitals and VB edge from the interaction between Pb-p and X-p orbitals. Mainly, the defects states are present near or within the VB edge due to very strong antibonding coupling between Pb–s and X-p orbitals, which results in raising the VB edge. Whereas, ionic character is shown by the CB edge due to the weak coupling between Pb-p and X-p orbitals

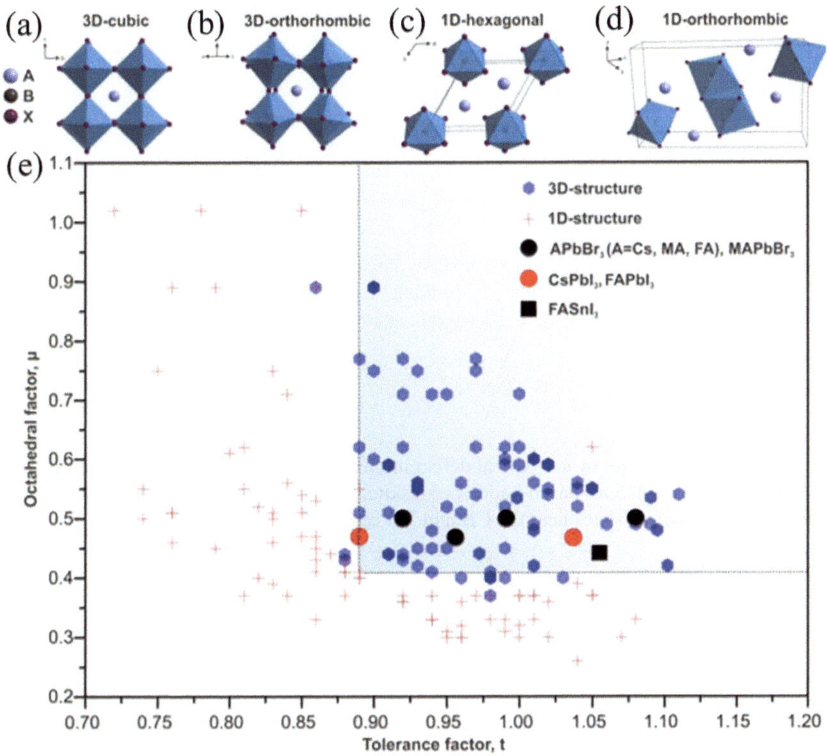

Fig. 1 Illustration of **a** α-$FAPbI_3$ with ideal 3D cubic structure, **b** $CsPbBr_3$, an orthorhombically distorted 3D structure, yellow phases of **c** $FAPbI_3$ and **d** $CsPbI_3$ with 1D hexagonal lattice and 1D orthorhombic structure, respectively. **e** Octahedral and Tetragonal factors for various 1D and 3D structures of MHPs (both all-inorganic as well as hybrid). The light blue squared area represents the area where the compositions adopt stable perovskite structure at room temperature and pressure [61]

which makes the energy states immune towards external defect levels [65]. Even though the defects states formed in MHPs are mostly benign, a careful investigation of defect physics for the perovskites needs to be carried out.

3 Synthesis Strategies/Methods for MHPs

Various synthesis strategies for high-quality MHP NCs are developed in order to have control over the shape, size, composition without compromising their quality. These strategies are basically divided into two categories, namely, "top-down" or "bottom-up" approaches as illustrated in Fig. 3. Top-down approaches involve ball-milling

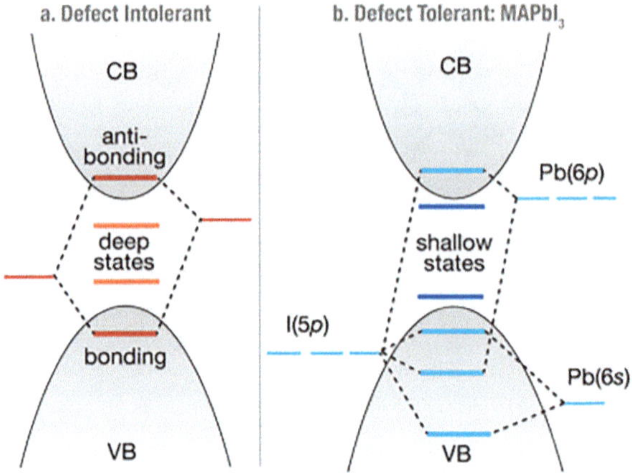

Fig. 2 Electronic structure of **a** a typical defect intolerant semiconductor belonging to III – V, II – VI or group IV and **b** lead-based halide perovskite, $MAPbI_3$, showing the defect tolerance due to the presence of defect states near the CB and VB edge [65]

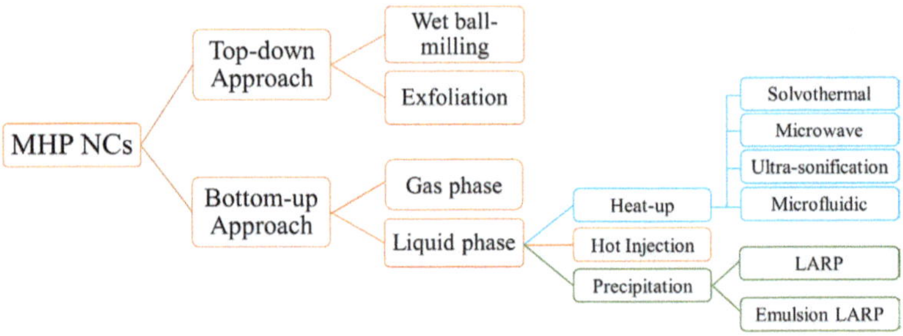

Fig. 3 Flow chart showing the various synthesis methods of nanocrystals of metal halide perovskites

(surfactants are used for size-control), chemical or mechanical (ligand-assisted) exfoliation [30, 66]. Bottom-up approaches mainly use gas- or liquid-phase chemical reactions and are further divided into three main sub-categories: (i) heat-up, (ii) re-precipitation and (iii) in-situ synthesis. In this chapter, we will mainly talk about the frequently used synthetic procedures viz., hot injection (HI) and ligand-assisted reprecipitation (LARP) method for colloidal MHP NCs.

3.1 Hot-Injection (HI) Method: Colloidal Synthesis

In 1993, Murray et al. first reported HI method for the preparation of high-quality cadmium chalcogenide (CdE; E = S, Se, Te) NCs [67]. For the synthesis, the organometallic reagents are heated along with other salts and surfactants in inert

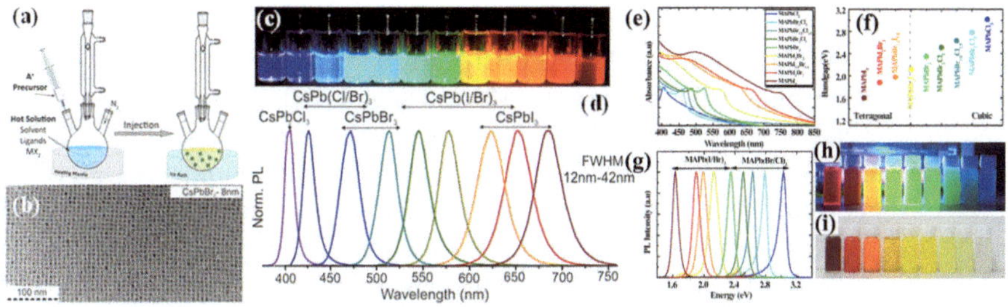

Fig. 4 **a** Hot-injection method for the formation of MHP NCs. **b** Recorded TEM image of uniformly distributed $CsPbBr_3$ NPs. **c** Digital image of dilute toluene dispersions of colloidal $CsPbBr_3$ NPs under UV lamp. **d** PL emission spectra of Cs-based perovskite NPs spanning the whole visible range (400–700 nm) obtained by adjusting the halogen ratios [56]. **e** Absorption spectra, **f** bandgap extracted using tau plot and **g** PL spectra of composition-controlled $MAPbX_3$ nanoparticles using HI method. Digital images of the corresponding samples **h** under UV lamp and **i** in normal light [69]

atmosphere (as illustrated in Fig. 4a). In this method, the nucleation is controlled to grow the highly monodispersed NCs [68]. Injection of precursors to the hot solution rapidly starts the nucleation process followed by the formation of small nuclei. With time, the nuclei start growing and finally contribute to the NC's population. The size, shape and the distribution of NCs can be controlled by varying the ratio of the precursors to the capping surfactants, or the reaction time or the temperature at which the precursor is introduced into the reaction medium. The main benefit of HI method is that it does not use any polar solvent to achieve the successful synthesis of nm-sized materials.

Protesescu et al. utilised the HI method for the successful synthesis of colloidal $CsPbX_3$, X = Cl, Br and I NCs in 2015 [56]. The method involves dissolving PbX_2 in octadecene (ODE) at high temperature (140–200 °C), where, oleylamine (OLA) and oleic acid (OA) are added in equimolar ratio to ensure the complete dissolution of PbX_2 with an added benefit of stabilizing the NCs in colloidal form. The reaction of PbX_2 in ODE with organic acids at high temperature results in the formation lead-oleate and halide ions. Finally, Cs ions (in the form of Cs-oleate, synthesized using Cs_2CO_3 and OA in the presence of ODE) are injected in the hot solution to obtain $CsPbX_3$ NCs. This instantaneous injection helps in attaining the uniform distribution of shape and size of the MHP NCs as illustrated in Fig. 4b. The size of the obtained NCs can also be tuned from 5–12 nm by varying temperature of the reaction [56]. The optical characteristics of the as-prepared NCs can be conveniently engineered across the entire visible range of spectrum from 400–700 nm using the HI method by merely adjusting the lead halide ratios ($PbCl_2/PbBr_2$ or $PbBr_2/PbI_2$) as observed from Fig. 4c–d. The amines and acids used in HI method play a crucial role in determining shape and size of perovskite NPs. Pan et al. studied the detailed structure and morphology of colloidal perovskite NCs by varying carboxylic acids and amines hydrocarbon chain length [70]. The HI method was modified further to synthesize hybrid organic–inorganic lead-based halide perovskite NCs by replacing Cs-oleate with MA solution or FA-oleate [71, 72]. Later, the process was utilised to obtain stable blue emitting mixed-halide $FAPb(Cl_{1-x}Br_x)_3$ NCs [73] and $FAPb(I_{1-x}Br_x)_3$ NCs with tunable emission ranging from 570 to 780 nm [74]. HI method was further used to fabricate various inorganic perovskites ranging from 3D-$CsPbX_3$ to 2D-$CsPb_2Br_5$ and 0D-Cs_4PbBr_6 [75, 76]. The high solubility of colloidal MHP NCs in polar solvents results in a deterioration of their stability in external environmental conditions such as humidity, light, heat, etc., resulting in the loss of their PLQY [53]. In this regard, Yuan et al. showed the sensitivity of perovskite NCs toward different antisolvents used for cleaning [77]. The change in the halide composition of $CsPbBr_{1-x}I_x$ NCs was observed in different antisolvents such as acetone, IPA, n-butanol etc., considerably affecting the optical properties in contrast with pristine $CsPbBr_3$ NCs. In order to compensate the loss of halide ions in $CsPbX_3$ NCs, Woo et al. reported fabrication of inorganically passivated (halide-rich condition) perovskite NCs via in situ route [78]. Additionally, the use of metal halide along with PbX_2 precursors resulted in the synthesis of halide-rich compositions which not only prevented the change in optical properties, but, enhanced the ambience stability as well.

The synthesis of composition-controlled MA-based halide perovskites using HI method started in 2015. Unlike Cs and FA precursors, the non-availability of MA-acetate as well as highly volatile nature of MA (boiling temperature of 4 °C) make the use of conventional HI method for their synthesis very difficult. This is the main reason behind storing this molecule in THF or alcohols like ethanol or methanol due to their boiling temperature of around 60–70 °C. Vybornyi et al. demonstrated a modified-HI method to synthesize $MAPbBr_3$ and $MAPbI_3$, however, the synthesis of $MAPbCl_3$ and the (I/Br and Br/Cl) mixed-halide counterparts was still a big challenge [71]. The approach included solubilizing PbX_2 in ODE with a minimal amount of OA and OLA ligands at 120 °C. Next, MA was injected rapidly into the solution after bringing the temperature down to 55 °C to avoid any decomposition. In 2019, Roy et al. utilized this method to obtain the whole series of composition-controlled $MAPbX_3$ nanoparticles, including $MAPbCl_3$ in gram scale [69].

3.1.1 Modified HI Method: Overcoming the Limitations of Conventional HI Method

The conventional HI method uses metal halide salt as a source for both cation and anion sites which restricts the use of this method in obtaining desirable ion stoichiometry [56]. The limitation of this method was overcome by introducing a modified, so called, "three-precursor" HI method. In this method, three separate precursors as sources for A cation, lead and halogen sites are taken which in turn helps in achieving and controlling the required lead to halogen ratio. Liu et al., for the first time, synthesised $CsPbX_3$ (X = halide) NCs using "three-precursor" HI method, where, instead of PbX_2 precursor as a source for lead and halogen, individual reagents NH_4X (halogen-source) and PbO (lead-source) were utilised [79]. Several reports suggest that the NCs synthesised in halide-rich conditions show improved optical properties as the halogen-stripping is somewhat compensated during the purification process [78]. For instance, $CsPbCl_3$ NCs synthesised under halide-rich conditions showed near-unity PLQY [28]. The three-precursor HI method was further utilised by Yassitepe et al. in order to eradicate the degradation of the perovskite QDs during the purification process [80]. The degradation is primarily ascribed to the rapid ligand loss due to the proton transfer between OA and OLA surfactants. The authors used quaternary alkylammonium halides as halogen sources (without any need for OLA), thus, synthesizing OA-capped perovskite QDs. This new synthetic route not only enhanced the colloidal stability of the QDs, but also substantially improved the PLQY to up to 70%. Further modification in the three-precursor HI method resulted in the successful synthesis of colloidal $FAPbX_3$ NCs [72]. The method utilized reaction of acetates of FA and Pb in OA and ODE with oleylammonium bromide, however, the impurity peaks corresponding to $NH_4Pb_2Br_5$ (probably due to the decomposition of FA^+ into NH_4^+) were also present in the final product.

While the three-precursor HI method offers the flexibility to fine-tune the quantity of the different precursors by using individual sources for metal and halide ions and subsequently, achieve the final composition of NCs, the method itself has a few limitations. The modified HI method failed to successfully synthesize $CsPbX_3$ (X = Cl and I) and $MAPbX_3$ (X = Br and I) NCs [74]. Additionally, as discussed earlier, due to the decomposition of alkylammonium halide during the synthesis process of $FAPbX_3$ NCs, substantial amount of impurities were present in the final phase [72]. In order to eliminate these drawbacks, Imran et al. reported the injection of highly reactive halides such as, benzoyl halides, into the solution of metal cations, primarily in carboxylate form for the formation of $APbX_3$ (A = Cs/MA/FA; X = halide) NCs (Fig. 5a) [81]. The injection induces immediate release of the halide ions which subsequently confirm the evolution of perovskite NCs. The NCs obtained using this method show extraordinary optical properties (narrow PL emission; Fig. 5b, c), uniform size distribution and excellent stability in air without any post-synthesis modifications. The process promotes the partial replacement of Cs ions with the oleylammonium ions, resulting in the lead halide-terminated surfaces, thereby, helping to improve their stability and photophysical properties. In addition, Creutz et al. extended this synthesis approach for the fabrication of lead-free colloidal Cs_2AgBiX_6 (X = halide) NCs using trimethylsilyl (TMS) halides as halogen sources (Fig. 5d–i) [82]. In short, these strategies allow to attain different compositions by adjusting cation or anion ratios, thereby helping in tuning the optical properties and enhancing the relative stability of the NCs.

3.1.2 HI Method: Regulating the Morphology of MHP Nanocrystals

The control on shape and size of both all-inorganic [25, 56] and hybrid organic–inorganic lead-based halide perovskite NCs [20] has been done by regulating reaction time, reaction temperature and compositions in colloidal-based synthesis procedures. After the synthesis of $MAPbBr_3$ ($MA = CH_3NH_3$) NCs as initiated by Schmidt et al. [20], the various reports on the synthesis of nanoplatelets of hybrid perovskites show

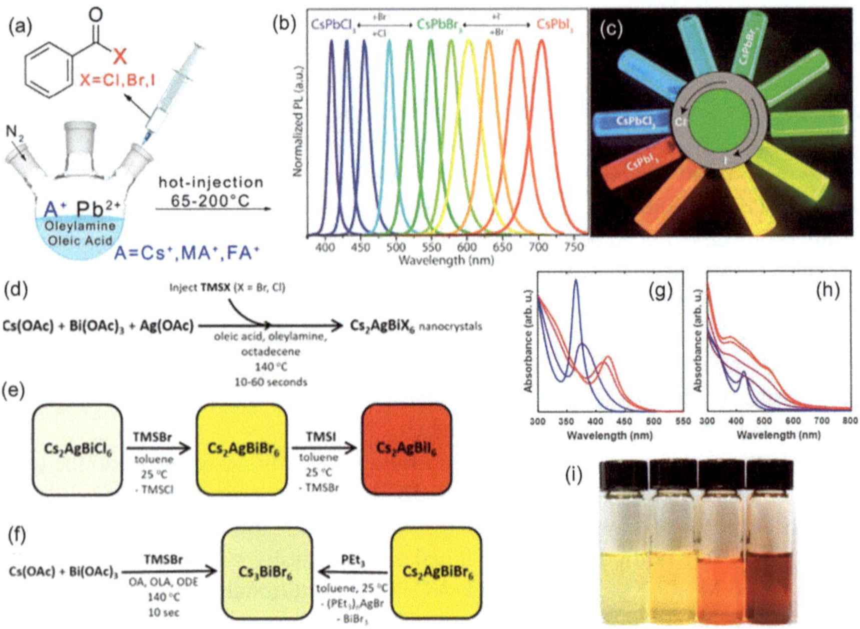

Fig. 5 **a** Schematics for the preparation of lead-based halide perovskite NCs using benzoyl halides as anion precursors utilizing colloidal-based HI method. **b** Evolution of PL spectra via anion-exchange achieved by adding benzoyl iodide or benzoyl chloride to already synthesized $CsPbBr_3$ NCs. (**c**) Digital images of different solutions prepared via anion-exchange reaction under UV lamp [81]. **d–f** Schematics for the synthesis route for Cs_2AgBiX_6 NCs. **g** Absorption spectrum of $Cs_2AgBiCl_6$ NCs (blue) converting to $Cs_2AgBiBr_6$ NCs (red) via anion-exchange mechanism (progressive addition of TMSBr). **h** Absorption spectrum of $Cs_2AgBiBr_6$ NCs (blue) converting to Cs_2AgBiI_6 NCs (red) via anion-exchange mechanism (progressive addition of TMSI). (**i**) Digital images of solutions of $Cs_2AgBiBr_6$, $Cs_2AgBiBr_{5.2}I_{0.8}$, $Cs_2AgBiBr_{1.6}I_{4.4}$, and Cs_2AgBiI_6 NCs in toluene (Left to Right) [82]

tuning of their thickness by varying the organic cations [83]. In case of all-inorganic perovskite NCs, Protesescu et al. reported the first successful synthesis of $CsPbX_3$ (X = Cl, Br, I) nanocubes using HI method followed by rapid cooling [31, 56]. Since then, significant advances in the development of MHP NCs resulted in the shape control ranging from nanowires (NWs) [84] to nanorods (NRs) [85] to nanoplatelets (NPLs) [27] and nanocubes [85]. Bekenstein et al. reported the tendency of formation of quasi-2D geometries (known as NPLs) for reactions using OA and OLA carried out at temperatures of 90–130 °C, symmetrical nanocubes at 150 °C, while NWs are preferred at relatively high temperatures of 170–200 °C [27, 84]. Further, Pan et al. studied in detail the effect of surfactants/ligands on $CsPbBr_3$ NC's morphology [70]. In order to understand this correlation, the chain lengths of the carboxylic acids and alkyl amines used in the reactions were systematically varied. It was observed that shorter length of carboxylic acids (added at 170 °C) favoured the formation of larger-sized nanocubes. Later, in another set of experiments, OA was kept fixed and

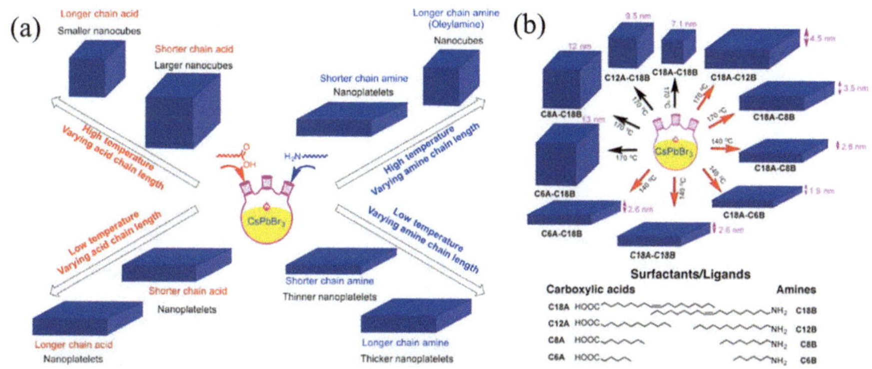

Fig. 6 **a** Schematics showing the correlation between the different morphologies (shape and size) and length of the chain of the amines and carboxylic acids. **b** Controlling the shape and size of NCs using HI method by using different ligands and at different temperatures. Nanocubes are formed at higher temperatures with short length carboxylic acids whereas, thin NPLs result from short length amines. At low temperatures, NPLs are formed using short length carboxylic acids [70].

different alkyl amines at 170 °C were used, resulting in the formation of NPLs (see Fig. 6).

The morphology of perovskite NCs depends significantly on the temperature at which the reaction is conducted. Generally, the precipitation of perovskite precursors leads to the formation of nanocubes at relatively high temperatures, while at low temperatures, formation of NPLs is more likely to occur. This dependence on temperature can be well understood by the protonation or deprotonation of the ligands (alkylamine) controlled by acid or base equilibria during the synthesis process while, Cs^+ ions compete for their incorporation to the facets during the growth of NCs [86]. Almeida et al. further elucidated the effect of relative concentration of carboxylic acids and alkylamines on the shape, size and distribuition of $CsPbBr_3$ NCs [86]. They reported an increase in the temperature for $PbBr_2$ precipitation from 195 °C to 290 °C with an increase in the concentration of the ligands. Additionally, the study revealed that low concentration of oleylammonium species tends to form nanocubes (size ranging from 4 nm to 16.4 nm with narrow size distribuition) while, the NCs grow anisotropically at higher concentrations (see Fig. 7).

Ultrathin LHP NWs (diameter < 3 nm; down to atomic level) are of current interest due to their strong confinement effects [87]. However, due to the difficulty in controlling the crystal growth at the atomic level, it is very hard to synthesize uniform ultrathin NWs. In 2015, Zhang et al. illustrated the synthesis of $CsPbBr_3$ NWs using conventional HI method, where, the reaction condition was altered drastically to room temperature (RT) at different intervals of time [84]. It was observed that initially at t < 10 min, nanocubes (size varying from 3 to 7 nm) dominated the reaction (Fig. 8). The structure evolved from nanocubes to square-shaped nanosheets (NSs) with fewer thin NWs after around 30 min. After one hour, the reaction was dominated with uniform NWs of diameter below 12 nm and lengths up to several microns.

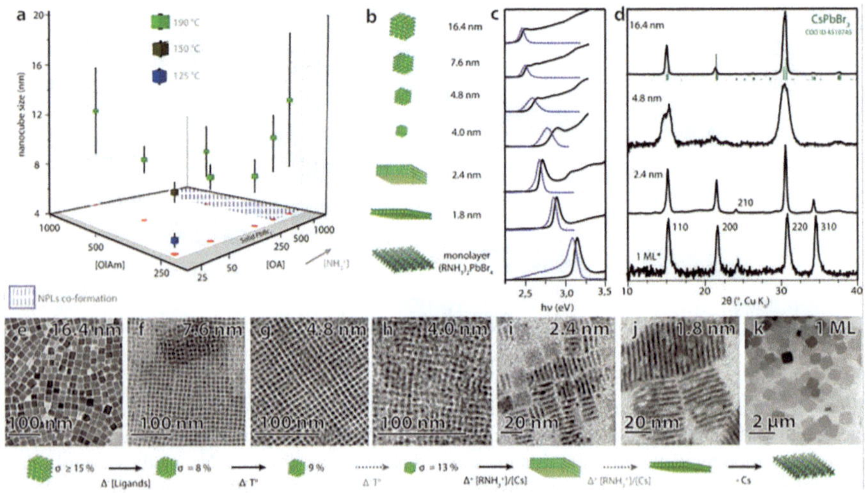

Fig. 7 **a** Change in OA and OLA concentration and temperature of reaction in order to vary the size of $CsPbBr_3$ nanocubes (vertical bars denote size distribution). **b** Drawing of various NCs of $CsPbBr_3$ attained using OA and OLA ligands only **c** corresponding absorbance (black lines) and photoluminescence spectra (blue lines) **d** X-ray diffraction (XRD) profiles and **e–k** TEM images [86]

Further in 2016, Zhang et al. worked on improving the quality and yield of ultrathin perovskite NWs (2.2 ± 0.2 nm width) showing a blue shift in PL (~465 nm) as compared to the bulk counterpart (~530 nm) [87]. Further reports on the modification of perovskite NWs (reduction in dimension up to quantum confinement regime) suggest the incorporation of carboxylic acids with short chain lengths in addition to OA and OLA [23].

Even though HI method has been used successfully for the fabrication as well as gaining control on the shape and size of LHP NCs, the method is tedious and requires inert conditions. Moreover, an additional step is needed in order to prepare Cs or FA oleate which are again heated just before the injection which makes the procedure arduous. Also, large scale production using HI method is difficult as precursors used in large quantities result in remarkable drop in the reaction temperature which encourages inhomogeneous nucleation [56]. Various alternative approaches

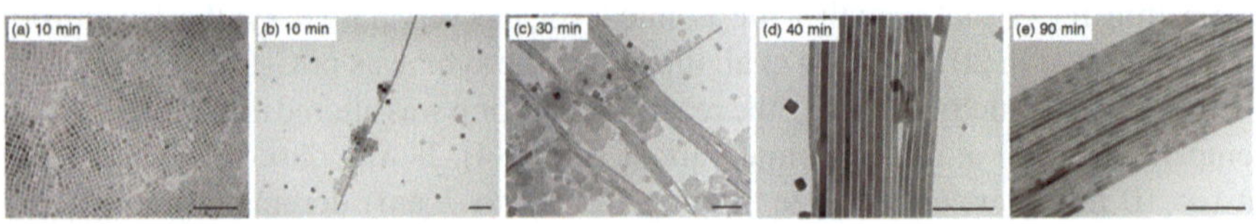

Fig. 8 Evolution of shape of $CsPbBr_3$ NCs as a result of different reaction times [84]

like ligand-assisted reprecipitation technique, ultrasonication, microwave irradiation, solvothermal method etc. have been employed in order to overcome the above mentioned drawbacks of HI method [88–91].

3.2 *Ligand-Assisted Reprecipitation (LARP) Method*

The process of reprecipitation stems from recrystallization of salts in clay pots back in 3500 BC in South Poland [85]. The process is simple and involves first dissolving all the precursor ions into a suitable solvent until an equilibrium state is reached. Then, the equilibrium state is disturbed by means of either evaporating the solvent or changing the temperature or by moving the solution into an anti-solvent. These conditions trigger a spontaneous precipitation and the crystallization happens until the equilibrium state is reached once again. The process can also be used to prepare nanoparticles, by carrying out the reaction in the presence of surfactants or ligands, that is why the name "ligand-assisted" reprecipitation is used, where, the crystals in the nm range can be grown (see Fig. 9).

The formation of MHP NCs using LARP method involves the dissolution of initial lead-halide (PbX_2) and BX (B = Cs, MA and FA) salts in a polar solvent like dimethylsulfoxide (DMSO), dimethylformamide (DMF) etc. In a different beaker, a poor (bad) solvent (in which the precursor ions are not soluble) such as hexane, toluene etc., is taken along with the ligands. When the two solutions are mixed together, a spontaneous supersaturation causes the nucleation and growth of LHP NCs. Unlike HI method, the whole reaction is carried out in open atmosphere and at room temperature. In HI method, the reaction temperature controls the size of nanocrystals and variation in temperature upon injection of the precursor solution hampers the reproducibility of NCs [92]. This affects the mass production (gram scale) of perovskite NCs and is resolved by LARP method, a RT-based synthetic

Fig. 9 Illustration of LARP synthesis method [85]

approach [92, 93]. In 2012, Papavassiliou utilized LARP method for the development of various lead-based perovskites which includes $MAPbX_3$ and other MA-based layered materials with $Pb_n(Br_xCl_{1-x})_{3n+1}$, $Pb_n(Br_xI_{1-x})_{3n+1}$, and $Pb_n(Cl_xI_{1-x})_{3n+1}$ ($n > 2$) units [94]. Only after a few years of rigorous efforts, LARP method was extended to prepare NCs of all-inorganic as well as hybrid halide perovskites, ABX_3, where B belongs to Pb, Bi, Sb or Sn [47, 85].

3.2.1 Synthesis of Hybrid and All-Inorganic Lead-Based Halide Perovskite NCs Using LARP Method

LARP method for the preparation of perovskite NCs in ambient conditions utilizes the inherent ionic nature of perovskites. In 2015, Zhang et al. prepared highly luminescent $MAPbBr_3$ NCs using LARP method for the first time (Fig. 10a–c) [22]. The NCs were obtained by preparing a clear precursor solution in DMF containing OLA and OA (surfactants). A suitable amount of this solution was added dropwise into toluene and vigorously stirred at RT resulting in the formation of a yellow-green colloidal solution which was further cleaned to obtain high-quality $MAPbBr_3$ nanoparticles. The work involved further exploration of the role of carboxylic acids and alkyl amines, where, they demonstrated the loss of control on the size of the crystals in the absence of amines. On the other hand, aggregated NCs were formed when carboxylic acids were not used. It was concluded that amines help in regulating the size of NCs (kinetics of crystallization) whereas, carboxylic acids help in preventing the aggregation of NCs and was followed by many other groups [95, 96]. Additionally, the heating of toluene controls the size of $MAPbBr_3$ NCs and their distribution, as suggested by Huang and his co-workers [95]. Further, the modified LARP method involves the formation of MAX salt in situ without the effort to make it individually [97]. In a conventional procedure, PbX_2 was dissolved in N-methylformamide (NMF; acting as a solvent as well as a source for MA^+ ions) instead of DMF along with OA and OLA at 100 °C for 10 min (Fig. 10d). Afterwards, the solution was dropped into chloroform at RT to yield $MAPbX_3$ ($X = $ Br and I) NCs. The bulk $MAPbX_3$ crystals can also be produced using this technique at RT with no antisolvent.

Levchuk and co-workers reported the synthesis of $FAPbX_3$ ($X = $ Cl, Br and I) NCs (Fig. 10e–f), where, the precursor solution of FAX and PbX_2 in DMF, OLA and OA were injected rapidly into chloroform instead of toluene at RT [98]. The study showed the prevention of formation of $FAPbI_3$ NCs and immediate agglomeration of $FAPbBr(Cl)_3$ NCs. The synthesis of colloidal all-inorganic $CsPbX_3$ NCs at RT using LARP method resembled that for $MAPbX_3$ NCs. Li et al. used the solution of inorganic salts PbX_2 and CsX in DMF, OA and OLA and mixed dropwise in toluene to obtain $CsPbX_3$ NCs (Fig. 10g–i) [89, 99].

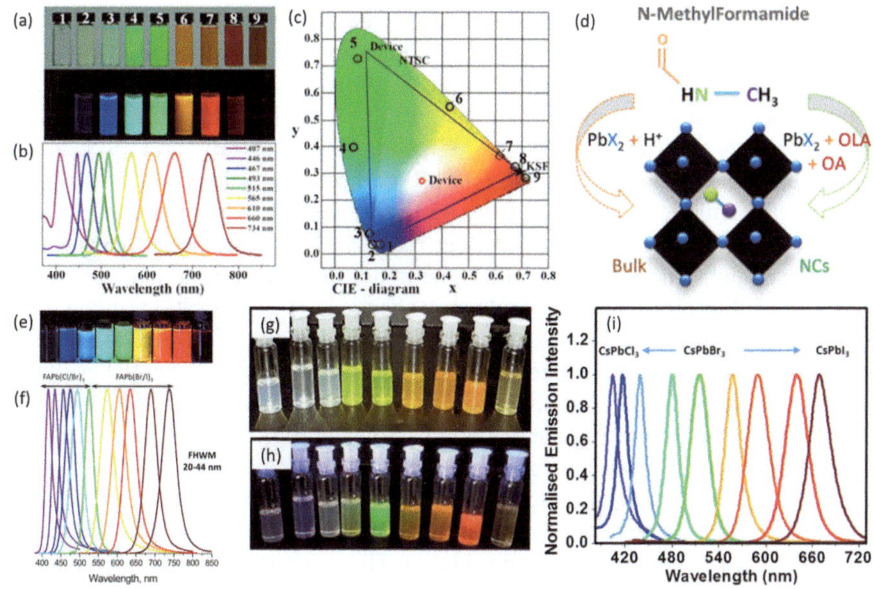

Fig. 10 **a** Digital images of $MAPbX_3$ (X = Cl/Br/I or their halide mixtures) NCs under ambient light and a UV lamp (λ = 365 nm). **b** Photoluminescence spectra taken for $MAPbX_3$ NCs. **c** Corresponding CIE color coordinates [22]. **d** N-methyl formamide (NMF) acts as a source of MA^+ ions and solvent for the synthesis of both bulk- and nano-crystals $MAPbX_3$. [97] **e** Digital images of colloidal solution of $FAPbX_3$ (X = Cl/Br/I or their mixture) NCs in toluene under a UV lamp. **f** Corresponding photoluminescence spectra showing the compositional tuning [98]. Digital images of all-inorganic $CsPbX_3$ NCs under **g** ambient light, **h** a UV lamp and **i** their corresponding PL spectra [99]

3.2.2 LARP Method for Controlling the Shape and Size of MHP NCs

After the successful synthesis of MHP NCs using LARP method, several research groups worked towards controlling their shape and size in order to optimize the PL emission. For instance, in 2015, Sichert and team demonstrated the control on the thickness of $MAPbBr_3$ NPLs which resulted in change of the PL emission by varying the ratio of the organic cations (octylammonium (OA^+) and MA^+) used [101]. The thickness of the NPLs decreases with the systematic increase in OA/MA ratio with "single-layered" NPLs obtained on using OABr only. Other reports also suggest the similar observation of obtaining monolayer thick NPLs using octylamine and OA [102]. Moreover, the transformation from 3D nanocubes to 2D nanostructures can be achieved by adding pyridine during the growth of NCs (Fig. 11a) [100]. The pyridine in the solution binds selectively to the $MAPbBr_3$ surface and promotes 2D growth due to the formation of native N $\rightarrow$ Pb bonds which restricts the vertical growth. Figure 11b–g shows the effect of variation of temperature on the thickness of these 2D-nanostructures resulting in the formation of n $\approx$ 2 (face-stacking NPLs), n $\approx$ 4 (thin sheets) and n $\approx$ 6 (quasi-2D nanostructures), where, n is the number of layers of the unit cell. Kirakosyan et al. achieved the lateral size reduction of the $MAPbBr_3$ NCs by slowly changing the quantity of the precursor solution at a constant rate to the anti-solvent which results in the reduction from 3 to 8 nm without altering the thickness (~3 nm) of the NCs [103]. Levchuk et al. varied OA to OLA ratio in order to fine tune the thickness of the $MAPbX_3$ (X = Br and I) NPLs down to the quantum confinement regime [104]. By adjusting the thickness of the NPLs only, the PL emission was tuned from 450 to 730 nm with pronounced quantum size effect (thickness $\leq$ 2.2 nm; 3D Bohr radius in $MAPbBr_3$ perovskites). Later, they utilised the same technique with some minor modifications to obtain $FAPbX_3$ NCs (use of chloroform instead of toluene); change in OA/OLA ratio resulted in the formation of different morphologies with different thickness (see Fig. 11h–l) [98]. Seth and co-workers demonstrated the formation as well as the sequential evolution of different morphologies (QDs, NPLs, nanocubes, NRs and NWs) of high-quality $CsPbX_3$ NCs just by varying the reaction parameters [99]. It was observed that on using ethyl acetate as anti-solvent, quasi-cubic $CsPbBr_3$ QDs (Fig. 11o–p) were formed immediately. On increasing the reaction time to 10 min, a few layered thick square-shaped NPLs were obtained which transformed to nanobars after increasing the reaction time further to 40 h (Fig. 11q–r). This is attributed to the high polarity of ethyl acetate as compared to toluene which enables it to work as a nucleophile causing the detachment of some OLA molecules from nuclei surface and results in an oriented attachment (Fig. 11m). The size of these nanocubes were further altered (edge length varying from ~12 nm to 34 nm) by using toluene as anti-solvent while keeping OA and OLA as ligands. However, at lower concentration of OLA, the formation of NRs (length = 800 nm and diameter = 70 nm) was observed which eventually transformed to NWs (diameter ~70 nm and length $\geq$ 15 μm) with increase in reaction time (Fig. 11s–v)). This is because in case of toluene, at lower concentration of OLA, the surface of NCs is not fully protected in all directions and the anisotropic growth of NCs (formation of NWs/NRs) happens when the reaction is carried out for prolonged duration, as seen from Fig. 11n.

3.2.3 Limitations of LARP Method

Apart from the advantage of direct synthesis of perovskite NCs in open environment without any heat treatment, the LARP method has certain limitations. LARP method makes use of the polar solvents like DMSO, DMF etc., which can possibly degrade or even solubilize the final product [92, 106]. The interaction between precursor and polar solvents played an important part in forming defective LHP NCs. In this contest, Zhang et al. studied these interactions as well as demonstrated the influence of various polar solvents on the crystallization of $MAPbI_3$ NCs [105]. It was observed that the use of coordinating solvents (forming strong intermediates with PbI_2) such as, DMF, DMSO and tetrahydrofuran (THF) led to the formation of $MAPbI_3$ NCs with residual solvent still present on the surface of these NCs. The removal of these solvents might create the intrinsic iodine vacancies resulting in the formation of defective NCs which

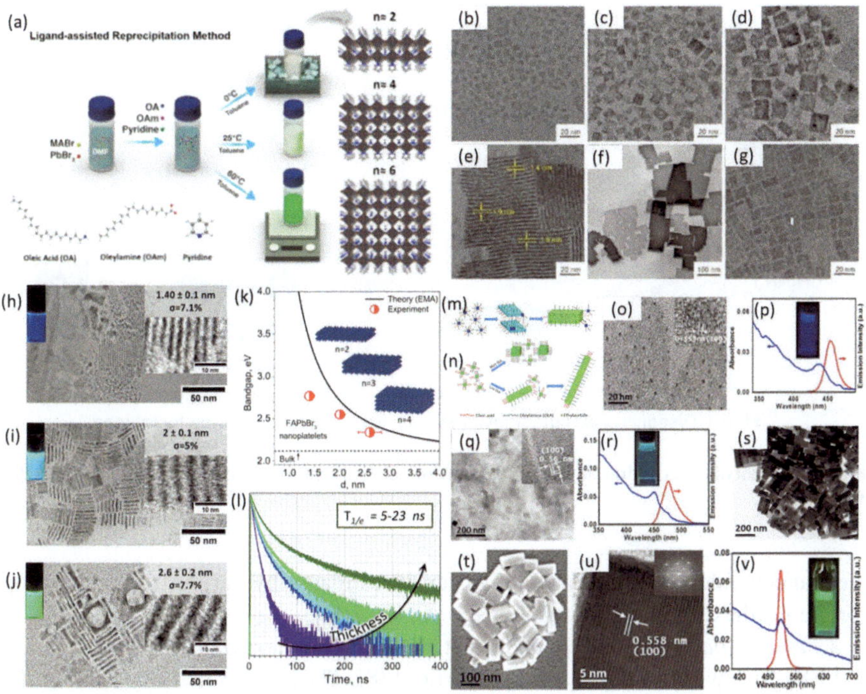

Fig. 11 **a** Illustration for the synthesis of $MAPbBr_3$ NCs using LARP method using OA, OLA and pyridine resulting in the reduction of 3D nanocubes to 2D nanostructures. TEM images of the corresponding nanostructures obtained at **b** 0 °C **c** 25 °C and **d** 60 °C without pyridine and at **e** 0 °C **f** 25 °C and **g** 60 °C after addition of pyridine [100]. TEM images of thickness-controlled $FAPbBr_3$ NCs prepared by varying OLA/OA ratio keeping amount of OLA fixed at 200 μl and varying OA to **h** 150 μl, **i** 80 μl and **j** 40 μl (inset shows the emission of different $FAPbBr_3$ NPLs; σ = thickness distribution). **k** Comparison between experimental and theoretical bandgap as a function of thickness, d. **l** PL decay [98]. Formation mechanism for (m) NPLs (cyan) and nanobars (green) and **n** larger nanocubes, NRs and NWs from smaller nanocubes in ethyl acetate and toluene, respectively. **o** TEM image and **p** absorption and PL spectra of $CsPbBr_3$ QDs. **q** TEM image and **r** absorption and PL spectra of $CsPbBr_3$ NPLs. **s** TEM image, **t** FESEM, **u** HR-TEM and **v** absorption and PL spectra of $CsPbBr_3$ nanobars [99]

probably convert to $MAPbI_3{\cdot}H_2O$ intermediate due to the water present in the solvent or air (turbid solution as seen in Fig. 12) [105].

3.3 *Alternate Approaches for the Synthesis of LHP NCs*

Various other chemical-based approaches such as solvothermal method, microwave irradiation, ultrasonication etc. as well as non-chemical-based approaches like

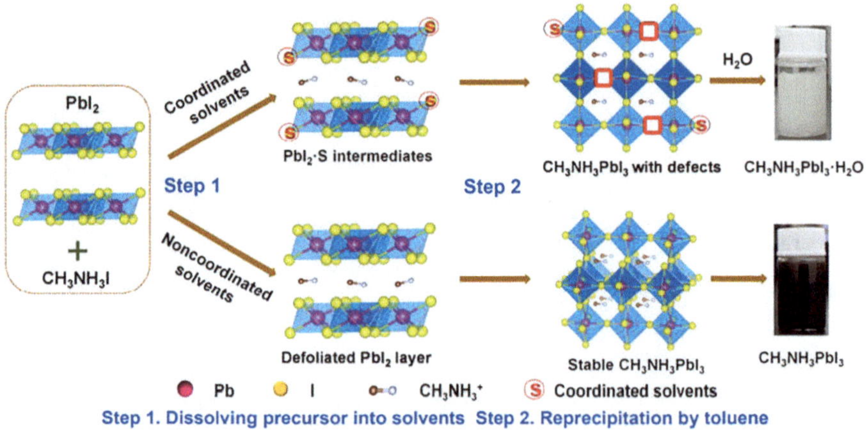

Fig. 12 Effect of the coordinating and non-coordinating solvents on the crystal structure of $MAPbI_3$ NCs [105]

mechanochemical method have also been investigated by various research groups for the synthesis of perovskite NCs [88, 90, 91, 107].

Chemical-based Approaches: Chen et al. used *solvothermal method* to synthesize $CsPbX_3$ NCs with tunable optical properties [88]. The process involved sealing stoichiometric amount of cesium acetate and PbX_2 salt dissolved in octadecene (ODE) along with ligands, OLA and OA in a stainless autoclave. The system is then heated at high temperatures of 140–200 °C in an oven for 30 min. Using this method, monodisperse nanocubes of $CsPbX_3$ NCs were prepared together with the successful control on the composition where, X is Cl, Br, I or their mixture and bright PL emission spanning the whole visible range. Furthermore, simply pre-dissolving the precursors in ODE using magnetic stirring, the morphology can surprisingly be changed from $CsPbBr_3$ nanocubes (green emission) to ultrathin $CsPbBr_3$ NWs with a diameter of $\approx$ 2.6 nm, giving a blue emission confirming the quantum confinement effect (Fig. 13a). The method was further extended by Praveen and co-workers to obtain the composition-controlled $MAPbX_3$ nanosheets with variable thickness resulting in a tunable emission down to the quantum confinement regime [108]. Another strategy, namely, *microwave irradiation* was also utilised by Pan et al. in order to control the halogen composition of $CsPbX_3$ NCs [90]. The process included mixing Cesium acetate, PbX_2, OA, OLA, trioctylphosphine oxide (TOPO) and ODE in a quartz tube which is then placed in a microwave reactor at 160 °C (28 °C/min) for 5 min. The use of TOPO helped in producing high-quality NCs by preventing the agglomeration of the precursors. The control on different morphologies viz., nanocubes, NPLs and NRs can also be achieved using this method (Fig. 13b). By reducing the reaction temperature from 160 °C down to 80 °C, the plate-like $CsPbBr_3$ NCs were obtained. Further, if the precursors were pre-dissolved before microwave treatment, just like in solvothermal method, rod-like morphology of $CsPbBr_3$ NCs was achieved. The minor modification of separating the nucleation step (addition of IPA to the solution of Cesium oleate and $PbBr_2$) from the growth step (when microwave irradiation is turned on) in the microwave-assisted method was done by Shamsi et al. in order to obtain the quantum-confined blue-emitting $CsPbBr_3$ NPLs [93]. Ultrasonic irradiation in place of a microwave treatment as a source of excitation was employed by Tong et al. to synthesize $CsPbX_3$ (X = Cl, Br, I or their mixture) perovskite NCs with successful control on the halogen composition, thickness and morphology of the NCs (Fig. 13c–g) [91]. Roy et al. utilized the ultrasonic vibration as an excitation source for the synthesis of hybrid mixed halide compound-phase perovskite nanoparticles, $MAPbX_{3-n}Y_n$ (X/Y = I, Br, or Cl), by mixing two pure-phase perovskites viz., $MAPbX_3$ and $MAPbY_3$ [107].

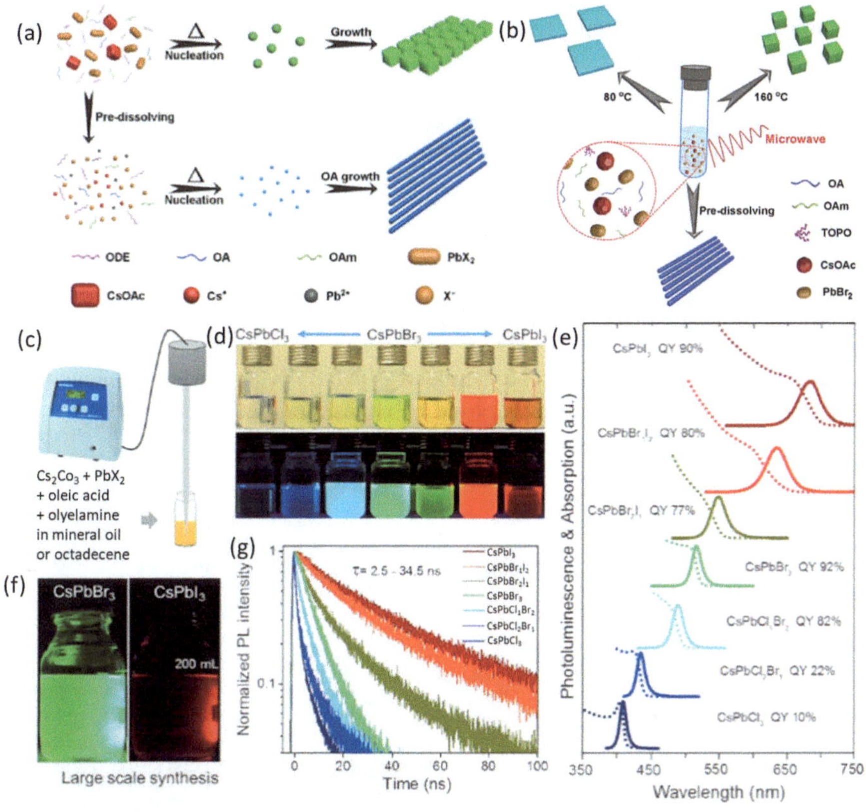

Fig. 13 **a** Schematics illustrating the growth process of NCs using solvothermal method with and without pre-dissolving the precursors [88]. **b** Proposed growth mechanism using microwave treatment. Different morphologies vis., NPLs and nanocubes of $CsPbBr_3$ are obtained at 80 °C and 160 °C, respectively without pre-dissolving the precursors however, pre-dissolving the precursors resulted in the formation of $CsPbBr_3$ NRs [90]. **c** Ultrasonication method for preparation of $CsPbX_3$ NCs. **d** Digital images of $CsPbX_3$ NCs with different halogen composition in ambient light (up) and under UV lamp (down) and their corresponding **e** optical absorption (dashed lines) and PL (solid lines) spectra. **f** Large-scale production using this technique (up to 10 times). **g** PL decay [91]

Non-chemical-based Approaches: Mechanochemical approach where, mechanical energy is utilized to trigger the chemical reagents, has several advantages over the conventional wet chemical methods. The process is completely solvent-free (volatile solvents are used), cost-effective and can be used for the mass-production of great quality and phase-pure perovskites. Furthermore, several researches show the perovskites synthesised using mechanochemical method show greater stability as compared with the polar-solvent based conventional methods [109, 110]. In a typical process, the weighed perovskite precursors PbX_2 and AX together with long-chain ammonium halide salts are mixed in a motar-pestle (hand-grinding) or ball-milling. Ball-milling is considered a better method for upscaling as hand-grinding is tedious and generally leads to impure phase. In ball-milling, precursor salts are taken in a stainless-steel jar, where, the mixing is done using the several balls (made of zirconia, agate or corundum). After sealing the jar, it is mounted on a motor where, two movements viz., rotation around the centre of the disk and around its own axis, are performed (Fig. 14a). Consequently, the tossing and collision with the reactant species provides enough energy for the synthesis of desired perovskite phase [66, 110]. The synthesis of NCs using mechanochemical synthesis (MS) approach is done by using surfactants and a volatile solvent in addition to the precursor salts for the reduction of the size as shown below:

$$\begin{aligned} &A(\text{solid}) + B(\text{solid}) + \text{surfactants} + \text{solvent}\, MS \\ &\rightarrow AB(\text{Colloidal}\, NCs) \end{aligned} \tag{1}$$

However, the first report on the synthesis of $MAPbX_3$ using mechanochemical approach by Hintermayr et al. was based on two-step process [30]. First, grinding

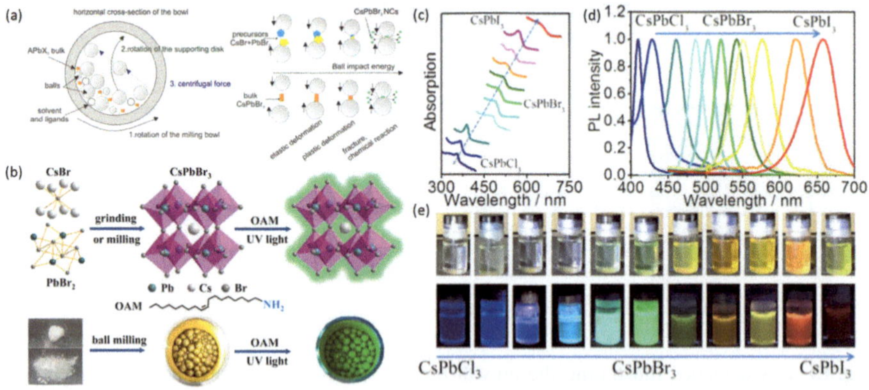

Fig. 14 **a** Illustration showing the working of a ball-mill where the horizontal cross-section of the jar is shown which undergoes two kinds of rotation; one about its own axis and another around the centre of the disk [66]. **b** Illustration for mechanochemical synthesis of $CsPbX_3$ QDs (top) and fluorescence shown by the samples under UV lamp (down). The corresponding **c** UV–Vis absorption and **d** Pl spectra of composition-controlled $CsPbX_3$ QDs and their **e** digital images in ambient light (top) and under UV lamp (down) [111]

of the equimolar precursor salts in motar-pestle for several mins was done to ensure the complete formation of bulk perovskite phase. Next, the perovskite powder was dispersed in toluene and OLA (ligands for size reduction) and sonicated for 30 min to obtain homogenous dispersion. Further, the excess ligands were removed using centrifugation which resulted in the formation of polydisperse rectangular NCs with size ranging from few tens to hundreds of nanometers. Similarly, a full range of $CsPbX_3$ NCs were obtained using this two-step process by Zhu and co-workers [111]. However, instead of using ultrasonic vibration for the second step, the bulk powder together with surfactants (without any solvent) was ground using ball-milling to obtain the composition-controlled all-inorganic perovskite NCs. Also, instead of bifurcating the process in two steps, perovskite NCs can be directly prepared by grinding the precursor salts in the presence of ligands and solvent, commonly known as liquid-assisted grinding (LAG) [112]. Protesescu et al. extended this method to synthesize hybrid $FAPbBr_3$, inorganic $CsPbBr_3$ and mixed halide $CsPb(Br/I)_3$ NCs via mechanically induced anion-exchange process [111]. The study also rendered the fact that prolonged milling produced quantum-confined perovskite nanoparticles.

4 Defect Physics in Halide Perovskites

4.1 Defects in Semiconductors: An Overview

Defects are present in all semiconductors, either by accident or by design. A Wide range of applications get affected because of the existence of defects such as, impurities or native defects introduced during the growth and processing of the semiconducting materials [113]. It is very important to understand the energy states for the occupation of electrons and the band structure of semiconductors in order to gain deep insight about the defects. Introduction of impurities during the fabrication or growth process and the lattice imperfections give rise to the energy states in the energy gap between VB and CB, commonly known as defect states. As discussed previously, shallow defects are mostly benign and correspond to the states present within a range of few tens of meV from the respective band edges. On the other hand, deep defects, as the name suggests, penetrate deeply into the energy bandgap and affect majorly the optical and electronic properties of the material [114]. In a typical semiconductor, charge carriers are generated after photo-excitation. Equation 2 shows the charge carriers' recombination as:

$$dt/dn = k_1 n_1 + k_2 n_2 + k_3 n_3 \tag{2}$$

Here, n represents the charge carrier density and k_1, k_2 and k_3 defines the rate constant for the three different kinds of recombination. The monomolecular recombination, as defined by k_1, , is mainly associated with either excitonic (where electrons and holes are bound by coulombic interactions) or trap-assisted recombination (when charge carrier gets captured in a trap) and is directly proportional to the charge carrier density, n. The rate constant k_1, , represents non-radiative recombination and is undesirable for photovoltaic application. The second kind of recombination is known as bimolecular recombination (rate constant, k_2) and is dependent on the density of both the charge carriers viz., electrons and holes, therefore is proportional to n [2] and corresponds to radiative recombination (resulting in luminescence). The rate constant k_3 is associated with Auger recombination (a three-particle process). It is a non-radiative process which includes the transfer of energy gained from charge carriers' recombination to electrons (or holes) which are then excited to higher states.

Shockley-Read Hall (SRH) theory describes the non-radiative recombination, R, mediated through defect states and is given by [115]:

$$R = \frac{n_e n_h - n_i^2}{T_h\left(n_e + N_c \exp\left(-\frac{E_c - E_T}{k_B T}\right)\right) + T_e\left(n_h + N_V \exp\left(-\frac{E_T - E_V}{k_B T}\right)\right)} \tag{3}$$

Here, n_e, n_h and n_i stands for electron, hole and intrinsic carrier density. T stands for lifetime, E_C (E_V) for the CB (VB) energy and N_C (N_V) for the effective density of states DOS of CB (VB). k_B is Boltzmann constant and E_T denotes defect energy level.

Assuming, $n_e \approx n_h >> n_i$,

$$R = n2T \tag{4}$$

$$T\alpha \frac{1}{N_T \sigma_T} \tag{5}$$

where, N_T and σ_T are number of traps and capture cross-section, respectively.

In case of lead-based halide perovskites, due to the high ability of $MAPbI_3$ to screen charge (high dielectric constant), σ_T is low and low N_T enables halide perovskites to have long carrier lifetime, making them suitable absorbing material for photovoltaics [116].

4.2 Halide Perovskites and Defects

In past few years, LHPs have revolutionized the photovoltaic and optoelectronic industry because of their excellent optoelectronic properties with added benefit of lower fabrication costs among other semiconductors. Particularly, Cs-based lead halide perovskite NCs, due to high stability and PLQY, are receiving great attention for their potential in display and light-emitting technologies [117, 118]. Though the LHPs are defect-tolerant, the synthesis conditions (reaction temperature, halide rich/poor conditions etc.) often promote the formation of defect states (both for bulk as well as NCs) as evident from their PLQY values which is far from unity

[28, 29]. The defects in crystalline semiconductors can mainly be characterized as crystallographic defects (imperfections in the crystal lattice such as vacancy, anti-site, interstitial, dislocation etc.) and impurities (addition of any foreign elements in the structure). The measured density of defects in bulk LHPs fall in the range of 10^{11}–10^{16} per cm^3 while, this value drops to one defect in ~10^2–10^7 for a 10 nm perovskite NC, which is too low to explain significant loss in PL efficiency in NCs [119, 120]. The formation of defects at nanoscale is quite different from that in bulk mainly because of the two reasons: (a) the presence of surfactants on the surface of NCs creates a different environment for the defects than in bulk and (b) due to large number of surface atoms, the defects are mostly present on the surface of the NCs than in the bulk [120]. During the washing of NCs, the surfactants can detach from the surface of NCs which might result in aggregation and precipitation of NCs and exposing the surface to defects [70, 120]. Furthermore, in case of NCs (with size in the range 8–10 nm), the defects formed in the bulk/core of the material might eventually travel to the surface, resulting in the formation of the trapping sites for the charge carriers at the surface. This is also seen in perovskite thin films based solar cells where, the loss in efficiency is majorly because of the carrier recombination at the grain boundaries. Also, the structural defects arising from the deviation of an atom(s) from the designated position in an ideal periodic crystal lattice introduces the disorder into the lattice affecting the mobility of free charge carriers as demonstrated in Fig. 15a [121]. Here, in addition to the carrier transport bands, localized electronic states are formed as a result of crystalline imperfections which act as trapping sites for photogenerated charge carriers (Fig. 15b). Depending upon the nature of the trap, the carriers can trace back to the CB by gaining energy either from further optical excitations or through thermal energy at RT ($k_BT \sim 26$ meV). As discussed earlier as well, the defect levels just above the CB and VB edges are benign (shallow defects), whereas, those residing in between the CB and VB edges (deep defects) are harmful as they can trap the charges. The difference in energy between the transport states and the defect levels is known as the trap depth (ΔE). As seen from Fig. 15c, for shallow traps, $\Delta E \leq k_BT$ (very small) and the mobility of free charge carriers are restricted only through trapping/detrapping processes. On the other hand, deep traps are found in the middle of the energy gap with $\Delta E > k_BT$ and accelerate non-radiative recombination by hindering detrapping processes. The formation of these disorders or defects affect adversely the PCE of photovoltaic or optoelectronic devices.

4.2.1 Theoretical Perspective on the Native Defects in MHPs

In order to gain further insights about the defect chemistry of NCs, DFT calculations turned out to be a powerful tool. Brinck et al. employed DFT simulation to compute the defect formation energies (DFEs) of vacancy, antisite and interstitial defects in different regions of NCs [120]. For this, the basic difference between the defect theory for bulk and NCs need to be investigated for better understanding of how these defect/trap states can be suppressed to prevent crystal degradation [45, 120]. Studies show that the interstitial defects are energetically more favourable than the

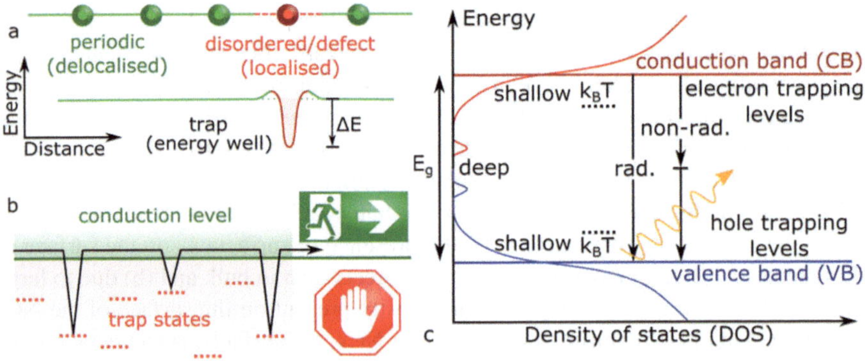

Fig. 15 **a** Schematic diagram of localized defects formed due to the structural defects in the semiconductor (top) and the energy levels introduced by these defects (bottom). **b** Trapping and detrapping processes slow down the transport of free charge carriers. **c** Density of states show the formation of shallow and deep defects with the corresponding radiative and non-radiative recombination in a disordered semiconductor [121]

vacancies and antisites which require large formation energies. However, interstitials only favour the formation of shallow trap states and do not have adverse effect on the material. On the other hand, vacancies result in the formation of deep traps. The density of these vacancies should be very high in order to actually transform them into trap states due to the high defect tolerance of perovskite structure. Experimentally, formation of these defect levels is evident due to the detachment of ligands/surfactants during the purification processes [122].

In this section, we will mainly focus on the defect formation in $CsPbBr_3$ NCs because of their high phase stability, optical purity and great tolerance against surface treatments [120]. For bulk systems, DFE is simulated using the following equation:

$$DFE(\alpha, q) = E(\alpha, q) - E(bulk) - \sum_A n_A \mu_A + q(E_F + E_{VBM} + \Delta V) + E^q \quad (6)$$

where, $E(bulk)$ and $E(\alpha, q)$ are the total energy of the pristine supercell and supercell with defect α and charge q, respectively; n_A and μ_A are the number and chemical potentials of the atom added/subtracted to form the defect; E_F is the Fermi energy; ΔV is the defect-induced electrostatic shift; E_{VBM} is the energy corresponding to the valence band maxima and E^q is the background electrostatic charge [123].

The accuracy of the DFE largely depends on choices of different exchange–correlation functionals e.g., semilocal, GW and hybrid along with or without spin–orbit coupling (soc) mostly for the correct estimation of the band edge energies [124]. However, this method is not effective for the NC systems as the perovskite NCs are dispersed in solution during the synthesis process. This results in the formation of the molecular ion pairs, viz., $PbBr_2$ and CsBr which further promotes the creation of the vacancy or interstitial defects in NCs. For instance, during dispersion, detachment of

$PbBr_2$ ion pair result in the formation of Pb vacancy ($V_{Pb}{}^{2-}$). Additionally, the Br^- ions are also removed from the lattice during the formation of lead vacancy therefore, the correct description for the defect is given by $V_{Pb}{}^{2-} + 2V_{Br}{}^{+}$. In NC model, the DFEs are calculated using the simplified equation given as follows:

$$DFE(\alpha) = E(NC(\alpha)) - E(NC) \pm E(s) \quad (7)$$

where, $E(s)$ is the energy of neutral ion pairs (e.g., $PbBr_2$) and $E(NC)$ and $E(NC(\alpha))$ are the total energies of the pristine NC and NC with defect α (referred to $V_{Pb}{}^{2-} + 2V_{Br}{}^{+}$ here). The DFE calculation in NC model take care of the charge neutrality as well as the background charge correction because of the explicit inclusion of the whole NC.

In the $CsPbBr_3$ NC, three different combination of defects/counterion (needed to neutralize the charge) are investigated: (a) core (c)/surface, (b) surface-centre (sc)/surface and (c) surface-edge (se)/surface. All the cases mentioned here show the energetically favourable formation of interstitial positions, as evident from Fig. 16. This implies that the addition of $CsBr/PbBr_2$ ion pairs in the lattice is energetically favourable and leads to the formation of interstitial sites (Fig. 17a–b). Experimentally, this infers the easy accommodation of metal salts from the environment, if available. Now, among the different interstitial sites, surface edge is usually more preferable than core. It is important to note that the addition of halide (counterion) results in the emergence of mid-band states irrespective of the cationic/anionic interstitial defects. Further, the formation of deep trap state is favoured by the occupancy of halide anion at the surface positions rather than the core. The creation of trap states is attributed to the dangling and non-bonding lone pairs of the excess halide present at the surface sites.

Removal of ions from the surface or the core results in the formation of vacancies, while the excess counterions need to be removed from the surface to neutralize the

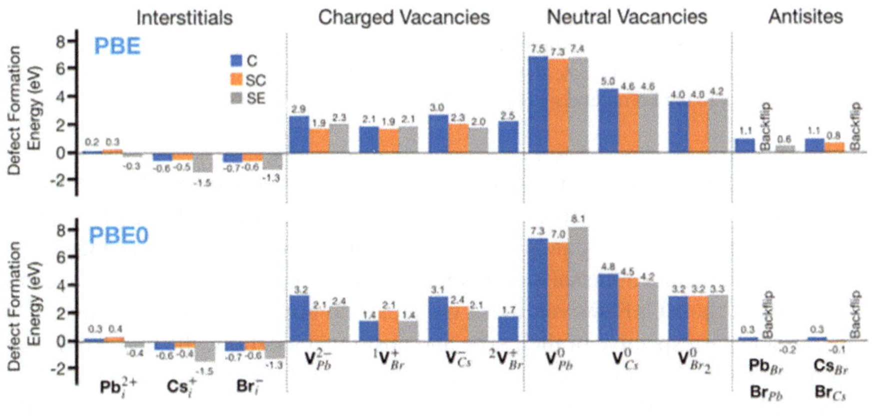

Fig. 16 DFE for interstitials, charged, neutral vacancies and antisite substitutions. All the DFEs are simulated using PBE/PBE0 (Perdew-Burke-Ernzerhof) exchange–correlation functionals [120]

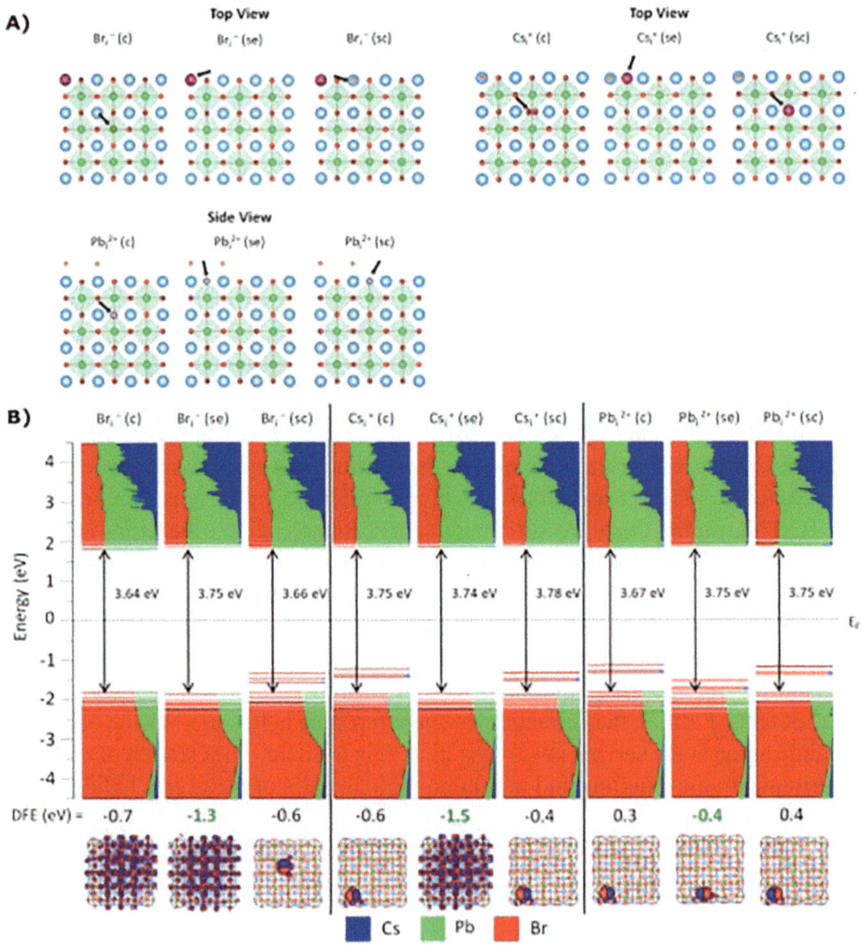

Fig. 17 **a** Illustration of the interstitial defects in the core, surface centre and surface edge of the NCs. **b** Density of states (DOS) calculated using PBE/PBE0 exchange–correlation functionals. The dotted line refers to the Fermi energy (E_F) level [120]

overall charge (Fig. 18a). From Fig. 18b, we can observe that to create these vacancies an external energy ranging from 1.4–2.3 eV is required which ultimately signifies the presence of very low vacancy densities. The creation of a neutral vacancy due to the removal of metal halide salts is energetically more favourable at the surface than in the core of the NC. However, these vacancies do not engage in the formation of any trap states irrespective of their type and position. Generally, the mid-band states originate when CsBr moieties, in large quantities, are removed from the outer surface of the NC lattice [125]. Moreover, attaining this kind of situation is very difficult due to the high DFE of these vacancies. Experimentally, the synthesized perovskite NCs are not ideal, where, the metalorganic salts for instance alkylammonium bromides are present at the surface than the simple CsBr as considered in the calculations. The moderate solubility of the surfactants in the organic solvent leads to the loss of these capping species. This results in the stripping of the metal halide salts during the purification process and promotes the formation of a large number of surface vacancies [122]. In this case, PL efficiency reduces due to the introduction of trap states. In contrast, the formation of the antisite i.e. the exchange of a cation position with an anion finally results in the reorganization of the structure to a non-antisite configuration (back-flipping of the corresponding ions to their original position). Due to this, the shallow traps are formed which do not affect electronic properties of the NCs significantly.

PL spectroscopy helps in gaining information about the excited states and charge dynamics of a semiconducting material. The technique, thus, can be used efficiently for the complete understanding of the electro-optical properties of MHP NCs. Generally, the high efficiency in halide perovskites is related to the fast generation of electron–hole pairs as a result of photo-illumination and long diffusion lengths. In addition to this, high PLQY as shown by these materials made possible the efficient use of MHP NCs for optoelectronic applications. For high emission intensity (PL), the efficient charge recombination is needed. However, charge trapping (electron/hole) after photogeneration in the mid-band defect states often results in non-radiative recombination. More importantly, the environmental factors such as air, O_2, moisture etc. affect significantly the density of the defects in the "soft" crystal lattices of MHPs majorly due to the trapped carriers [126, 127]. The trap-mediated carrier dynamics in MHP NCs can be effectively controlled once the reasons for optical instability in these materials under various environmental conditions are well understood. In this context, it is imperative to study the PL intermittency i.e., blinking in MHP NCs under ambient conditions to understand the nature of the defects.

4.2.2 Defect-Mediated Carrier Recombination in MHP NCs Favoured by Atmospheric Constituents

Initially, LHPs grabbed much attention for their efficient use in photovoltaic devices, however, over the past few years, their use in optoelectronic applications such as LEDs or lasers has been promising [55]. It is important to note that the emissive properties of MHP crystals might change with reduction in their size. Roy et al. reported the phase transition of $MAPbBr_3$ NPLs from quasi-2D to 3D under photo-illumination, typically referred as photo-driven phase transformation (PDPT) process [128]. Here, the red shift (78 nm) as well as PL quenching of emission spectra was observed under continuous photo-illumination as a result of increase in the layer thickness of the perovskite platelets. This can be explained in two steps: first involves the removal of the barrier layer i.e. the surfactants (OLm) from the surface of the NPLs due to the external energy supplied to the system in the form of photon irradiation. Due to the low binding energy of the surfactants, the ligands easily get released, resulting in the formation of surface defects promoting carrier trapping. The trap states, thus formed, stimulate the non-radiative recombination leading to quenching

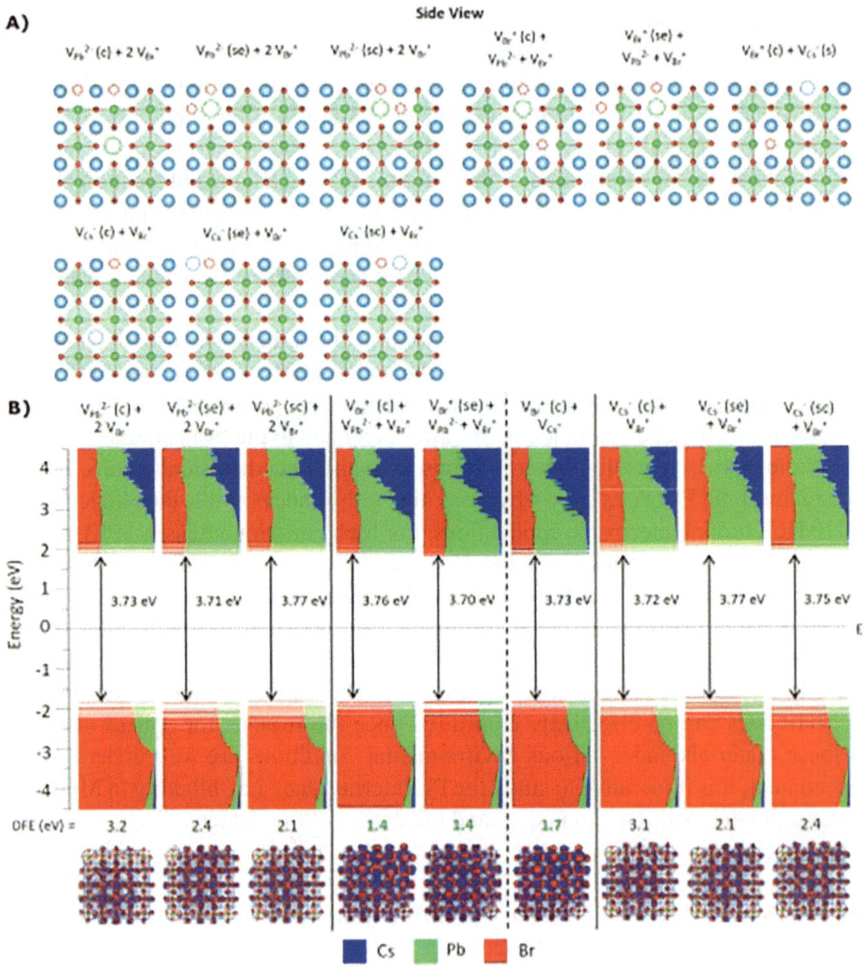

Fig. 18 **a** Illustration of the vacancy defects in the core, surface centre and surface edge of the NCs. **b** Density of states (DOS) calculated using PBE/PBE0 exchange–correlation functionals. The dotted line refers to the Fermi energy (E_F) level [120]

of the emission spectra. Second is the coalescence of the perovskite layers due to the strong ion mobilization between them to form a 3D structure. Recently, various groups also reported the observation of intermittency or blinking in PL when MHP NCs are subjected to continuous light illumination [129, 130]. In case of hybrid organic–inorganic lead-based halide perovskites, the intermittency or blinking is observed in nanostructures with size well above the Bohr's exciton radius [131]. Mukherjee et al. studied the interaction of the organo-metallic halide perovskite (OHP) NCs with surrounding gases in order to probe the nature of the active defects present in these NCs and their influence on the carrier dynamics [132]. For this,

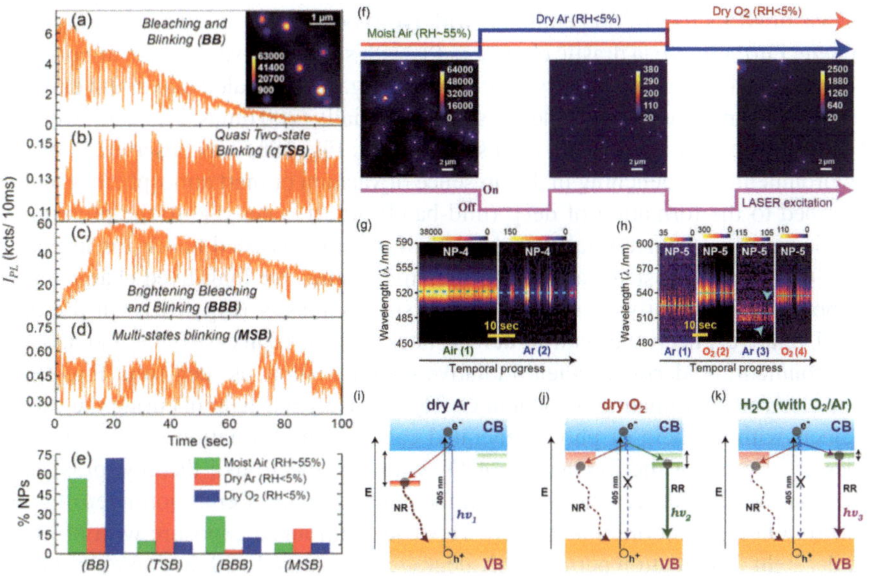

Fig. 19 **a–d** Different PL intermittency and intensity (I_{PL}) trajectories of various individual $MAPbBr_3$ NPs under similar environmental conditions. **e** PL blinking frequency under moist air (investigated for 221 emitters), dry Ar (investigated for 504 emitters) and dry O_2 (investigated for 604 emitters). **f** Inhomogeneous emission intensity of $MAPbBr_3$ NPs as shown by PL images under varied atmospheres. **g** Maximum emission intensity in ambient air decreases substantially by about 200 times in dry Ar whereas, **h** further tenfold increment in emission intensity was observed in dry O_2 environment. Illustration representing the pathways for charge carrier recombination under **i** dry Ar, **j** dry O_2 and **k** moist O_2 or Ar. The straight and dashed lines represent the radiative (RR) and non-radiative (NR) recombination, respectively. The thickness of lines represents the efficiencies qualitatively [132].

the authors first investigated the temporal instabilities of PL spectra of $MAPbBr_3$ NPLs under different environment conditions such as dry and humid Argon (Ar) and O_2 as well as moist (RH $\sim$ 55%) ambient environment (Air). Figure 18a–d shows the various PL intensity (I_{PL}) trajectories of different individual NPs under similar conditions (ambient air). Based on these observations, the temporal fluctuations or blinking of single NPs can be classified into *as-stated* four categories: "(a) Bleaching with blinking (BB) (high initial intensity), (b) Quasi two-state blinking (qTSB) (low to moderate intensity) (c) Photo-brightening and bleaching with blinking (BBB) (moderate to high intensity) and (d) Multistate blinking (MSB) (low to moderate intensity)" [132]. Further, on changing the surrounding environment, the remarkable variation in the behaviour of the emitters was noticed. As evident from Fig. 16e, when the measurements were done in moist air, BB ($\sim$57%) was the dominant type of blinking among the emitters (221 individual NPs were investigated) however, in an inert atmosphere (dry Ar), there was a significant drop in the PL intensity with majority of qTSB ($\sim$60%) type blinking along with minor MSB behaviour ($\sim$20%). On the other hand, a considerable improvement in PL intensity was observed for

~75% of NPLs (out of 605 emitters) with BB type blinking in case of dry O_2 atmosphere. Furthermore, a drastic change in PL with sequential variation in atmospheric constituents i.e., moist air to dry Ar to dry O_2 (Fig. 18f) revealed maximum emission intensity in ambient air which decreased substantially (200 times; Fig. 18g) in dry Ar. Further increment in emission intensity (tenfold; Fig. 18h) was observed in dry O_2 environment. PL quenching in the presence of vacuum or inert atmosphere is mainly ascribed to the formation of deep (mid-band) defect states due to the presence of halide vacancy or interstitial sites as a result of photo-assisted halide ion migration [131]. These defect states act as non-radiative centres for long-lived charge carrier trapping, subsequently quenching the emission intensity (Fig. 18i). On the other hand, relatively deep and shallow defect states are formed in case of O_2 and moisture environment, rendering efficient radiative recombination, as seen from Fig. 18j, k. The electron coordinating oxygen in O_2/H_2O constituents help in curing the surface defects with the help of photo-generated carriers (the formation of O_2^- superoxides as a result of the interaction of O_2 with free electrons), thereby, passivating the non-radiative defects and result in the enhancement of emission intensity with mostly On-type intermittency [133]. However, prolonged exposure with moist O_2 under light illumination permanently degrade the material with irreversible spectral changes whereas, PL intensity could be restored while going from moist to dry conditions in inert (Ar) atmosphere. This concludes that individually O_2 or moisture does not have adverse effect on the optical properties of MHP NCs as the combination of two i.e., moist O_2 has on the spectral stability of these materials.

Tian et al. reported that the chemical/crystal defects attributing to the formation of trapping sites are often present on the surfaces (corners or edges) with large geometrical distortions exposed to the ambient conditions [129]. The defect site can act as a PL quencher or an emitter based on the chemical structure which leads to the formation of different local energy levels [134]. The PL blinking in MHP NCs might be due to the activation and deactivation of these defect states under light illumination. Additionally, as seen earlier, emission intensity gets adversely affected due to the combined effect of O_2 with moisture which increases the density of such defects into the material and further leads to an enhanced non-radiative recombination. This has a detrimental effect on the radiative efficiency and might affect the charge transport properties of the material as well. Therefore, it is very crucial to control the population of these defects in order to make efficient use of these perovskite NCs.

5 Summary

To summarize, we have presented several syntheses methods and their advantages/disadvantages for the preparation of halides perovskite nanocrystals. A deeper insight about the possible defects and their role in dictating optoelectronic properties is portrayed in detail both from theoretical as well as experimental perspective. HI method can be used successfully in the preparation of monodisperse and high-quality LHP NCs with a good control over their composition, shape (QDs, NPLs,

NRs, nanocubes and NWs), size and distribution. The use of halide salts both as a source for metal cation and halogen result in the formation of halide deficit NCs. The undercoordinated lead atoms/NPs thus formed on the surface lead to the non-radiative recombination thereby, quenching the PL emission. This encouraged the development of the modified-HI synthesis method which employs the use of benzoyl or silyl halides (halide-rich conditions) and provided the flexibility to change the metal to halide ratio for a precise composition control. However, the stripping of the metal halide salts during the purification process promotes the formation of large number of surface vacancies. Theoretically calculated defect formation energies (DFEs) for the interstitial, vacancy and antisite defects at different sites i.e., core, surface-centre and surface-edge of the NC showed the surface to be the most stable position for defect formation in NCs. Interestingly, only the excessive detachment of capping species from the surface of NCs leave behind the undercoordinated X^- (halide) ions which might lead to the formation of deep trap states. Additionally, the vacancies at the surface of NCs trap moisture or O_2 present in the environment and accelerate degradation of the perovskite NCs.

References

1. Wells HL (1893) Z Anorg Chem 3:195–210
2. Mitzi DB, Chondroudis K, Kagan CR (2001) IBM J Res Dev 45:29–45
3. Mitzi DB, Feild CA, Harrison WTA, Guloy AM (1994) Nature 369:467–469
4. Kojima A, Teshima K, Shirai Y, Miyasaka T (2009) J Am Chem Soc 131:6050–6051
5. Kim H-S, Lee C-R, Im J-H, Lee K-B, Moehl T, Marchioro A, Moon S-J, Humphry-Baker R, Yum J-H, Moser JE, Grätzel M, Park N-G (2012) Sci Rep 2:591
6. Chung I, Lee B, He J, Chang RPH, Kanatzidis MG (2012) Nature 485:486–489
7. N T Energy J (2020)
8. Xing G, Mathews N, Sun S, Lim Swee S, Lam Yeng M, Grätzel M, Mhaisalkar S, Sum Tze C (2013) Science 342:344–347
9. Manser JS, Christians JA, Kamat PV (2016) Chem Rev 116:12956–13008
10. Tan Z-K, Moghaddam RS, Lai ML, Docampo P, Higler R, Deschler F, Price M, Sadhanala A, Pazos LM, Credgington D, Hanusch F, Bein T, Snaith HJ, Friend RH (2014) Nat Nanotechnol 9:687–692
11. Deschler F, Price M, Pathak S, Klintberg LE, Jarausch D-D, Higler R, Hüttner S, Leijtens T, Stranks SD, Snaith HJ, Atatüre M, Phillips RT, Friend RH (2014) J Phys Chem Lett 5:1421–1426
12. D'Innocenzo V, Srimath Kandada AR, De Bastiani M, Gandini M, Petrozza A (2014) J Am Chem Soc 136:17730–17733
13. Zhang L, Yang X, Jiang Q, Wang P, Yin Z, Zhang X, Tan H, Yang YM, Wei M, Sutherland BR, Sargent EH, You J (2017) Nat Commun 8:15640
14. de Quilettes Dane W, Vorpahl Sarah M, Stranks Samuel D, Nagaoka H, Eperon Giles E, Ziffer Mark E, Snaith Henry J, Ginger David S (2015) Science 348:683–686
15. Uratani H, Yamashita K (2017) J Phys Chem Lett 8:742–746
16. Soe CMM, Nagabhushana GP, Shivaramaiah R, Tsai H, Nie W, Blancon J-C, Melkonyan F, Cao DH, Traoré B, Pedesseau L, Kepenekian M, Katan C, Even J, Marks TJ, Navrotsky A, Mohite AD, Stoumpos CC, Kanatzidis MG (2019) Proc Natl Acad Sci 116:58
17. Bohn BJ, Tong Y, Gramlich M, Lai ML, Döblinger M, Wang K, Hoye RLZ, Müller-Buschbaum P, Stranks SD, Urban AS, Polavarapu L, Feldmann J (2018) Nano Lett 18:5231–5238

18. Zheng Y, Niu T, Ran X, Qiu J, Li B, Xia Y, Chen Y, Huang W (2019) J Mater Chem A 7:13860–13872
19. Li C, Li J, Li Z, Zhang H, Dang Y, Kong F (2021) CrystEngComm 23:3619–3630
20. Schmidt LC, Pertegás A, González-Carrero S, Malinkiewicz O, Agouram S, Mínguez Espallargas G, Bolink HJ, Galian RE, Pérez-Prieto J (2014) J Am Chem Soc 136:850–853
21. Gonzalez-Carrero S, Galian RE, Pérez-Prieto J (2015) J Mater Chem A 3:9187–9193
22. Zhang F, Zhong H, Chen C, Wu X-G, Hu X, Huang H, Han J, Zou B, Dong Y (2015) ACS Nano 9:4533–4542
23. Imran M, Di Stasio F, Dang Z, Canale C, Khan AH, Shamsi J, Brescia R, Prato M, Manna L (2016) Chem Mater 28:6450–6454
24. Zhang D, Yang Y, Bekenstein Y, Yu Y, Gibson NA, Wong AB, Eaton SW, Kornienko N, Kong Q, Lai M, Alivisatos AP, Leone SR, Yang P (2016) J Am Chem Soc 138:7236–7239
25. Akkerman QA, Motti SG, Srimath Kandada AR, Mosconi E, D'Innocenzo V, Bertoni G, Marras S, Kamino BA, Miranda L, De Angelis F, Petrozza A, Prato M, Manna L (2016) J Am Chem Soc 138:1010–1016
26. Huang H, Polavarapu L, Sichert JA, Susha AS, Urban AS, Rogach AL (2016) NPG Asia Mater 8:e328–e328
27. Bekenstein Y, Koscher BA, Eaton SW, Yang P, Alivisatos AP (2015) J Am Chem Soc 137:16008–16011
28. Dutta A, Behera RK, Pal P, Baitalik S, Pradhan N (2019) Angew Chem Int Ed 58:5552–5556
29. Zhong Q, Cao M, Xu Y, Li P, Zhang Y, Hu H, Yang D, Xu Y, Wang L, Li Y, Zhang X, Zhang Q (2019) Nano Lett 19:4151–4157
30. Hintermayr VA, Richter AF, Ehrat F, Döblinger M, Vanderlinden W, Sichert JA, Tong Y, Polavarapu L, Feldmann J, Urban AS (2016) Adv Mater 28:9478–9485
31. Nedelcu G, Protesescu L, Yakunin S, Bodnarchuk MI, Grotevent MJ, Kovalenko MV (2015) Nano Lett 15:5635–5640
32. De Roo J, Ibáñez M, Geiregat P, Nedelcu G, Walravens W, Maes J, Martins JC, Van Driessche I, Kovalenko MV, Hens Z (2016) ACS Nano 10:2071–2081
33. Behera RK, Das Adhikari S, Dutta SK, Dutta A, Pradhan N (2018) J Phys Chem Lett **9**:6884–6891
34. Koscher BA, Swabeck JK, Bronstein ND, Alivisatos AP (2017) J Am Chem Soc 139:6566–6569
35. Manspeaker C, Venkatesan S, Zakhidov A, Martirosyan KS (2017) Curr Opin Chem Eng 15:1–7
36. Kim S, Bae S, Lee S-W, Cho K, Lee KD, Kim H, Park S, Kwon G, Ahn S-W, Lee H-M, Kang Y, Lee H-S, Kim D (2017) Sci Rep 7:1200
37. Li J, Dong Q, Li N, Wang L (2017) Adv Energy Mater 7:1602922
38. McClure ET, Ball MR, Windl W, Woodward PM (2016) Chem Mater 28:1348–1354
39. Slavney AH, Hu T, Lindenberg AM, Karunadasa HI (2016) J Am Chem Soc 138:2138–2141
40. Volonakis G, Filip MR, Haghighirad AA, Sakai N, Wenger B, Snaith HJ, Giustino F (2016) J Phys Chem Lett 7:1254–1259
41. Zhang C, Gao L, Teo S, Guo Z, Xu Z, Zhao S, Ma T (2018) Sustain Energy Fuels 2:2419–2428
42. Volonakis G, Haghighirad AA, Milot RL, Sio WH, Filip MR, Wenger B, Johnston MB, Herz LM, Snaith HJ, Giustino F (2017) J Phys Chem Lett 8:772–778
43. Bhawna, Roy M, Vikram, Borkar H, Alam A, Aslam M (2021) Chem Commun 57:1478–1481
44. Bhawna V, Roy M, Ghorui S, Alam A, Aslam M (2022) J Mater Chem C. https://doi.org/10.1039/D1TC05877K
45. Kangsabanik J, Ghorui S, Aslam M, Alam A (2020) Phys Rev Appl 13:014005
46. Roy M, Ghorui S, Bhawna, Kangsabanik J, Yadav R, Alam A, Aslam M (2020) J Phys Chem C 124:19484–19491
47. Dey A, Ye J, De A, Debroye E, Ha SK, Bladt E, Kshirsagar AS, Wang Z, Yin J, Wang Y, Quan LN, Yan F, Gao M, Li X, Shamsi J, Debnath T, Cao M, Scheel MA, Kumar S, Steele JA, Gerhard M, Chouhan L, Xu K, Wu X-G, Li Y, Zhang Y, Dutta A, Han C, Vincon I, Rogach AL, Nag A, Samanta A, Korgel BA, Shih C-J, Gamelin DR, Son DH, Zeng H, Zhong H, Sun

H, Demir HV, Scheblykin IG, Mora-Seró I, Stolarczyk JK, Zhang JZ, Feldmann J, Hofkens J, Luther JM, Pérez-Prieto J, Li L, Manna L, Bodnarchuk MI, Kovalenko MV, Roeffaers MBJ, Pradhan N, Mohammed OF, Bakr OM, Yang P, Müller-Buschbaum P, Kamat PV, Bao Q, Zhang Q, Krahne R, Galian RE, Stranks SD, Bals S, Biju V, Tisdale WA, Yan Y, Hoye RLZ, Polavarapu L (2021) ACS Nano 15:10775–10981
48. Zheng X, Hou Y, Sun H-T, Mohammed OF, Sargent EH, Bakr OM (2019) J Phys Chem Lett 10:2629–2640
49. Stranks SD, Burlakov VM, Leijtens T, Ball JM, Goriely A, Snaith HJ (2014) Phys Rev Appl 2:034007
50. Srimath Kandada AR, Neutzner S, D'Innocenzo V, Tassone F, Gandini M, Akkerman QA, Prato M, Manna L, Petrozza A, Lanzani G (2016) J Am Chem Soc 138:13604–13611
51. Rossetti R, Nakahara S, Brus LE (1983) J Chem Phys 79:1086–1088
52. Brus L (1986) J Phys Chem 90:2555–2560
53. Li X, Cao F, Yu D, Chen J, Sun Z, Shen Y, Zhu Y, Wang L, Wei Y, Wu Y, Zeng H (2017) Small 13:1603996
54. Yang D, Cao M, Zhong Q, Li P, Zhang X, Zhang Q (2019) J Mater Chem C 7:757–789
55. Veldhuis SA, Boix PP, Yantara N, Li M, Sum TC, Mathews N, Mhaisalkar SG (2016) Adv Mater 28:6804–6834
56. Protesescu L, Yakunin S, Bodnarchuk MI, Krieg F, Caputo R, Hendon CH, Yang RX, Walsh A, Kovalenko MV (2015) Nano Lett 15:3692–3696
57. Giustino F, Snaith HJ (2016) ACS Energy Lett 1:1233–1240
58. Cortecchia D, Dewi HA, Yin J, Bruno A, Chen S, Baikie T, Boix PP, Grätzel M, Mhaisalkar S, Soci C, Mathews N (2016) Inorg Chem 55:1044–1052
59. Pilania G, Balachandran PV, Kim C, Lookman T (2016) Front Mater 3:19
60. Correa-Baena J-P, Saliba M, Buonassisi T, Grätzel M, Abate A, Tress W, Hagfeldt A (2017) Science 358:739–744
61. Protesescu L, Yakunin S, Kumar S, Bär J, Bertolotti F, Masciocchi N, Guagliardi A, Grotevent M, Shorubalko I, Bodnarchuk MI, Shih CJ, Kovalenko MV (2017) ACS Nano 11:3119–3134
62. Snaith HJ (2018) Nat Mater 17:372–376
63. Kim G-W, Petrozza A (2020) Adv Energy Mater 10:2001959
64. Zakutayev A, Caskey CM, Fioretti AN, Ginley DS, Vidal J, Stevanovic V, Tea E, Lany S (2014) J Phys Chem Lett 5:1117–1125
65. Brandt RE, Poindexter JR, Gorai P, Kurchin RC, Hoye RLZ, Nienhaus L, Wilson MWB, Polizzotti JA, Sereika R, Žaltauskas R, Lee LC, MacManus-Driscoll JL, Bawendi M, Stevanović V, Buonassisi T (2017) Chem Mater 29:4667–4674
66. Protesescu L, Yakunin S, Nazarenko O, Dirin DN, Kovalenko MV (2018) ACS Appl Nano Mater 1:1300–1308
67. Murray CB, Norris DJ, Bawendi MG (1993) J Am Chem Soc 115:8706–8715
68. Bullen CR, Mulvaney P (2004) Nano Lett 4:2303–2307
69. Roy M, Vikram, Banerjee S, Mitra A, Alam A, Aslam M (2019) Chem—A European J 25:9892–9901
70. Pan A, He B, Fan X, Liu Z, Urban JJ, Alivisatos AP, He L, Liu Y (2016) ACS Nano 10:7943–7954
71. Vybornyi O, Yakunin S, Kovalenko MV (2016) Nanoscale 8:6278–6283
72. Protesescu L, Yakunin S, Bodnarchuk MI, Bertolotti F, Masciocchi N, Guagliardi A, Kovalenko MV (2016) J Am Chem Soc 138:14202–14205
73. Lignos I, Protesescu L, Emiroglu DB, Maceiczyk R, Schneider S, Kovalenko MV, deMello AJ (2018) Nano Lett 18:1246–1252
74. Protesescu L, Yakunin S, Kumar S, Bär J, Bertolotti F, Masciocchi N, Guagliardi A, Grotevent M, Shorubalko I, Bodnarchuk MI, Shih C-J, Kovalenko MV (2017) ACS Nano 11:3119–3134
75. Akkerman QA, Park S, Radicchi E, Nunzi F, Mosconi E, De Angelis F, Brescia R, Rastogi P, Prato M, Manna L (2017) Nano Lett 17:1924–1930
76. Han C, Li C, Zang Z, Wang M, Sun K, Tang X, Du J (2017) Photon Res 5:473–480

77. Yuan L, Patterson R, Wen X, Zhang Z, Conibeer G, Huang S (2017) J Colloid Interface Sci 504:586–592
78. Woo JY, Kim Y, Bae J, Kim TG, Kim JW, Lee DC, Jeong S (2017) Chem Mater 29:7088–7092
79. Liu P, Chen W, Wang W, Xu B, Wu D, Hao J, Cao W, Fang F, Li Y, Zeng Y, Pan R, Chen S, Cao W, Sun XW, Wang K (2017) Chem Mater 29:5168–5173
80. Yassitepe E, Yang Z, Voznyy O, Kim Y, Walters G, Castañeda JA, Kanjanaboos P, Yuan M, Gong X, Fan F, Pan J, Hoogland S, Comin R, Bakr OM, Padilha LA, Nogueira AF, Sargent EH (2016) Adv Func Mater 26:8757–8763
81. Imran M, Caligiuri V, Wang M, Goldoni L, Prato M, Krahne R, De Trizio L, Manna L (2018) J Am Chem Soc 140:2656–2664
82. Creutz SE, Crites EN, De Siena MC, Gamelin DR (2018) Nano Lett 18:1118–1123
83. Yuan Z, Shu Y, Xin Y, Ma B (2016) Chem Commun 52:3887–3890
84. Zhang D, Eaton SW, Yu Y, Dou L, Yang P (2015) J Am Chem Soc 137:9230–9233
85. Shamsi J, Urban AS, Imran M, De Trizio L, Manna L (2019) Chem Rev 119:3296–3348
86. Almeida G, Goldoni L, Akkerman Q, Dang Z, Khan AH, Marras S, Moreels I, Manna L (2018) ACS Nano 12:1704–1711
87. Zhang D, Yu Y, Bekenstein Y, Wong AB, Alivisatos AP, Yang P (2016) J Am Chem Soc 138:13155–13158
88. Chen M, Zou Y, Wu L, Pan Q, Yang D, Hu H, Tan Y, Zhong Q, Xu Y, Liu H, Sun B, Zhang Q (2017) Adv Func Mater 27:1701121
89. Li X, Wu Y, Zhang S, Cai B, Gu Y, Song J, Zeng H (2016) Adv Func Mater 26:2435–2445
90. Pan Q, Hu H, Zou Y, Chen M, Wu L, Yang D, Yuan X, Fan J, Sun B, Zhang Q (2017) J Mater Chem C 5:10947–10954
91. Tong Y, Bladt E, Aygüler MF, Manzi A, Milowska KZ, Hintermayr VA, Docampo P, Bals S, Urban AS, Polavarapu L, Feldmann J (2016) Angew Chem Int Ed 55:13887–13892
92. Wei S, Yang Y, Kang X, Wang L, Huang L, Pan D (2016) Chem Commun 52:7265–7268
93. Shamsi J, Rastogi P, Caligiuri V, Abdelhady AL, Spirito D, Manna L, Krahne R (2017) ACS Nano 11:10206–10213
94. Papavassiliou GC, Pagona G, Karousis N, Mousdis GA, Koutselas I, Vassilakopoulou A (2012) J Mater Chem 22:8271–8280
95. Huang H, Susha AS, Kershaw SV, Hung TF, Rogach AL (2015) Adv Sci 2:1500194
96. Zhao Y, Xu X, You X (2016) Sci Rep 6:35931
97. Shamsi J, Abdelhady AL, Accornero S, Arciniegas M, Goldoni L, Kandada ARS, Petrozza A, Manna L (2016) ACS Energy Lett 1:1042–1048
98. Levchuk I, Osvet A, Tang X, Brandl M, Perea JD, Hoegl F, Matt GJ, Hock R, Batentschuk M, Brabec CJ (2017) Nano Lett 17:2765–2770
99. Seth S, Samanta A (2016) Sci Rep 6:37693
100. Ahmed GH, Yin J, Bose R, Sinatra L, Alarousu E, Yengel E, AlYami NM, Saidaminov MI, Zhang Y, Hedhili MN, Bakr OM, Brédas J-L, Mohammed OF (2017) Chem Mater 29:4393–4400
101. Sichert JA, Tong Y, Mutz N, Vollmer M, Fischer S, Milowska KZ, García Cortadella R, Nickel B, Cardenas-Daw C, Stolarczyk JK, Urban AS, Feldmann J (2015) Nano Lett 15:6521–6527
102. Kumar S, Jagielski J, Yakunin S, Rice P, Chiu Y-C, Wang M, Nedelcu G, Kim Y, Lin S, Santos EJG, Kovalenko MV, Shih C-J (2016) ACS Nano 10:9720–9729
103. Kirakosyan A, Kim J, Lee SW, Swathi I, Yoon S-G, Choi J (2017) Cryst Growth Des 17:794–799
104. Levchuk I, Herre P, Brandl M, Osvet A, Hock R, Peukert W, Schweizer P, Spiecker E, Batentschuk M, Brabec CJ (2017) Chem Commun 53:244–247
105. Zhang F, Huang S, Wang P, Chen X, Zhao S, Dong Y, Zhong H (2017) Chem Mater 29:3793–3799
106. Li X, Yu D, Cao F, Gu Y, Wei Y, Wu Y, Song J, Zeng H (2016) Adv Func Mater 26:5903–5912
107. Roy M, Vikram B, Dedhia U, Alam A, Aslam M (2021) J Phys Chem Lett 12:1189–1194
108. Parveen S, Paul KK, Giri PK (2020) ACS Appl Mater Interfaces 12:6283–6297
109. Jang DM, Kim DH, Park K, Park J, Lee JW, Song JK (2016) J Mater Chem C 4:10625–10629

110. Leupold N, Schötz K, Cacovich S, Bauer I, Schultz M, Daubinger M, Kaiser L, Rebai A, Rousset J, Köhler A, Schulz P, Moos R, Panzer F (2019) ACS Appl Mater Interfaces 11:30259–30268
111. Zhu Z-Y, Yang Q-Q, Gao L-F, Zhang L, Shi A-Y, Sun C-L, Wang Q, Zhang H-L (2017) J Phys Chem Lett 8:1610–1614
112. Baláž P, Achimovičová M, Baláž M, Billik P, Cherkezova-Zheleva Z, Criado JM, Delogu F, Dutková E, Gaffet E, Gotor FJ, Kumar R, Mitov I, Rojac T, Senna M, Streletskii A, Wieczorek-Ciurowa K (2013) Chem Soc Rev 42:7571–7637
113. McCluskey MD, Janotti A (2020) J Appl Phys 127:190401
114. Queisser Hans J, Haller Eugene E (1998) Science 281:945–950
115. Shockley W, Read WT (1952) Phys Rev 87:835–842
116. Tress W (2016). https://doi.org/10.1007/978-3-319-35114-8_3
117. Shi W, Zhang X, Matras-Postolek K, Yang P (2021) ACS Appl Nano Mater 4:9391–9400
118. Paul S, Ahmed T, Das S, Samanta A (2021) J Phys Chem C 125:23539–23547
119. Adinolfi V, Yuan M, Comin R, Thibau ES, Shi D, Saidaminov MI, Kanjanaboos P, Kopilovic D, Hoogland S, Lu Z-H, Bakr OM, Sargent EH (2016) Adv Mater 28:3406–3410
120. ten Brinck S, Zaccaria F, Infante I (2019) ACS Energy Lett 4:2739–2747
121. Jin H, Debroye E, Keshavarz M, Scheblykin IG, Roeffaers MBJ, Hofkens J, Steele JA (2020) Mater Horiz 7:397–410
122. Krieg F, Ochsenbein ST, Yakunin S, ten Brinck S, Aellen P, Süess A, Clerc B, Guggisberg D, Nazarenko O, Shynkarenko Y, Kumar S, Shih C-J, Infante I, Kovalenko MV (2018) ACS Energy Lett 3:641–646
123. Freysoldt C, Neugebauer J, Van de Walle CG (2009) Phys Rev Lett 102:016402
124. Meggiolaro D, Mosconi E, De Angelis F (2018) ACS Energy Lett 3:447–451
125. Bodnarchuk MI, Boehme SC, ten Brinck S, Bernasconi C, Shynkarenko Y, Krieg F, Widmer R, Aeschlimann B, Günther D, Kovalenko MV, Infante I (2019) ACS Energy Lett 4:63–74
126. Chouhan L, Ghimire S, Biju V (2019) Angew Chem 131:4796–4796
127. Yang J, Yuan Z, Liu X, Braun S, Li Y, Tang J, Gao F, Duan C, Fahlman M, Bao Q (2018) ACS Appl Mater Interfaces 10:16225–16230
128. Roy M, Vikram B, Alam A, Aslam M (2021) Phys Chem Chem Phys 23:27355–27364
129. Tian Y, Merdasa A, Peter M, Abdellah M, Zheng K, Ponseca CS, Pullerits T, Yartsev A, Sundström V, Scheblykin IG (2015) Nano Lett 15:1603–1608
130. Yuan H, Debroye E, Caliandro G, Janssen KPF, van Loon J, Kirschhock CEA, Martens JA, Hofkens J, Roeffaers MBJ (2016) ACS Omega 1:148–159
131. Galisteo-López JF, Calvo ME, Rojas TC, Míguez H (2019) ACS Appl Mater Interfaces 11:6344–6349
132. Mukherjee A, Roy M, Pathoor N, Aslam M, Chowdhury A (2021) J Phys Chem C 125:17133–17143
133. Feng X, Su H, Wu Y, Wu H, Xie J, Liu X, Fan J, Dai J, He Z (2017) J Mater Chem A 5:12048–12053
134. Tian Y, Merdasa A, Unger E, Abdellah M, Zheng K, McKibbin S, Mikkelsen A, Pullerits T, Yartsev A, Sundström V, Scheblykin IG (2015) J Phys Chem Lett 6:4171–4177

Synthesis and Applications of Graphene and Its Nanocomposites

Mohd Asif and Irfan Ahmad

Abstract Graphene has been a material of interest, especially since the discovery of its free-standing form in 2003. The discovery provided hope to researchers looking for breakthroughs in the field that had not seen significant growth for long. Incremental improvements are not enough to meet the exponentially growing demands for cheap, convenient, and high-performing technologies. Graphene has the potential to provide new ways of achieving goals that previously seemed impossible by redefining the frontiers of science. It is because of the unprecedented material properties of graphene that were never demonstrated before by any other material. Novel and better material properties open up doors to new technologies and advancements in existing ones. However, it is imperative to obtain the material of suitable quality at a reasonable cost for it to compete with prevailing alternatives. In this chapter, various methods for synthesizing graphene have been discussed with a particular focus on the liquid-phase exfoliation (LPE) of graphite. Characterization with Raman spectroscopy, electron diffraction, and microscopy-based tools have been explored. The chapter also reviews applications of graphene in a few emerging areas. Graphene-based composites, with emphasis on their syntheses and applications, will be discussed.

Keywords Graphene synthesis · Liquid-phase exfoliation · Graphene characterization · Graphene applications

M. Asif
Interdisciplinary Nanotechnology Centre (INC), ZHCET, Aligarh Muslim University (AMU), Aligarh UP-202002, India

Department of Electrical Engineering, Indian Institute of Technology Delhi (IITD), New Delhi-110016, India

I. Ahmad (✉)
School of Engineering Science and Technology (SEST), Jamia Hamdard, New Delhi-110062, India
e-mail: i.ahmad@alientt.com

I. Uddin and I. Ahmad (eds.), *Synthesis and Applications of Nanomaterials and Nanocomposites*, Composites Science and Technology,
https://doi.org/10.1007/978-981-99-1350-3_2

1 Introduction

Carbon is one of the critical elements for sustaining life on earth and has so many compounds that a complete branch of chemistry has been dedicated to it. Pure carbon itself has several naturally occurring allotropes, and numerous others can be synthetically made in a laboratory. One such naturally occurring allotrope is graphite, in which atomically thin layers are stacked together by van der Waals forces. These individual layers are called graphene, where the ending term 'ene' marks the existence of sp^2 bonding within the sheet plane. Atoms in graphene are bound together in a honeycomb-like structure, with six carbon atoms forming a hexagon. Three out of four valance electrons of each carbon atom make strong covalent σ-bonds with other carbon atoms. The remaining one electron per atom contributes to π-bonding, responsible for graphene's aromaticity and unique electronic properties [1]. It has been found that two π-electrons per ring are delocalized in graphene, making it different from other aromatic compounds such as benzene [2]. Graphene has been theoretically explored by many researchers for more than the past seven decades. The first theoretical explanation of graphene is dated back to 1947, when Canadian theoretical physicist P.R. Wallace studied the band structure of graphite [3]. Though Wallace did not use the term 'graphene', his contribution is regarded as the basis of theoretical studies of graphene.

For numerous decades, researchers tried to obtain free-standing single-layer graphene, while another section of theirs believed that it was practically impossible. The discrepancy came to an end in 2004, when K.S. Novoselov et al. reported successful isolation of single-layer graphene (SLG) from graphite [4]. The discovery of free-standing graphene and subsequent experiments by K.S. Novoselov and A. Geim won them a Nobel in physics in 2010. Controversy followed the award and W.A. de Heer of Georgia Tech., who had also contributed significantly to graphene science, wrote a letter to the Nobel committee stating that "The Nobel Prize committee did not do its homework" [5]. Nevertheless, graphene had been studied extensively and had demonstrated tremendous potential in multiple applications by that time because of its exceptional properties.

The exotic properties arise broadly due to single-layer thickness and sp^2 hybridization of electronic orbitals in graphene. The most interesting ones are high electrical conductivity ($>10^6$ S/m) 6] very high thermal conductivity (up to ~ 5300Wm^{-1} K^{-1}), [7] highest ever measured mechanical strength (~130GPa) and Young's modulus (~1TPa), [8] extremely high experimentally calculated carrier mobility (~2×10^5 cm^2V^{-1} s^{-1} for suspended graphene), [9] and very high theoretical specific surface area of (~2630 m^2g^{-1}) [10].

The high carrier mobility is due to the conical band structure of SLG, in which the highest occupied molecular orbital (HOMO) and lowest unoccupied molecular orbital (LUMO) meet each other at Dirac points (vertices of the Dirac cones) [3]. It is a well-known fact that the effective mass of charge carriers is inversely proportional to the curvature of the band, and at the Dirac points, the curvature is very high (ideally infinite), making electrons massless fermions (ideally). Carrier mobility is inversely

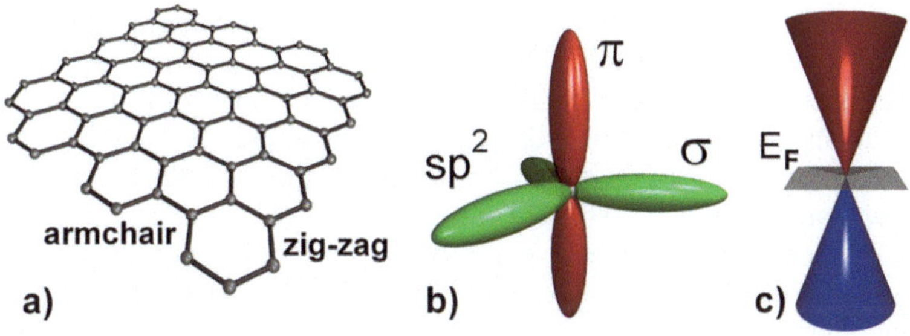

Fig. 1 Representative images of (**a**) hexagonal structure of graphene and chirality, (**b**) sp^2 hybridization with sigma and pi bonding electrons, and (**c**) valence and conduction bands meeting each other at a Dirac point [11]

related to its effective mass hence such high mobility is observed in graphene. Due to the unique band-structure of graphene, it has been called by multiple names, such as zero-bandgap semiconductor and zero-overlap semimetal. The high electrical conductivity of graphene is due to delocalized π-electrons that are free to roam throughout the sheet [11]. The magnificent mechanical properties arise due to sp^2 hybridization of orbitals, resulting in stronger bonding between hexagonally organized carbon atoms [12]. Representative images of hexagonal structure, σ and π bonds, and Dirac cones for graphene are shown in Fig. 1.

The thermal conductivity of materials increases with bond strength between their constituent atoms [13]. The outstanding thermal conductivity of graphene is, thus, attributed to very strong sp^2 bonding between carbon atoms [14]. High thermal conductivity and very high mechanical strength is an excellent combination for heat sinks in electronic circuits and devices [15]. Another compelling property of an SLG is its light absorption, measured to be 2.3% [16]. A single layer with such a high conductivity and good transparency makes it a suitable candidate for transparent electrodes for optoelectronic devices [17].

The properties of materials degrade with their quality, and for a material with such unique properties, high-quality synthesis is challenging. Often, researchers seek an optimization between quality and cost (quantity). Various methods have been developed to obtain graphene, and graphene's quality varies with its synthesis route. Some commonly used methods are epitaxial growth, [18] chemical vapor deposition (CVD), [19] and exfoliation of graphite [20]. Graphite exfoliation-based techniques are more promising for producing large quantities with reasonable quality. The quality assessment is the next step after the synthesis, for which various tools and methods are employed. Raman spectroscopy is one of the most reliable techniques for acquiring scores of information on synthesized graphene. It provides information regarding the number of layers, [21] defects, [22, 23] doping, [23] chirality, [24] etc., as well as effects of external parameters such as pressure, [25] and temperature [26]. Other tools, for instance, selected area electron diffraction (SAED), and high-resolution transmission electron microscopy (HRTEM), are competent for analyzing crystallinity, defects, and the number of layers [27]. Atomic force microscopy (AFM) is often used for the determination of electrical [28] and mechanical [29] properties in addition to quality evaluation.

In the recent past, graphene has become a material of interest in a vast spectrum of applications for commercialization. It has been, therefore, forecasted in a research report by an apex market experts group named IDTechEx that the graphene market will grow from < 100 M USD in 2020 to ~ 700 M USD in 2031 [30]. Graphene has been commercialized with several products currently available in the market such as automobile coating products by 'Turtle Wax', [31] sporting products by a leading sports goods manufacturer 'Head', [32] shoes by 'inov-8', [33] many products, especially face masks, by 'G1 Wonders', [34] tailor-made solutions for a range of industrial requirements by 'Graphene XT' [35]. There are numerous sectors where graphene is being explored, and the resulting products are expected to hit the market in the coming years.

Even though graphene itself is a wonder material with a wide range of applications, still in a large number of applications, graphene-based nanocomposites (NCs) are a better choice. Graphene has been investigated as the continuous phase (matrix) as well as dispersed phase (filler) when hybridized with diverse classes of materials, including metals, their oxides, and sulfides, and organic polymers. In-situ polymerization is the most common route for making polymer-graphene NCs. For metal-graphene NCs coatings, electrodeposition is commonly employed, while for their powdered forms, simultaneous reduction of metal ions and graphene oxide (GO) has been frequently reported. Graphene NCs with metal oxides and sulfides are usually produced by solvo-/hydro-thermal methods.

This chapter is dedicated to discussing various synthesis protocols, characterization techniques, and applications of graphene. Synthesis using CVD, epitaxial growth, and exfoliation of graphite will be briefly introduced. It will be followed by a detailed discussion on the LPE of graphite. Evaluation of the quality of graphene through characterization techniques, especially with Raman spectroscopy, will be elaborated. For the sake of completeness, crucial methods such as SAED and microscopic techniques will also be discussed. Potential applications of graphene in electronics, energy conversion and storage, environmental remediation, and healthcare sectors will be reviewed. NCs of graphene with polymers, metals, metal oxides, and metal sulfides will also be within the scope of this chapter. Finally, the chapter will be summarized, and future scopes will be highlighted to conclude the chapter.

2 Synthesis of Graphene

Properties of the synthesized graphene are dependent upon the preparation route it has undergone. Therefore, it is imperative to discuss major synthesis methods in order to understand the variation that may arise due to the preparation method. Commonly utilized synthesis routes are discussed in this section.

2.1 *Epitaxial Growth*

Epitaxial growth involves the deposition of a material layer onto a well-oriented substrate. Thermal decomposition of silicon carbide and molecular beam epitaxy (MBE) are the two most commonly employed methods for the epitaxial growth of graphene. The decomposition of SiC, when heated to high temperatures, and its structural studies were reported by D. Badami in 1962 [36]. In this approach, a crystal of hexagonal SiC is annealed at high temperatures of around 1000°C in an ultra-high vacuum (UHV) [37]. The high temperatures allow silicon atoms to sublimate into the vacuum while carbon atoms, having lower vapor pressure than Si, [38] remain on the surface. The carbon atoms at the surface then rearrange into a hexagonal structure which is nothing but a graphene sheet. Two layers of SiC decompose to form a single layer of graphene [39]. Both SLG and few-layer graphene (FLG) can be grown with this method. Walt de Heer and his team studied the 2D electron gas of epitaxially grown graphene on the (0001) face of a 6H-SiC crystal [40]. This method is advantageous for the semiconductor industry because graphene inherently grows on an insulating SiC substrate. Disadvantages of the process include the limitation of synthesized graphene area by the SiC crystal size and the control over the number of graphene layers synthesized. Decomposition of SiC and subsequent graphene formation is represented schematically in Fig. 2.

Another frequently reported technique for growing graphene epitaxially is MBE. In this technique, carbon deposition occurs atom by atom on a substrate from the source material in the growth chamber under UHV (as shown in Fig. 3). The source of carbon atoms could be solid such as highly ordered pyrolytic graphite (HOPG), [41] or could even be a gas, such as acetylene (C_2H_2) [42]. One significant advantage of this method is that graphene can be grown onto a variety of substrates. Researchers have demonstrated the use of metals, [43] insulators, [44] elemental semiconductors, [45] and metal oxides [46] as substrates. Films of another material (especially metals) can also be deposited onto the substrate before exposure to the beam of carbon atoms and the subsequent growth of graphene. The choice of substrate affects the quality of the graphene, and thus high-quality graphene can be grown with this approach [47]. Furthermore, MBE offers possibilities for in-situ growth monitoring and characterization of graphene sheets [48]. The growth process, however, is slow and takes tens of minutes to hours depending on the product requirements and thus

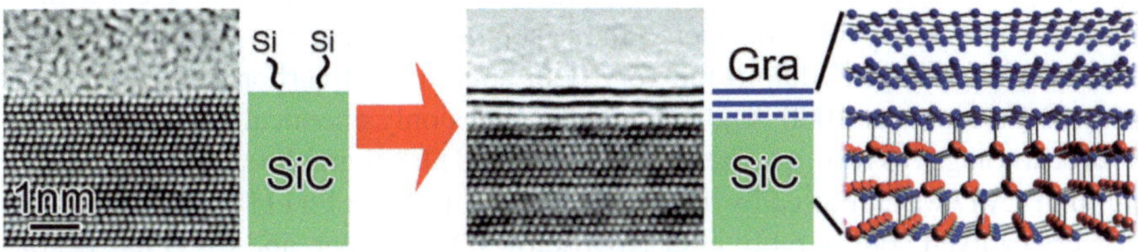

Fig. 2 Formation of graphene via decomposition of SiC at {0001}surface; HRTEM images and schematic representation [37]

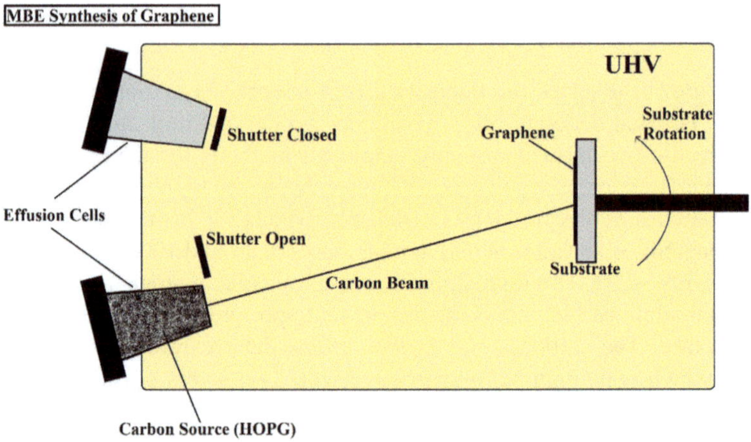

Fig. 3 A schematic representation of the setup for MBE synthesis of graphene

low throughput, which is a major disadvantage of this method. Also, the process is costly, and therefore, it is generally reserved for applications where high-quality (pristine) graphene is the prime requirement.

2.2 *Chemical Vapour Deposition (CVD)*

CVD technique is the most commonly used technique for growing high-quality and large-area graphene sheets [49]. The process typically involves the adsorption of a carbon-containing (hydrocarbon) gas on a transition metal surface, its decomposition and removal of unwanted groups, and finally, the rearrangement of carbon atoms to form the honeycomb structure (Fig. 4 shows the schematics). Commonly used hydrocarbon gas precursors are ethylene, [50] methane, [51] and acetylene [52]. Several transition metals have been reported to perform as catalytic substrates such as Pt, [50, 53] Ni, [51, 54] Fe, [52] Cu [54]. The solubility of carbon in these metals at high temperatures affects the process. Dissolved carbon tends to segregate on the surface when cooled, and thus, an additional number of graphene layers are formed. Therefore, it becomes challenging to control the number of layers with the CVD process. Copper is considered the best amongst the metal substrates demonstrated so far for SLG due to the very poor solubility of carbon in it (<1 C atom per 10^5 Cu atoms at 1000 °C), [55] which allows only chemical decomposition of hydrocarbon gas to form graphene. The first 'single layer graphite' (as they called it) through CVD was reported by T.A. Land et al. in 1992 on Pt(111) surface [53]. The process evolved considerably after Novoselov's 2004 paper, and CVD has become one of the most promising techniques for obtaining large-area graphene. The crystallinity of a substrate is another critical parameter. Smooth and well-oriented single

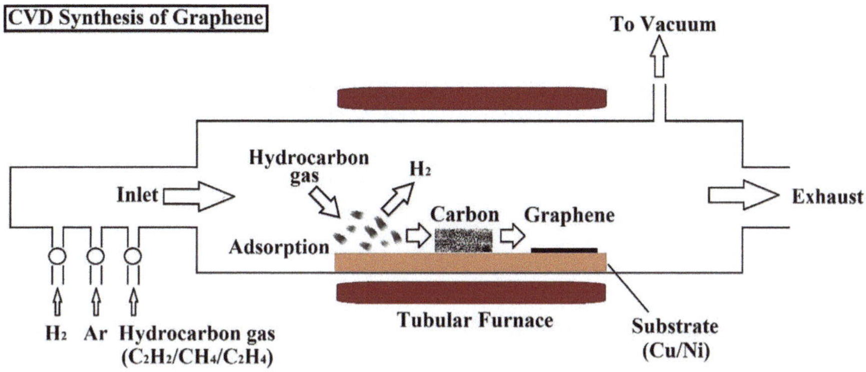

Fig. 4 Schematic representation of a typical CVD setup and process

crystal substrates are preferred for monolayer graphene because grain boundaries and other defects act as nucleation sites and thus stimulate the formation of multi-layer graphene (MLG) [56]. Hydrogen is generally used in the annealing step to remove the oxide layer, if any, from the catalyst surface. It is crucial to note that increasing the hydrogen:methane ratio in the feedstock reduces the graphene growth rate on Cu substrate while boosting the growth rate when Ni substrate is used [54]. It can be ascribed to the high hydrogen solubility and low carbon solubility in Cu, while the opposite is true for Ni. The quality of the produced graphene depends on the flow rate of the precursor gas, the transition metal used as the catalytic substrate, and chamber temperature. Growth can take place at relatively higher pressures with argon introduced into the system.

Two configurations, namely Thermal CVD (TCVD) and Plasma Enhanced CVD (PECVD) are commonly employed for graphene synthesis. TCVD process uses high temperatures of around 1000°C (close to the melting point of substrate metals) for the decomposition of hydrocarbon gas [51]. High temperatures, generally, favor the formation of highly crystalline wrinkle-free graphene sheets [50]. Hot wall CVD and cold wall CVD are two subclasses of TCVD. The entire growth chamber is heated in a hot-wall setup while only the substrate is heated in a cold-wall configuration. High throughput is an inherent advantage of cold wall CVD because the reactants are not deposited on the reactor walls, and thus it is more suitable for scale-up production. PECVD, on the other hand, is a deposition technique of substantially lower temperatures. The PECVD process for graphene synthesis has been reported at temperatures as low as 240°C [57]. The elimination of heating and cooling of the substrate in the PECVD process significantly reduces the synthesis time but at the cost of quality degradation of product material. In a typical setup, plasma is created, which reacts with the hydrocarbon gas to form more reactive radicalized species capable of forming a honeycomb lattice at relatively low temperatures. Li et al. studied the growth of graphene through the PECVD process on Cu substrate at different temperatures [58].

The CVD process is a reliable technique for synthesizing high-quality large-area graphene sheets. However, the transfer of the prepared graphene from metal substrates to other substrates is necessary for most applications. The transfer process is critical and leads to the degradation of the quality of graphene by introducing imperfections such as kinks, wrinkles, cracks, etc. [59]. Some techniques have also been developed for synthesizing graphene on non-metallic substrates such as glass, which can avoid transfer steps for multiple applications [60].

2.3 Exfoliation of Graphite

Graphite is a meticulous stack of graphene layers, and thus, free-standing graphene can be obtained by unstacking these layers. Exfoliation of graphite is just another term for unstacking it to obtain the graphene sheets. The energy required for exfoliation (or separation) of two $1 nm^2$ graphene sheets has been calculated to be around 2eV [61]. Several methods have been explored by researchers for providing this energy to exfoliate graphite. Mechanical, liquid-phase, and electrochemical exfoliation methods have been reported more frequently in the literature.

Mechanical exfoliation was the first method used by Novoselov and Geim to obtain free-standing graphene [4]. They used scotch tape and a crystal of graphite for the purpose, and thus the technique is referred to as the "scotch-tape-method" in layman's terms. In this experiment, a piece of tape is first pressed against a well-oriented HOPG crystal face and removed carefully. Due to this peeling process, many layers detach from the crystal and come off sticking onto the scotch tape. Another piece of scotch tape is then pressed against the first piece and then carefully separated, which results in the distribution of the detached layers onto these two pieces. The process is repeated multiple times until the desired number of layers remains on the tape. The pieces of scotch tape are then either put in a solution to remove epoxy and obtain free-standing graphene or on a silica slide to get graphene transferred onto a substrate.

The electrochemical exfoliation process generally involves a setup of two electrodes, one of which is graphite which could either be an anode or a cathode in a liquid electrolyte solution. When a voltage is applied across the two electrodes, ionic species intercalate between the layers of graphite, causing weakening of van der Walls attraction and thus dissociation of mono- and few-layer graphene from the graphite electrode into the solution [62]. Various geometries of graphite such as foils, flakes, rods, and discs have been studied for this purpose. Electrodes made up of HOPG were observed to be providing better quality graphene than naturally occurring forms [63]. DC voltages between 1 to 30V and low-frequency AC voltages up to 20V have been explored and reported [64]. Sulfuric acid is a commonly used electrolyte, which dissociates into H^+ and $SO_4{}^{2-}$ ions in an aqueous solution. Sulfate ions, compared to other anions, easily intercalate between the graphite layers causing them to exfoliate into graphene layers more efficiently [65]. High yield of around 50% was reported by J. Liu et al. by using vertical cell configuration [66]. Even better results (65%) were obtained by T.C. Achee et al. with the help of a permeable and expandable container to make a compressed graphite electrode from graphite powder [67].

Another fundamental method is the oxidation of graphite flakes to obtain exfoliated and oxidized graphene called graphene oxide (GO), which is useful for numerous applications. It can also be used as a precursor for obtaining graphene through suitable reduction processes. Hummer's method is the most popular protocol for oxidation and exfoliation of graphite to convert it into GO [68]. The process is an age-old technique used for graphite oxide synthesis, but it kept on modifying over time. In a typical procedure, sulfuric acid is used for intercalation and potassium permanganate as the oxidizing agent, while hydrogen peroxide is used at a later stage of the process to eliminate excess permanganate. The material is then washed and dried to obtain yellowish water-soluble GO. Several methods have been developed to convert GO into graphene; chemical [69] and thermal [70] reduction are the most common amongst them. The obtained graphene is not pure or pristine as some functional groups remain attached to it. Thus, it is referred to as reduced graphene oxide (rGO).

Liquid phase methods use solvents such as water (with and without surfactants) and organic solvents to exfoliate graphene sheets from graphite flakes. Ultrasonication in a suitable solvent is the primary method used in this category. Other improvised methods wield a combination of mechanical techniques such as high energy ball milling (HEBM), shear mixing, and centrifugation, in addition to ultrasonication. Biological substances instead of synthetic surfactants have also been reported for LPE synthesis of graphene [71]. LPE methods will be discussed in more detail in the next section. Exfoliation synthesis, especially LPE, is one of the best techniques to obtain large quantities of quality graphene at an economical cost.

3 Liquid-Phase Exfoliation (LPE) Synthesis

Liquid phase exfoliation method has tremendous potential for scale-up production of graphene. It is a simple technique where mechanical energy is applied to exfoliate graphite flakes in a liquid medium. The energy could be provided through acoustic waves, shear force, or centrifugal force. Acoustic waves such as ultrasonic waves are applied through a liquid to transfer energy to graphite flakes which can stimulate the separation process and lead to the formation of graphene sheets. Surfactants are generally used to assist the exfoliation process by creating an electrostatic repulsive force between layers. Surfactant-free approaches, in which organic solvents are used instead of water for dispersing the graphite powder, also recurrently appear in the literature. Graphene is not dispersible in water but exhibits a high dispersion in many organic solvents, which provides a better possibility of its stability after exfoliation and hence a better yield. Organic solvents such as ortho-dichlorobenzene (ODCB) [72] and N-methyl-pyrrolidone (NMP) [73] have been reported quite often for synthesizing graphene through the LPE method. Further improvements were reported by the addition and tuning of the concentration of n-octylbenzene in the former two organic solvents [74]. The primary stabilization mechanism, with or without surfactants, mainly involves a charge transfer between unstacked graphene and the stabilizing agents. The charge transfer, most of the time, takes place from electron-rich graphene to the stabilizer molecules, but the reverse can also happen (such as in the case of pyridine) [75].

M. Telkhozhayeva et al. studied in detail the ultrasonication-assisted exfoliation of graphite [76]. They explored the effect of the frequency of sonic vibrations on the quality of synthesized graphene and the efficiency of synthesis. They reported that good quality graphene (~30% monolayer and ~ 45% bi- and tri-layer) with an average lateral size of 13μm could be obtained in ethanolic bath sonication by increasing the frequency from the usually used 40kHz to a higher value of 80kHz. They demonstrated that increasing the frequency positively affects the exfoliation process and reduces the chances of inducing lattice defects in the synthesized graphene. The group also reported that bath sonication is a more efficient technique than probe sonication.

M. Monajjemi theoretically explored the LPE of graphite by ultrasonication [77]. Various types of surfactants and their effects on the efficiency of exfoliation were studied, and it was found that ionic surfactants are more efficient than others. The exfoliation efficiency in any dispersion medium can be written in descending order as cationic, anionic, zwitterionic, and non-ionic. It was further reported that the sulfonic group is very effective in assisting the exfoliation of graphite and stabilizing the produced graphene in solvents to prevent restacking.

Water-insoluble stabilizers have also shown potential for increasing the efficiency of LPE [78]. Y. Shin and group have recently demonstrated the use of pyrene derivatives such as bis-pyrene stabilizers functionalized with pyrrolidine, which can substantially improve the quality of obtained graphene. The use of toxic substances, however, limits its applications to the areas where biocompatibility is not a requirement.

A detailed experimental as well as theoretical study on LPE of graphite was performed by Coleman's group, which has greatly contributed to LPE research [79]. The study reveals three stages of the conversion of graphite into graphene through the sonication-assisted LPE process, namely: *Flake rupture and kink band formation; peeling off of thin graphite strips; and exfoliation to thin flakes.* In the first stage, acoustic waves travel through the graphite piece due to applied sonic energy, resulting in ridges forming on the graphite structure. These ridges have a large number of defects which increase the reactivity and thus the possibility of attachment of various functional groups. The ridges then crack in the second stage, and the exposed edges become oxygen-rich due to the addition of functional groups. The third and final stage mainly involves intercalation and peeling off the layers from the graphite structure and graphene dispersion into the solvent. The study has led to a very significant improvement in the basic understanding of the LPE process and is expected to improve the overall process as well. The three stages are schematically represented in Fig. 5.

A large number of organic and inorganic solvents have been reported for LPE of graphite since the discovery of free-standing graphene. A.B. Bourlinos et al., in 2009,

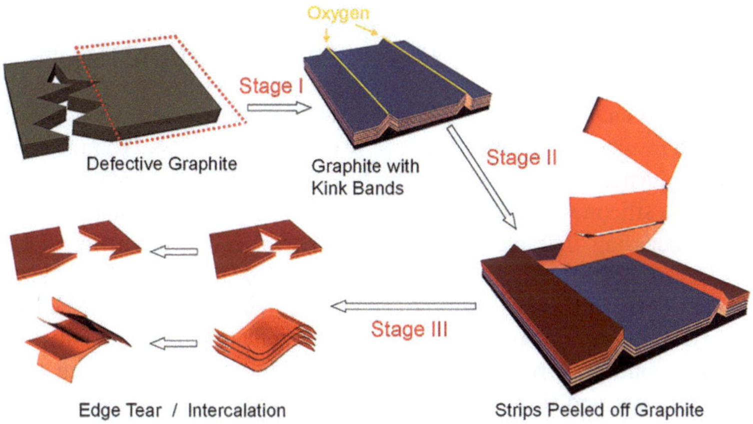

Fig. 5 Three stages of LPE process [79]

added multiple new solvents to the list of already known solvents for LPE [75]. They experimented LPE with pyridine and some perfluorinated aromatic solvents such as hexafluoro-benzene (C_6F_6), octafluoro-toluene ($C_6F_5CF_3$), pentafluoro-benzonitrile (C_6F_5CN), and pentafluoro-pyridine (C_5F_5N).

A mixed solvent strategy was proposed by M. Yi et al. for exfoliating graphite powder to obtain graphene nanosheets [80]. They performed two experiments with mixtures of alcohol and water. Mixture-1 was prepared with ethanol in water, and mixture-2 with isopropyl alcohol in water. Mild sonication was employed to exfoliate nanosheets from graphite powder. The concentration of alcohols was varied to optimize the processes for maximum yield and better quality. The optimum concentration was reported to be 40% for mixture-1 and 55% for mixture-2, with a yield up to 10%. Simple strategies like this one could benefit upscale graphene production through LPE in the future.

D. Nuvoli and the team first used ionic liquids as solvents for graphite dispersion [81]. The team used grinding in mortar followed by ultrasonication and finally centrifugation for obtaining FLG. Ionic solvent 1-hexyl-3-methyl-imidazolium hexafluorophosphate was used with no other functionalizing compounds. They reported a very high dispersivity of graphene (5.33 mg/ml) in the solvent. One more advantage of this method is that the solvent can be processed for reuse, significantly reducing the production cost.

The addition of common organic salts during the LPE process has also been shown to have positive impacts. A group led by X. Jiang proposed using organic salts such as edetate disodium, sodium tartrate, potassium sodium tartrate, and sodium citrate in organic solvents to improve the exfoliation efficiency [82]. The group reported more than a hundred times improvement in efficiency with a mixture of dimethyl sulfoxide (DMSO) and sodium citrate.

A more sophisticated but effective 'lab on a chip' LPE was proposed by X. Qiu and associates [83]. They exploited a phenomenon called hydrodynamic cavitation, which involves bubble generation and implosion in flowing liquid due to sudden changes in pressure, to induce the exfoliation of graphite. The setup, a microfluidic channel with a smaller orifice, was fabricated on a silicon wafer. A mixture of water, sodium cholate (surfactant), and graphite powder were forced to flow through the orifice, which resulted in a change in pressure of 10 bar. This local change in pressure stimulates the unstacking of graphene layers. A series of these microchannels can be fabricated on a silicon wafer for upscale production of graphene nanosheets.

HEBM processing of graphite powder is another crucial method within LPE genre for cost-effective high quality synthesis of graphene sheets. It has been reported by several researchers that the blending of inorganic materials such as NaCl [84] and sulfur [85] during the milling process improves the flake size and the quality of the synthesized graphene, owing to the intercalation of these particles acting as a wedge between the layers and thus assisting the exfoliation. However, M.F. Alam et al. reported that the blending of salt during the ball milling process results in the creation of defects in the synthesized graphene and the residual salt impurities degrade the overall quality of thus synthesized graphene [86]. The group used a combination of three techniques viz. ultrasonication, HEBM, and centrifugation, with and without inorganic salts (NaCl, KCl, and LiI). They compared the results obtained from these experiments and reported HEBM followed by three times centrifugation without salts to be the best combination for synthesizing high-quality sheets. The results are arranged in Table 1 as shown below:

The above discussed results seem counter-intuitive as the earlier discussion in this section pointed out the intercalation of ions between the sheets acting as tiny wedges during LPE, thus making it easier to exfoliate the graphene sheets. The results in this part are with HEBM processing before the LPE step. With HEBM processing, blending of salt negatively impacts graphene synthesis.

4 Characterization of Graphene

4.1 Raman Spectroscopy

Raman (inelastic) scattering is so tiny compared to Rayleigh scattering (elastic) that its detection was practically impossible to be utilized in a probing technique until high-intensity lasers were invented. Raman spectroscopy has become a tool with extraordinary capabilities to probe molecules and crystals based on the Raman scattering phenomenon which takes place due to the vibrational and rotational modes of the material. The modes that change the polarizability of material are called Raman active modes, and the resulting output spectrum consists of bands of different intensities. A typical Raman instrument's output, a plot between intensity and Raman shift, gives information about the atoms and interatomic bondings that are signatures of a molecule, and thus, the technique acts as a molecular fingerprinting. This tool has

Table 1 Summary of results by M.F. Alam et al. [86]

Preparation Method	Halide Used	Results (TEM and SAED)	Remarks
3 × Ultrasonication	None	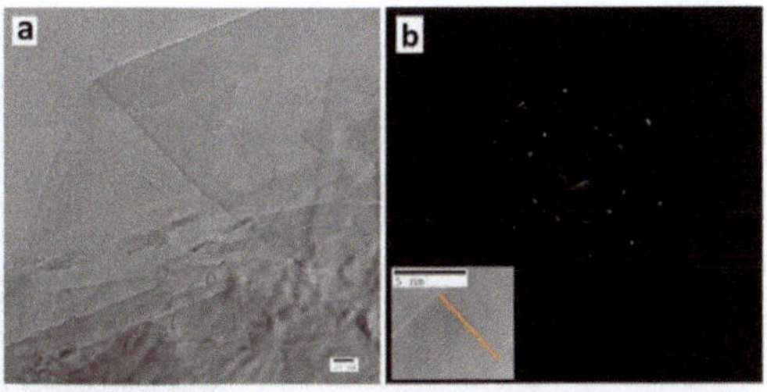	Few 100 nm size, Fringes 0.33 nm, 2 prominent rings in SAED
HEBM + 3 × Ultrasonication	None	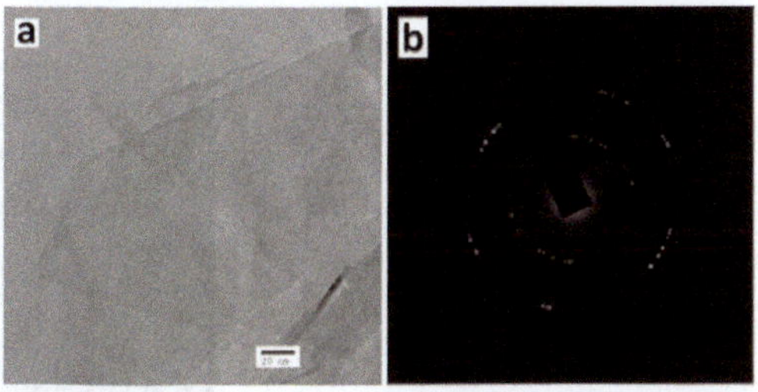	Few 100 nm size, 2 prominent 1 feeble rings in SAED, Lattice defects
HEBM + 3 × Centrifugation	None	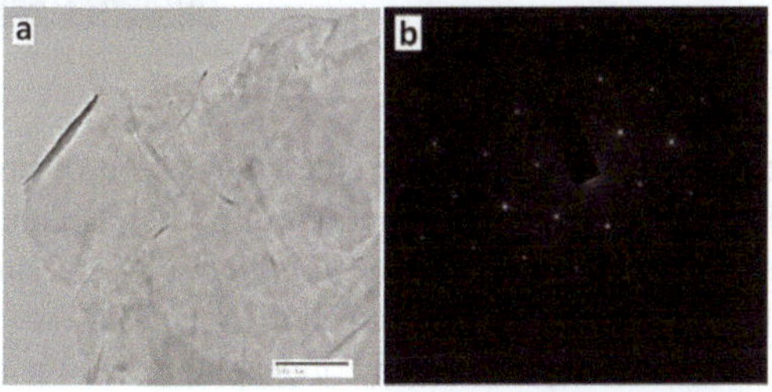	Size in μm, sixfold symmetry in SAED, Defect-free lattice
HEBM + 3 × Centrifugation	NaCl (40%)	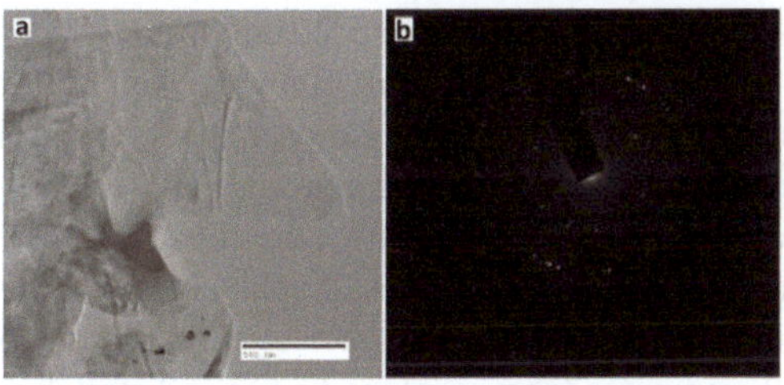	Size in μm, 2 prominent 1 feeble rings in SAED, multi-layer, Lattice defects

(continued)

Table 1 (continued)

Preparation Method	Halide Used	Results (TEM and SAED)	Remarks
HEBM + 3 × Centrifugation	KCl (40%)	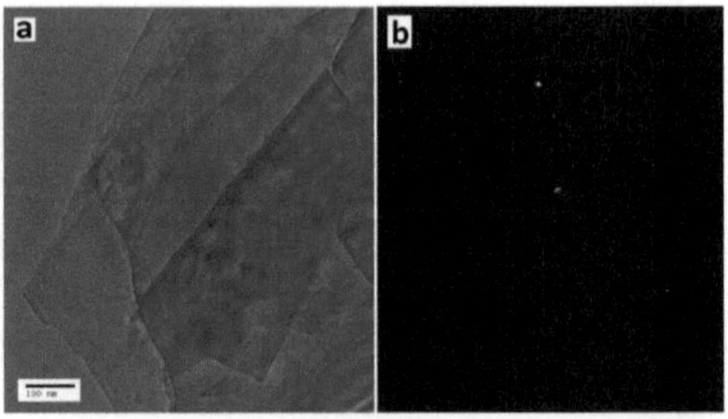	Multiple overlapping sheets, Asymmetric pattern in SAED, Inefficient method
HEBM + 3 × Centrifugation	LiI (40%)	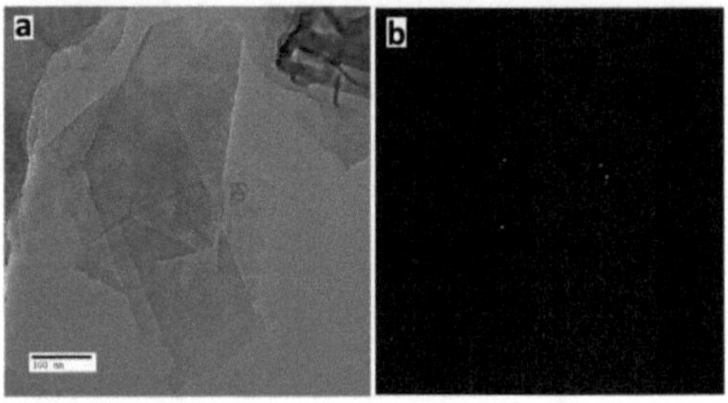	Multiple overlapping sheets, 1 prominent 1 feeble ring in SAED, Lattice defects

been proven extremely useful for the characterization of honeycomb carbon materials. The technique is so sensitive that it can be used to analyze the electronic and phononic behavior in a single sheet of graphene, [87] a single CNT, [88] a single fullerene, [89] and any changes introduced in their structure. This section will mainly be focused on the Raman spectroscopy of graphene and the effects of the number of layers, defects, and doping on the spectrum.

Figure 6a shows a typical Raman spectrum of an SLG. Many bands arise due to Raman active phonons, but most result in a weak signal. It is, therefore, a common practice to study only four prominent bands, namely G, D, G', and D', for obtaining relevant information on the underlying material. These bands originate due to various phenomena occurring in graphene, which are illustrated in Fig. 6b. The information about these bands in the following paragraphs is for laser photon energy of 2.41eV (514nm wavelength).

The G band originates from in-plane C–C bond stretching vibrations and occurs at around 1585 cm^{-1} due to first-order Raman scattering involving one phonon. The peak of the G band is an outcome of resonance and is a characteristic peak for honeycomb carbon materials. It is related to doubly degenerate phonon modes, namely, longitudinal optical (LO) and in-plane transverse optical (ITO) at the center of the first Brillouin zone. The G band is the only non-dispersive band (having no correlation between band position and photon energy) in the Raman spectrum

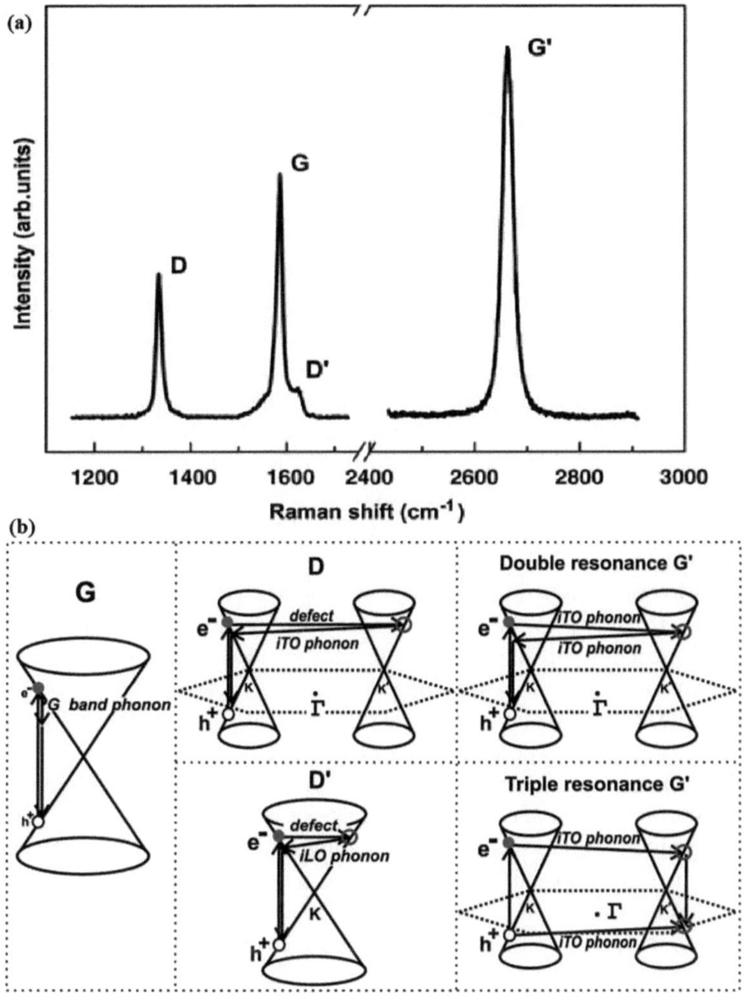

Fig. 6 Raman spectrum of an SLG showing: various bands (**a**) and origin of these bands (**b**) [87]

of graphene. The G' band (also known as 2D band because it occurs at twice the Raman shift of D band) at around 2700 cm^{-1} results from a second-order process involving two phonons participating in the Raman scattering of the excited electron. A double resonance process arising from inter/intra-valley scattering gives rise to this band in the spectrum. Triple resonance processes are also sometimes involved and are responsible for G' band. The G' band is a signature of sp^2 hybridized carbon materials in Raman spectra. This band exhibits dispersion, and the slope of the dispersion curve has been estimated to be ~ 90 cm^{-1}/eV [87, 90]. The D band (D, here, is generally referred to as defects), seen at around 1350 cm^{-1}, is again a second-order process, but it arises due to inelastic scattering of the excited electron by a phonon followed by elastic scattering by a defect. This band is an outcome of the so-called 'breathing of carbon hexagons'. Similar to G' band, the D band also shows dispersive nature with the slope of dispersion curve ~ 50 cm^{-1}/eV [87, 90]. It is important to note here that the G' band, being the overtone of the D band, is twice as dispersive as the D band. Sometimes other bands associated with the double resonance mechanism in which a defect instead of a second phonon is responsible for the conservation of momentum are also observed, such as D' band at ~ 1620 cm^{-1} and D" band at ~ 1100 cm^{-1} [91].

Y. You et al. studied the correlation between the intensity of D band and crystal orientation (edge chirality) of graphene [92]. Their study revealed that for the zigzag edge of graphene, the D band is either absent or is of very low intensity, while the armchair structure produces a much intense D band. G band has also been studied for determining the orientation of the graphene sheet [24]. The D band intensity, being a more prominent Raman signature of defects than the G' band, is a measure of the number of defects in the graphene crystal. The nature of defects also plays a significant role in altering the Raman spectrum of graphene. A detailed study in this regard was performed by A. Eckmann et al. [22]. The study discloses the dependence of the intensity ratio of D band and D' band ($I_D/I_{D'}$) on the type of defects such as a change in hybridization at some locations to sp3 (instead of sp2) and the creation of vacancies. In the study, the intensity ratio was found to be ~ 13 for sp3 defects and ~ 7 for vacancy type defects. A more careful study on the sub-bands of the D band for suspended graphene was done by Z. Luo et al. [93]. They mainly focused on whether the scattering first takes place through a phonon or a defect. The research demonstrates that the two processes give rise to two D sub-bands.

A relative intensity, taking the intensity of the G band (I_G) as a reference, rather than the intensity of the band under investigation, is generally estimated for quantitative and qualitative analysis of defects, doping, and the number of graphene layers in the system. G.S. Papanai et al. studied the intensity of the G' band as a function of the number of graphene layers on Si/SiO_2 substrate [94]. The G' band of an SLG consists of one Lorentzian peak of high intensity ($I_{G'}/I_G = 24$) resulting from one double resonance process. For two layers stacked in the Bernal AB pattern, the system becomes more complex, and four double resonance processes take place, due to which the G' band becomes a superposition of four Lorentzian peaks. The intensity of the G' band compared to the G band is significantly reduced ($I_{G'}/I_G \leq 1$). Scattering possibilities keep increasing with the number of layers, and the G' band becomes a combination of multiple Lorentzian peaks, each representing a scattering phenomenon. The relative intensity ($I_{G'}/I_G$) also decreases, which can be estimated to calculate the number of graphene layers. Figure 7 shows the Raman spectra for 1–5 graphene layers and HOPG.

The positions of G and G' bands change with doping because of coupling between electron and phonon and are, therefore, studied to estimate the doping concentration in graphene [95]. The G band shifts towards higher values of Raman shift, and its width is also reduced with increasing dopant concentration. However, the dependency of the G' band is different for electron concentration and hole concentration. The

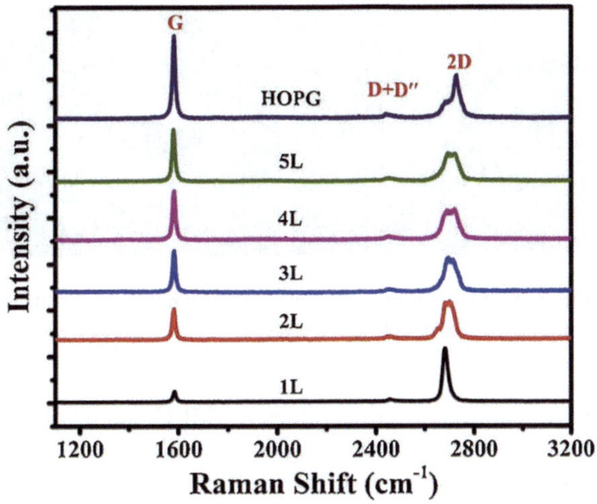

Fig. 7 Raman spectra for different numbers of graphene layers on Si/SiO_2 substrate [94]

position of the G' band shifts upwards with increasing hole concentration while in the reverse direction for increasing electron concentration [23].

The temperature change significantly affects the position of the G band. I. Calizo and his group calculated the temperature coefficient of the G band peak position for SLG and two-layer graphene [26]. For SLG, the value was measured to be -1.62×10^{-2} cm^{-1}/°C, while for the two layers, it was found to be -1.54×10^{-2} cm^{-1}/°C. The negative value indicates a blue shift in the scattering frequency when the temperature is increased. Various other parameters such as pressure, [25] strain, [96] etc. have also been found to be affecting the Raman spectrum of graphene, making it one of the most valuable tools to study graphene and related carbon materials.

4.2 *Selected Area Electron Diffraction (SAED)*

Diffraction is simply a change in a wave's path around the corners of and reaching in geometrical shadow regions of an obstacle. The electrons, behaving as waves, are diffracted by atoms in a crystal, and a pattern of diffraction maxima and minima emerges. This pattern, obtained for a small area, is called SAED and has long been used to study the crystallinity and crystal structure of materials. It is another valuable tool to analyze the quality of synthesized graphene. First of all, the technique is used for assessing the crystallinity of graphene sheets by visual analysis of the electron diffraction pattern. Distinct and sharp spots arranged in a hexagonal pattern with six-fold symmetry indicate the crystalline nature of graphene [97]. Furthermore, by finding the intensity ratios of various spots corresponding to different planes, the number of graphene layers present in the material can be estimated. The intensity ratio of (1–210) and (1–100) points (Miller-Bravais indices) is measured, and a value of less than unity indicates monolayer graphene. On the other hand, research groups

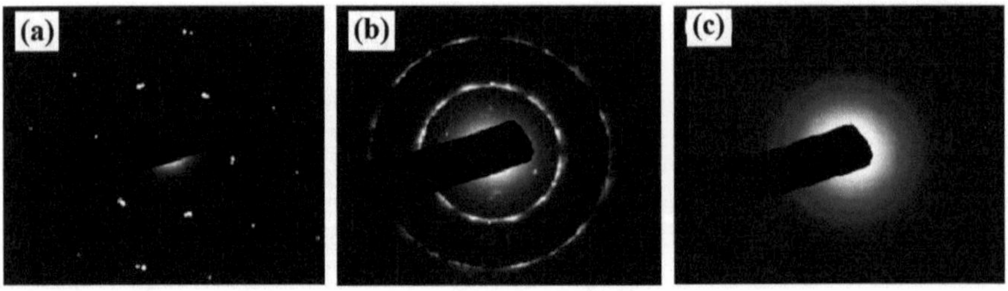

Fig. 8 SAED patterns attributed to **a** Graphene (with some overlapping/folding/rotational stacking), **b** MLG with some amorphous carbon, and **c** only amorphous carbon [102]

have reported different values for MLG, but a value close to 2 for bilayer and 4 for trilayer Bernal AB stacking is generally observed [98–100]. All these groups have also confirmed the results with Raman spectroscopy.

Sharp concentric rings appear, in general, for polycrystalline materials with randomly oriented crystallites. Similarly, a random orientation of sheets in MLG (instead of Bernal AB structure) results in a sharp diffraction pattern with rings and some intense points on the rings [101]. Similar multiple ring patterns with maxima points can also appear due to the presence of amorphous carbon along with graphene sheets, as can be seen in Fig. 8b. In contrast, only the halo ring pattern similar to Fig. 8c infers the absence of crystallinity and, thus, of graphene (only amorphous carbon is present) [102]. MLG, with slight misalignment between layers, produces a cometic aberration-like pattern (Fig. 8a) instead of sharp points [103].

It is necessary to point out here that the electron beam is focused in a small area, and thus, the results obtained give information only on that particular area of the material. For thorough investigation, data should be obtained from different regions of the sample separated by a fixed distance through SAED mapping [97, 99]. Some researchers have proposed more sophisticated but faster methods to circumvent this issue for large area sheets [104].

4.3 Microscopic Techniques

Non-microscopic techniques for the characterization of graphene provide data based on the material's properties, but a visual perspective makes an entirely different impact on the understanding of a material. Optical microscopy is one of the oldest in this category and is also employed for studying graphene in the micro regime, but the information it reveals is very limited [105]. Transmission electron microscopy (TEM), especially with high resolution (HRTEM), is a powerful tool to probe the materials in nano and even sub-nano regimes. TEM is widely used to obtain information on surface topographies, growth patterns, and discontinuities. Sheet-level defects such as wrinkles, folds, kinks, rippling, and twists can be easily observed in TEM micrographs [27, 98, 103]. It is a highly recommended tool to study the potentially preferred sites of the attachment [102, 106]. Information on a much deeper level, such as lattice defects, atomic planes, and interplanar spacing, can be obtained from the HRTEM [98]. It can also be utilized to measure the number of graphene layers by carefully examining high-resolution images of the edges [98, 99, 107].

Atomic force microscope (AFM) is different from other techniques because it uses a material cantilever tip, rather than electromagnetic (EM) waves or electrons, to probe the underlying material. It is another useful tool in the microscopy group to obtain crucial information about the synthesized graphene such as morphology, [108] imperfections, [109] functionalization, [110] mechanical, [29, 111] and electrical properties [28, 112]. The technique can also estimate the number of layers by measuring thickness, but results are not very reliable unless confirmed with other techniques. C.J. Shearer et al. conducted experiments to accurately measure the thickness of graphene with the help of AFM [105]. Apart from characterization, AFM can also be modified for controlled thermal reduction of GO to draw graphene patterns for the fabrication of electronic devices [113].

5 Applications

5.1 *Electronics*

One of the major problems that modern transistors face is power scaling with size. Dennard scaling law states that the power density of transistors remains the same, [114] but it no longer holds because of high leakage currents in devices with feature sizes less than 65 nm [115]. M.C. Lemme et al. first demonstrated the use of graphene as the channel material in field effect devices [116]. The power consumption was significantly reduced due to the high conductivity of graphene, and they were called graphene FETs or GFETs. A typical configuration of graphene channel field effect device is presented in Fig. 9. Graphene, being a zero bandgap semiconductor, causes turn-off issues in GFETs and makes them unsuitable for high-frequency applications, particularly for digital circuits [117]. Several proposed solutions such as electrical double layer gating, [118] uniaxially strained graphene, [119] graphene nanoribbons (GNRs), [120] high-k dielectrics, [121] and wrapped channel [122] have demonstrated potential. The performance was substantially enhanced, and up to 10^7 I_{DS} ON/OFF ratio was achieved by using GNRs [120]. A novel configuration, MSIS-FET, was proposed by P. Li et al. with which they reported an even better ON/OFF ratio than GNRs [123].

Graphene has shown massive potential for light emitting diode (LED) applications. Graphene can be used for engineering various components such as electrodes and active material of the device. F. Withers et al. prepared van der Waals heterostructures using graphene, boron nitride, and metal chalcogenides to fabricate flexible LEDs with an extrinsic quantum efficiency of nearly 10% [124]. They also proposed

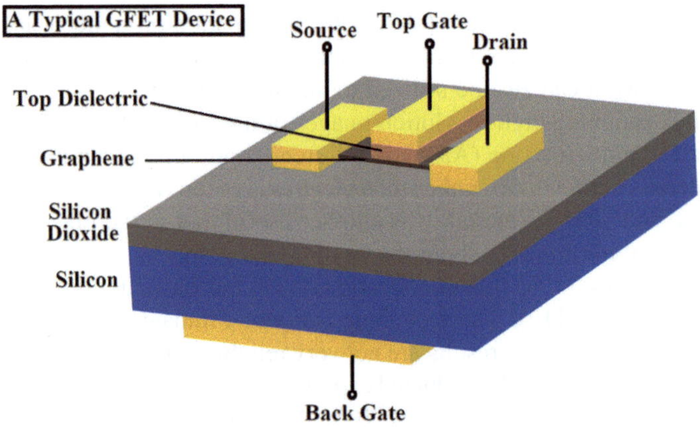

Fig. 9 A typical graphene-based field effect device (GFET)

further improvements by introducing 2D layers of other chalcogenides. F. Rodríguez-Mas and his group introduced rGO in a previously explored structure for LED and reported better intensity with an unaffected output spectrum [125]. D. Yu et al. fabricated GaN micro-rod arrays on graphene films for flexible micro-LEDs emitting blue light [126]. Y.-X. Chen with his team prepared functionalized graphene quantum dots (QDs) from fullerenes and blended them with polyvinyl alcohol to form photoluminescent films that can transform ultraviolet LED into white [127]. Graphene is not only used for improving the existing LED devices but has also demonstrated the potential for novel all-graphene devices. X. Wang et al. demonstrated an all-graphene, wavelength-tunable LED device by using GO/rGO interface for producing light ranging from blue to red [128]. Graphene can also be used outside of the principal device to enhance its overall performance and stability [129].

5.2 Healthcare

The biocompatibility of graphene has been widely studied, [130, 131] which provides opportunities for its use in healthcare applications. Graphene has been explored for its potential in diagnostics, therapeutics, medicine, water purification, etc. It has provided technological advances in smart diagnostic tools for diseases and general health monitoring. A group led by W. Gao made a diagnosis and telemonitoring platform for COVID-19 [132]. A recent study shows that septicemia, which causes the immune system to damage its own tissues, can be diagnosed with the help of magnetic graphene-based micromotors [133]. Graphene-Based Transistors have also been applied to disease diagnosis [134]. Graphene and its derivatives have exhibited potential for detection of common bacterial [135] and viral infections [136]. A large number of reports and reviews on graphene-based diagnostic tools for cancer indicate its enormous potential for the sector. E.G. Afshar et al. have reviewed applications of graphene in the diagnosis and treatment of one of the most menacing cancers, Glioblastoma multiform [137]. R. Majidi and M. Nadafan theoretically studied the application of twin graphene to diagnose lung cancer by sensing exhaled gases in human breath with the help of density functional theory (DFT) [138]. An immune-sensor for detecting a biomarker (HER-2) for breast cancer diagnosis was developed by H. Nasrollahpour with his team [139]. The group exploited the electroluminescence property of rGO/chitosan nanocomposite for the purpose.

Graphene is not only suitable for diagnostics but also has properties that are desirable in various therapies. Graphene-based therapeutics is one of the hot topics in modern healthcare for superior quality products. Several medical scientists have proposed and demonstrated ideas for the evolution of existing approaches and innovation of techniques by exploiting the unmatched properties of graphene. Among other methods, graphene-based materials for photothermal therapy (PTT) have gained significant popularity for cancer treatment in the past decade. PTT involves the delivery of a photothermal agent to the infection site (infected cells) and its subsequent heating through exposure to the near-infrared (NIR) radiation resulting in the killing of infected cells (Fig. 10 represents the scheme). C.C. Barrera et al. studied the interaction between rGO/Fe_3O_4 NC and cell membrane and the effects of PTT on the viability of the cells [140]. The results show an excellent potential of the material for PTT of cancer. Another iron oxide and rGO-based NC, Fe_3O_4/Au/rGO, was demonstrated by T.S. Ardakani et al. as an effective material for synergistic radio-/photo-therapy on the oral squamous carcinoma cell line [141]. The material was reported to have good biocompatibility for healthy cells while having cytotoxic effects on the infected ones. X. Jia et al. used a hybrid of rGO, Au nanostars, and lipid bilayer for PTT of pancreatic cancer [142]. The material also exhibits an improved targeting capability and hence a better efficacy. R. Lima-Sousa et al. demonstrated, for the first time, the use of injectable GO and rGO incorporated hydrogels for the treatment of breast cancer [143]. The system shows a chemo-photothermal effect to reduce the cancer cell viability to 34% with Doxorubicin:Ibuprofen. Another group reported functionalized rGO as drug loading platforms for targeted delivery of the drug, doxorubicin hydrochloride, and subsequent chemo-photothermal treatment [144]. Various other therapies utilizing graphene and derived materials, such as sonodynamic, [145] photodynamic, [146] magnetothermodynamic [147] have also been reported for tumor inhibition. Many researchers are exploring graphene-based scaffolds for regenerative medicines, which help enhance the repair process of damaged tissues, and have reported impressive results [148, 149].

5.3 Energy

The ever-growing energy demand requires continuous improvements in existing technologies for energy conversion and storage and the creation of new paths that can meet the needs of the future. Graphene has been extensively studied for solar energy

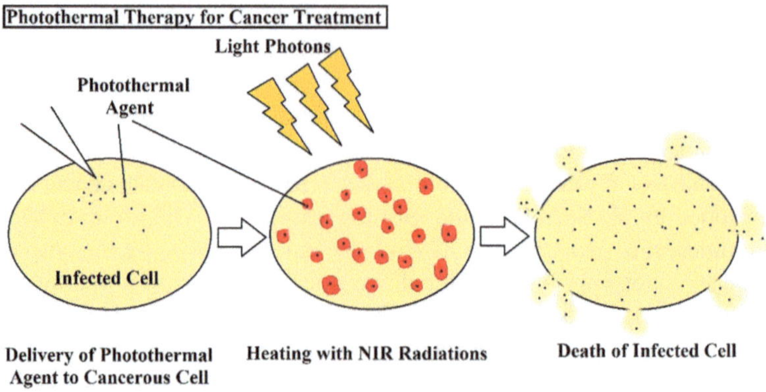

Fig. 10 Step-by-step procedure of PTT for the treatment of cancer

conversion, irrespective of the type of device. However, substantial improvements in the performance of dye-sensitized solar cells (DSSCs) were observed with graphene introduced in photoanode or used at the counter electrode. One such device was fabricated and extensively characterized by F.W. Low et al. (shown in Fig. 11) [150]. They studied the effects of TiO_2 content in rGO as photoanode material on the open-circuit voltage and short circuit current density of the device. S. Sun et al. also demonstrated the use of graphene/TiO_2 composite for DSSC photoanodes and reported 59% enhanced efficiency compared to TiO_2 photoanodes [151]. K. Basu et al. used graphene incorporated SnO_2/TiO_2 composite as photoanode material and reported better efficiency and durability [152]. M.N. Mustafa and Y. Sulaiman prepared photoanode with graphene QDs decorated titania and identified the QDs as light scatters responsible for better performance [153]. The enhanced efficiency has been attributed to the enhanced dye loading due to increased surface area, which improves the carrier generation process. The addition of graphene also improves charge transfer by providing shorter paths to the electrode, reducing the carrier recombination rate. Pt has been the most widely used counter electrode by far; however, its high cost led many researchers to explore alternative options. Counter electrodes made by graphene ink spray deposited onto transparent conductive oxide (TCO) were reported by S. Casaluci et al. for large area DSSCs [154]. W.C. Oh et al. demonstrated the use of a rather complicated material, graphene composite with $Cu_2ZnNiSe_4$-WO_3 nanorods, for efficient (even better than Pt) counter electrode material [155].

Apart from DSSCs, graphene has also exhibited potential for diverse applications in most other types of solar cells. For example, X. Miao et al. used SLG/n-Si for Schottky junction solar cells; [156]. J. Wu et al. made graphene transparent electrodes applied to organic solar cells; [157]. N.F. Ramali et al. inserted graphene passivation layer in perovskite solar cells; [158] graphene nanohills on silicon were illustrated for solar cell applications by M.A. Rehman et al. [159]. T. Lin and team used pristine and boron-doped graphene as the back electrode for CdTe solar cells [160].

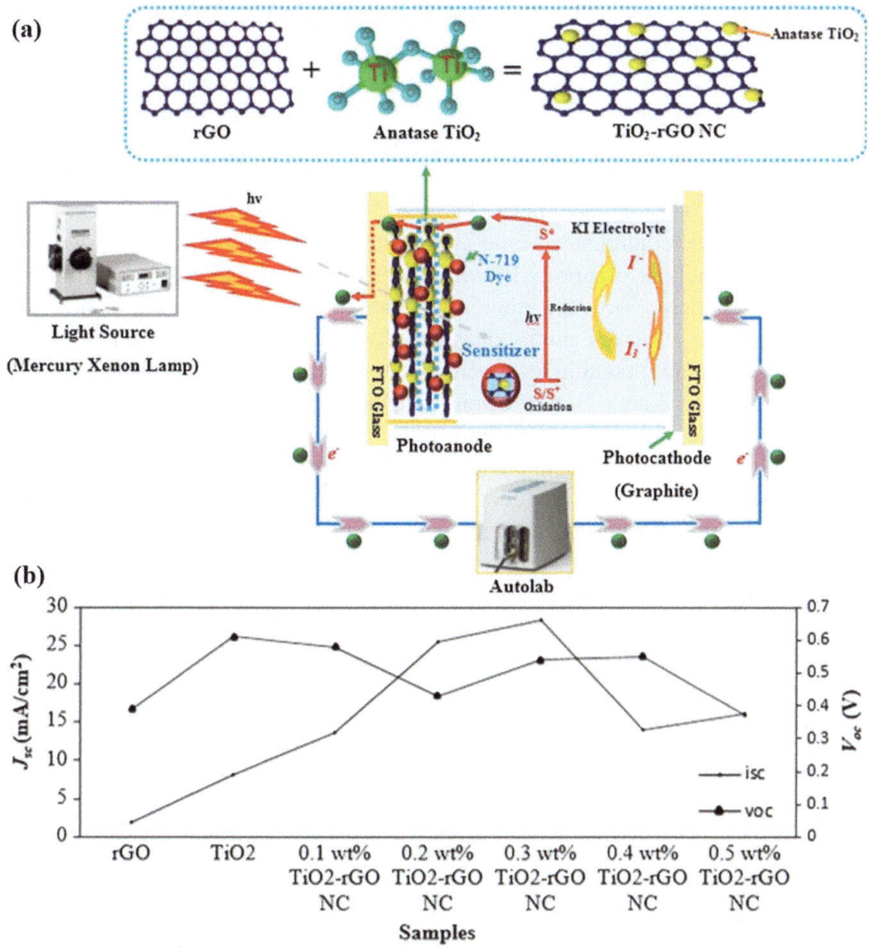

Fig. 11 Schematic of the device fabricated by Low et al. (**a**) and results obtained from the device using different content of TiO_2 with graphene as photoanode material (**b**) [150]

The scope of energy generation, particularly through renewable sources, is limited unless a device can store it for later use. Batteries have long been used for energy storage and are the most prominent technology at present for the purpose. Amongst them, lithium-based batteries have been commercially available in the market for a long time. A large fraction of them uses graphite electrodes. As the technology is evolving, new materials are being explored to achieve higher charge storage capacities and longer cycle life. C. Wang et al. demonstrated the use of graphene paper as cathode for Li batteries [161]. N-doped graphene films have shown better Li-ion intercalation in an article by A.L. Mohana Reddy et al. [162]. J. Xiao et al. fabricated functionalized porous graphene structures for very high capacity (15,000 mAh/g) Li-air batteries [163]. Graphene-incorporated polymer-derived SiOC aerogels have demonstrated stunning results as anode material for Li batteries [164]. X. Han et al. illustrated the use of core–shell structures (Si core, graphene shell) for Li-ion battery anodes [165]. Using a novel strategy with rGO/Ag-Li scaffolds as anode for Li-metal batteries was developed by T. Ma et al. [166]. A group from Zhejiang University led by C. Gao proposed a defect-free principle for high-performance Al-graphene batteries in 2017 [167]. He, leading another group, reported an exceptionally long cycle life of 250,000 for Al-graphene batteries later that same year [168].

Another prime requirement for an energy storage device is fast charging. Electrochemical capacitors, otherwise known as supercapacitors, are the most promising alternative to batteries with the inherent advantage of speedy charging. They offer high power density, but the energy density is low. Graphene-based electrodes have exhibited tremendous potential for supercapacitor applications because of graphene's high specific surface area. Supercapacitors made by RuO_2 modified rGO as anode and polyaniline modified rGO as cathode have been demonstrated by J. Zhang et al. [169]. L.T. Le et al. prepared supercapacitor electrodes by inkjet printing of GO on Ti and its subsequent thermal reduction [170]. Y. Fang et al. fabricated supercapacitor electrodes with functionalized graphene and demonstrated that the technology could bridge the gap between supercapacitor and battery with further optimizations [171]. H. Kim et al. proposed an all-graphene synergistic system for energy storage [172]. They also reported that the system works on both battery and supercapacitor mechanisms and thus creates a link between the two.

5.4 Environmental Remediation

Pollution, whether air, soil, water, or radiation, has caused much harm to the environment and human health. Scientists are looking for materials and devices that can be engineered for providing tools to reduce/eliminate pollution-causing agents. Many properties of graphene, such as corrosion resistance and mechanical and chemical stability, are ideal for environmental remediation applications. It has, therefore, been explored for removing pollutants, especially from water and air. Several reports have shown that graphene and derived materials are excellent for organic dye removal applications for wastewater treatment. H. Guo et al. reported GO/polyethylenimine (PEI) hydrogel for large-scale and efficient dye, Methylene Blue (MB) and Rhodamine B (RB), removal from wastewater [173]. They attributed the adsorption of dye to GO while PEI assisted the gelation process of GO. T. Jiao et al., however, synthesized rGO/Ag/PEI hydrogel via simultaneous reduction of GO and Ag^+ and suggested its application in catalytic degradation of MB and RB [174]. Y. Zhong and the group used dopamine GO composite decorated membrane to adsorb Congo Red and Basic Blue dyes [175]. They also demonstrated that treatment of membrane with a minute quantity of reducing agent ($NaBH_4$) to reduce a small part of GO makes it useful for removal of small concentrations of methyl orange (MO), MB, RB, and 4-Nitrophenol. Photocatalytic dye (MB) degradation ability of Ag decorated rGO was exhibited by M. Ikram et al. [176]. Heavy metals in water are another primary class of pollutants that are toxic to aquatic as well as non-aquatic life. T.A. Tabish et al. prepared highly porous graphene and demonstrated its application for efficiently removing MB and RB [177]. They further demonstrated the removal of oils (vegetable, engine, and pump oil), heavy metal (arsenic ion), and other harmful ions (fluoride and nitrate) from water. Many other groups also reported graphene-based techniques as promising candidates for heavy metal removal from wastewater. G. Zhao et al. prepared GO with the help of modified Hummer's method and displayed its application for sorption of Cd(II) and Co(II) from water [178]. K. Zhang et al. anchored TiO_2 nanoparticles on graphene sheets for effective reduction of Cr(VI) to Cr(III) [179]. A. Marjani et al. prepared GO dispersed in polyethersulfone membrane for degradation of MB and MO dyes and removal of heavy metal ions (Cu^{2+}, Cd^{2+}, and Zn^{2+}) [180]. J. Zhang et al. studied, for the first time, the molecular interaction mechanism experimentally through single-molecule force spectroscopy and theoretically with DFT [181]. The study revealed critical aspects of molecular interactions that will help design graphene-based materials for wastewater treatment.

Though graphene has been extensively studied for the removal of water pollutants, it has also shown potential for removing contaminants from the air. V. Kumar et al. reviewed potential applications of graphene and related materials for adsorption and removal of volatile organic compounds [182]. W. Jung et al. fabricated a robust system by sandwiching a condenser between two layers of rGO filters to remove particulate matter from the air [183]. The group demonstrated the removal of both filterable and condensable particulate matter with the help of the fabricated system.

5.5 Sensing

Graphene, having a high surface-to-volume ratio, is an excellent material for sensing applications. Graphene has been applied to a range of sensors for physical parameters like pressure, temperature, strain, humidity, etc., and chemical species like gases, analytes, heavy metals, etc. Due to extraordinary mechanical and electrical properties, graphene has been studied extensively for pressure and strain sensors that are key to modern non-invasive techniques for health monitoring. Multiple innovative graphene-based wearable pressure and strain sensors were demonstrated by T.-L. Ren and associates for the purpose. In one report, they used tissue paper soaked in GO solution, which was later given thermal treatment to reduce GO for making pressure sensors for human motion detection [184]. Another report explains the use of Ag nanoparticles linked to graphene sheets over PDMS substrate for strain sensors that can monitor human motion [185]. In this work, the group used the laser scribing technique to reduce the GO. In another innovative work, an abrasive paper was used to make patterns on PDMS over which GO was deposited and subsequently reduced [186]. The rGO was then sandwiched by placing another layer of PDMS, forming a structure imitating the epidermis. The system was used to fabricate a wearable pressure sensor for human motion detection. A strain sensor for motion detection was made by dipping a polyester fabric in GO followed by thermal treatment to reduce

GO [187]. In a more recent work by T.-L. Ren and associates, a breathable graphene electronic skin, susceptible to strain and vibrations, and thus, useful for monitoring electrocardiogram, body motion, and respiration, was demonstrated [188]. A wearable pulse monitoring device with high sensitivity and long-range linear response graphene pressure sensor was reported by J. He et al. [189]. A graphene strain sensor for real-time pulse monitoring systems was explored by T. Yang et al. [190]. W. Liu et al. used wrinkled graphene for making piezoresistivity-based flexible, susceptible, and reliable pressure sensors [191]. Another class of graphene-based sensors, useful for healthcare applications, detect particular analytes from body fluids such as non-enzymatic glucose sensors [192, 193].

Z. Zheng and H. Wang performed first-principles calculations through DFT on doped and undoped graphene-based gas (CO_2) sensors [194]. The results show a better performance with Al-doped than B-, N-, and P-doped graphene. Y. Seekaew and C. Wongchoosuk reported a novel CO_2 sensor configuration utilizing electroluminescence with graphene film deposited over a patterned Ag electrode on a phosphor material [195]. G. Liu et al. fabricated a temperature sensor using rGO and demonstrated its use in robot skin for internet of things (IoT) [196]. B. Davaji et al. demonstrated the use of suspended SLG, deposited on SiO_2/Si substrate and SiN substrate for a resistance temperature sensor [197]. Devices formed on SiN substrates showed better performance. J. Yun et al. illustrated using graphene electrodes to build high-performance capacitive pressure sensors [198]. Graphene-based sensors for other physical quantities such as humidity, [199] refractive index, [200] magnetic field, [201] etc. have also been demonstrated.

5.6 Protective Coatings

Corrosion and biofouling are the major concerns for components used in marine applications and wastewater treatment. Graphene has displayed a great potential for protecting metals and alloys in corrosive conditions due to its hydrophobicity and impenetrability. F. Yu et al. demonstrated the use of the CVD process for graphene deposition on Al alloys for long-term corrosion protection [202]. A comparative study on GO and rGO based epoxy coatings for corrosion protection was performed by F.A. Ghauri et al. [203]. GO was found to be better protecting mild steel than rGO when exposed to a 3.5% NaCl solution. Y. Ye et al. reported better corrosion resistance with salinized aniline trimer functionalized graphene [204]. They reported that graphene forms an impermeable coating while salinized trianiline promotes self-healing of the underlying metal; thus, these synergistic effects provide better protection from the corrosive environment. S. Qiu et al. reported similar self-healing properties with polypyrrole intercalation in graphene [205].

Even with numerous reports of graphene providing protection from corrosive fluids, researchers have also shown that graphene coating can have degrading effects due to it being cathodic to most metals [206]. They have proposed that graphene inhibits corrosive solutions from reacting with underlying metal, but when a small

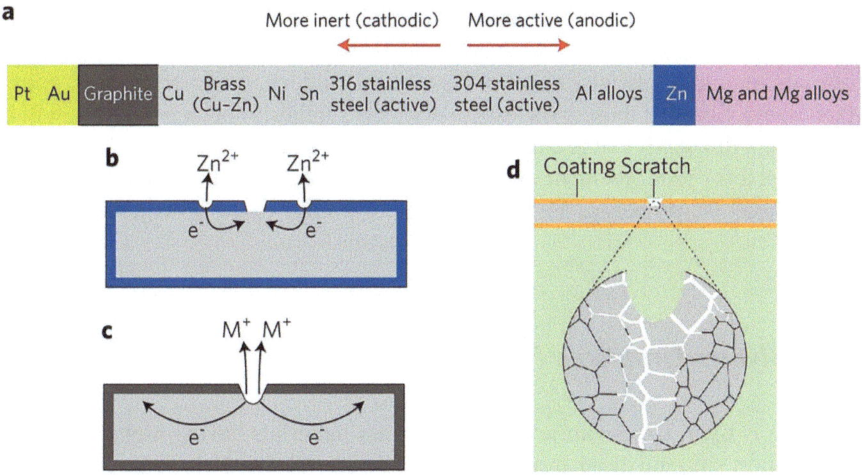

Fig. 12 The chemistry of anticorrosion coatings: **a** the corrosion tendency of various metals and graphite (galvanic sequence), **b** anodic coating of Zn over steel, **c** cathodic coating promoting local oxidation, and **d** corrosion causing weakening of underlying metal/alloy by etching through grain boundaries. [206]

portion of the metal is exposed due to scratch or some other reason, graphene, being conductive, acts as the cathode and promotes localized oxidation (Fig. 12). The topic is debatable, and some material scientists have reported methods to circumvent this problem. X. Xu et al. demonstrated defect-free graphene coating on Cu(111) surface for durable protection from humid air [207].

Graphene/Si based antifouling coatings have shown vast potential, especially for maritime applications. H. Jin et al. proposed a novel graphene/Si rubber-based antifouling coatings and its working mechanism [208]. They proposed a mechanism inspired by sea animals having soft and flexible skin like dolphins due to which a dynamic surface is created, and they named it 'harmonic motion effect'. They further reported, in another study, the effects of elastic modulus and color of the material on its antifouling properties [209]. M.S. Selim and the group also explored graphene/Si based materials for antifouling properties [210]. Antifouling coatings have also been proven advantageous for protecting sensors used in marine technology [211]. Multifunction coatings, such as antifouling coatings with anticorrosion [212] and antibacterial [213] effects, have also been proposed and tested.

6 Graphene Nanocomposites

Pristine graphene itself is a material with wonderful qualities and thus, has been studied for a range of applications and has been commercially employed in various industries. Other forms of graphene such as functionalized graphene, GNRs, GO, and graphene NCs can have desired properties for several applications that pristine graphene cannot provide. Graphene has been utilized to make NC with many materials, including metals, metal oxides, metal sulfides, polymers, CNTs, epoxy, Mxenes, etc. Some of these NCs, especially their synthesis and applications, will be discussed in this section. It is crucial to point out that in some of these NCs, graphene will be used for reinforcement, while in others, graphene will be the matrix phase. Nevertheless, the properties are generally improved due to the synergistic effects of both phases.

6.1 Polymer-Graphene NCs

Graphene is hybridized with polymers to obtain materials having high mechanical strength, porosity, and thermal stability. Porous materials have a high specific surface area, one of the key requirements for electrochemical applications. X. Li et al. demonstrated the synthesis of highly porous polyaniline (PANI)/graphene NC through electropolymerization of PANI on graphene for electrochemical capacitors [214]. They reported up to 1209 F/gm specific capacitance and long cyclic stability with the synthesized materials. S. Fazli-Shokouhi et al. used in situ polymerization to prepare PANI-graphene NC, which demonstrated good anticorrosion and antifouling properties when incorporated in epoxy.[212] PANI- graphene NCs have also been used for EM interference mitigation [215]. In-situ polymerization, as presented schematically in Fig. 13, is one of the most appreciated methods for polymer-graphene NCs.

Poly-methyl methacrylate (PMMA) has been used for making composite materials with graphene through techniques such as in-situ emulsion polymerization, [216] precipitation polymerization, [217] solution blending method, [218] etc. The synthesized materials were characterized to find mechanical, thermal, and electrical properties [216] and dielectric and rheological properties [217]. Z. Zabihi et al.

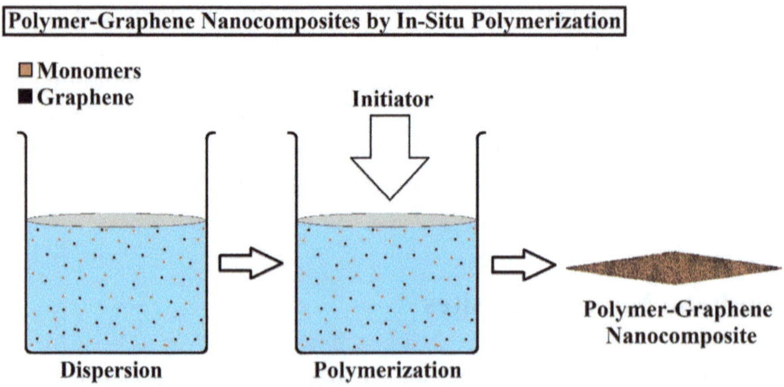

Fig. 13 Steps involved in the in-situ polymerization process for the synthesis of polymer-graphene NCs

studied the interaction between the two components, graphene and PMMA, through theoretical simulations and their potential for detecting volatile organic compounds. [218].

Small quantities of graphene significantly improve the mechanical and thermal properties of polymers such as polyvinyl alcohol (PVA). The majority of researchers illustrated blending GO with PVA and then reduction of GO through a suitable method to obtain the desired rGO/PVA NC. X. Yuan reported improved tensile strength and thermal stability of PVA when 0.8% rGO was water blended into the polymer matrix [219]. X. Zhao et al. have reported 150% increased tensile strength and more than ten times enhanced Young's modulus with 1.8 vol % graphene addition via facial aqueous solution [220]. X. Wang et al. also reported improved mechanical and thermal properties of PVA with much smaller (0.1 and 0.3%) fractions of graphene [221]. They attributed strong interfacial interactions between the components to the improved tensile strength and restriction of motion of polymer chains by graphene sheets to the enhanced thermal stability. T. Zhou et al. obtained mechanically more robust and electrically conductive rGO/PVA NC with the help of a sodium hydrosulfite reducing agent [222].

Graphene has also been reported for making NC with polydopamine for strain sensor;[[223]] with poly(p-phenylenediamine) for detection of protein biomarkers; [224] with poly(isobutylene-co-isoprene) for barrier, dielectric, and sensing applications; [225] with polyimide [226] and poly(ethylene-co-methyl acrylate) [227] for EM interference shielding; with poly(butylene succinate) [228] and polyvinylbutyral [229] for improving thermal and mechanical properties; with poly(styrene sulfonate) [230] and poly vinyl pyrrolidone [231] for ascorbic acid sensing.

6.2 *Metal-Graphene NCs*

Metal-graphene NCs have shown potential in a large spectrum of applications ranging from electronics to biomedicine. Nickel, which also acts as a catalyst for the synthesis of graphene, can quickly form NCs with graphene via electrodeposition technique, which are widely employed in tribological applications [232–236]. It was reported by all these groups that incorporation of graphene in Ni increases hardness and, thus, increases wear resistance of the coating as compared to Ni coating. Surfactants, such as sodium dodecyl sulfate, were also used in deposition baths to further improve the synthesis process and hence the resultant properties of the coatings [232]. A. Jabbar et al. explored the effects of bath temperature in the range from 15°C to 60°C on deposited coatings and reported that moderate temperatures around 45°C are suitable for obtaining refined grain size and better microhardness [233]. A. Algul et al. extensively studied the effects of graphene content on wear mechanism [234]. A theoretical nanoindentation simulation study on Ni-graphene NCs was performed by S.-W. Chang et al. [237]. Their study reveals that an increasing number of graphene layers negatively impact the hardness but improve elastic deformability. The same group also performed atomistic simulations on Ni- and Cu-based graphene NCs to record the effects of temperature and geometry on thermal interface conductance [238]. They reported that the effects are similar for both Ni and Cu. They also demonstrated that the SLG provides a high thermal interface conductance of ~ 500MWm^{-2} K^{-1}, which decreases to ~ 300MWm^{-2} K^{-1} for bilayer and 150MWm^{-2} K^{-1} for trilayer. For more than three layers, the value (~100MWm^{-2} K^{-1}) becomes independent of the number of graphene layers.

Similar to Ni, Cu-graphene NCs have also been used for tribofilm formation. N. Khobragade et al. fabricated graphene-reinforced Cu NCs through the high-pressure torsion method and found that the synthesized NC with 10wt% graphene is twice as hard as pure copper [239]. S. Wang et al. also reported similar improvements in mechanical properties when graphene layers were directly deposited over Cu nanoparticles [240]. C.L.P. Pavithra et al. synthesized Cu-graphene NC foil via electrochemical route and reported ~ 250Gpa hardness, which is also close to the above-discussed results [241]. Not only tribological applications but Cu-graphene NCs have also been employed for sensing applications such as EC immune-sensors for detection of newcastle disease [242] and glucose sensor for food [243].

Noble metals such as Ag and Au have also been utilized to make NCs with graphene and have found applications in numerous areas. The synthesis protocols are quite similar but differ in the reduction methods applied for Ag^{+}/Au^{3+} ions and GO. In a typical procedure, an aqueous solution of silver/gold salt is mixed with an aqueous solution of GO, followed by simultaneous reduction of ions and GO [244–251]. Some researchers also reported sequential, instead of simultaneous, reduction of metal ions and GO [251]. The reducing agents may be chemical (e.g., sodium citrate,[244] hydrazine, [247, 251] sodium borohydride [246]) and biological (e.g., electrochemically active biofilms), [245, 249], or it can be done with other means such as electrochemistry [250] and exposure to ultraviolet radiations [248]. Bimetallic-graphene NCs have also been frequently reported in the literature. F. Tahernejad-Javazmi et al. studied rGO/$FeNi_3$ NCs applied to the detection of tert-butylhydroquinone in the presence of folic acid [252]. M.R. Vengatesan et al. used Ag–Cu/graphene NC for hybrid capacitance deionization application [253].

6.3 *Metal Oxide-Graphene NCs*

Metal oxides are one of the most crucial inorganic materials because they exhibit a range of properties with a distinct variation within each one. For example, metal oxides can have bandgaps such that some of them are narrow bandgap semiconductors or wide bandgap semiconductors, and others are insulators. NCs made by incorporating graphene into metal oxide matrix, or the other way around, offer novel properties desirable for many applications. Solvothermal/hydrothermal routes are generally employed for synthesizing NCs of graphene with metal oxides as well as sulfides. A typical procedure, as schematically represented in Fig. 14, involves the treatment of precursor material at high pressure and a temperature above the boiling

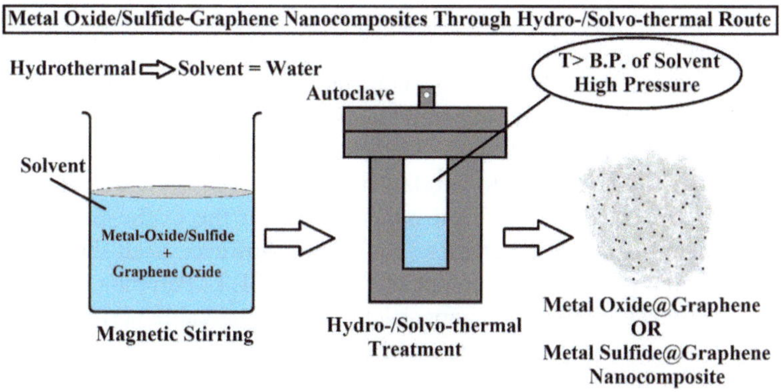

Fig. 14 Solvo-/hydro-thermal route for the preparation of graphene-metal oxide/sulfide NCs

point (BP) of the solvent for 24 h. Titanium dioxide (TiO_2), a wide bandgap material, makes NC with graphene, and the NC thus formed has been widely explored for solar cells, especially DSSCs, as discussed in section VI [151–153]. A group led by Y.-J. Xu has studied the photocatalytic applications of TiO_2/graphene NC for the degradation of volatile aromatic compounds [254] and oxidation of alcohols [255]. Both these reports compared the photocatalytic activity of TiO_2/graphene NC with TiO_2/CNT NC and reported better performance with the former. TiO_2/graphene NCs have also been employed for several other applications such as battery, [256] pseudocapacitor, [257] and sensing [258]. Solvo-/hydro-thermal techniques are often used to synthesize graphene-metal oxide NCs, especially graphene-TiO_2 NCs. [254–256, 258].

Iron oxide, with its various oxidation states, has been reported to make NCs with graphene, which has found applications in multiple areas. S. Lee et al. prepared Fe_2O_3/graphene NC via a solventless route and demonstrated its use in detecting heavy metal ions (Zn^{2+}, Pb^{2+}, and Cd^{2+}) [259]. H. Su et al. prepared FeO_x/GO NC using the co-precipitation method and demonstrated its use for highly efficient arsenic removal [260]. Y. Yuan et al. employed a hydrothermal method with hydrazine to reduce GO and prepare Fe_2O_3/graphene NC applied as a catalyst for the thermal decomposition of NH_4ClO_4 [261]. Numerous research papers on iron oxide/graphene NC as anode material for batteries stipulate its potential for energy storage devices. L. Xiao et al. used direct self-assembly of iron oxide and GO followed by hydrothermal treatment to reduce GO to synthesize Fe_2O_3/rGO NC [262]. They demonstrated its use as anode material in lithium-ion batteries with superior performance. W. Jiang et al. reported preparing a novel FeO_x/graphene NC through a solid-state method and exhibited its use as anode material for nickel–iron batteries [263]. They ball-milled ferrous oxalate dihydrate, GO, and glucose (reducing agent) followed by a series of heating processes to obtain the final product. H. Ren et al. used a simple hydrothermal method for obtaining Mn-doped Fe_3O_4/graphene NC for Na-ion batteries [264]. Fe_3O_4/graphene NCs have also attracted attention for mitigating EM pollution [265, 266].

Copper oxide/graphene NCs for Li-ion batteries have also been well documented in the literature. B. Wang et al. prepared CuO/graphene NC through simple stirring and centrifugation processes [267]. They reported much enhanced cyclic performance with the NC compared to anodes made by using only CuO nanoparticles. A.K. Rai also reported a significant enhancement in the cyclic performance of CuO/graphene NC for Li-ion batteries [268]. They, however, used short time spex-milling for obtaining the desired composite materials. L.L. Perreault et al. prepared graphene-wrapped Cu-Ni oxide nanoparticles with remarkable performance as Li-ion battery anode [269]. The group spray-dried the oxide nanoparticles and graphene mixture to form the NC. The use of CuO/graphene NC as an interfacial layer of metal/interlayer/semiconductor junction diode was demonstrated by Z. Orhan et al. [270]. They also studied gamma radiation resistance properties of the layer to demonstrate the device's ability to work in a radiation-rich environment. L. Luo et al. discussed CuO/graphene NC for non-enzymatic highly sensitive glucose sensor applications [271].

Several materials engineers have exploited ZnO/graphene NCs for photocatalysis, [272, 273] sensing, [274, 275] solar energy conversion, [276] and pollution mitigation[[277]] applications. Preparation methods and thus the synthesized NCs differ in properties, with the most common route again being the solvothermal method [273, 277]. Other methods that have also been explored oftentimes include combinations of stirring, ultrasonication, and centrifugation techniques; [272, 274] ball milling; [273] and in-situ reduction of zinc acetate and GO [275, 276]. Tin oxide (SnO_2) is also hybridized with graphene to form NC materials that have found applications in many interesting areas. Na/Li-ion batteries have seen much advancement due to the synergistic effects of graphene and tin oxide on the anode [278, 279]. D. Zhang et al. developed highly sensitive humidity sensors using SnO_2/graphene NCs [279].

6.4 Metal Sulfide-Graphene NCs

Metal sulfides have been materials of interest for the scientific community working in diverse areas, from energy to biomedicine. Some of them, such as MoS_2 and SnS, have layered structures similar to graphite, making them suitable for high surface area applications such as energy storage. The addition of graphene to these metal sulfides allowed researchers to push the performance limits in these research areas.

Graphene NCs with tin sulfides (SnS and SnS_2) have been researched for multiple applications, energy conversion and storage remain the dominant ones. M. Zhang et al. introduced a novel solvothermal technique to synthesize SnS_x/graphene NC and demonstrated its application as anode material for Li-ion batteries [280]. Researchers have been following the route with few or no modifications to synthesize metal sulfide/graphene NCs. T. Ma and the team also used the solvothermal method to prepare SnS/graphene NC, which was later covered by N-doped carbon coating [281].

They demonstrated the use of this material to make anodes for Li-ion batteries. J. Shi et al. used graphene sheets hybridized with N, S co-doped carbon-coated SnS nanoflakes, synthesized again via solvothermal route, for Na-ion battery applications [282]. B. Yang et al. used solvothermal technique for the preparation of SnS/graphene NC applied to counter electrodes for DSSCs, replacing the commonly employed Pt electrode [283]. Another group used the hydrothermal route for the fabrication of flexible SnS_2/graphene sensors for NO_2 detection [284]. J. Johny et al. made SnS/graphene NCs for solar energy conversion and electrochemical applications, but they followed a rather different synthesis protocol. The procedure involved spraying a mixture of SnS and GO onto moderately (120 and 250 °C) heated substrates to obtain the desired material [285].

Cadmium sulfide (CdS), with its various forms (films, nanoparticles, etc.), is a popular material for solar cell applications [286]. When blended with conducting graphene, CdS can also be employed for scores of other applications. Hydrogen production is one such sector where CdS/graphene NCs have been extensively studied [287–289]. A group led by T. Peng used GO, cadmium acetate, and dimethyl sulfoxide as precursor materials which were given solvothermal treatment for CdS/rGO NC preparation [287]. The same group also reported using unreduced GO to form NC with CdS and its application in H_2 storage [288]. In this work, they just stirred the GO and cadmium acetate mixture and added Na_2S dropwise, skipping the thermal treatment step with the autoclave. S. Hammadi et al. used 7–8 layer graphene (purchased), cadmium sulfate, and thiourea as precursors for NC synthesis [289]. They performed experiments to display its potential for solar cells in addition to H_2 production applications. CdS/graphene NC has also exhibited potential for photoelectrochemical sensing applications in a report by L. Ge et al. [290]. They used an entirely different approach for NC synthesis directly onto an indium tin oxide (ITO) coated glass with the help of a direct-laser writing technique. The same article also demonstrated the fabrication of sensing devices with PbS/graphene NC through a similar approach. Other researchers have also investigated PbS/graphene NCs for sensing applications [291].

Cobalt sulfide (CoS)/graphene NCs have found applications in supercapacitors [292–294] and, like other metal-sulfide/graphene NCs, are generally prepared by solvothermal route [292, 293, 295]. R. Ramachandran et al., however, demonstrated the use of oil bath technique to synthesize the material [294]. G. Huang et al. described their cobalt sulfide phase as a mixture of CoS, CoS_2, and Co_9S_8 and demonstrated the use of NC thus formed for Li storage with excellent cyclic performance [295]. Graphene NCs with several other metal sulfides have also been investigated, but the list is long, and thus we will restrict this discussion by mentioning them with a few of their applications. MoS_2, for instance, has been studied for EM pollution removal, [296] hydrogen evolution reaction, [297] desalination, [298] etc. Copper sulfide (CuS)/graphene NCs have been explored for hydrazine and hydrogen peroxide sensors, [299] biosensor to detect trichloroacetic acid [300] etc. Various reports on the supercapacitor application of nickel sulfide (NiS)/graphene NC can be found in the literature [301, 302]. Sulfides of tungsten (WS_2) [303] and vanadium (VS_2) [304]

have been investigated for sensing dihydroxybenzene isomers (catechol, resorcinol, hydroquinone) and Li-ion batteries, respectively.

7 Summary and Future Prospects

In this chapter, graphene and its nanocomposites with several crucial organic and inorganic materials have been discussed. Researchers from across the globe unanimously believe that graphene is a material of unmatched properties and, therefore, has a bright future. This chapter is designed to encourage this faith to further add to what researchers have achieved in the past two decades with graphene. The challenges that current technologies face for the scaled-up production of graphene and the currently available options to circumvent these issues have been explored. The discussion has an emphasis on the liquid phase exfoliation (LPE) process as it is a simple and one of the most promising techniques for cost-effective production of commercial-grade graphene in larger quantities.

Characterization tools are required for the quality assessment of graphene. Few of them, which reveal critical information, have been discussed in the chapter. Basics of Raman spectroscopy and various aspects of it, when implemented for characterization of graphene, have been critically discoursed. The origin of different bands appearing in the Raman spectrum of graphene, their broadening and shifting due to structural changes introduced by imperfections have been discussed. SAED was introduced, within the scope of graphene, for information on the crystal structure and stacking of layers. A brief discussion of microscopic tools such as TEM and AFM has also been presented.

The remarkable properties of graphene made curious minds wonder about their exploitation in real-world applications. In this chapter, applications of graphene are reviewed in electronics, healthcare, energy, environmental remediation, sensing, and protective coating sectors. Graphene nanocomposites are generally prepared to optimize properties for application-specific requirements. For example, graphene NCs with metals have excellent tribological properties, while NCs with metal oxides and sulfides have shown much potential for energy storage devices, especially batteries and supercapacitors. A discussion on the nanocomposites of graphene with polymers, metals, metal-oxides, and sulfides has been included in the chapter.

References

1. Yazyev OV (2016) Theory of magnetism in graphitic materials. In: Esquinazi P (eds) Basic physics of functionalized graphite. Springer Series in Materials Science, Springer, Cham. 244
2. Popov IA, Bozhenko KV, Boldyrev AI (2012) Is Graphene Aromatic? Nano Res 5(2):117–123
3. Wallace PR (1947) The Band Theory of Graphite. Phys Rev 71:622
4. Novoselov KS, Geim AK, Morozov SV, Jiang D, Zhang Y, Dubonos SV, Grigorieva IV, Firsov AA (2004) Electric field effect in atomically thin carbon films. Science 306:666

5. E. Reich, Nobel prize committee under fire. Nature (2010)
6. Kundhikanjana W, Lai K, Wang H, Dai H, Kelly MA, Shen Z (2009) Hierarchy of electronic properties of chemically derived and pristine graphene probed by microwave imaging. Nano Lett 9(11):3762–3765
7. Ghosh S, Nika DL, Pokatilov EP, Balandin AA (2009) Heat conduction in graphene: experimental study and theoretical interpretation. New J Phys 11:095012
8. Ovid'ko IA (2013) Mechanical properties of graphene. Rev Adv Mater Sci 34, 1–11
9. Bolotin KI, Sikes KJ, Jiang Z, Klima M, Fudenberg G, Hone J, Kim P, Stormer HL (2008) Ultrahigh electron mobility in suspended graphene. Solid State Commun 146:351–355
10. McAllister MJ, Li JL, Adamson DH, Schniepp HC, Abdala AA, Liu J, Herrera-Alonso M, Milius DL, Car R, Prud'homme RK, Aksay IA (2007) Single sheet functionalized graphene by oxidation and thermal expansion of graphite. Chem Mater 19:4396–4404
11. Lemme MC (2009) Current status of graphene transistors. Solid State Phenom 156–158:499–509
12. Atalaya J, Isacsson A, Kinaret JM (2008) Continuum elastic modeling of graphene resonators. Nano Lett 8(12):4196–4200
13. Wei Z, Chen Y, Dames C (2013) Negative correlation between in-plane bonding strength and cross-plane thermal conductivity in a model layered material. Appl Phys Lett 102:011901
14. Pop E, Varshney V, Roy AK (2012) Thermal properties of graphene: Fundamentals and applications. MRS Bull 37:1273–1281
15. Rho H, Jang YS, Bae H, Cha A-N, Lee SH, Ha J-S (2021) Fanless, porous graphene-copper composite heat sink for micro devices. Sci Rep 11:17607
16. Pirruccio G, Moreno LM, Lozano G, Rivas JG (2013) Coherent and broadband enhanced optical absorption in graphene. ACS Nano 7(6):4810–4817
17. Pang S, Hernandez Y, Feng X, Müllen K (2011) Graphene as transparent electrode material for organic electronics. Adv Mater 23(25):2779–2795
18. Tetlow H, Posthuma de Boer J, Ford IJ, Vvedensky DD, Coraux J, Kantorovich L (2014) Growth of epitaxial graphene: Theory and experiment. Phys Rep 542:195–295
19. Muñoz R, Gómez-Aleixandre C (2013) Review of CVD synthesis of graphene. Chem Vap Deposition 19:297–322
20. Cai M, Thorpe D, Adamson DH, Schniepp HC (2012) Methods of graphite exfoliation. J Mater Chem 22:24992–25002
21. Niilisk A, Kozlova J, Alles H, Aarik J, Sammelselg V (2016) Raman characterization of stacking in multi-layer graphene grown on Ni. Carbon 98:658–665
22. Eckmann A, Felten A, Mishchenko A, Britnell L, Ralph Krupke, Novoselov KS, Casiraghi C (2012) Probing the nature of defects in graphene by raman spectroscopy. Nano Lett. 12, 3925–3930
23. Beams R, Cancado LG, Novotny L (2015) Raman characterization of defects and dopants in graphene. J Phys: Condens Matter 27, 083002
24. Cong C, Yu T, Wang H (2010) Raman study on the G mode of graphene for determination of edge orientation. ACS Nano 4(6):3175–3180
25. Proctor JE, Gregoryanz E, Novoselov KS, Lotya M, Coleman JN, Halsall MP (2009) High-pressure Raman spectroscopy of graphene. Phys Rev B 80:073408
26. Calizo I, Balandin AA, Bao W, Miao F, Lau CN (2007) Temperature dependence of the raman spectra of graphene and graphene multilayers. Nano Lett 7(9):2645–2649
27. Suh YJ, Park SY, Kim MJ (2009) High resolution TEM and electron diffraction study of graphene layers. Microsc Microanal 15(S2):1168–1169
28. Yu Z, Song A, Sun L, Li Y, Gao L, Peng H, Ma T, Liu Z, Luo J (2020) Understanding interlayer contact conductance in twisted bilayer graphene. Small 16:1902844
29. Teklu A, Barry C, Palumbo M, Weiwadel C, Kuthirummal N, Flagg J (2019) Mechanical characterization of reduced graphene oxide using AFM. Adv Condens Matter Phys 2019:8713965
30. Collins R (30 Dec, 2021) Graphene market & 2D materials assessment 2021–2031, IDTechEx, https://www.idtechex.com/en/research-report/graphene-market-and-2d-materials-assessment-2021-2031/789

31. Hybrid Solutions Graphene, Turtle Wax, (30 Dec, 2021) https://www.turtlewax.com/collections/hybrid-solutions-graphene
32. Graphene 360+, Head, (30 Dec, 2021) https://www.head.com/en/sports/tennis/technology/graphene360plus
33. G-Series, inov-8, (30 Dec, 2021) https://www.inov-8.com/row/g-series
34. G1 wonders, (30 Dec, 2021) https://g1wonders.com/shop/
35. Graphene XT, (30 Dec, 2021) https://www.graphene-xt.com/en/
36. Badami D (1962) Graphitization of α-Silicon carbide. Nature 193:569–570
37. Norimatsu W, Kusunoki M (2014) Growth of graphene from SiC{0001} surfaces and its mechanisms. Semicond Sci Technol 29:064009
38. Rohbeck N, Xiao P (2014) Effects of thermal treatment on the mechanical integrity of silicon carbide in HTR fuel up to 2200°C. J Nucl Mater 451:168–178
39. Badami DV (1965) X-Ray studies of graphite formed by decomposing silicon carbide. Carbon 3(1):53–57
40. Berger C, Song Z, Li T, Li X, Ogbazghi AY, Feng R, Dai Z, Marchenkov AN, Conrad EH, First PN, de Heer WA (2004) Ultrathin epitaxial graphite: 2D electron gas properties and a route toward graphene-based nanoelectronics. J Phys Chem B 108(52):19912–19916
41. Wofford JM, Oliveira MH Jr, Schumann T, Jenichen B, Ramsteiner M, Jahn U, Fölsch S, Lopes JMJ, Riechert H (2014) Molecular beam epitaxy of graphene on ultra-smooth nickel: growth mode and substrate interactions. New J Phys 16:093055
42. Zheng R, Xu Z, Khanaki A, Tian H, Zuo Z, Zheng J-G, Liu J (2017) Low-temperature growth of graphene on iron substrate by molecular beam epitaxy. Thin Solid Films 627:39–43
43. Garcia JM, He R, Jiang MP, Yan J, Pinczuk A, Zuev YM, Kim KS, Kim P, Baldwin K, West KW, Pfeiffer LN (2010) Multilayer graphene films grown by molecular beam deposition. Solid State Commun 150:809–811
44. Cheng TS, Davies A, Summerfield A, Cho YJ, Cebula I, Hill RJA, Mellor CJ, Khlobystov AN, Taniguchi T, Watanabe K, Beton PH, Foxon CT, Eaves L, Novikova SV (2016) High temperature MBE of graphene on sapphire and hexagonal boron nitride flakes on sapphire. J Vac Sci Technol B 34:02L101
45. Dabrowski J, Lippert G, Avila J, Baringhaus J, Colambo I, Dedkov YuS, Herziger F, Lupina G, Maultzsch J, Schaffus T, Schroeder T, Kot M, Tegenkamp C, Vignaud D, Asensio M-C (2016) Understanding the growth mechanism of graphene on Ge/Si(001) surfaces. Sci Rep 6:31639
46. Dillemans L, Tran T, Bhuiyan NK, Smets T, Menghini M, Lieten R, Seo JW, Locquet J-P (2011) Epitaxial growth of V2O3 on Al2O3 by reactive MBE. MRS Online Proc Libr 1292:61–66
47. Schumann T, Lopes JMJ, Wofford JM, Oliveira MH Jr, Dubslaff M, Hanke M, Jahn U, Geelhaar L, Riechert H (2015) The impact of substrate selection for the controlled growth of graphene by molecular beam epitaxy. J Cryst Growth 425:274–278
48. Hernández-Rodríguez I, García JM, Martín-Gago JA, de Andrés PL, Méndez J (2015) Graphene growth on Pt(111) and Au(111) using a MBE carbon solid-source. Diam Relat Mater 57:58–62
49. Wang M, Luo D, Wang B, Ruoff RS (2021) Synthesis of Large-Area Single-Crystal Graphene. Trends Chem. 3(1):15–33
50. Cushing GW, Johanek V, Navin JK, Harrison I (2015) Graphene Growth on Pt(111) by ethylene chemical vapor deposition at surface temperatures near 1000 K. J Phys Chem C 119(9):4759–4768
51. Chung WR, Zhao Y, Oye M, Nguyen C (2011) Graphene synthesis by thermal-CVD method. In: 2011 11th IEEE International Conference on Nanotechnology, Portland Marriott, August 15–18
52. An H, Lee W-J, Jung J (2011) Graphene synthesis on Fe foil using thermal CVD. Curr Appl Phys 11:S81–S85
53. Land TA, Michely T, Behm RJ, Hemminger JC, Comsa G (1992) STM investigation of single layer graphite structures produced on Pt(111) by hydrocarbon decomposition. Surf Sci 264:261–270

54. Losurdo M, Giangregorio MM, Capezzuto P, Bruno G (2011) Graphene CVD growth on copper and nickel: role of hydrogen in kinetics and structure. Phys Chem Chem Phys 13:20836–20843
55. McLellan RB (1969) The solubility of carbon in solid gold, copper, and silver. Scr Mater 3:389–392
56. Wood JD, Schmucker SW, Lyons AS, Pop E, Lyding JW (2011) Effects of polycrystalline Cu substrate on graphene growth by chemical vapor deposition. Nano Lett 11(11):4547–4554
57. Kalita G, Wakita K, Umeno M (2012) Low temperature growth of graphene film by microwave assisted surface wave plasma CVD for transparent electrode application. RSC Adv 2:2815–2820
58. Li N, Zhen Z, Zhang R, Xu Z, Zheng Z, He L (2021) Nucleation and growth dynamics of graphene grown by radio frequency plasma-enhanced chemical vapor deposition. Sci Rep 11:6007
59. Huet B, Raskin J-P, Snyder DW, Redwing JM (2020) Fundamental limitations in transferred CVD graphene caused by Cu catalyst surface morphology. Carbon 163:95–104
60. Chen Z, Qi Y, Chen X, Zhang Y, Liu Z (2019) Direct CVD growth of graphene on traditional glass: methods and mechanisms. Adv Mater 31:1803639
61. Niyogi S, Bekyarova E, Itkis ME, McWilliams JJ, Hamon MA, Haddon RC (2006) Solution properties of graphite and graphene. J Am Chem Soc 128:7720–7721
62. Yu P, Lowe SE, Simon GP, Zhong YL (2015) Electrochemical exfoliation of graphite and production of functional graphene. Curr Opin Colloid Interface Sci 20:329–338
63. Munuera JM, Paredes JI, Villar-Rodil S, Ayán-Varela M, Pagán A, Aznar-Cervantes SD, Cenis JL, Martínez-Alonso A, Tascón JMD (2015) High quality, low oxygen content and biocompatible graphene nanosheets obtained by anodic exfoliation of different graphite types. Carbon 94:729–739
64. Liu F, Wang C, Sui X, Riaz MA, Xu M, Wei L, Chen Y (2019) Synthesis of graphene materials by electrochemical exfoliation: Recent progress and future potential. Carbon Energy 1(2):173–199
65. Parvez K, Wu Z-S, Li R, Liu X, Graf R, Feng X, Müllen K (2014) Exfoliation of graphite into graphene in aqueous solutions of inorganic salts. J Am Chem Soc 136(16):6083–6091
66. Liu J, Poh CK, Zhan D, Lai L, Lim SH, Wang L, Liu X, Sahoo NG, Li C, Shen Z, Lin J (2013) Improved synthesis of graphene flakes from the multiple electrochemical exfoliation of graphite rod. Nano Energy 2:377–386
67. Achee TC, Sun W, Hope JT, Quitzau SG, Sweeney CB, Shah SA, Habib T, Green MJ (2018) High-yield scalable graphene nanosheet production from compressed graphite using electrochemical exfoliation. Sci Rep 8:14525
68. Hummers WS, Offeman RE (1958) Preparation of graphitic oxide. J Am Chem Soc 80:1339
69. Chua CK, Pumera M (2014) Chemical reduction of graphene oxide: a synthetic chemistry viewpoint. Chem Soc Rev 43:291–312
70. Sengupta I, Chakraborty S, Talukdar M, Pal S, Chakraborty S (2018) Thermal reduction of graphene oxide: How temperature influences purity. J Mater Res 33(23):4113–4122
71. Salunke BK, Kim BS (2016) Facile synthesis of graphene using a biological method. RSC Adv 6:17158–17162
72. Hamilton CE, Lomeda JR, Sun Z, Tour JM, Barron AR (2009) High-Yield organic dispersions of unfunctionalized graphene. Nano Lett 9(10):3460–3462
73. Hernandez Y, Nicolosi V, Lotya M, Blighe FM, Sun Z, De S, Mcgovern IT, Holland B, Byrne M, Gun'ko YK, Boland JJ, Niraj P, Duesberg G, Krishnamurthy S, Goodhue R, Hutchison J, Scardaci V, Ferrari AC, Coleman JN (2008) High-yield production of graphene by liquid-phase exfoliation of graphite. Nat Nanotechnol. 3, 563–568
74. Haar S, El Gemayel M, Shin Y, Melinte G, Squillaci MA, Ersen O, Casiraghi C, Ciesielski A, Samorì P (2015) Enhancing the liquid-phase exfoliation of graphene in organic solvents upon addition of n-Octylbenzene. Sci Rep 5:16684
75. Bourlinos AB, Georgakilas V, Zboril R, Steriotis TA, Stubos AK (2009) Liquid-Phase exfoliation of graphite towards solubilized graphenes. Small 5(16):1841–1845

76. Telkhozhayeva M, Teblum E, Konar R, Girshevitz O, Perelshtein I, Aviv H, Tischler YR, Nessim GD (2021) Higher ultrasonic frequency liquid phase exfoliation leads to larger and monolayer to Few-Layer flakes of 2D layered materials. Langmuir 37(15):4504–4514
77. Monajjemi M (2017) Liquid-phase exfoliation (LPE) of graphite towards graphene: An ab initio study. J Mol Liq 230:461–472
78. Shin Y, Just-Baringo X, Boyes M, Panigrahi A, Zarattini M, Chen Y, Liu X, Morris G, Prestat E, Kostarelos K, Vranic S, Larrosa I, Casiraghi C (2021) Enhanced liquid phase exfoliation of graphene in water using an insoluble bis-pyrene stabiliser. Faraday Discuss 227:46–60
79. Li Z, Young RJ, Backes C, Zhao W, Zhang X, Zhukov AA, Tillotson E, Conlan AP, Ding F, Haigh SJ, Novoselov KS, Coleman JN (2020) Mechanisms of liquid-phase exfoliation for the production of graphene. ACS Nano 14(9):10976–10985
80. Yi M, Shen Z, Ma S, Zhang X (2012) A mixed-solvent strategy for facile and green preparation of graphene by liquid-phase exfoliation of graphite. J Nanopart Res 14:1003
81. Nuvoli D, Valentini L, Alzari V, Scognamillo S, Bon SB, Piccinini M, Illescas J, Mariani A (2011) High concentration few-layer graphene sheets obtained by liquid phase exfoliation of graphite in ionic liquid. J Mater Chem 21:3428–3431
82. Du W, Lu J, Sun P, Zhu Y, Jiang X (2013) Organic salt-assisted liquid-phase exfoliation of graphite to produce high-quality graphene. Chem Phys Lett 568–569:198–201
83. Qiu X, Bouchiat V, Colombet D, Ayela F (2019) Liquid-phase exfoliation of graphite into graphene nanosheets in a hydrocavitating 'lab-on-a-chip.' RSC Adv 9:3232–3238
84. Alinejad B, Mahmoodi K (2017) Synthesis of graphene nanoflakes by grinding natural graphite together with NaCl in a planetary ball mill. Funct Mater Lett 10:1750047
85. Lin T, Tang Y, Wang Y, Bi H, Liu Z, Huang F, Xie X, Jiang M (2013) Scotch tape-like exfoliation of graphite assisted with elemental sulfur and graphene–sulphur composites for high-performance lithium-sulfur batteries. Energy Environ Sci 6:1283
86. Alam MF, Khan MS, Uddin I, Khan SN, Ahmad I (2020) Exfoliation synthesis of graphene and optimization with alkali halides salts. Surf Interfaces 20:100548
87. Malard LM, Pimenta MA, Dresselhaus G, Dresselhaus MS (2009) Raman spectroscopy in graphene. Phys Rep 473(5–6):51–87
88. Dresselhaus MS, Dresselhaus G, Jorio A, Souza Filho AG, Saito R, Raman (2002) spectroscopy on isolated single wall carbon nanotubes. Carbon 40, 2043–2061
89. Artur CG, Miller R, Meyer M, Le Ru EC, Etchegoin PG (2012) Single-molecule SERS detection of C60. Phys Chem Chem Phys 14:3219–3225
90. Jorio A (2012) Raman spectroscopy in graphene-based systems: prototypes for nanoscience and nanometrology. ISRN Nanomater. 2012:234216
91. Jorio A, Dresselhaus MS, Saito R, Dresselhaus G (2011) Raman spectroscopy in graphene related systems. Wiley-VCH, Weinheim, Germany
92. You Y, Ni Z, Yu T, Shen Z (2008) Edge chirality determination of graphene by Raman spectroscopy. Appl Phys Lett 93:163112
93. Luo Z, Cong C, Zhang J, Xiong Q, Yu T (2012) The origin of sub-bands in the Raman D-band of graphene. Carbon 50:4252–4258
94. Papanai GS, Sharma I, Gupta BK (2020) Probing number of layers and quality assessment of mechanically exfoliated graphene via Raman fingerprint. Mater Today Commun 22:100795
95. Tang B, Guoxin H, Gao H (2010) Raman spectroscopic characterization of graphene. Appl Spectrosc Rev 45(5):369–407
96. Popov VN, Lambin P (2013) Theoretical Raman intensity of the G and 2D bands of strained graphene. Carbon 54:86–93
97. Geng D, Wu B, Guo Y, Huang L, Xue Y, Chen J, Yu G, Jiang L, Hu W, Liu Y (2012) Uniform hexagonal graphene flakes and films grown on liquid copper surface. PNAS 21:7992–7996
98. Shen Y, Lua AC (2013) A facile method for the large-scale continuous synthesis of graphene sheets using a novel catalyst. Sci Rep 3:3037
99. Li J, Ji H, Zhang X, Wang X, Jin Z, Wang D, Wan L-J (2014) Controllable atmospheric pressure growth of mono-layer, bi-layer and tri-layer graphene. Chem Commun 50:11012–11015

100. Luo B, Chen B, Wang A, Geng D, Xu J, Wang H, Zhang Z, Peng L, Xu Z, Yu G (2016) Chemical vapor deposition of bilayer graphene with layer-resolved growth through dynamic pressure control. J. Mater. Chem. C 4:7464–7471
101. Sengupta J, Das K, Nandi UN, Jacob C (2019) Substrate free synthesis of graphene nanoflakes by atmospheric pressure chemical vapour deposition using Ni powder as a catalyst. Bull Mater Sci 42:136
102. Cui T, Lv R, Huang Z-H, Zhu H, Jia Y, Chen S, Wang K, Wu D, Kang F (2012) Low-temperature synthesis of multilayer graphene/amorphous carbon hybrid films and their potential application in solar cells. Nanoscale Res Lett 7:453
103. Kaur A, Kaur J, Singh RC (2018) Green exfoliation of graphene nanosheets based on freezing induced volumetric expansion of carbonated water. Mater Res Express 5:085601
104. Zhao W, Xia B, Lin L, Xiao X, Liu P, Lin X, Peng H, Zhu Y, Yu R, Lei P, Wang J, Zhang L, Xu Y, Zhao M, Peng L, Li Q, Duan W, Liu Z, Fan S, Jiang K (2017) Low-energy transmission electron diffraction and imaging of large-area graphene. Sci Adv 3:e1603231
105. Shearer CJ, Slattery AD, Stapleton AJ, Shapter JG, Gibson CT (2016) Accurate thickness measurement of graphene. Nanotechnology 27:125704
106. Kim HJ, Lee S-M, Oh Y-S, Yang Y-H, Lim YS, Yoon DH, Lee C, Kim J-Y, Ruoff RS (2014) Unoxidized graphene/alumina nanocomposite: fracture- and wear-resistance effects of graphene on alumina matrix. Sci Rep 4:5176
107. Yan Z, Lin J, Peng Z, Sun Z, Zhu Y, Li L, Xiang C, Samuel EL, Kittrell C, Tour JM (2012) Toward the synthesis of wafer-scale single-crystal graphene on copper foils. ACS Nano 6(10):9110–9117
108. Haghighian N, Convertino D, Miseikis V, Bisio F, Morgante A, Coletti C, Canepa M, Cavalleri O (2018) Rippling of graphitic surfaces: a comparison between few-layer graphene and HOPG. Phys Chem Chem Phys 20:13322–13330
109. Prakash G, Capano MA, Bolen ML, Zemlyanov D, Reifenberger RG (2010) AFM study of ridges in few-layer epitaxial graphene grown on the carbon-face of 4H–SiC(000–1). Carbon 48(9):2383–2393
110. Ramphal IA, Hagerman ME (2019) Nanoscale morphology, tribology and electrical properties of polyaniline/graphene oxide/LAPONITE composites investigated using atomic force microscopy. Nanoscale 11:20876–20883
111. Lina L-Y, Kim D-E, Kim W-K, Jun S-C (2011) Friction and wear characteristics of multi-layer graphene films investigated by atomic force microscopy. Surf Coat Technol 205(20):4864–4869
112. Hauquier F, Alamarguy D, Viel P, Noël S, Filoramo A, Huc V, Houzé F, Palacin S (2012) Conductive-probe AFM characterization of graphene sheets bonded to gold surfaces. Appl Surf Sci 258(7):2920–2926
113. Wei Z, Wang D, Kim S, Kim S-Y, Hu Y, Yakes MK, Laracuente AR, Dai Z, Marder SR, Berger C, King WP, De Heer WA, Sheehan PE, Riedo E (2010) Nanoscale tunable reduction of graphene oxide for graphene electronics. Science 328(5984):1373–1376
114. Dennard RH, Gaensslen FH, Yu H, Rideout VL, Bassous E, LeBlanc AR (1974) Design of ion-implanted MOSFET's with very small physical dimensions. IEEE J Solid-State Circuits 9(5):256–268
115. Xue J, Li T, Deng Y, Yu Z (2010) Full-chip leakage analysis for 65nm CMOS technology and beyond. Integration 43:353–364
116. Lemme MC, Echtermeyer TJ, Baus M, Kurz H (2007) A graphene field-effect device. IEEE Electron Device Lett 28(4):282–284
117. Schwierz F (2010) Graphene transistors. Nat Nanotechnol 5:487–496
118. Hayashi CK, Garmire DG, Yamauchi TJ, Torres CM, Ordonez RC (2020) High on-off ratio graphene switch via electrical double layer gating. IEEE Access 8:92314–92321
119. Ni ZH, Yu T, Lu YH, Wang YY, Feng YP, Shen ZX (2008) Uniaxial strain on graphene: Raman spectroscopy study and bandgap opening. ACS Nano 2:2301–2305
120. Li X, Wang X, Zhang L, Lee S, Dai H (2008) Chemically derived, ultrasmooth graphene nanoribbon semiconductors. Science 319:1229–1232

121. Robinson JA, LaBella M III, Trumbull KA, Weng X, Cavelero R, Daniels T, Hughes Z, Hollander M, Fanton M, Snyder D (2010) Epitaxial graphene materials integration: Effects of dielectric overlayers on structural and electronic properties. ACS Nano 4:2667–2672
122. Ahmed T, Biswas A, Subrina S (2021) Enhanced performance of strained graphene wrapped channel cylindrical FET. In: 2021 IEEE Region 10 Symposium (TENSYMP), 1–4
123. Li P, Zeng RZ, Liao YB, Zhang QW, Zhou JH (2019) A novel graphene metal semi-insulator semiconductor transistor and its new super-low power mechanism. Sci Rep 9:3642
124. Withers F, Del Pozo-Zamudio O, Mishchenko A, Rooney AP, Gholinia A, Watanabe K, Taniguchi T, Haigh SJ, Geim AK, Tartakovskii AI, Novoselov KS (2015) Light-emitting diodes by band-structure engineering in van der Waals heterostructures. Nat Mater 14:301–306
125. Rodríguez-Mas F, Ferrer JC, Alonso JL, de Ávila SF, Valiente D (2021) Reduced graphene oxide inserted into PEDOT:PSS layer to enhance the electrical behaviour of light-emitting diodes. Nanomaterials 11:645
126. Yoo D, Lee K, Tchoe Y, Guha P, Ali A, Saroj RK, Lee S, Hamidul Islam ABM, Kim M, Yi G-C (2021) Dimension—and position-controlled growth of GaN microstructure arrays on graphene films for flexible device applications. Sci Rep 11, 17524
127. Chen Y-X, Lu D, Wang G-G, Huangfu J, Wu Q-B, Wang X-F, Liu L-F, Ye D-M, Yan B, Han J (2020) Highly efficient orange emissive graphene quantum dots prepared by acid-free method for white LEDs. ACS Sustain Chem Eng 8(17):6657–6666
128. Wang X, Tian H, Mohammad MA, Li C, Wu C, Yang Y, Ren T-L (2015) A spectrally tunable all-graphene-based flexible field-effect light-emitting device. Nat Commun 6:7767
129. Lim YS, Hung YM (2021) Anomalously enhanced light-emitting diode cooling via nucleate boiling using graphene-nanoplatelets coatings. Energy Convers Manag 244:114522
130. Bullock CJ, Bussy C (2019) Biocompatibility considerations in the design of graphene biomedical materials. Adv Mater Interfaces 6:1900229
131. Syama S, Mohanan PV (2016) Safety and biocompatibility of graphene: A new generation nanomaterial for biomedical application. Int J Biol Macromol 86:546–555
132. Torrente-Rodrı´guez RM, Lukas H, Tu J, Min J, Yang Y, Xu C, Rossiter HB, Gao W (2020) SARS-CoV-2 RapidPlex: A graphene-based multiplexed telemedicine platform for rapid and Low-Cost COVID-19 Diagnosis and Monitoring. Matter 3, 1981–1998
133. Molinero-Fernandez A, Arruza L, Lopez MA, Escarpa A (2020) On-the-fly rapid immunoassay for neonatal sepsis diagnosis: C-reactive protein accurate determination using magnetic graphene-based micromotors. Biosens Bioelectron 158:112156
134. Li G, Lu Z, Luan X, Wang Z, Liu F, Liu L (2021) Measurement method of akkermansia muciniphila by graphene-based transistor for diseases diagnosis. IEEE Trans Nanotechnol 20(1):332–337
135. Jeon J, Lee J, -in So J, Lee JB, Lee H, Chang Y, Shin S, Jo J, Ban C (2020) Homogeneous fluorescent aptasensor for active tuberculosis diagnosis by direct quantification of circulating TB7.7 based on aptamer beacon with graphene oxide. Sens Actuators B Chem 317, 128126
136. Siew QY, Pang EL, Loh H-S, Tan MTT (2021) Highly sensitive and specific graphene/TiO2 impedimetric immunosensor based on plant-derived tetravalent envelope glycoprotein domain III (EDIII) probe antigen for dengue diagnosis. Biosens Bioelectron 176:112895
137. Afshar EG, Zarrabi A, Dehshahri A, Ashrafizadeh M, Dehghannoudeh G, Behnam B, Mandegary A, Pardakhty A, Mohammadinejad R, Tavakol S (2020) Graphene as a promising multifunctional nanoplatform for glioblastoma theranostic applications. FlatChem 22:100173
138. Majidi R, Nadafan M (2020) Detection of exhaled gas by γ -graphyne and twin-graphene for early diagnosis of lung cancer: A density functional theory study. Phys Lett A 384:126036
139. Nasrollahpour H, Isildak I, Rashidi M-R, Hashemi EA, Naseri A, Khalilzadeh B (2021) Ultrasensitive bioassaying of HER-2 protein for diagnosis of breast cancer using reduced graphene oxide/chitosan as nanobiocompatible platform. Cancer Nano 12:10
140. Barrera CC, Groot H, Vargas WL, Narváez DM (2020) Efficacy and molecular effects of a reduced graphene Oxide/Fe3O4 nanocomposite in photothermal therapy against cancer. Int J Nanomedicine 15

141. Ardakania TS, Meidanchia A, Shokria A, Shakeri-Zadeh A (2020) Fe3O4@Au/reduced graphene oxide nanostructures: Combinatorial effects of radiotherapy and photothermal therapy on oral squamous carcinoma KB cell line. Ceram Int 46:28676–28685
142. Jia X, Xu W, Ye Z, Wang Y, Dong Q, Wang E, Li D, Wang J (2020) Functionalized graphene@gold nanostar/lipid for pancreatic cancer gene and photothermal synergistic therapy under photoacoustic/photothermal imaging dual-modal guidance. Small 16:2003707
143. Lima-Sousa R, de Melo-Diogo D, Alves CG, Cabral CSD, Miguel SP, Mendonça AG, Correia IJ (2020) Injectable in situ forming thermo-responsive graphene based hydrogels for cancer chemo-photothermal therapy and NIR light-enhanced antibacterial applications. Mater Sci Eng C 117:11129
144. Hao L, Song H, Zhan Z, Lv Y (2020) Multifunctional reduced graphene oxide-based nanoplatform for synergistic targeted chemo-photothermal therapy. ACS Appl Bio Mater 3:5213–5222
145. Yang S, Wang X, He P, Xu A, Wang G, Duan J, Shi Y, Ding G (2021) Graphene quantum dots with pyrrole N and pyridine N: Superior reactive oxygen species generation efficiency for metal-free sonodynamic tumor therapy. Small 17:2004867
146. Li Z, Wang D, Xu M, Wang J, Hu X, Anwar S, Tedesco AC, Moraisde PC, Bi H (2020) Fluorine-containing graphene quantum dots with a high singlet oxygen generation applied for photodynamic therapy. J Mater Chem B 8, 2598, 2606
147. Liu X, Yan B, Li Y, Ma X, Jiao W, Shi K, Zhang T, Chen S, He Y, Liang X-J, Fan H (2020) Graphene Oxide-Grafted Magnetic Nanorings Mediated Magnetothermodynamic Therapy Favoring Reactive Oxygen Species-Related Immune Response for Enhanced Antitumor Efficacy. ACS Nano 14:1936–1950
148. Raslan A, del Burgo LS, Ciriza J, Pedraz JL (2020) Graphene oxide and reduced graphene oxide-based scaffolds in regenerative medicine. Int J Pharm 580:119226
149. Bellet P, Gasparotto M, Pressi S, Fortunato A, Scapin G, Mba M, Menna E, Filippini F (2021) Graphene-Based scaffolds for regenerative medicine. Nanomaterials 11:404
150. Low FW, Lai CW, Abd Hamid SB (2017) Study of reduced graphene oxide film incorporated of TiO2 species for efficient visible light driven dye-sensitized solar cell. J Mater Sci: Mater Electron. 28, 3819–3836
151. Sun S, Gao L, Liu Y (2010) Enhanced dye-sensitized solar cell using graphene- photoanode prepared by heterogeneous coagulation. Appl Phys Lett 96:083113
152. Basu K, Selopal GS, Mohammadnezad M, Akilimali R, Wang ZM, Zhao H, Vetrone F, Rosei F (2020) Hybrid graphene/metal oxide anodes for efficient and stable dye sensitized solar cell. Electrochim Acta 349:136409
153. Mustafa MN, Sulaiman Y (2020) Optimization of titanium dioxide decorated by graphene quantum dot as a light scatterer for enhanced dye-sensitized solar cell performance. J Electroanal Chem 876:114516
154. Casaluci S, Gemmi M, Pellegrini V, Carloa AD, Bonaccorso F (2016) Graphene-based large area dye-sensitized solar cell modules. Nanoscale 8:5368–5378
155. Oh WC, Cho KY, Jung CH, Areerob Y (2020) Hybrid of Graphene based on quaternary Cu2ZnNiSe4–WO3 Nanorods for Counter Electrode in Dye-sensitized Solar Cell Application. Sci Rep 10:4738
156. Miao X, Tongay S, Petterson MK, Berke K, Rinzler AG, Appleton BR, Hebard AF (2012) High efficiency graphene solar cells by chemical doping. Nano Lett 12:2745–2750
157. Wu J, Becerril HA, Bao Z, Liu Z, Chen Y, Peumans P (2008) Organic solar cells with solution-processed graphene transparent electrodes. Appl Phys Lett 92:263302
158. Ramli NF, Fahsyar PNA, Ludin NA, Teridi MAM, Ibrahim MA, Sepeai S (2021) Graphene dispersion as a passivation layer for the enhancement of perovskite solar cell stability. Mater Chem Phys 257:123798
159. Rehman MA, Roy SB, Gwak D, Akhtar I, Nasir N, Kumar S, Khan MF, Heo K, Chun S-H, Seo Y (2020) Solar cell based on vertical graphene nano hills directly grown on Silicon. Carbon 164:235–243

160. Lin T, Huang F, Lianga J, Wang Y (2011) A facile preparation route for boron-doped graphene, and its CdTe solar cell application. Energy Environ Sci 4:862–865
161. Wang C, Li D, Too CO, Wallace GG (2009) Electrochemical Properties of Graphene Paper Electrodes Used in Lithium Batteries. Chem Mater 21:2604–2606
162. Reddy ALM, Srivastava A, Gowda SR, Gullapalli H, Dubey M, Ajayan PM (2010) Synthesis of Nitrogen-Doped graphene films for lithium battery application. ACS Nano 4(11):6337–6342
163. Xiao J, Mei D, Li X, Xu W, Wang D, Graff GL, Bennett WD, Nie Z, Saraf LV, Aksay IA, Liu J, Zhang J-G (2011) Hierarchically porous graphene as a lithium-air battery electrode. Nano Lett 11:5071–5078
164. Shao G, Hanaor DAH, Wang J, Kober D, Li S, Wang X, Shen X, Bekheet MF, Gurlo A (2020) Polymer-Derived SiOC integrated with a graphene aerogel as a highly stable li-ion battery anode. ACS Appl Mater Interfaces 12:46045–46056
165. Han X, Zhang Z, Chen H, Zhang Q, Chen S, Yang Y (2020) On the interface design of si and multilayer graphene for a high-performance li-ion battery anode. ACS Appl Mater Interfaces 12:44840–44849
166. Ma T, Su TY, Zhang L, Yang J-W, Yao H-B, Lu L-L, Liu Y-F, He C, Yu S-H (2021) Scallion-Inspired graphene scaffold enabled high rate lithium metal battery. Nano Lett 21:2347–2355
167. Chen H, Guo F, Liu Y, Huang T, Zheng B, Ananth N, Xu Z, Gao W, Gao C (2017) A defect-free principle for advanced graphene cathode of aluminum-ion battery. Adv Mater 29:1605958
168. Chen H, Xu H, Wang S, Huang T, Xi J, Cai S, Guo F, Xu Z, Gao W, Gao C (2017) Ultrafast all-climate aluminum-graphene battery with quarter-million cycle life. Sci Adv 3, 12, eaao7233
169. Zhang J, Jiang J, Lib H, Zhao XS (2011) A high-performance asymmetric supercapacitor fabricated with graphene-based electrodes. Energy Environ Sci 4:4009–4015
170. Le LT, Ervin MH, Qiu H, Fuchs BE, Lee WY (2011) Graphene supercapacitor electrodes fabricated by inkjet printing and thermal reduction of graphene oxide. Electrochem commun 13:355–358
171. Fang Y, Luo B, Jia Y, Li X, Wang B, Song Q, Kang F, Zhi L (2012) Renewing Functionalized Graphene as Electrodes for High-Performance Supercapacitors. Adv Mater 24:6348–6355
172. Kim H, Park K-Y, Hong J, Kang K (2014) All-graphene-battery: bridging the gap between supercapacitors and lithium ion batteries. Sci Rep 4:5278
173. Guo H, Jiao1 T, Zhang Q, Guo W, Peng Q, Yan X. (2015) Preparation of graphene oxide-based hydrogels as efficient dye adsorbents for wastewater treatment. Nanoscale Res Lett 10, 272
174. Jiao T, Guo H, Zhang Q, Peng Q, Tang Y, Yan X, Li B (2015) Reduced graphene oxide-based silver nanoparticle-containing composite hydrogel as highly efficient dye catalysts for wastewater treatment. Sci Rep 5:11873
175. Zhong Y, Mahmud S, He Z, Yang Y, Zhang Z, Guo F, Chen Z, Xiong Z, Zhao Y (2020) Graphene oxide modified membrane for highly efficient wastewater treatment by dynamic combination of nanofiltration and catalysis. J Hazard Mater 397:122774
176. Ikram M, Raza A, Imran M, Ul-Hamid A, Shahbaz A, Ali S (2020) Hydrothermal synthesis of silver decorated reduced graphene oxide (rGO) nanoflakes with effective photocatalytic activity for wastewater treatment. Nanoscale Res Lett 15:95
177. Tabish TA, Memon FA, Gomez DE, Horsell DW, Zhang S (2018) A facile synthesis of porous graphene for efcient water and wastewater treatment. Sci Rep 8:1817
178. Zhao G, Li J, Ren X, Chen C, Wang X (2011) Few-Layered graphene oxide nanosheets as superior sorbents for heavy metal ion pollution management. Environ Sci Technol 45:10454–10462
179. Zhang K, Kemp KC, Chandra V (2012) Homogeneous anchoring of TiO2 nanoparticles on graphene sheets for waste water treatment. Mater Lett 81:127–130
180. Marjani A, Nakhjiri AT, Adimi M, Jirandehi HF, Shirazian S (2020) Effect of graphene oxide on modifying polyethersulfone membrane performance and its application in wastewater treatment. Sci Rep 10:2049
181. Zhang J, Lu X, Shi C, Yan B, Gong L, Chen J, Xiang L, Xu H, Liu Q, Zeng H (2019) Unraveling the molecular interaction mechanism between graphene oxide and aromatic organic compounds with implications on wastewater treatment. Chem Eng J 358:842–849

182. Kumar V, Lee Y-S, Shin J-W, Kim K-H, Kukkar D, Tsang YF (2020) Potential applications of graphene-based nanomaterials as adsorbent for removal of volatile organic compounds. Environ Int 135:105356
183. Jung W, Jeong MH, Ahn KH, Kim T, Kim YH (2020) Reduced graphene-oxide filter system for removing filterable and condensable particulate matter from source. J Hazard Mater 391:122223
184. Tao L-Q, Zhang K-N, Tian H, Liu Y, Wang D-Y, Chen Y-Q, Yang Y, Ren T-L (2017) Graphene-paper pressure sensor for detecting human motions. ACS Nano 11:8790–8795
185. Pang Y, Zhang K, Yang Z, Jiang S, Ju Z, Li Y, Wang X, Wang D, Jian M, Zhang Y, Liang R, Tian H, Yang Y, Ren T-L (2018) Epidermis microstructure inspired graphene pressure sensor with random distributed spinosum for high sensitivity and large linearity. ACS Nano 12:2346–2354
186. Yang Z, Wang D-Y, Pang Y, Li Y-X, Wang Q, Zhang T-Y, Wang J-B, Liu X, Yang Y-Y, Jian J-M, Jian M-Q, Zhang Y-Y, Yang Y, Ren T-L (2018) Simultaneously detecting subtle and intensive human motions based on a silver nanoparticles bridged graphene strain sensor. ACS Appl Mater Interfaces 10:3948–3954
187. Yang Z, Pang Y, Han X-L, Yang Y, Ling J, Jian M, Zhang Y, Yang Y, Ren T-L (2018) Graphene textile strain sensor with negative resistance variation for human motion detection. ACS Nano 12:9134–9141
188. Qiao Y, Li X, Jian J, Wu Q, Wei Y, Shuai H, Hirtz T, Zhi Y, Deng G, Wang Y, Gou G, Xu J, Cui T, Tian H, Yang Y, Ren T-L (2020) Substrate-Free multilayer graphene electronic skin for intelligent diagnosis. ACS Appl Mater Interfaces 12:49945–49956
189. He J, Xiao P, Lu W, Shi J, Zhang L, Liang Y, Pan C, Kuo S-W, Chen T (2019) A Universal high accuracy wearable pulse monitoring system via high sensitivity and large linearity graphene pressure sensor. Nano Energy 59:422–433
190. Yang T, Jiang X, Zhong Y, Zhao X, Lin S, Li J, Li X, Xu J, Li Z, Zhu H (2017) A wearable and highly sensitive graphene strain sensor for precise home-based pulse wave monitoring. ACS Sens. 2:967–974
191. Liu W, Liu N, Yue Y, Rao J, Cheng F, Su J, Liu Z, Gao Y (2018) Piezoresistive pressure sensor based on synergistical innerconnect polyvinyl alcohol nanowires/wrinkled graphene film. Small 14:1704149
192. Prabhakaran A, Nayak P (2020) Surface engineering of laser-scribed graphene sensor enables nonenzymatic glucose detection in human body fluids. ACS Appl Nano Mater 3:391–398
193. Zhang Y, Li N, Xiang Y, Wang D, Zhang P, Wang Y, Lu S, Xu R, Zhao J (2020) A flexible non-enzymatic glucose sensor based on copper nanoparticles anchored on laser-induced graphene. Carbon 156:506–513
194. Zhenga Z, Wang H (2019) Different elements doped graphene sensor for CO2 greenhouse gases detection: The DFT study. Chem Phys Lett 721:33–37
195. Seekaew Y, Wongchoosuk C (2019) A novel graphene-based electroluminescent gas sensor for carbon dioxide detection. Appl Surf Sci 479:525–531
196. Liu G, Tan Q, Kou H, Zhang L, Wang J, Lv W, Dong H, Xiong J (2018) A flexible temperature sensor based on reduced graphene oxide for robot skin used in internet of things. Sensors 18:1400
197. Davaji B, Cho HD, Malakoutian M, Lee J-K, Panin G, Kang TW, Lee CH (2017) A patterned single layer graphene resistance temperature sensor. Sci Rep 7, 8811
198. Yang J, Luo S, Zhou X, Li J, Fu J, Yang W, Wei D (2019) Flexible, tunable, and ultrasensitive capacitive pressure sensor with microconformal graphene electrodes. ACS Appl Mater Interfaces 11:14997–15006
199. Pang Y, Jian J, Tu T, Yang Z, Ling J, Li Y, Wang X, Qiao Y, Tian H, Yang Y, Ren T-L (2018) Wearable humidity sensor based on porous graphene network for respiration monitoring. Biosens Bioelectron 116:123–129
200. Liang C, Niu G, Chen X, Zhou Z, Yi Z, Ye X, Duan T, Yi Y, Xiao S (2019) Tunable triple-band graphene refractive index sensor with good angle-polarization tolerance. Opt Commun 436:57–62

201. Kaidarova A, Liu W, Swanepoel L, Almansouri A, Geraldi NR, Duarte CM, Kosel (2021) Flexible hall sensor made of laser-scribed graphene. npj Flex. Electron. 5, 2
202. Yu F, Camilli L, Wang T, Mackenzie DMA, Curioni M, Akid R, Bøggild P (2018) Complete long-term corrosion protection with chemical vapour deposited graphene. Carbon 132:78–84
203. Ghauri FA, Raza MA, Baig MS, Ibrahim S (2017) Corrosion study of the graphene oxide and reduced graphene oxide-based epoxy coatings. Mater Res Express 4:125601
204. Ye Y, Zhang D, Liu T, Liu Z, Pu J, Liu W, Zhao H, Li X, Wang L (2019) Superior corrosion resistance and self-healable epoxy coating pigmented with silanzied trianiline-intercalated graphene. Carbon 142:164–176
205. Qiu S, Li W, Zheng W, Zhao H, Wang L (2017) Synergistic effect of polypyrrole-intercalated graphene for enhanced corrosion protection of aqueous coating in 3.5% NaCl Solution. ACS Appl Mater Interfaces 9, 34294–34304
206. Cui C, Lim ATO, Huang J (2017) A cautionary note on graphene anti-corrosion coatings. Nat Nanotechnol 12:834–835
207. Xu X, Yi D, Wang Z, Yu J, Zhang Z, Qiao R, Sun Z, Hu Z, Gao P, Peng H, Liu Z, Yu D, Wang E, Jiang Y, Ding F, Liu K (2018) Greatly enhanced anticorrosion of cu by commensurate graphene coating. Adv Mater 30:1702944
208. Jin H, Zhang T, Bing W, Dong S, Tian L (2019) Antifouling performance and mechanism of elastic graphene–silicone rubber composite membranes. J Mater Chem B 7:488–497
209. Jin H, Bing W, Tian L, Wang P, Zhao J (2019) Combined effects of color and elastic modulus on antifouling performance: a study of graphene Oxide/Silicone Rubber Composite Membranes. Materials 12, 2608
210. Selim MS, Fatthallah NA, Higazy SA, Hao Z, Mo PJ (2021) A comparative study between two novel silicone/graphene-based nanostructured surfaces for maritime antifouling. J Colloid Interface Sci 606:367–383
211. Jiang T, Qi L, Qin W (2019) Improving the environmental compatibility of marine sensors by surface functionalization with graphene oxide. Anal Chem 91:13268–13274
212. Fazli-Shokouhi S, Nasirpouri F, Khatamian M (2019) Polyaniline-modified graphene oxide nanocomposites in epoxy coatings for enhancing the anticorrosion and antifouling properties. J Coat Technol Res 16(4):983–997
213. Zhang W, Cheng W, Ziemann E, Be'er A, Lu X, Elimelech M, Bernstein R (2018) Functionalization of ultrafiltration membrane with polyampholyte hydrogel and graphene oxide to achieve dual antifouling and antibacterial properties. J Membr Sci 565, 293–302
214. Li X, Zhao W, Yin R, Huang X, Qian L (2018) A highly porous polyaniline-graphene composite used for electrochemical supercapacitors. Eng Sci 3, 89–95
215. Singh K, Ohlan A, Pham VH, Balasubramaniyan R, Varshney S, Jang J, Hur SH, Choi WM, Kumar M, Dhawan SK, Kong BS, Chung JS (2013) Nanostructured graphene/Fe3O4 incorporated polyaniline as a high performance shield against electromagnetic pollution. Nanoscale 5, 2411
216. Kuila T, Bose S, Khanra P, Kim NH, Rhee KY, Lee JH (2011) Characterization and properties of in situ emulsion polymerized poly(methyl methacrylate)/graphene nanocomposites. Compos Part A 42:1856–1861
217. Thomassin J-M, Trifkovic M, Alkarmo W, Detrembleur C, Jerome C, Macosko C (2014) Poly(methyl methacrylate)/Graphene oxide nanocomposites by a precipitation polymerization process and their dielectric and rheological characterization. Macromolecules 47:2149–2155
218. Zabihi Z, Araghi H, Rodriguez PEDS, Boujakhrout A, Villalonga R (2019) Vapor sensing and interface properties of reduced graphene oxide–poly(methyl methacrylate) nanocomposite. J Mater Sci: Mater Electron 30, 2908–2919
219. Yuan X (2011) Enhanced interfacial interaction for effective reinforcement of poly(vinyl alcohol) nanocomposites at low loading of graphene. Polym Bull 67:1785–1797
220. Zhao X, Zhang Q, Chen D (2010) Enhanced mechanical properties of graphene-based poly(vinyl alcohol) composites. Macromolecules 43:2357–2363
221. Wang X, Liu X, Yuan H, Liu H, Liu C, Li T, Yan C, Yan X, Shen C, Guo Z (2018) Non-covalently functionalized graphene strengthened poly(vinyl alcohol). Mater Des 139:372–379

222. Zhou T, Chen F, Tang C, Bai H, Zhang Q, Deng H, Fu Q (2011) The preparation of high performance and conductive poly (vinyl alcohol)/graphene nanocomposite via reducing graphite oxide with sodium hydrosulfite. Compos Sci Technol 71:1266–1270
223. Li B, Luo J, Huang X, Lin L, Wang L, Hu M, Tang L, Xue H, Gao J, Mai Y-W (2020) A highly stretchable, super-hydrophobic strain sensor based on polydopamine and graphene reinforced nanofiber composite for human motion monitoring. Compos Part B 181:107580
224. Amani J, Maleki M, Khoshroo A, Sobhani-Nasab A, Rahimi-Nasrabadi M (2018) An electrochemical immunosensor based on poly p-phenylenediamine and graphene nanocomposite for detection of neuron-specific enolase via electrochemically amplified detection. Anal Biochem 548:53–59
225. Kumar SK, Castro M, Saiter A, Delbreilh L, Feller JF, Thomas S, Grohens Y (2013) Development of poly(isobutylene-co-isoprene)/reduced graphene oxide nanocomposites for barrier, dielectric and sensing applications. Mater Lett 96:109–112
226. Yu Z, Dai T, Yuan S, Zou H, Liu P (2020) Electromagnetic interference shielding performance of anisotropic polyimide/graphene composite aerogels. ACS Appl Mater Interfaces 12:30990–31001
227. Ganguly S, Ghosh S, Das P, Das TK, Ghosh SK, Das NC (2020) Poly(N-vinylpyrrolidone)-stabilized colloidal graphene-reinforced poly(ethylene-co-methyl acrylate) to mitigate electromagnetic radiation pollution. Polym Bull 77:2923–2943
228. Wang X, Yang H, Song L, Hu Y, Xing W, Lu H (2011) Morphology, mechanical and thermal properties of graphene-reinforced poly(butylene succinate) nanocomposites. Compos Sci Technol 72, 1–6
229. Hajian M, Reisi MR, Koohmareh GA, Jam ARZ (2012) Preparation and characterization of Polyvinylbutyral/Graphene Nanocomposite. J Polym Res 19:9966
230. Chen J, Ge J, Zhang L, Li Z, Li J, Sun Y, Qu L (2016) Reduced graphene oxide nanosheets functionalized with poly(styrene sulfonate) as a peroxidase mimetic in a colorimetric assay for ascorbic acid. Microchim Acta 183:1847–1853
231. Wu Y, Deng P, Tian Y, Feng J, Xiao J, Li J, Liu J, Li G, He Q (2020) Simultaneous and sensitive determination of ascorbic acid, dopamine and uric acid via an electrochemical sensor based on PVP-graphene composite. J Nanobiotechnol 18:112
232. Yasin G, Arif M, Nizam MN, Shakeel M, Khan MA, Khan WQ, Hassan TM, Abbas Z, Farahbakhsh I, Zuo Y (2018) Effect of surfactant concentration in electrolyte on the fabrication and properties of nickel-graphene nanocomposite coating synthesized by electrochemical co-deposition. RSC Adv 8:20039–20047
233. Jabbar A, Yasin G, Khan WQ, Anwar MY, Korai RM, Nizam MN, Muhyodin G (2017) Electrochemical deposition of nickel graphene composite coatings: effect of deposition temperature on its surface morphology and corrosion resistance. RSC Adv 7:31100–31109
234. Algul H, Tokur M, Ozcan S, Uysal M, Cetinkaya T, Akbulut H, Alp A (2015) The effect of graphene content and sliding speed on the wear mechanism of nickel–graphene nanocomposites. Appl Surf Sci 359:340–348
235. Xiang L, Shen Q, Zhang Y, Bai W, Nie C (2019) One-step electrodeposited Ni-graphene composite coating with excellent tribological properties. Surf Coat Technol 373:38–46
236. Hau TV, Trinh PV, Nam NPH, Tu NV, Lam VD, Phuong DD, Minh PN, Thang BH (2020) Electrodeposited nickel–graphene nanocomposite coating: effect of graphene nanoplatelet size on its microstructure and hardness. RSC Adv 10:22080–22090
237. Chang S-W, Nair AK, Buehler MJ (2013) Nanoindentation study of size effects in nickel–graphene nanocomposites. Philos Mag Lett 93(4):196–203
238. Chang S-W, Nair AK, Buehler MJ (2012) Geometry and temperature effects of the interfacial thermal conductance in copper—and nickel–graphene nanocomposites. J Phys: Condens Matter 24, 245301
239. Khobragade N, Sikdar K, Kumar B, Bera S, Roy D (2019) Mechanical and electrical properties of copper-graphene nanocomposite fabricated by high pressure torsion. J Alloys Compd 776:123–132

240. Wang S, Han S, Xin G, Lin J, Wei R, Lian J, Sun K, Zu X, Yu Q (2018) High-quality graphene directly grown on Cu nanoparticles for Cu-graphene nanocomposites. Mater Des 139:181–187
241. Pavithra CLP, Sarada BV, Rajulapati KV, Rao TN, Sundararajan G (2014) A new electrochemical approach for the synthesis of copper-graphene nanocomposite foils with high hardness. Sci Rep 4:4049
242. Huang J, Xie Z, Huang Y, Xie L, Luo S, Fan Q, Zeng T, Zhang Y, Wang S, Zhang M, Xie Z, Deng X (2020) Electrochemical immunosensor with Cu(I)/Cu(II)-chitosan-graphene nanocomposite-based signal amplifcation for the detection of newcastle disease virus. Sci Rep 10:13869
243. Shabnam L, Faisal SN, Roy AK, Haque E, Minett AI, Gomes VG (2017) Doped graphene/Cu nanocomposite: A high sensitivity non-enzymatic glucose sensor for food. Food Chem 221:751–759
244. Yuan W, Gu Y, Li L (2012) Green synthesis of graphene/Ag nanocomposites. Appl Surf Sci 261:753–758
245. Khan ME, Khan MM, Cho MH (2015) Biogenic synthesis of a Ag–graphene nanocomposite with efficient photocatalytic degradation, electrical conductivity and photoelectrochemical performance. New J Chem 39:8121–8129
246. Shen J, Shi M, Li N, Yan B, Ma H, Hu Y, Ye M (2010) Facile synthesis and application of ag-chemically converted graphene nanocomposite. Nano Res 3:339–349
247. Pasricha R, Gupta S, Srivastava AK (2009) A facile and novel synthesis of ag–graphene-based nanocomposites. Small 5(20):2253–2259
248. Wang P, Liu Z-G, Chen X, Meng F-L, Liu J-H, Huang X-J (2013) UV irradiation synthesis of an Au–graphene nanocomposite with enhanced electrochemical sensing properties. J. Mater. Chem. A 1:9189
249. Khan ME, Khan MM, Cho MH (2015) Green synthesis, photocatalytic and photoelectrochemical performance of an Au–Graphene nanocomposite. RSC Adv 5:26897–26904
250. Govindhan M, Amiri M, Chen A (2015) Au nanoparticle/graphene nanocomposite as a platform for the sensitive detection of NADH in human urine. Biosens Bioelectron 66:474–480
251. Çiplak Z, Yildiz N, Çalimli Ay (2015) Investigation of graphene/ag nanocomposites synthesis parameters for two different synthesis methods. Fuller Nanotub Carbon Nanostructures 23, 4, 361–370
252. Tahernejad-Javazmi F, Shabani-Nooshabadi M, Karimi-Maleh H (2019) 3D reduced graphene oxide/FeNi3-ionic liquid nanocomposite modified sensor; an electrical synergic effect for development of tert-butylhydroquinone and folic acid sensor. Compos Part B 172:666–670
253. Vengatesan MR, Darawsheh IFF, Govindan B, Alhseinat E, Banat F (2019) Ag-Cu bimetallic nanoparticle decorated graphene nanocomposite as an effective anode material for hybrid capacitive deionization (HCDI) System. Electrochim Acta 297:1052–1062
254. Zhang Y, Tang Z-R, Fu X, Xu Y-J (2010) TiO2-Graphene nanocomposites for gas-phase photocatalytic degradation of volatile aromatic pollutant: Is TiO2-Graphene truly different from other TiO2-Carbon composite materials? ACS Nano 4(12):7303–7314
255. Zhang Y, Tang Z-R, Fu X, Xu Y-J (2011) Engineering the unique 2D mat of graphene to achieve graphene-TiO2 nanocomposite for photocatalytic selective transformation: what advantage does graphene have over its forebear carbon nanotube? ACS Nano 5(9):7426–7435
256. Qiu J, Lai C, Wang Y, Li S, Zhang S (2014) Resilient mesoporous TiO2/graphene nanocomposite for high rate performance lithium-ion batteries. Chem Eng J 256:247–254
257. Le Z, Liu F, Nie P, Li X, Liu X, Bian Z, Chen G, Wu HB, Lu Y (2017) Pseudocapacitive sodium storage in mesoporous single-crystal-like TiO2−Graphene nanocomposite enables high performance sodium-ion capacitors. ACS Nano 11:2952–2960
258. Fan Y, Lu H-T, Liu J-H, Yang C-P, Jing Q-S, Zhang Y-X, Yang X-K, Huang K-J (2011) Hydrothermal preparation and electrochemical sensing properties of TiO2–graphene nanocomposite. Colloids Surf B: Biointerfaces 83:78–82
259. Lee S, Oh J, Kim D, Piao Y (2016) A sensitive electrochemical sensor using an iron oxide/graphene composite for the simultaneous detection of heavy metal ions. Talanta 160:528–536

260. Sua H, Ye Z, Hmidi N (2017) High-performance iron oxide–graphene oxide nanocomposite adsorbents for arsenic removal. Colloids and Surfaces A: Physicochem. Eng. Aspects 522, 161–172
261. Yuan Y, Jiang W, Wang Y, Shen P, Li F, Li P, Zhao F, Gao H (2014) Hydrothermal preparation of Fe2O3/graphene nanocomposite and its enhanced catalytic activity on the thermal decomposition of ammonium perchlorate. Appl Surf Sci 303:354–359
262. Xiao L, Schroeder M, Kluge S, Balducci A, Hagemann U, Schulz C, Wiggers H (2015) Direct self-assembly of Fe2O3/reduced graphene oxide nanocomposite for high-performance lithium-ion batteries. J Mater Chem A 3:11566–11574
263. Jiang W, Liang F, Wang J, Su L, Wu Y, Wang L (2014) Enhanced electrochemical performances of FeOx–graphene nanocomposites as anode materials for alkaline nickel–iron batteries. RSC Adv 4:15394–15399
264. Ren H, Bai Y, Wang X, Ni Q, Wang Z, Li Y, Chen G, Wu F, Xu H, Wu C (2019) High-Capacity interstitial Mn-Incorporated MnxFe3−xO4/Graphene nanocomposite for sodium-ion battery anodes. ACS Appl Mater Interfaces 11:37812–37821
265. Liang C, Song P, Ma A, Shi X, Gub H, Wang L, Qiu H, Kong J, Gu J (2019) Highly oriented three-dimensional structures of Fe3O4 decorated CNTs/reduced graphene oxide foam/epoxy nanocomposites against electromagnetic pollution. Compos Sci Technol 181:107683
266. Shen B, Zhai W, Tao M, Ling J, Zheng W (2013) Lightweight, multifunctional polyetherimide/graphene@Fe3O4 composite foams for shielding of electromagnetic pollution. ACS Appl Mater Interfaces 5:11383–11391
267. Wang B, Wu X-L, Shu C-Y, Guo Y-G, Wang C-R (2010) Synthesis of CuO/graphene nanocomposite as a high-performance anode material for lithium-ion batteries. J Mater Chem 20:10661–10664
268. Rai AK, Anh LT, Gim J, Mathew V, Kang J, Paul BJ, Singh NK, Song J, Kim J (2013) Facile approach to synthesize CuO/reduced graphene oxide nanocomposite as anode materials for lithium-ion battery. J Power Sources 244:435–441
269. Perreault LL, Colò F, Meligrana G, Kim K, Fiorilli S, Bella F, Nair JR, Vitale-Brovarone C, Florek J, Kleitz F, Gerbaldi C (2018) Spray-Dried mesoporous mixed Cu-Ni Oxide@Graphene nanocomposite microspheres for high power and durable li-ion battery anodes. Adv Energy Mater 8:1802438
270. Orhan Z, Cinan E, Çaldıran Z, Kurucu Y, Daş E (2020) Synthesis of CuO–graphene nanocomposite material and the effect of gamma radiation on CuO–graphene/p-Si junction diode. J Mater Sci: Mater Electron 31, 12715–12724
271. Luo L, Zhu Li, Wang Z (2012) Nonenzymatic amperometric determination of glucose by CuO nanocubes–graphene nanocomposite modified electrode. Bioelectrochemistry 88, 156–163
272. Gayathri S, Jayabal P, Kottaisamy M, Ramakrishnan V (2014) Synthesis of ZnO decorated graphene nanocomposite for enhanced photocatalytic properties. J Appl Phys 115:173504
273. Lonkar SP, Pillai V, Abdala A (2019) Solvent-free synthesis of ZnO-graphene nanocomposite with superior photocatalytic activity. Appl Surf Sci 465:1107–1113
274. Yukird J, Kongsittikul P, Qin J, Chailapakul O, Rodthongkum N (2018) ZnO@graphene nanocomposite modified electrode for sensitive and simultaneous detection of Cd (II) and Pb (II). Synth Met 245:251–259
275. Anand K, Singh O, Singh MP, Kaur J, Singh RC (2014) Hydrogen sensor based on graphene/ZnO nanocomposite. Sens. Actuator B 195:409–415
276. Jayabal P, Gayathri S, Sasirekha V, Mayandi J, Ramakrishnan V (2014) Preparation and characterization of ZnO/graphene nanocomposite for improved photovoltaic performance. J Nanopart Res 16:2640
277. Moradi S, Sobhgol SA, Hayati F, Isari AA, Kakavandi B, Bashardoust P, Anvaripour B (2020) Performance and reaction mechanism of MgO/ZnO/Graphene ternary nanocomposite in coupling with LED and ultrasound waves for the degradation of sulfamethoxazole and pharmaceutical wastewater. Sep Purif Technol 251:117373
278. Chen W, Song K, Mi L, Feng X, Zhang J, Cui S, Liu C (2017) Synergistic effect induced ultrafine SnO2/graphene nanocomposite as an advanced lithium/sodium-ion batteries anode. J. Mater. Chem. A 5, 10027-10038

279. Su D, Ahn H-J, Wang G (2013) SnO2@graphene nanocomposites as anode materials for Na-ion batteries with superior electrochemical performance. Chem Commun 49:3131–3133
280. Zhang D, Chang H, Li P, Liu R, Xue Q (2016) Fabrication and characterization of an ultrasensitive humidity sensor based on metal oxide/graphene hybrid nanocomposite. Sens Actuators B 225:233–240
281. Zhang M, Lei D, Yu X, Chen L, Li Q, Wang Y, Wang T, Cao G (2012) Graphene oxide oxidizes stannous ions to synthesize tin sulfide–graphene nanocomposites with small crystal size for high performance lithium ion batteries. J Mater Chem 22:23091–23097
282. Ma T, Sun L, Niu Q, Xu Y, Zhu K, Liu X, Guo X, Zhang J (2019) N-doped carbon-coated Tin sulfide/graphene nanocomposite for enhanced lithium storage. Electrochim Acta 300:131–137
283. Shi J, Wang Y, Su Q, Cheng F, Kong X, Lin J, Zhu T, Liang S, Pan A (2018) N-S co-doped C@SnS nanoflakes/graphene composite as advanced anode for sodium-ion batteries. Chem Eng J 353:606–614
284. Yang B, Zuo X, Chen P, Zhou L, Yang X, Zhang H, Li G, Wu M, Ma Y, Jin S, Chen X (2015) Nanocomposite of tin sulfide nanoparticles with reduced graphene oxide in high-efficiency dye-sensitized solar cells. ACS Appl Mater Interfaces 7:137–143
285. Wu J, Wu Z, Ding H, Wei Y, Huang W, Yang X, Li Z, Qiu L, Wang X (2020) Flexible, 3D SnS2/Reduced graphene oxide heterostructured NO2 sensor. Sens Actuators B: Chem 305:127445
286. Johny J, Sepulveda-Guzman S, Krishnan B, Avellaneda DA, Aguilar Martinez JA, Anantharaman MR, Shaji S (2019) Tin sulfide: Reduced graphene oxide nanocomposites for photovoltaic and electrochemical applications. Sol Energy Mater Sol Cells 189, 53–62
287. Kapadnis RS, Bansode SB, Supekar AT, Bhujbal PK, Kale SS, Jadkar SR, Pathan HM (2020) Cadmium telluride/cadmium sulfide thin films solar cells: a review. ES Energy Environ 10:3–12
288. Zeng P, Zhang Q, Peng T, Zhang X (2011) One-pot synthesis of reduced graphene oxide–cadmium sulfide nanocomposite and its photocatalytic hydrogen production. Phys Chem Chem Phys 13:21496–21502
289. Peng T, Li K, Zeng P, Zhang Q, Zhang X (2012) Enhanced photocatalytic hydrogen production over graphene oxide—cadmium sulfide nanocomposite under visible light irradiation. J Phys Chem C 116:22720–22726
290. Alhammadi S, Minnam Reddy VR, Gedi S, Park H, Sayed MS, Shim J-J, Kim WK (2020) Performance of Graphene–CdS hybrid nanocomposite thin film for applications in Cu(In,Ga)Se2 Solar Cell and H2 Production. Nanomaterials 10, 245
291. Ge L, Hong Q, Li H, Liu C, Li F (2019) Direct-Laser-Writing of metal sulfide-graphene nanocomposite photoelectrode toward sensitive photoelectrochemical sensing. Adv Funct Mater 29:1904000
292. Roshan H, Sheikhi MH, Faramarzi Haghighi MK, Padidar P (2020) High-Performance room temperature methane gas sensor based on lead sulfide/reduced graphene oxide nanocomposite. IEEE Sens. J. 20, 5
293. Qu B, Chen Y, Zhang M, Hu L, Lei D, Lu B, Li Q, Wang Y, Chen L, Wang T (2012) β-Cobalt sulfide nanoparticles decorated graphene composite electrodes for high capacity and power supercapacitors. Nanoscale 4:7810–7816
294. Zhu J, Zhou W, Zhou Y, Cheng X, Yang J (2019) Cobalt sulfide/reduced graphene oxide nanocomposite with enhanced performance for supercapacitors. J Electron Mater 48:1531–1539
295. Ramachandran R, Felix S, Saranya M, Santhosh C, Velmurugan V, Chakkravarthy Ragupathy BP, Jeong SK, Grace AN (2013) Synthesis of cobalt sulfide-graphene (CoS/G) nanocomposites for supercapacitor applications. IEEE Trans. Nanotechnol. 12, 6, 985–990
296. Huang G, Chen T, Wang Z, Chang K, Chen W (2013) Synthesis and electrochemical performances of cobalt sulfides/graphene nanocomposite as anode material of Li-ion battery. J Power Sources 235:122–128
297. Prasad J, Singh AK, Yadav AN, Kumar A, Tomar M, Srivastava A, Kumar P, Gupta V, Singh K (2020) Molybdenum disulfide-wrapped carbon nanotube-reduced graphene oxide

(CNT/MoS2-rGO) nanohybrids for excellent and fast removal of electromagnetic interference pollution. ACS Appl Mater Interfaces 12:40828–40837
298. Khan M, Yousaf AB, Chen M, Wei C, Wu X, Huang N, Qi Z, Li L (2016) Molybdenum sulfide/graphene-carbon nanotube nanocomposite material for electrocatalytic applications in hydrogen evolution reactions. Nano Res 9(3):837–848
299. Peng W, Wang W, Han G, Huang Y, Zhang Y (2020) Fabrication of 3D flower-like MoS2/graphene composite as high-performance electrode for capacitive deionization. Desalination 473:114191
300. Yang YJ, Li W, Wu X (2014) Copper sulfide/reduced graphene oxide nanocomposite for detection of hydrazine and hydrogen peroxide at low potential in neutral medium. Electrochim Acta 123:260–267
301. Shi F, Zheng W, Wang W, Hou F, Lei B, Sun Z, Sun W (2015) Application of graphene-copper sulfide nanocomposite modified electrode for electrochemistry and electrocatalysis of haemoglobin. Biosens Bioelectron 64:131–137
302. Wang A, Wang H, Zhang S, Mao C, Song J, Niu H, Jin B, Tian Y (2013) Controlled synthesis of nickel sulfide/graphene oxide nanocomposite for high-performance supercapacitor. Appl Surf Sci 282:704–708
303. Xing Z, Chu Q, Ren X, Tian J, Asiri AM, Alamry KA, Al-Youbi AO, Sun X (2013) Biomolecule-assisted synthesis of nickel sulfides/reduced graphene oxide nanocomposites as electrode materials for supercapacitors. Electrochem commun 32:9–13
304. Huang K-J, Wang L, Liu Y-J, Gan T, Liu Y-M, Wang L-L, Fan Y (2013) Synthesis and electrochemical performances of layered tungsten sulfide-graphene nanocomposite as a sensing platform for catechol, resorcinol and hydroquinone. Electrochim Acta 107:379–387
305. Wu D, Wang C, Wu M, Chao Y, He P, Ma J (2020) Porous bowl-shaped VS2 nanosheets/graphene composite for high-rate lithium-ion storage. J Energy Chem 43:24–32

Synthesis of Graphene Oxide and Its Metal Composites

Ahmad Khalilullah, Ramsha Anwer, and Imran Uddin

Abstract Because of the unique features of a derivative of allotropic carbon graphite that has been known as graphene oxide, a number of unique optical, electrical and thermal breakthroughs has come into limelight. This has made graphene oxide as the material having the most intriguing nature which is still under investigation. Furthermore, apart from just a precursor for the manufacturing of graphene, researchers have discovered a plethora of unique optical, electrical, and chemical characteristics of graphene oxide that may be used in a variety of applications. The synthesis of GO, its structure and characterisation along with its functionalization and GO applications are the subject of this chapter. Additionally, we have discussed the use of GO in environmental, medicinal, and biological applications, freestanding membranes, and diverse composite systems. The synthesis of graphene oxide and its nanocomposite based on novel nanoparticles will be covered in this chapter. A brief overview has also been provided, with a focus on the use of graphene oxide and its nanocomposite in various fields, particularly waste water treatment of water.

Keywords Graphene oxide · Nanocomposite · Metals · Synthesis · Economical aspects

1 Introduction

Graphene oxide is the oxidised form of graphene with disrupted lattice as compared to the lattice of the pristine graphene. For the first time oxidation of graphite has been studied in the mid-nineteenth century by Brodie in 1860 when graphite was treated with potassium chlorate ($KClO_3$) and fuming nitric acid (HNO_3) [5]. Later in 1898 Staudenmaier improved this approach for oxidising graphite by slowly adding the

A. Khalilullah
Aligarh Muslim University (AMU), Aligarh 202002, India

R. Anwer · I. Uddin (✉)
Research Institute of Biomolecular and Chemical Engineering, Nanolab, University of Pannonia, Veszprém 8200, Hungary
e-mail: uddin.imran@mk.uni-pannon.hu

I. Uddin and I. Ahmad (eds.), *Synthesis and Applications of Nanomaterials and Nanocomposites*, Composites Science and Technology,
https://doi.org/10.1007/978-981-99-1350-3_3

potassium chlorate to a solution containing concentrated sulphuric acid, concentrated nitric acid (63%) and graphite [59]. Researchers, however, deemed this approach unsafe as the mass ratio of potassium chlorate with respect to graphite is quite high, along with that the method is also time consuming. After more than half-century Hummers and Offeman in 1958 gave a safe method of graphene oxide synthesis, based on a water-free mixture of concentrated sulphuric acid, sodium nitrate, and potassium permanganate known as the Hummers method [49]. Temperatures of only 45 °C are necessary for this approach, and the entire reaction took only two hours to finish. Literature-based on graphene oxide states that Brodie, Staudenmaier, and Hummer proposed three methodologies which were later considered to be the main approaches for the synthesis of graphene oxide via the oxidation of graphite, resulting in the formation of exfoliated graphite sheets with oxygen based functional groups attached to it. The structure of graphene oxide constitutes nonstoichiometric geometry having various oxygen based functional groups such as hydroxyl, epoxy, carboxyl groups and their bonding to carbon atoms in the graphene layer. Since the functional groups are having electron pairs which makes graphene oxide as negatively charged in nature. Thus, graphene oxide sheets disperse well in water due to the negative surface charges on the sheets that arise from the phenols, epoxy, hydroxyl, carbonyl and carboxylic acid groups present on graphene oxide and keep it from aggregating [23]. Due to the hydrophilic nature of graphene oxide, intersheet separations range from 0.6 nm to 1.2 nm [6]. Not only in water, but Graphene oxide is also dispersible in a wide range of organic solvents, such as N,N dimethylformamide, N-methyl-2-pyrrolidone, Tetrahydrofuran and ethylene glycol [45].

Initially there have been five proposed models of Graphene oxide but at last the final model of graphene oxide comes out to be the collective model of Scholz-Boehm's and Ruess' models. Graphene oxide on reduction gives rise to graphene thus it is used to synthesise graphene by means of a reducing agent like sodium borohydride ($NaBH_4$), hydrazine and dimethylhydrazine [55, 65]. In pure anhydrous hydrazine, maximum percentile carboxyl groups attached with the lattice carbon atoms of graphene oxide are reduced which makes graphene formation by anhydrous hydrazine an effective method. However, the former use of pure hydrazine for the formation of graphene shows effective and impressive results but the nature of hydrazine in this methodology should be anhydrous which requires a dry-box that act as a limitation that hinders the large-scale production. Graphene oxide can also be used to form graphene by reducing graphene oxide via electrochemical method in which a sharp increase in current is employed that give rise to electrochemical reduction of the graphene oxide.

Exfoliation of graphite is not only done by chemical means but also by thermal means. In case of thermal expansion of graphite oxide, the decomposition rate of epoxy and hydroxyl sites of graphene oxide becomes larger than the rate of diffusion of the evolved gases and this causes a build-up of pressure which overcomes the van der Waals forces that bind the graphite sheets together and exfoliation occurs resulting in the formation of graphene oxide [40]. The carbon atom has different hybridization due to which graphene and graphene oxide exhibit different electrical nature. The functionalised graphene sheets are electrically conductive due to sp^2

hybridization while graphene oxide act as an insulator because of sp^3 hybridization [52].

Graphene oxide is a two-dimensional crystal structure formed by a flat monolayer of carbon atoms arranged in a hexagonal lattice [42]. Graphene oxide has been a source of contention in recent years, as it represents a reasonable starting material for the mass manufacture of graphene. Additionally, Graphene oxide Nano sheets have recently attracted a great deal of attention due to their unique chemical and physical properties that highlight its potential as a promising material for biological applications such as bio functionalization [74]. Graphene oxide nanosheets are interesting materials for enzyme immobilization due to their large specific surface area and abundant functional groups [37]. Depending on the degree of oxidation Graphene oxide acts as a semiconductor or insulator. This behaviour of graphene oxide attracted attention towards its chemical structure [47], electronic properties [74], reductive nature [46] and chemical modification or functionalization [7]. The electronic properties of graphene oxide mainly depend on the oxidation level and chemical composition which can be tailored by removal or addition of certain oxygen groups to adjust the proportion of sp^2 and sp^3 carbon [30]. The electrical, mechanical, and electrochemical characteristics of graphene oxide are all influenced by the oxygenated groups present in it. Because of the polar oxygen functional groups, graphene oxide exhibit hydrophilic nature due to which graphene oxide shows dispensability in water and hence form stable graphene oxide dispersion, making it possible to make thin conductive films by means of drop-casting, spraying, or spin-coating [31]. Additionally, these functional groups act as sites responsible for functionalization of graphene oxide. Thus, the chemical composition of graphene oxide makes it possible to tune its physicochemical properties and makes graphene oxide a promising nanomaterial to fabricate electrochemical and electro analysis sensors.

Graphene consists of a one-atom-thick planar sheet comprising sp^2 bonded carbon structure with exceptionally high crystal and electronic quality [20]. Graphene has been making a profound impact in many areas of science and technology due to its remarkable physicochemical properties like high specific surface area [64] strong mechanical strength and excellent thermal and electrical conductivities [4]. These unique physicochemical properties make graphene useful for critical improvements in the field of electrochemistry.

Apart from graphene oxide and graphene there exist a partially reduced form of graphene oxide that has layered lattice of carbon atoms similar to that of graphene oxide but the all the carbon atoms of lattice plane in reduced graphene oxide are not attached with oxygen-containing groups, rather of all the carbon atoms of lattice plane, some of them are sp^2 hybridised while some of them are sp^3 hybridised. Therefore, based on the observations it has been concluded that graphene oxide is a single atomic layered nanomaterial synthesized by the oxidation of graphite [18], and has property of water dispersibility. However, the partly reduced graphene oxide consist of carbon atoms having sp^3 as well as sp^2 hybridisation. Some other names have also been given to reduced graphene oxide, such as functionalized graphene, chemically modified graphene, chemically converted graphene, or reduced graphene [16].

Graphene oxide can be synthesized using graphite, an allotrope of carbon that act as raw material through cost effective chemical methods with a high yield. Since, graphene oxide is hydrophilic in nature therefore it forms stable water colloids and makes it possible for the construction of macroscopic structures by economical processes. The graphene oxide sheets consist of sp3 carbon atoms which are attached with oxygen based functional groups that are oriented either above or below the lattice plane. Because of the structure deformation and the presence of covalently bonded functional groups, graphene oxide sheets are atomically rough. Several researchers have studied the surface of graphene oxide and observed highly defective regions, probably due to the presence of oxygen, and other areas are nearly intact [11]. According to a report the carbon atoms of graphene oxide attached to functional groups are slightly displaced, but the overall size of the unit cell in graphene oxide remains similar to that of graphene [44].

2 Synthesis of Graphene Oxide

The synthesis methods of graphene oxide include two main techniques that are termed as, Top-down approach & Bottom-up approach.

2.1 *Top-Down Approach*

Synthesis of graphene oxide based on top-down approach includes the principle of separation of various graphite layers by overcoming the van der Walls forces of attraction between them. A schematic representation showing top down approach is depicted here in Fig. 1.

The major synthesis routes which come under this category are chemical exfoliation and electrochemical exfoliation. Chemical exfoliation employs appropriate

Fig. 1 Schematic representation of top-down approach of nanoparticles

liquid environment and temperature to exfoliate graphite into individual layers via oxidation. This technique can be considered to be divided into three steps:

(a) Dispersion of graphite
(b) Exfoliation of graphite into single and multi-layer graphene oxide
(c) Purification which separates single and multilayer graphene oxide generally by centrifugation.

The graphite exfoliation by chemical means is economical and appropriate for the large scale synthesis. Further for applications like composites [24] thin films [22] and ink [64], this method gained wide popularity. The major limitations of this technique are the limited size of synthesized graphene oxide sheets and the presence of defects and non-exfoliated graphene oxide sheets [43]. Based on the principle of chemical exfoliation, graphene oxide can be synthesized by using methods developed by Brodie, Park and Hummers. The quality of graphene oxide can be improved by means of thermal exfoliation under constant pressure [40]. Graphene oxide synthesized by chemical oxidation method contains many functional groups which make them ideal for making composites [58]. This technique produces large size flakes, controllable solution processing and can be scaled up for industrial synthesis. Graphene oxide synthesised from top down approach is used in different applications in its actual as well as in its reduced form like electrochemical glucose sensor, transparent conducting films, electrodes, solar cells, and ultrafast lasers [17, 66]. The time-consuming nature of this technique and implication of hazardous chemicals like hydrazine hydrate, sulphuric acid during the reduction and oxidation process makes this approach non eco-friendly.

2.2 Bottom-Up Approach

This type of approach refers to the building of material from constituent particle by constituent particle where, constituent particle includes atom, molecule or cluster. In the bottom-up approach, one starts with employing carbon molecules usually obtained from different sources as precursors [68]. The schematic representation showing concept of bottom up approach is shown below in Fig. 2.

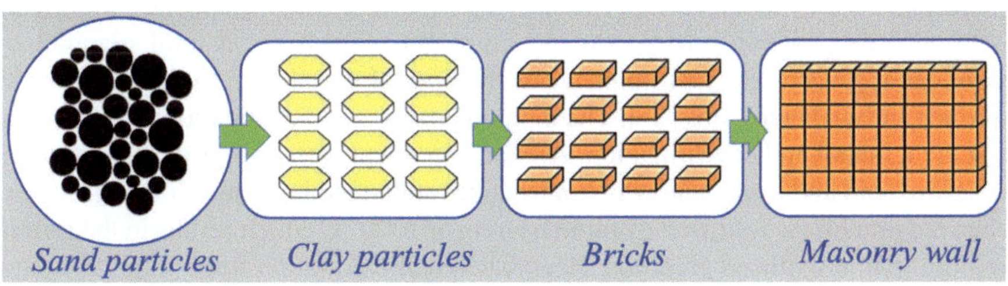

Fig. 2 Schematic representation of bottom-up approach of nanoparticles

Graphene oxide is usually prepared by various methods that are based on this approach, and are known as Hummers' method or modified Hummers' methods. Since, top-down approaches employs strong oxidizing agents, to overcome this drawback, synthesis of graphene oxide is done by the bottom-up approach in such a way that graphene oxide so formed will be of required thickness (as per the need) ranging from 1 to 1500 nm. The lateral sizes of the monolayer and few-layer graphene oxide are about 20 and 100 mm respectively [27].

In bottom up approach the graphene oxide is prepared by a hydrothermal method. In this method glucose is used as a sole reagent, that utilizes the bottom-up approach to synthesize graphene oxide with controllable number of layers ranging from monolayer to multilayers, having tuneable properties that have been achieved by thermal annealing of sample using Rapid Thermal Processor so that the amount of oxygen present in graphene oxide can be controlled. This method is advantageous over other methods of top-down approaches as it is environmentally friendly, facile, low-cost as well as capable of scaling up for mass production [63].

Since graphene oxide is an emerging area of research, a large number of articles as well as reviews have been published exploring different synthesis approaches to graphene oxide. Each synthesis technique specializes in various properties of graphene oxide like dimensions, layers, conductivity, quality, cost-effectiveness, and so on. The methods used for the synthesis of graphene oxide include chemical and electrochemical exfoliation of graphene oxide and by means of a hydrothermal method based on bottom-up technique using glucose as a sole reagent [33, 63].

Since, functionalisation of graphene oxide has a risk on specificity, greater loading capacity, solubility, stability, and biocompatibility. Therefore, researchers have implemented a various level of interactions like covalent bonding of the functional molecules with the graphene oxide basal planes, defects and edges of graphene oxide sheets, noncovalent adsorption via hydrogen bonding, π-π stacking, electrostatic interactions, and van der Waals attractions in order to make graphene oxide synthesis more efficient [12]. Considering the above factors various chemical and electrochemical methods have been proposed for the synthesis of graphene oxide which have been discussed in the following section below.

2.2.1 Chemical Methods

There are various chemical approaches reported to extract graphene oxide from graphite. The possibility of producing graphene oxide from natural graphite by chemical approach was first pointed out by Horiuchi et al. during their attempt to produce carbon Nano films [25]. The method consists of the oxidation process, dilution in methanol and centrifugation. Another chemical method was developed using sulphuric acid and nitric acid which introduce intercalate into the layers of graphite. Then the graphite is followed by rapid heating at 1000 °C which results in the formation of the thin oxidised graphitic sheet. Chemical exfoliation from graphite crystal is a widely accepted technique to produce graphene oxide. This method is capable of producing low-cost graphene oxide in large quantities. Typically, the method includes oxidation followed by thermal expansion of oxidised graphite [39]. This results in the formation of graphene oxide having covalently bonded functional groups namely hydroxyl, epoxide groups, that are bonded to the basal plane of the graphene oxide sheets, and carboxyl groups that occupy the edges of the lattice planes. Due to the presence of such functional groups graphene oxide sheets exhibit hydrophilic nature and hence form negatively charged colloidal suspension.

Reduction of graphene oxide can be reduced to form graphene. The method is advantageous because of its scalability, large volume of production and effectiveness in multipurpose functionalization, but the higher number of defects is the main limitation of the method. In the chemical methodology, one of the reliable techniques is used which is called a modified Hummers' method. Hummers' method is a chemical process in which graphite is subjected to oxidation, with strong oxidising agents namely nitric acid, potassium permanganate and sulphuric acid, which results in an increment of spacing between two consecutive layers of graphite from 0.335 nm for to more than 0.625, nm resulting in the synthesis of large quantities of one atom thick graphene oxide. B. C. Brodie for the very first time synthesized graphene oxide in 1859 by oxidising slurry of graphite in the presence of fuming nitric acid and potassium chlorate. Later, in 1898, Staudenmaier developed this method by introducing concentrated sulphuric acid to the above-mentioned reaction system and this work opened a new systematic approach towards graphene oxide synthesis. In 1958, the scientist William S. Hummers reported an alternative method for the synthesis of graphene oxide by using $KMnO_4$ and $NaNO_3$ in concentrated H_2SO_4, which came to be known as modified Hummers' method [55]. This method is economically favourable, but a significant number of defects is the limitation of the method. The typical experimental procedure of modified Hummer's method is as follows:

1. Five grams of graphite powder and 2.5 g of $NaNO_3$ dispersed in 115 ml H_2SO_4 (98%) as an intercalating agent and stirred for 2 h.
2. Then 15 gm of ground $KMnO_4$ added gradually to the above solution. The temperature of the solution was kept less than 20 °C by keeping it in the ice bath then the solution is a deep oily green colour.
3. Next, the mixture was stirred at 35–37 °C for 2 h. The solution colour in the medium temperature stage is like thick brown paste.
4. The resulting solution is diluted gradually by adding 200 ml of deionized water under vigorous stirring to get a reddish-brown suspension of graphite oxide solution with the temperature at this stage being 95 °C keeping constant stirring for 1 h.
5. The suspension is then treated further by adding 30 ml of 30% H_2O_2 solution and 350 ml of distilled water. The bright orange colour of graphene oxide suspension will appear after adding the peroxide solution (H_2O_2 convert the residual permanganate to soluble manganese sulphate and the solution colour changed to yellowish-brown indicating the oxidation of graphite).
6. The suspension was first washed with 5% hydrochloric acid (HCl) in order to remove the sulphate ions attached with the graphene oxide and then with DI until the pH of suspension became neutral by removing the attached salt impurities.

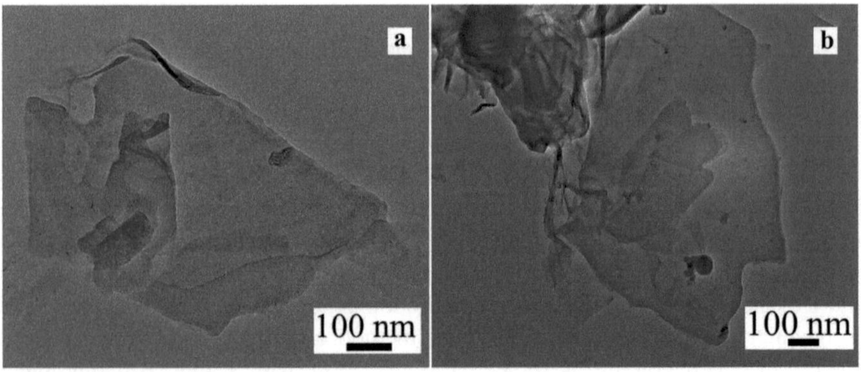

Fig. 3 **a, b** Transmission electron micrograph image (TEM) of graphene oxide nanosheets

7. After maintaining the pH at 7 the thick brown sticky precipitate is dried in temperature not exceeding the limit of 60 °C.
8. After drying, the precipitate is ground and finally, the silver-brown powder of graphene oxide is ready to use.

Transmission Electron Microscopy is used to determine the structure of graphene oxide that has been found similar to that of the structure discussed in the former. Figure 3 exhibits TEM image of graphene oxide along with SAED pattern representing its crystalline nature.

2.2.2 Electrochemical Methods

Electrolysis is an important phenomenon occurring due to the chemical effect of electric current. The effect produced on metallic conductors due to the flow of electric current heavily differs from the effect produced by the electric current in an ionic solution called electrolyte. The conduction of electric current through a metallic conductor is due to the drifting of free electrons and hence chemically or physically no change occurs except the generation of heat. At the same time, the conduction of electric current through an electrolyte is due to the movement of ions and hence that will be associated with chemical changes. In a typical electrolytic experiment, two metal rods or graphite rods are used as electrodes which are connected to the positive and negative terminal of the battery and are known as anode and cathode respectively. These two electrodes are immersed in an ionic solution called electrolyte. An appropriate dc voltage is provided across the electrodes by connecting the battery. The dissociation process of an electrolyte occurs and the dissociated ions namely cation and anion will move towards the electrodes.

Electrolytic synthesis is a reliable, cost-effective and eco-friendly method for the synthesis of nanomaterials. A wide range of nanomaterials can be synthesized using this technique. An electrochemical approach for the synthesis of nanomaterials was first effectively introduced and studied in detail by Reetz and Helbig in 1994. For the formation of metallic particles, Reetz et al. dissolved a metal sheet (anodic reaction) to form an intermediate metal salt which was reduced at the cathode. In the reaction, Tetraalkylammonium salts were used as a stabilizer. Later as motivation from above work, Rodigrigous Sanchez et al. successfully synthesized silver nanoparticles using acetonitrile containing tetrabutylammonium salt. The synthesis of high purity nanoparticles with controlled size without requiring any expensive equipment or vacuum is considered as the main advantage of the electrochemical route. The features like reliable operation, possibly in size and shape control, high yield and absence of unwanted side products, etc. increased the acceptance of electrolysis method among researchers. The method is especially advanced in the synthesis of metallic nanostructured materials and is also used for the synthesis of graphene sheets [34]. The electrochemical synthesis of graphene oxide includes two graphite rods as electrodes, the ionic solution as an electrolyte and a constant dc voltage as a power supply. When a voltage is applied across the graphite rods, the anode gets structural expansion due to the intercalation of anions and hence exfoliated graphene oxide sheets form by this oxidation process [57]. Thus, on electrolysis, the graphene oxide sheets are obtained at graphite made anode electrode.

3 Properties of Graphene Oxide

Graphene oxide sheets disperse well in water due to the negative surface charges on the sheets and keep it from aggregating [23]. Also due to the hydrophilic nature of graphene oxide, it shows variable inter sheet separations ranging from 0.6 to 1.2 nm [6]. Inspite of having hydrophilic nature, graphene oxide can dissolve in a wide range of organic solvents such as N-methyl-2-pyrrolidone, THF, N, N dimethylformamide and ethylene glycol. Graphene oxide can act as a semiconductor or in some cases as an insulator depending upon the parameters of the system and their optical as well as electronic properties, as per their requirement, can be modified.

The structure of graphene oxide is of nonstoichiometric nature [28]. Graphene oxide has a carbon network consisting of two kinds of regions i.e. trans-linked cyclohexane chairs and ribbons of flat hexagons with C–C double bonds as well as functional groups [62]. Being a 2D structure with various oxygenated functional groups, graphene oxide exhibits various excellent qualities in the electrochemical, optical, mechanical, thermal and chemical reactivity fields. Depending on the chemical & atomic structures, structural defects and presence of sp^3 hybridised carbon atoms, graphene oxide shows electronic properties such as conductivity. Graphene oxide shows insulating nature with a sheet resistance of 1012 $\Omega\ sq^{-1}$ or higher due to the sp^3 hybridised C–O bonding [54]. However, the reduction of graphene oxide results in a decrease in sheet resistance by several orders of magnitude and hence transforms the material into a semiconductor and ultimately into a carbonaceous semimetal [21]. The conductivity of the reduced graphene oxide is approximately equal to 1000 S/m and the activation energy of the graphene oxide has been estimated to be 32 $\pm$

5 kcal/mol. In addition to these interesting electronic properties, graphene oxide also exhibit optical properties due to which it shows photoluminescence in Uv–Visible wavelength range which changes with reduction of graphene oxide and makes it useful for biosensing, and optoelectronic applications [14]. Graphene oxide shows electrochemical properties due to its electron mobility and unique surface properties which make it be used efficiently in electrical wires of the redox centres of several heme-containing metalloproteins [56]. Graphene oxide also possesses excellent electrocatalytic properties toward oxygen reduction and certain biomolecules [38]. Graphene oxide exhibits high electrochemical capacitance with excellent cycle performance and hence has potential application in ultra-capacitors [8]. Due to the presence of a large number of oxygen-containing functional groups and structural defects, graphene oxide exhibits enhanced chemical activity [15]. One of the most important reactions of graphene oxide is its reduction which can be done using various reducing agents like hydrazine, sodium borohydride or hydroquinone in liquid medium [50]. One more significant chemical reactivity of graphene oxide is its capacity regarding chemical functionalization that involves introducing other groups to graphene oxide sheets via chemical reactions based on in-situ technique and ex-situ technique. For chemical functionalization, various oxygen-based groups attached to graphene oxide are subjected to orthogonal reactions. Typically, graphene oxide can be covalently functionalized using specially selected small molecules or polymers through activation and amidation/esterification of either the carboxyls or hydroxyls in graphene oxide via coupling reactions [69]. The noncovalent functionalization of graphene oxide is achieved by π–π stacking, van der Waals interactions, or hydrogen bonding graphene oxide can also be used as carbocatalyst for facilitating oxidation and hydration reactions and hence can be used to catalyse the oxidation of alcohols and alkenes and for the hydration of various alkynes into their respective aldehydes and ketones [48, 19]. Also graphene oxide shows strong oxidizing properties which makes it liable to use for oxidation in a broad range of reactions like oxidation of olefins to diones, methyl benzenes to aldehydes, diarylethenes to ketones, and various dehydrogenations [29].

4 Applications of Graphene Oxide

Due to antimicrobial properties and low toxicity graphene oxide has been proposed for the use as antimicrobial agents. The functionalized graphene oxide could inhibit Herpes Simplex Virus Type-1 infection through cell attachment inhibition at low concentrations [51]. A graphene oxide/carbon nanotube-based nanocomposite has been created and is being used in the treatment of trichomoniasis sickness. This due to the fact that nanocomposite formed a stable colloidal solution and interacted with the cell membrane of the protozoan trichomonas foetus strongly. Therefore, the hybrid nanocomposite might be an excellent candidate for a drug carrier, which could be used to deliver therapeutic agents to treat trichomoniasis disease [71]. Graphene oxide sheets has high potential for biological applications which makes it possible to use graphene oxide biofunctionalization like enzyme immobilization. For example, cross-linking of amino groups on pectinase and functional groups and site-specific interactions can be used to immobilize acid pectinase and Chloroperoxidase on graphene oxide Nanosheets acting as the matrix. The system showed higher activity, better thermal stability, and remained active in basic medium and at higher concentrations of oxidant in comparison with the free enzyme [14]. α-Amylase immobilized on graphene oxide sheets also exhibits enhanced thermal and storage stability. The application fields of graphene oxide are mainly focused on sensor and drug delivery [56]. A number of graphene oxide based electronic devices have been fabricated that are now implemented in the medicinal field for diagnosis, sensing and monitoring of drugs. One such device is a graphene-based field-effect transistor (FET) [26]. Another example is fluorescent-based biosensors that have been employed for the detection of DNA and proteins, with the potential for improved HIV diagnosis. One more remarkable application of graphene oxide is that folic acid-functionalized graphene oxide helps in detecting human cervical cancer and human breast cancer cells [69]. Graphene oxide is also used to make glucose sensors that use dc power. One of the primary ways by which graphene oxide is anticipated to be used is to create conductive transparent films that can be put on any surface. Such coatings could be used in flexible electronics, solar cells, liquid crystal devices, chemical sensors, and touch screen devices [8]. One such example is the transparent electrode of graphene oxide that has been used as a hole transport layer in polymer solar cells and LEDs [50]. Graphene oxide has an extremely high surface area which makes it useful as an electrode in batteries and double-layered capacitors, as well as in fuel cells and solar cells [2]. Due to its ability to store hydrogen, graphene oxide based nanocomposite can also be employed in fabricating high-capacity energy storage devices like lithium ion batteries. Anticancer medications have been delivered using functionalized graphene oxide. Furthermore, since, the interlayer distance of dried Hummers graphite oxide was 6.35 A°, but it rose to 11.6 A° in liquid water. This swelling of graphene oxide structures allows a water penetration channel between individual graphene oxide layers, which is attributed to water permeation and water diffusion across the membrane at a rate of 0.1 mg/min/cm^2 and 1 cm/h respectively. These property of graphene oxide makes it suitable for using as a potentially active nanomaterial for water desalination in 1960, as the membranes of graphene oxide in a reverse osmosis process shows 90% retention rate for NaCl solutions, and as the cation exchange membrane especially for the case of massive alkaloid ions. Graphene oxide based polymer nanocomposites exhibit enhanced properties like elasticity, tenacity, thermal stability and conductivity as compared to the properties of the original polymer that is used as the matrix for nanocomposite. Keeping in view of this extraordinary behaviour of nanocomposite, it is used in the form of stabilized paper-like structures specifically for hydrogen storage applications, ion conductors and Nanofiltration membranes [48].

5 Economical Aspects of Graphene Oxide

Due to the unique and exceptional electrical, optical, mechanical, and chemical capabilities of graphene and graphene oxide, these materials have received a lot of attention in the last decade [70]. Keeping in view of their applications the economical aspect of graphene oxide has been a great centre of concern. Graphene Oxide used in high-performance energy generation and storage technologies. Researchers have harnessed graphene oxide's unusual capabilities to build revolutionary electronic materials such as transparent conductors and ultrafast transistors. Understanding graphene oxide's numerous chemical characteristics has recently facilitated its use in high-performance energy generation and storage technologies [13]. The excellent 2D planar structure, large surface area, mechanical as well as chemical stability, superconductivity and biocompatibility of graphene oxide, have been extensively investigated as some of the most promising in biomedical applications. These characteristics suggest that they could be useful in the development of sophisticated drug delivery systems and the delivery of a wide range of treatments [35]. A graphene oxide based drug delivery system can also achieve co-distribution of numerous drugs or genes with improved chemotherapeutic efficacy, thanks to the specialised design of molecular structure of matrices. It was discovered when two anticancer medicines, doxorubicin (DOX) and camptothecin (CPT), were simultaneously loaded onto the Folic Acid-conjugated graphene oxide, via p–p stacking and hydrophobic interactions. Folic acid—graphene oxide delivered two medicines simultaneously that had superior targeting efficacy and cytotoxicity than graphene oxide that simply delivered DOX or CPT [72]. It indicates that the usage of functionalized graphene oxide for controlled loading and targeted delivery of multiple drugs, can improve the therapeutic efficacy because of the increased surface area and surface functionality of graphene oxide. Also, the pi-pi stacking interactions of graphene oxide GO could be useful in successfully creating reactive oxygen species (ROS) that can destroy cancer cells when exposed to radiation [41].

The graphene oxide nano-walls generated by a chemical exfoliation technique by electrophoretic deposition of Mg^{2+} on graphene oxide nanosheets are found to have antibacterial activity. The cell membrane injury of the bacteria generated by direct contact of the bacteria with the extremely sharp edges of the nanowalls was discovered to be the effective mechanism in the bacterial inactivation based on monitoring the outflow of cytoplasmic components of the bacteria. Gram-negative *Escherichia coli* bacteria with an outer membrane were more resistant to nanowall-induced cell membrane disruption than Gram-positive *Staphylococcus aureus* bacteria without an outside membrane. Furthermore, hydrazine-reduced graphene oxide nanowalls were more hazardous to bacteria than unreduced graphene oxide nanowalls. The better charge transfer between the bacteria and the more sharpened edges of the reduced nanowalls during the contact interaction was attributed to the reduced nanowall's enhanced antibacterial activity [1]. An experiment conducted on antibacterial activity of four types of graphene-based materials graphite, graphite oxide, graphene oxide, and reduced graphene oxide against *Escherichia coli* as a bacterial model. Graphene

Oxide dispersion had the strongest antibacterial activity (under similar concentration and incubation circumstances), followed by reduced graphene oxide, graphite and graphite oxide. Both membrane and oxidative stress may play a role in bacterial cytotoxicity as graphene materials with a higher density of functional groups and smaller sizes have a greater likelihood of interacting with bacteria, leading in cell deposition. Graphene oxide nanosheets can cause membrane stress by disrupting and destroying cell membranes, resulting in cell death when they come into close contact [36].

Graphene oxide can be used as a catalyst as it has an extraordinary catalytic capability on its own and in combination with another substance. It's an ideal environment for molecular engineering. Graphene oxide can act as an oxidant during anaerobic oxidation and is decreased at the end of the first catalytic cycle due to its many oxygen atoms. During aerobic oxidation, however, partly reduced graphene oxide can continue to activate molecular oxygen. Organic dyes or organo-catalysts can also be used to hybridise graphene oxide. Dye-induced photosensitization and easy charge transfer across the graphene contact have a synergistic effect that improves catalytic conversion [60].

The single-layer graphene oxide sheets sized down to a few nanometres in lateral width have been used to develop techniques for cell imaging in biological systems. The functionalization chemistry is implemented in order to impart solubility and compatibility of nano-graphene oxide in biological environments. The obtained size separated pegylated nano-graphene oxide sheets are soluble in buffers and serum without agglomeration. The nano-graphene oxide sheets are found to be photoluminescent in the visible and infrared regions. The intrinsic photoluminescence (PL) of nano-graphene oxide is used for live cell imaging in the near-infrared (NIR). One technique for producing photoluminescence in graphene materials is by the induction of energy band gaps by reducing graphene to finite sizes or by forming sp^2 islands, while the other involves the production of defects [61]. In the photoluminescence emission of nanoscale graphene materials, both processes are usually active at the same time. This is due to the fact that these materials are often made by extensive oxidising graphene cutting, followed by reduction to reduced graphene oxide. Chemical treatments invariably result in structural defects. Any sites that are not excellent sp^2 domains are referred to as defects. The defect-derived photoluminescence is similar to that of other carbon materials, such as carbon dots, in terms of mechanism [10]. Owing to its small size, intrinsic optical properties, large specific surface area, low cost, and useful non-covalent interaction they are effective in imaging [61].

Due to the unique properties of graphene oxide, the graphene oxide based materials have been used to generate a number of biosensors using a variety of sensing modes, including electrochemical and optical signalling. The electrochemical approach has received a lot of interest for biomolecule detection as one of the best methods due to its ease of operation, quick response, low cost, and high sensitivity. Graphene oxide based electrodes for detection fields can produce high-sensitivity electrochemical sensors because graphene has shown significant efficiency [3].

6 Functionalisation of Graphene Oxide with Metal Nanoparticles

The synthesis of graphene oxide/nanoparticle hybrids is categorized into two basic categories i.e. In-situ technique and Ex-situ technique. In the former approach, the crystalline nanoparticles are formed in the presence of graphene oxide and direct development or growth of nanostructures on graphene oxide sheets is obtained. On the other hand, in the ex-situ technique, nanostructures with desired shape and size are first synthesised, followed by the functionalisation of the graphene oxide sheets with nanoparticles to form a graphene oxide-based nanocomposite [34]. Figure 4 represents the TEM image of graphene oxide nanocomposite which is functionalised by gold nanoparticles.

There are many synthesis techniques reported under these approaches. However, the two main synthesis techniques that have been used for the functionalisation of graphene oxide to form graphene oxide-based nanocomposites are; hydrothermal method and electrochemical method.

The hydrothermal technique adapted from mineral formation occurring in nature has been used as an efficient method for the synthesis of inorganic nanomaterials. In nature generally, minerals are formed under specific pressure and temperature in the presence of water. The possibilities of a hydrothermal method for the artificial synthesis of inorganic compounds were first commercially explored by Karl Josef Bayer in 1892. Bayer converted aluminium hydroxide to Al_2O_3 by hydrothermal method. Today, the hydrothermal method has conquered the major fields of science such as nanotechnology, medical research, etc. due to its attractive properties like reduced contamination, the low temperature of synthesis, etc. The hydrothermal

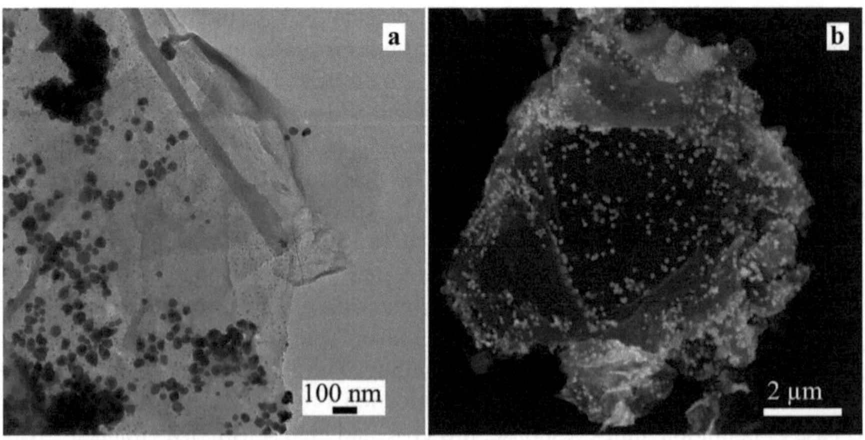

Fig. 4 **a** Transmission electron micrograph (TEM) and **b** scanning electron micrograph (SEM) of graphene oxide nanosheets

method has found its application in the field of crystal growth, thin-film preparation, material processing, etc.

Typically, hydrothermal reactions are carried out in an autoclave. The simple scale type autoclave uses a 50–100 ml Teflon lined vessel for carrying the precursors and is designed to work at low pressure and low temperature (around 300 °C and 1000 bars).

The hydrothermal method is capable of reducing graphene oxide. Moreover, in the hydrothermal method, formation of nanoparticles from its precursors and functionalization of the nanoparticles on reduced form of graphene oxide, both are simultaneously possible. Superheated H_2O facilitates acid-catalyzed reactions during the hydrothermal process, which results in an increase in the restoration of the -conjugation network inside the graphene oxide sheets owing to the availability of suitably high H+ concentration. This results in the formation of reduced graphene oxide from graphene oxide. The main advantages of the hydrothermal method are that it has high production efficiency and the possibility for synthesising nanoparticles with a high crystalline degree. The requirement of high temperature and need of a long time is the limitations of the hydrothermal method [53, 67].

The electrochemical approach is also used for the functionalization of nanoparticles with graphene oxide by using different electrolytes. This method is considered as advantageous as it is reliable and cost-effective. Although the in-situ functionalization is advantageous because of its efficiency, simplicity and cost-effectiveness, it has certain limitations, one of which is tuning the morphology of the nanoparticles. Keeping in view the above merits and limitations of in situ technique, ex situ technique of graphene oxide functionalization for the fabrication of nanocomposite have been proposed. In ex situ technique the noble metal based nanocomposites are prepared using nanoparticles which are not dispersed directly into the graphene oxide matrix. Although the ex situ approach is an excellent one, its limitation acts as a challenge during nanocomposite synthesis especially for nanoparticles having higher dispersibility in the matrix of graphene oxide. In order to solve these problems sonication methods if needed have been employed as per the requirement, to get complete and stable dispersion of the nanoparticles in the matrix.

7 Graphene Oxide Nanocomposites

Nanotechnology is an emerging technology, which deals with synthesis, characterization, and manipulations in the nanometre scale, which leads to various applications of material science, engineering, and devices. The main attraction of nanotechnology is its tendency to approach material to explore the unique physical and chemical properties of a material. Generally, for nanoparticles, at least one phase of the particle must be in a range of size 1–100 nm. The larger surface-volume ratio compared to bulk materials makes nanoparticles as authentic structures to fascinate the chemical transformations. The nanomaterials are bulkily classified as environmental and engineered. Metal oxides and metal sulphides are commonly found as minerals in nature and are termed as environmental. Artificially synthesized nanoparticles from their precursors and other basic units of the materials are called engineered nanoparticles. Recent trends in research are mainly based on four different types of engineered nanoparticles. They are metal-based nanoparticles, carbon-based and Nano polymers (constructed from pieces of different Nano molecules—also called dendrimers). Today, advanced research is concentrated on a combination of various engineered nanoparticles to form nanocomposites. Basically, nanocomposites are a mixture of different nanoparticles in which one type of nanoparticles (size in a range of 1–100 nm) will be dispersed on or attached to larger, continuous matrices. Hence the nanocomposites contain two parts, one is the continuous phase and the other is discontinuous dispersed phase [32]. An enhancement in the properties of a nanomaterial is indented by the formation of its nanocomposite. Any combinations of materials can be designed for the fabrication of nanocomposites and they are bulkily classified into three basic building blocks; they are metals (metal oxides), polymers and ceramics. The composites can have different dimensions such as core–shell (0D), nanowires (1D), lamellar or sheet-like structures (2D) and metal matrix composites (3D) [9]. The schematic representation of nanocomposite is shown as below in Fig. 5.

To utilize the full potential of unique properties of graphene oxide, one possible route is by making its composites with other materials. The major requirement of making composites is the availability of processable graphene oxide in large amount and the two popular ways to achieve it are:

1. Chemical and Electrochemical exfoliation of graphite.

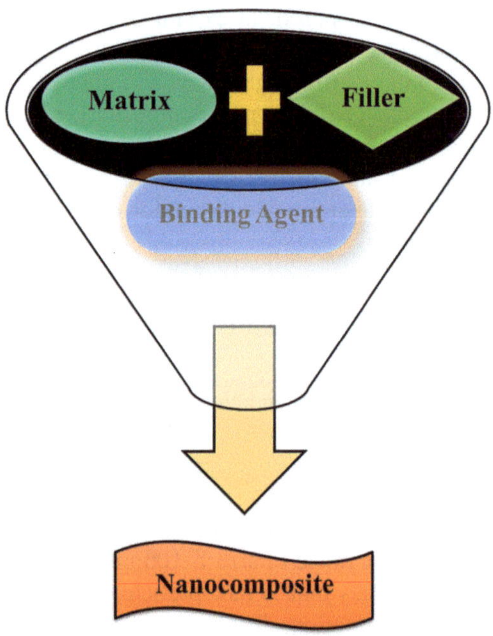

Fig. 5 Schematic representation of nanocomposite

2. By utilizing the bottom-up assembly technique and glucose as a sole reagent in the hydrothermal method.

These methods have advantages of functionalization and tuning of properties of graphene oxide depending on the desired applications. Graphene oxide composites are broadly classified into two categories:

1. Graphene oxide/inorganic composites
2. Graphene oxide/polymer composites.

Depending on the material other than graphene oxide, graphene oxide/inorganic composites can be further classified into graphene oxide/metal, graphene oxide/carbon and graphene oxide/noble metal nanocomposites.

Noble metal nanoparticles have unique optical and chemical properties which are completely different from their bulk counterparts. These properties are a function of size, shape, composition and structure of nanoparticles. They have found a wide variety of applications viz. catalysis, electronics, sensors and biomedical [73]. Incorporation of noble metal nanoparticles in carbonaceous materials further enhances their chemical reactivity, catalytic behaviour and electrical properties. Among all carbonaceous materials, graphene oxide has emerged as the most promising base matrix. The synthesis of graphene oxide/noble metal nanocomposites generally follows the reduction of metal salt in the presence of graphene oxide Nanosheets. Further graphene oxide, being two dimensional in nature, besides providing a base for nanoparticles also increases the effective surface area, which is helpful in enhancing the properties of noble metal nanostructures. The graphene oxide-based metal nanocomposites have found applications in biosensors, intracellular analysis, bio-distribution, bio-imaging, and gene mapping [2].

Although graphite and its intercalated compounds have been known and studied for more than a century it is only after the experimental discovery of wonder material graphene that carbon-based materials have created a cornucopia for many potential applications and new physics. There are a number of reports describing the utilization of graphene oxide-based composites in many applications. The present status of the ongoing research in the field graphene oxide-based bioinspired nanocomposite has exhibited a great potential. Both top-down and bottom-up approaches are available for the synthesis of noble metal nanoparticles composites but still, the non biocompatibility, self-assembly, aggregation, reproducibility, low through-output, control over size and shape of nanoparticles are major obstacles in their implication with respect to their reliability. Therefore a reliable substrate for the synthesis of nanocomposite is demanded till now. Two-dimensional nature, cost-effectiveness, bio-compatibility and the ability to absorb or form complexes with target molecules make graphene oxide an attractive candidate. So, it would be interesting to study the noble metal nanoparticles decorated with graphene oxide as composite [73].

References

1. Akhavan O, Ghaderi E (2010) Toxicity of graphene and graphene oxide nanowalls against bacteria. ACS Nano 4(10):5731–5736
2. Ali MA, Singh C, Srivastava S, Admane P, Agrawal VV, Sumana G, John R, Panda A, Dong L, Malhotra BD (2017) Graphene oxide-metal nanocomposites for cancer biomarker detection. RSC Adv 7(57):35982–35991 https://doi.org/10.1039/C7RA05491B
3. Behbudi G (2020) Mini review of graphene oxide for medical detection and applications. Adv Appl NanoBio-Technol 1(3):63–66
4. Bonaccorso F, Lombardo A, Hasan T, Sun Z, Colombo L, Ferrari AC (2012) Production and processing of graphene and 2d crystals. Mater Today 15(12):564–589. https://doi.org/10.1016/S1369-7021(13)70014-2
5. Brodie BC (1860) Sur le poids atomique du graphite. Ann Chim Phys 59(466):e472
6. Buchsteiner A, Lerf A, Pieper J (2006) Water dynamics in graphite oxide investigated with neutron scattering. J Phys Chem B 110(45):22328–22338
7. Burress JW, Gadipelli S, Ford J, Simmons JM, Zhou W, Yildirim T (2010) Graphene oxide framework materials: theoretical predictions and experimental results. Angew Chem Int Ed 49(47):8902–8904
8. Cai B, Wang S, Huang L, Ning Y, Zhang Z, Zhang G-J (2014) Ultrasensitive label-free detection of PNA–DNA hybridization by reduced graphene oxide field-effect transistor biosensor. ACS Nano 8(3):2632–2638
9. Cury CPH, Gundappa SK, Fernando W (2009) Nanocomposites : synthesis, structure, properties and new application opportunities. Mater Res 12(1):1–39. https://doi.org/10.1590/S1516-14392009000100002
10. Cao LI, Meziani MJ, Sahu S, Sun Y-P (2013) Photoluminescence properties of graphene versus other carbon nanomaterials. Acc Chem Res 46(1):171–180
11. Castro Neto AH, Guinea F, Peres NMR, Novoselov KS, Geim AK (2009) The electronic properties of graphene. Rev Mod Phys 81(1):109–162. https://doi.org/10.1103/RevModPhys.81.109
12. Chen D, Feng H, Li J (2012) Graphene oxide: preparation, functionalization, and electrochemical applications. Chem Rev 112(11):6027–6053 https://doi.org/10.1021/cr300115g
13. Chung C, Kim Y-K, Shin D, Ryoo S-R, Hong BH, Min D-H (2013) Biomedical applications of graphene and graphene oxide. Acc Chem Res 46(10):2211–2224
14. Ding Y, Cui R, Hu M, Li S, Zhai Q, Jiang Y (2017) Well-oriented bioarchitecture for immobilization of chloroperoxidase on graphene oxide nanosheets by site-specific interactions and its catalytic performance. J Mater Sci 52(17):10001–10012
15. Dreyer DR, Park S, Bielawski CW, Ruoff RS (2010) The chemistry of graphene oxide. Chem Soc Rev 39(1):228–240
16. Eda G, Chhowalla M (2010) Chemically derived graphene oxide: towards large-area thin-film electronics and optoelectronics. Adv Mater 22(22):2392–2415
17. Eda G, Lin Y-Y, Mattevi C, Yamaguchi H, Chen H-A, Chen I-S, Chen C-W, Chhowalla M (2010) Blue photoluminescence from chemically derived graphene oxide. Adv Mater 22(4):505–509
18. Geim AK, Novoselov KS (2010) The rise of graphene. In: Nanoscience and technology: a collection of reviews from nature journals. World Scientific, pp 11–19
19. Ghorbani M, Abdizadeh H, Golobostanfard MR (2015) reduction of graphene oxide via modified hydrothermal method. Procedia Mater Sci 11:326–330. https://doi.org/10.1016/j.mspro.2015.11.104
20. Gómez-Navarro C, Meyer JC, Sundaram RS, Chuvilin A, Kurasch S, Burghard M, Kern K, Kaiser U (2010) Atomic structure of reduced graphene oxide. Nano Lett 10(4):1144–1148
21. Han HJ, Chen YN, Wang ZJ (2015) Effect of microwave irradiation on reduction of graphene oxide films. RSC Adv 5(113):92940–92946
22. Hasan T, Torrisi F, Sun Z, Popa D, Nicolosi V, Privitera G, Bonaccorso F, Ferrari AC (2010) Solution-phase exfoliation of graphite for ultrafast photonics. Phys Status Solidi (B) Basic Res 247(11–12):2953–2957. https://doi.org/10.1002/pssb.201000339

23. He H, Klinowski J, Forster M, Lerf A (1998) A new structural model for graphite oxide. Chem Phys Lett 287(1–2):53–56
24. Hernandez Y, Nicolosi V, Lotya M, Blighe FM, Sun Z, De Sukanta I, McGovern T et al (2008) High-yield production of graphene by liquid-phase exfoliation of graphite. Nat Nanotechnol 3(9):563–568. https://doi.org/10.1038/nnano.2008.215
25. Horiuchi S, Gotou T, Fujiwara M, Sotoaka R, Hirata M, Kimoto K, Asaka T, Yokosawa T, Matsui Y, Watanabe K (2003) Carbon nanofilm with a new structure and property. Jpn J Appl Phys 42(9A):L1073
26. Hu H, Onyebueke L, Abatan A (2015) Characterizing and modeling mechanical properties of nanocomposites-review and evaluation. J Miner Mater Charact Eng. https://doi.org/10.4236/jmmce.2010.94022
27. Hummers WS, Offeman RE (1958) Preparation of graphitic oxide. J Am Chem Soc 80(6):1339. https://doi.org/10.1021/ja01539a017
28. Inagaki M, Kang F (2014) Engineering and applications of carbon materials. materials science and engineering of carbon: fundamentals. https://doi.org/10.1016/b978-0-12-800858-4.00003-6
29. Jia H-P, Dreyer DR, Bielawski CW (2011) C–H oxidation using graphite oxide. Tetrahedron 67(24):4431–4434
30. Johari P, Shenoy VB (2011) Modulating optical properties of graphene oxide: role of prominent functional groups. ACS Nano 5(9):7640–7647
31. Kim F, Cote LJ, Huang J (2010) Graphene oxide: surface activity and two-dimensional assembly. Adv Mater 22(17):1954–1958
32. Lateef A, Nazir R (2017) Metal nanocomposites : synthesis , characterization and their applications. Sci Appl Tailored Nanostruct 239–56
33. Lee G, Kim KS, Cho K (2011) Theoretical study of the electron transport in graphene with vacancy and residual oxygen defects after high-temperature reduction. J Phys Chem C 115(19):9719–9725
34. Liu F, Choi JY, Seo TS (2010) Graphene oxide arrays for detecting specific DNA hybridization by fluorescence resonance energy transfer. Biosens Bioelectron 25(10):2361–2365
35. Liu J, Cui L, Losic D (2013) Graphene and graphene oxide as new nanocarriers for drug delivery applications. Acta Biomater 9(12):9243–9257
36. Liu S, Zeng TH, Hofmann M, Burcombe E, Wei J, Jiang R, Kong J, Chen Y (2011) Antibacterial activity of graphite, graphite oxide, graphene oxide, and reduced graphene oxide: membrane and oxidative stress. ACS Nano 5(9):6971–6980
37. Liu Y, Li Q, Feng Y-Y, Ji G-S, Li T-C, Jie T, Gu X-D (2014) Immobilisation of acid pectinase on graphene oxide nanosheets. Chem Pap 68(6):732–738
38. Liu Y, Yu D, Zeng C, Miao Z, Dai L (2010) Biocompatible graphene oxide-based glucose biosensors. Langmuir 26(9):6158–6160
39. Mattevi C, Eda G, Agnoli S, Steve Miller K, Mkhoyan A, Celik O, Mastrogiovanni D, Granozzi G, Carfunkel E, Chhowalla M (2009) Evolution of electrical, chemical, and structural properties of transparent and conducting chemically derived graphene thin films. Adv Func Mater 19(16):2577–2583. https://doi.org/10.1002/adfm.200900166
40. McAllister MJ, Li J-L, Adamson DH, Schniepp HC, Abdala AA, Liu J, Herrera-Alonso M, Milius DL, Car R, Robert K, Prud'homme RK, Aksay IA (2007) Single sheet functionalized graphene by oxidation and thermal expansion of graphite. Chem Mater 19(18):4396–4404
41. Mousavi S, Zarei M, Hashemi S (2018) polydopamine for biomedical application and drug delivery system. Med Chem (Los Angeles) 8:218–229
42. Novoselov KS (2011) Nobel lecture: graphene: materials in the flatland. Rev Mod Phys 83(3):837–849. https://doi.org/10.1103/RevModPhys.83.837
43. O'Neill A, Khan U, Nirmalraj PN, Boland J, Coleman JN (2011) Graphene dispersion and exfoliation in low boiling point solvents. The J Phys Chem C 115(13):5422–5428
44. Pandey D, Reifenberger R, Piner R (2008) Scanning probe microscopy study of exfoliated oxidized graphene sheets. Surf Sci 602(9):1607–1613

45. Paredes JI, Villar-Rodil S, Martínez-Alonso A, Tascon JMD (2008) Graphene oxide dispersions in organic solvents. Langmuir 24(19):10560–10564
46. Pei S, Cheng H-M (2012) The reduction of graphene oxide. Carbon 50(9):3210–3228
47. Rasheed M, Shihab S, Wissam Sabah O (2021) An investigation of the structural, electrical and optical properties of graphene-oxide thin films using different solvents. J Phys Conf Ser 1795:12052. IOP Publishing
48. Ray SC (2015) Application and uses of graphene oxide and reduced graphene oxide. Elsevier Inc., Applications of graphene and graphene-oxide based nanomaterials. https://doi.org/10.1016/b978-0-323-37521-4.00002-9
49. Sabzevari M, Cree D, Wilson L (2018) Preparation and characterization of graphene oxide cross-linked composites 2014. https://doi.org/10.25071/10315/35427
50. Saha SK, Bhaumik S, Maji T, Mandal TK, Pal AJ (2014) Solution-processed reduced graphene oxide in light-emitting diodes and photovoltaic devices with the same pair of active materials. RSC Adv 4(67):35493–35499
51. Sametband M, Kalt I, Gedanken A, Sarid R (2014) Herpes simplex virus type-1 attachment inhibition by functionalized graphene oxide. ACS Appl Mater Interfaces 6(2):1228–1235
52. Schniepp HC, Li J-L, McAllister MJ, Sai H, Herrera-Alonso M, Adamson DH, Prud'homme RK, Car R, Saville DA, Aksay IA (2006) Functionalized single graphene sheets derived from splitting graphite oxide. J Phys Chem B 110(17):8535–8539
53. Shao Y, Wang J, Engelhard M, Wang C, Lin Y (2010) Facile and controllable electrochemical reduction of graphene oxide and its applications. J Mater Chem 20(4):743–748
54. Shin H-J, Kim KK, Benayad A, Yoon S-M, Park HK, Jung I-S, Jin MH, Jeong H-K, Kim JM, Choi J-Y (2009) efficient reduction of graphite oxide by sodium borohydride and its effect on electrical conductance. Adv Func Mater 19(12):1987–1992
55. Si Y, Samulski ET (2008) Synthesis of water soluble graphene. Nano Lett 8(6):1679–1682
56. Singh K, Srivastava G, Talat M, Srivastava ON, Kayastha AM (2015) α-amylase immobilization onto functionalized graphene nanosheets as scaffolds: its characterization, kinetics and potential applications in starch based industries. Biochem Biophys Rep 3:18–25
57. Singh PK, Singh U, Bhattacharya B, Rhee HW (2014) Electrochemical synthesis of graphene oxide and its application as counter electrode in dye sensitized solar cell. J Renew Sustain Energy 6(1). https://doi.org/10.1063/1.4863834
58. Stankovich S, Dikin DA, Dommett GHB, Kohlhaas KM, Zimney EJ, Stach EA, Piner RD, Nguyen ST, Ruoff RS (2006) Graphene-based composite materials. Nature 442(7100):282–286
59. Staudenmaier L (1898) Verfahren Zur Darstellung Der Graphitsäure. Ber Dtsch Chem Ges 31:1481–1487
60. Su C, Loh KP (2013) Carbocatalysts: graphene oxide and its derivatives. Acc Chem Res 46(10):2275–2285
61. Sun X, Liu Z, Welsher K, Robinson JT, Goodwin A, Zaric S, Dai H (2008) Nano-graphene oxide for cellular imaging and drug delivery. Nano Res 1(3):203–212
62. Szabó T, Berkesi O, Forgó P, Josepovits K, Sanakis Y, Petridis D, Dékány I (2006) Evolution of surface functional groups in a series of progressively oxidized graphite oxides. Chem Mater 18(11):2740–2749
63. Tang L, Li X, Ji R, Teng KS, Tai G, Ye J, Wei C, Lau SP (2012) Bottom-up synthesis of large-scale graphene oxide nanosheets. J Mater Chem 22(12):5676–5683. https://doi.org/10.1039/c2jm15944a
64. Torrisi F, Hasan T, Wu W, Sun Z, Lombardo A, Kulmala TS, Hsieh G-W et al (2012) Inkjet-printed graphene electronics 4:2992–3006
65. Tung VC, Allen MJ, Yang Y, Kaner RB (2009) High-Throughput solution processing of large-scale graphene. Nat Nanotechnol 4(1):25
66. Wang X, Zhi L, Tsao N, Tomović Ž, Li J, Müllen K (2008) Transparent carbon films as electrodes in organic solar cells. Angewandte Chemie—Int Edition 47(16):2990–2992. https://doi.org/10.1002/anie.200704909
67. Wang Y, Wan Y, Zhang D (2010) Reduced graphene sheets modified glassy carbon electrode for electrocatalytic oxidation of hydrazine in alkaline media. ElectrochemCommun 12(2):187–190

68. Warner JH, Schäffel F, Bachmatiuk A, Rümmeli MH (2013) Properties of graphene. In Graphene. https://doi.org/10.1016/B978-0-12-394593-8.00003-5
69. Yang X, Wang Y, Huang X, Ma Y, Huang Y, Yang R, Duan H, Chen Y (2011) Multi-functionalized graphene oxide based anticancer drug-carrier with dual-targeting function and PH-sensitivity. J Mater Chem 21(10):3448–3454
70. Yu W, Sisi L, Haiyan Y, Jie L (2020) Progress in the functional modification of graphene/graphene oxide: a review. RSC Adv 10(26):15328–15345
71. Zanin H, Margraf-Ferreira A, Da Silva NS, Marciano FR, Corat EJ, Lobo AO (2014) Graphene and carbon nanotube composite enabling a new prospective treatment for trichomoniasis disease. Mater Sci Eng, C 41:65–69
72. Zhang L, Xia J, Zhao Q, Liu L, Zhang Z (2010) Functional graphene oxide as a nanocarrier for controlled loading and targeted delivery of mixed anticancer drugs. Small 6(4):537–544
73. Zhang X, Xie H, Liu Z, Tan C, Luo Z, Li H, Lin J et al (2015) Black phosphorus quantum dots. Angewandte Chemie—Int Edition 54(12):3653–3657. https://doi.org/10.1002/anie.201409400
74. Zhu Y, Murali S, Cai W, Li X, Suk JW, Potts JR, Ruoff RS (2010) Graphene and graphene oxide: synthesis, properties, and applications. Adv Mater 22(35):3906–3924

Synthesis of Self-stabilized Metal-Oxide and Metal-Hydroxide Nanorods

Mohd Asif, Vinayak Pundir, and Irfan Ahmad

Abstract Nanorods (NRs) have been a subject of profound interest because of a wide variation in their electronic properties with confinement. In this article, a single-step, self-stabilizing, two-electrode electrochemical synthesis method is demonstrated for growing metal–oxide and metal-hydroxide NRs at room temperature. Barium hydroxide NRs were fabricated using a simple electrochemical reduction of Ba^{2+} ions from a barium chloride precursor solution without using any capping agent. The synthesized NRs were characterized using Scanning Electron Microscopy (SEM), Transmission Electron Microscopy (TEM), X-ray Diffraction (XRD), Optical Microscopy (OM), Energy Dispersive X-Ray Spectroscopy (EDS), Selected Area Electron Diffraction (SAED), and UV–Vis. absorption spectroscopy. Effects of temperature and potential differences across the electrodes on the shape and size of the synthesized NRs were also investigated. NRs of diameters in the range of 80 to 300 nm were fabricated with different growth parameters. Furthermore, the synthesis of silver oxide NRs at room temperature is also demonstrated with the method.

Keywords Barium hydroxide nanorods · Silver-oxide nanorods · Electrochemical synthesis · Surfactant-free synthesis

---

M. Asif · V. Pundir · I. Ahmad (✉)
Interdisciplinary Nanotechnology Centre, ZHCET, Aligarh Muslim University (AMU), Aligarh-202002, India
e-mail: irfanplsr@gmail.com; i.ahmad@alientt.com

M. Asif
Department of Electrical Engineering, Indian Institute of Technology Delhi (IITD), New Delhi-110016, India

V. Pundir
Material Science Program, Indian Institute of Technology Kanpur (IITK), Kanpur-208016, India

I. Ahmad
School of Engineering Sciences and Technology (SEST), Jamia Hamdard, New Delhi-110062, India

I. Uddin and I. Ahmad (eds.), *Synthesis and Applications of Nanomaterials and Nanocomposites*, Composites Science and Technology,
https://doi.org/10.1007/978-981-99-1350-3_4

1 Introduction

Nanomaterials (NMs) are a very exciting class of materials offering properties that their larger siblings usually could not. One of the very obvious is a larger surface area which is advantageous in the applications where the surface becomes a playground for atoms and molecules such as catalysis and Electrochemical (EC) capacitors. NMs can be engineered in numerous shapes, most common being sphere, prism, rod, cube, and sheet, with every method of their preparation providing different dimensional variations to them. With different shapes, the properties NMs exhibit could be different. [1–4] Researchers could achieve technological advancements that previously seemed impossible by exploiting these properties. Nanorods (NRs), due to their unusual physical properties, originating out of the confinement in two dimensions, [5] have a wide range of applications in the fields of piezoelectric devices, [6, 7] solar cells, [8] LEDs, [9] gas-sensors, [10] etc. Properties of NRs may vary with size, surface density, and alignment [11] which are tuneable and thus can be utilized to control (enhance) the performance of NRs in many applications.

Not only the shape and size are crucial, but their preparation route also plays a vital role in the efficacy of NMs for many applications. EC methods have been used to synthesize metal [12] and metal-oxide [13] NMs. In the EC synthesis method, electrons are provided from a low-voltage electric power supply through the cathode for the reduction of the positively charged metal cations. Metals with more negative reduction potential tend to oxidize in water by giving up electrons to the dissolved oxygen and forming their respective oxides. This redox process primarily results in the formation of nanosheets of metal oxides, but with a controlled process other shapes such as NRs are also possible to synthesize.

Capping agents/surfactants are crucial for the chemical and the biological synthesis of NMs, [14] but adversely impact their effectiveness, especially in applications such as catalysis, as they make it difficult for reactants to access the surface of the catalyst. [15] Therefore, the capping-agent-free synthesis approach is desirable in several applications. Several researchers have previously demonstrated surfactant-free techniques for nanostructure fabrication, but a high temperature or some other harsh conditions are usually required. [16] Some attempts have also been made with the EC method for surfactant-free synthesis of nanomaterials, [17] but the approach involves stochastic collisions of aqueous nanodroplet reactors on microelectrodes and has thus far produced only metallic nanoparticles.

Barium hydroxide is an alkaline-earth metal hydroxide with a diverse line of applications such as catalysis, [18, 19] cathode buffer layer of polymer solar cells, [20] heat storage, [21] etc. The shape and size of nanomaterials play essential roles in determining their properties. [22] Barium hydroxide NRs and nanoparticles have attracted a great deal of attention in the conservation of cultural heritage. [23, 24] In the past, approaches such as microwave-assisted synthesis in aqueous solution [25] have been used for the synthesis of barium hydroxide NRs.

In this work, we demonstrate a simple, economical, and effective EC method to synthesize metal-oxide and metal-hydroxide NRs at room temperature by reducing metal ions directly from the precursor solution without using any other chemical in the synthesis. Effects of varying growth parameters (voltage and temperature) on the synthesized NRs were also investigated. To the best of our knowledge, this work on barium hydroxide NRs synthesis is the first report with a two-electrode electrochemical synthesis method at room temperature without using any surfactants. Silver oxide nanomaterials have also attracted interest due to applications in electronics, [26] gas sensing, [27] antimicrobial [28] and biomedical [29] research. To exhibit the versatility of the proposed method, synthesis of silver oxide NRs is also demonstrated with the method.

2 Electrochemical Synthesis of Metal Hydroxide Nanorods

Barium chloride salt ($BaCl_2.2H_2O$, purity 99%) was purchased from Qualigens Fine Chemicals. The salt weighing 0.98 gm was dissolved in 20 ml of double distilled water to make a 200 mM barium chloride solution. A two-electrode electrolytic cell was formed by using an ITO coated glass (100 Ω/sq.,) purchased from Techinstro® as the cathode (2.5 cm × 1.25 cm) and a small aluminium sheet (2.5 cm × 1.25 cm) as the anode. The two electrodes were held 3 cm apart in 100 ml of double distilled water with the ITO coating side facing the other electrode. The temperature of the solution was measured by using a digital thermometer and controlled with the help of an ice bath and a hot plate. A DC voltage was applied across the two electrodes by using a regulated power supply. 1 ml of the as-prepared solution was added gradually in a dropwise manner to the cell system. With each drop, the current from the power supply spiked and gradually reduced as the ions were consumed by respective electrodes. After a few drops, coatings on both the electrodes were visible, marking the deposition of respective ions in the solution on the electrodes. Initially, the electrodes were only about 10% submerged (along the longer edge) in the water. After a few drops, the reaction was slowing down, marking nearly complete coverage of non-conductive coating on at least one electrode. A few drops of the distilled water were added as needed to raise the level of solution to about 2 mm and thus expose fresh areas of electrodes for the reaction. The reaction continued in this manner until about 75% area of the electrodes was visibly coated. The ITO electrode was taken out of the setup for characterization, and the Al electrode was discarded. Some detachment of coating off of the cathode (ITO) was observed that inevitably mixed in the solution. The left-over solution in the EC setup was also dried to obtain the remainder of the material as a residual powder for characterization. SEM images were captured for morphological characterization of the material deposited on ITO as well as in the powdered form by using a JEOL-JSM 6510 LV electron microscope with 15 kV accelerating voltage. EDS was performed by using the EDS attachment with SEM to analyze the chemical composition and impurities in the samples. Carbon tape was used for holding the powder sample inside the chamber. JEOL-JEM 2100 TEM was used with 200 kV accelerating voltage to obtain information about the size and shape of the synthesized nanomaterial. A carbon-coated copper grid was used for holding the material in the vacuum chamber. SAED attachment with JEM 2100 was used for structural characterization of the material. XRD data of the residual powder sample containing barium hydroxide NRs (Fig. 5) was obtained using Rigaku Miniflex-II X-Ray diffractometer with Cu-Kα radiation (wavelength = 1.54 Å) and scan step of 0.02°.

2.1 Mechanism

When an ionic salt is dissolved in water, it creates positively charged metal cations and negatively charged anions. In the presence of an applied electric field, the metal cations accelerate towards the cathode and impinge on it. At the cathode, the cations are reduced from M^+ to M^O state and form a thin coating. In the case of alkaline earth metals, a swift hydroxide formation occurs in an aqueous medium at the cathode. In the case of barium, barium hydroxide with monohydrate $\{Ba(OH)_2 \cdot H_2O\}$, trihydrate $\{Ba(OH)_2 \cdot 3H_2O\}$ and octahydrate $\{Ba(OH)_2 \cdot 8H_2O\}$ formation occur, the latter two being more common in an aqueous medium. [30] In the case of precursor solutions containing ions of noble metals, no reaction with water occurs to form hydroxide after the coating of the metal at the ITO cathode. However, since with every drop of dilute precursor solution a very thin coating of metal forms on the cathode, oxidation is possible. Nanomaterials of different shapes and sizes can be fabricated with the proposed method and synthesis parameters can be tuned to provide good control over the synthesized material.

2.2 Results and Discussion

In Fig. 1a, synthesized barium hydroxide NRs are presented. NRs seem coalesced and formed a dense network. Several randomly oriented NRs of diameters in the range of 100 to 300 nm and lengths in the range of 1500–2500 nm are observed. A closer view of NRs is shown in Fig. 1b. A NR of diameter about 150 nm and length of about 2000 nm is observed in the central region of the image. Another NR of diameter about 300 nm and length of about 3000 nm is observed in the lower central region, which at the other end is branching out into 4 NRs of average 150 nm diameter. Another NR of diameter about 120 nm and length of about 800 nm is observed in the central top region of the image.

EDS spectrum of the fabricated $Ba(OH)_2$ NRs on ITO is displayed in Fig. 2. Atomic percentages of the elements measured on the surface of the sample are listed on the inset of the figure. A large amount of oxygen (55%) and carbon (32%) were present on the surface. A substantial amount of barium (8.41%) and small amounts of silicon (2.27%), chlorine (1.51%), and indium (1.05%) were measured. It is not uncommon to find about 30% carbon on ITO substrate, due to hydrocarbons present in usual workplaces [31]. While about 5% oxygen is being contributed from

Fig. 1 Barium hydroxide nanorods synthesized on ITO cathode (**a**) and a magnified view of NRs (**b**)

the glass substrate (SiO_2), about 2% of oxygen is being contributed from the ITO coating and about 38% (~7.7% × 5) is originating from barium hydroxide trihydrate {$Ba(OH)_2.3H_2O$} NRs. A small amount of chlorine was measured on the sample, due to the presence of unreacted chlorine ions in the precursor solution.

TEM image of barium hydroxide NRs grown on ITO is shown in Fig. 3a. A network of NRs of different aspect ratios was formed. A NR of diameter about 12 nm and length of about 100 nm is observed in the left central region of the image. Another NR of diameter about 20 nm and length of about 200 nm is observed in the left central region.

It is crucial to mention that performing XRD measurement to identify the exact phases of the synthesized NRs was not practical on the coated film of NRs on ITO substrate. The EC reaction on the cathode is self-limiting after a thin coating of barium hydroxide forms on top of the ITO substrate. The film is not thick enough for XRD measurement. To mitigate the issue, we performed SAED measurement, which is compatible with this sample. However, the XRD measurements were performed on the residual powder samples containing NRs and are discussed later in the article.

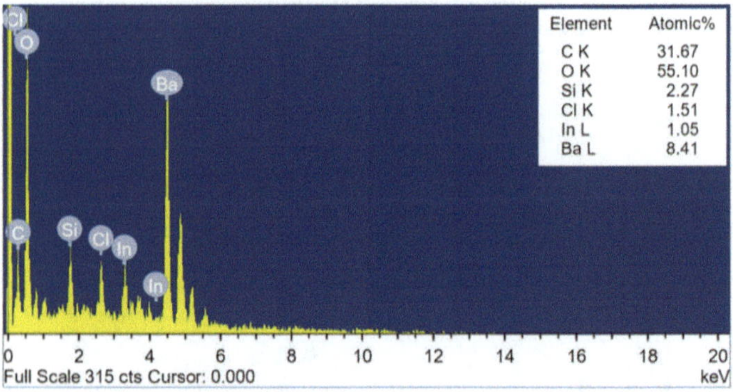

Fig. 2 EDS of $Ba(OH)_2$ nanorods on ITO with inset showing the atomic percentage of constituent elements

SAED image of the sample is presented in Fig. 3b. Discrete diffraction spots are aligned in a distinct rectangular grid fashion, without diffraction rings. The pattern strongly indicates a highly crystalline nature of the synthesized NRs. It is important to note that the diffraction pattern of unreacted barium would reveal a body centre cubic (bcc) pattern, [32] which would result in two square grids pattern, symmetrically intersecting each other with one grid of significantly higher intensity than the second grid. The coated barium layer would react with the water present in the medium to form barium hydroxide with trihydrate. The structure of barium hydroxide octahydrate is monoclinic with $\beta = 99°$, [33] partly eliminating the possibility of its presence since the observed SAED has a rectangular grid pattern. It is also crucial to point out that commercial ITO coating on glass substrates is mostly performed with the sputtering technique. Thus prepared ITO coatings are known to have very limited crystallinity. Vaufrey et al. have reported SAED of sputtered ITO with multiple diffraction rings and occasional discrete diffraction maxima and have termed it nearly an amorphous phase [34]. Therefore, the diffraction pattern in the image has no significant contribution from the ITO substrate. Since no other chemical was used in the synthesis, the SAED pattern could not have originated from any other material than barium hydroxide trihydrate $\{Ba(OH)_2.3H_2O\}$ deposited on the ITO electrode, which is also consistent with the analysis of the EDS data.

Some of the coated material from the ITO cathode was detached-off. In order to investigate the material, we dried the left-over solution as a residual powder and characterized it. In Fig. 4, an SEM image of the residual powder is presented. Multiple NRs were found in the residual sample. Some NRs seem to coalesce, while a few are isolated. Two NRs are visible in the central-right region of the image with diameters in the range of 150–200 nm and lengths between 2000–2500 nm.

In Fig. 5, XRD data of the residual powder of barium chloride (with some barium hydroxide NRs) is presented. Multiple peaks are observed in the data, indicating several growth planes. Major peaks are measured at 44.6°, 43.3°, 35.0°, 30.4°,

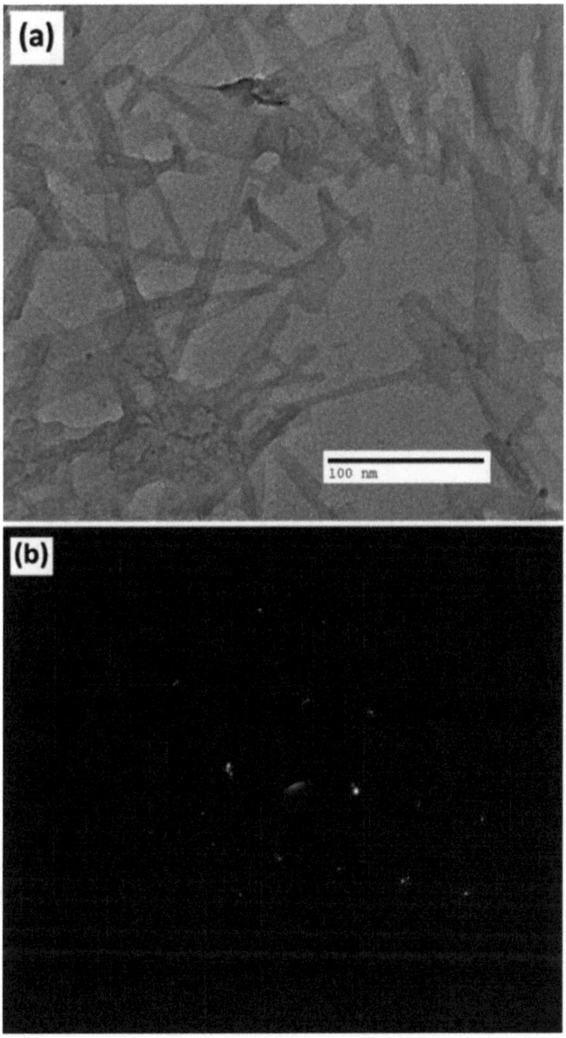

Fig. 3 TEM image of barium hydroxide nanorods on ITO (**a**) and corresponding SAED image (**b**)

19.9°, 63.9°, 36.9°, 16.1°, 37.6°, 26.2°, 60.5°, and 40.5°, in decreasing peak intensity order. These peak positions match well with the reported XRD data of barium chloride dihydrate {$BaCl_2 \cdot 2H_2O$} in the literature. [35, 36] Peaks at 30.4°, 63.9°, 36.9°, and 60.5° matches with the reported XRD peaks of barium hydroxide trihydrate {$Ba(OH)_2 \cdot 3H_2O$}. [37] The peak at 36.9° is better matched with the barium hydroxide trihydrate data. The findings are consistent with SEM and EDS data of the residual powder.

Fig. 4 SEM of residual powder. Barium hydroxide trihydrate nanorods with $BaCl_2$ agglomerated particles

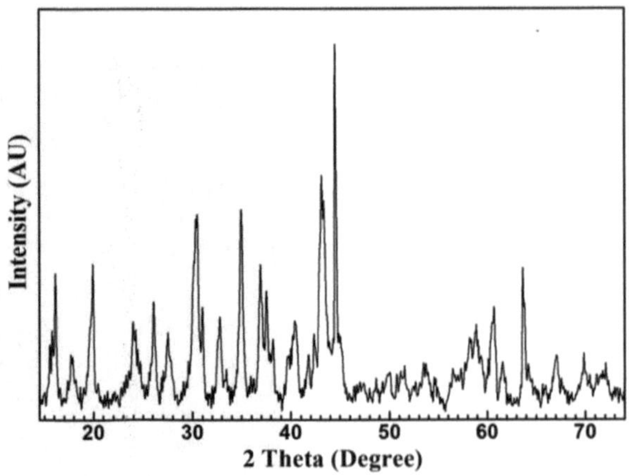

Fig. 5 XRD data of the residual powder (barium chloride dihydrate containing some barium hydroxide trihydrate NRs)

3 Electrochemical Synthesis of Metal Oxide Nanorods

In order to establish the versatility of the synthesis method, the growth of silver oxide NRs is also demonstrated in this section. The synthesis was carried out in a similar manner as before, just the synthesis setup was deliberately placed inside an ultra-sonicator to assist detachment of the synthesized NRs from the cathode to increase NRs/salt ratio in the residual powder. AgNO3 salt (from RFCL Ltd., India) was utilized for the precursor solution. No other chemical or capping agent was used in the synthesis. After the reaction, partial detachment of the ITO coating was also observed on the cathode. The residual powder was characterized to investigate the synthesized material.

A digital optical microscope (OM) from Metzer (Model: Vision Plus 5000DTM) was used to capture images of the structures formed on ITO electrodes. Agilent Cary 5000 UV–Vis.-NIR spectrometer was used to record the absorption spectrum of the

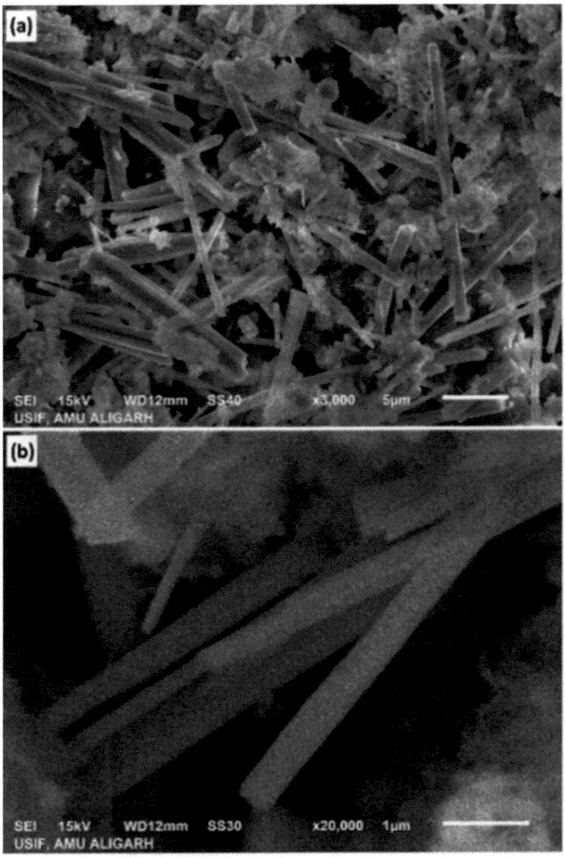

Fig. 6 SEM image of residual powder containing silver oxide nanorods and agglomerated $AgNO_3$ particles (**a**) and a magnified view of silver oxide nanorods (**b**)

materials. The residue powder from the specimen was dissolved in double-distilled water and taken in quartz cuvette for the purpose.

Presented in Fig. 6a, b are the SEM images of the residual powder containing silver oxide NRs and agglomerated $AgNO_3$ particles that did not take part in the electrochemical reaction. In the image (a) multiple NRs with no preferred growth direction are observed. Since in this case there was a deliberate attempt to detach NRs from the cathode, the density of NRs is much higher in comparison to that observed in the residual powder of barium chloride (with barium hydroxide NRs Fig. 4. In Fig. 6b, a magnified section of the specimen is shown. Multiple NRs with diameters in the range of 80–250 nm and lengths of 1200–4000 nm are observed in the image.

UV–Vis. absorption spectrum of the synthesized material is presented in Fig. 8. The lower (red) and the upper (black) curves are from the absorption spectrum of $AgNO_3$ (precursor) and the residual powder, respectively. The absorption spectrum of the residue powder has a broad absorption ramp that saturates beyond 400 nm. Haram et al. have reported UV–Vis spectrum of silver oxide nanoparticles with a broad peak around 400 nm, which is in reasonable agreement with our results. [38]

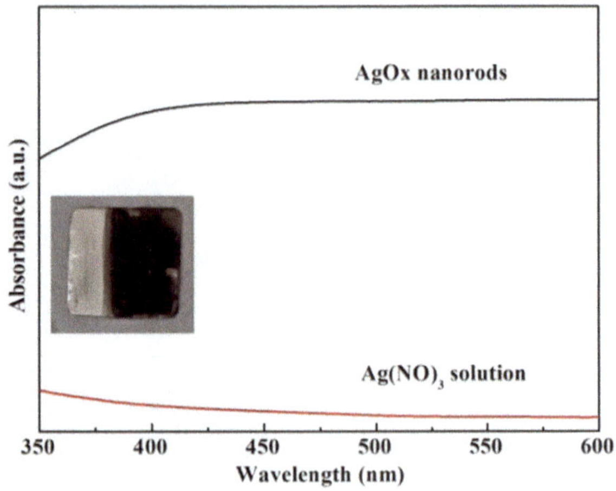

Fig. 7 UV–Vis. absorption spectrum of silver oxide nanorods (black) and silver nitrate solution (red). Inset: dark black colour of the coated material

It is also pertinent to point out that UV–Vis spectra of Ag nanoparticles display a much sharper characteristic peak around 410 nm. [39] Thus, the UV–Vis. spectrum also indicates the growth of silver oxide NRs. In the inset, an image of the coated material on another cathode is shown that was grown in a similar setup. The dark black colour of the material also indicates the presence of silver oxide in the residual powder.

4 Effects of Parameter Variations on the Aspect Ratio of NRs

In Fig. 8, optical micrographs of barium hydroxide nanostructures grown at different temperatures (10 °C, 25 °C, and 60 °C) are displayed. The voltage in this investigation was maintained at 5 V. At lower temperature (10 °C), a dense dendritic network of rods is observed with a relatively lower aspect ratio. At room temperature (25 °C) rods became isolated with a higher aspect ratio than that synthesized at the previous temperature. Further increase in the growth temperature (60 °C) resulted in an even higher aspect ratio of the fabricated rods. There are multiple rods in Fig. 8c that are twice the length of the reference bar of 10 μm in the image. It is straightforward to conclude from the investigation that an increase in the growth temperature would lead to longer rods, at least up to the temperature range investigated here. This observation is consistent with the articles reporting a lower surface sticking coefficient on the substrate with increased temperature [40, 41]. Lower sticking coefficient at higher temperature results in lower surface density of nucleation points [42] (per unit area on the glass/ITO substrate) on which nanorods eventually grow. The lower density of nucleation points leads to dispersed nanorods at higher temperatures. Furthermore,

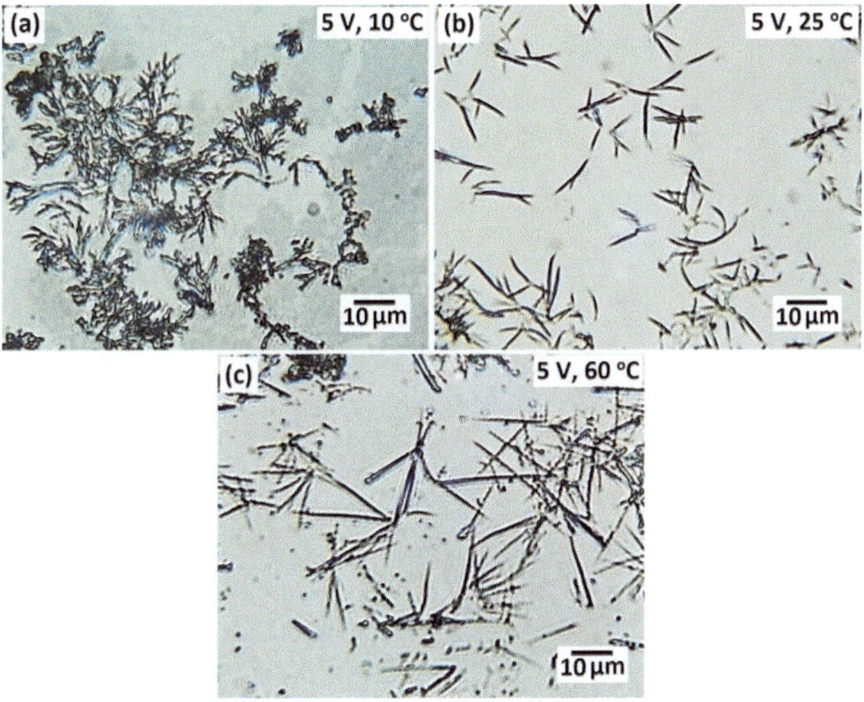

Fig. 8 Optical micrographs showing effects of temperature variation: **a** 10 °C, **b** 25 °C and **c** 60 °C

since the current density (or the influx of Ba^{+2} ions on the cathode) has a feeble dependence on the substrate temperature, a preferential/channelized deposition would take place, resulting in a higher aspect ratio of nanorods at higher temperatures.

The effect of voltage variation on the synthesized barium hydroxide nanostructures was also studied and is displayed in Fig. 9. The temperature was maintained at around 25 °C during the study and samples were investigated at 1, 2.5, 5, and 10 V. At 1 V, a network of small rods was formed. The network became denser at 2.5 V. At 5 V, the growth of isolated well-formed rod-like structures was visible. Further increase in the voltage resulted in a massive reduction in the aspect ratio. The study demonstrates the optimum voltage of around 5 V for the synthesis of NRs.

5 Conclusions

An efficient and cost-effective two-electrode electrochemical synthesis method is presented for the fabrication of self-stabilized metal-oxide and metal-hydroxide nanorods directly from the precursor solution at room temperature. The demonstrated method does not require any surfactant or any other chemical for the synthesis of

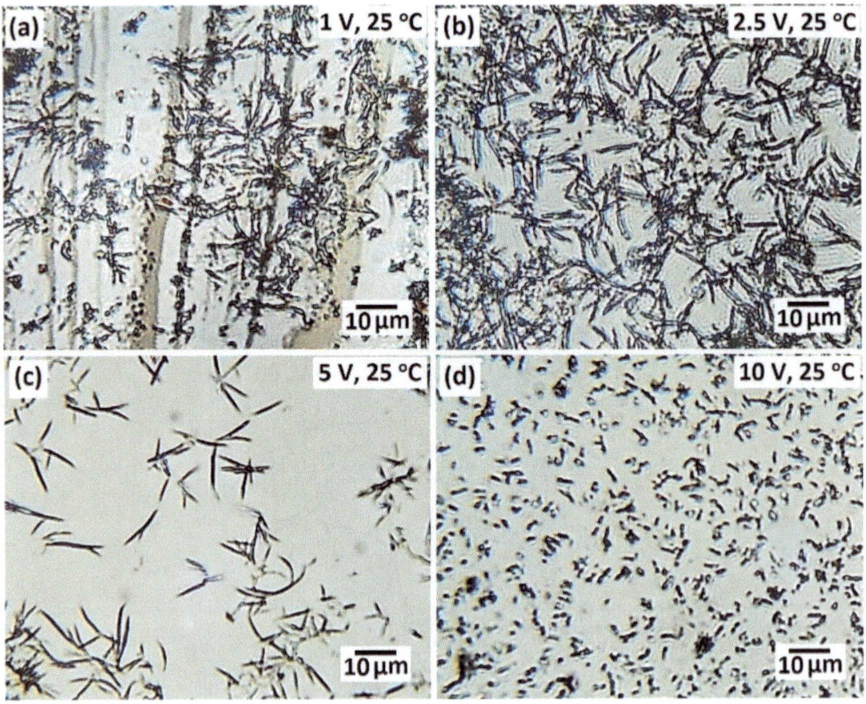

Fig. 9 Optical micrographs showing effects of voltage variation: **a** 1 V, **b** 2.5 V, **c** 5 V, and **d** 10 V

nanorods. Barium hydroxide trihydrate and silver oxide nanorods were successfully fabricated using this synthesis method. Based on SEM images, $Ba(OH)_2{\cdot}3H_2O$ NRs of diameters in the range of 100 to 300 nm and AgO_x NRs in the range of 80–250 nm were synthesized with different parameters. EDS, SAED, and XRD results confirmed the formation of barium hydroxide trihydrate. The effect of temperature variation revealed that a higher growth temperature favors a higher aspect ratio of the fabricated nanorods. The voltage variation study disclosed an optimum applied voltage of 5 V at 3 cm electrode separation. UV–Vis results indicate the formation of silver oxide NRs. The versatility of the method is demonstrated in the present work and the synthesis method can be utilized to synthesize other metal-oxide and metal-hydroxide nanorods.

Acknowledgements The authors gratefully acknowledge the support of INC departmental funds for TEM grids, ITO substrates, chemicals, etc. used in this work. The authors acknowledge critical comments from an expert electrochemist/editor in improving the manuscript. The authors acknowledge the help of staff members of USIF-AMU for SEM, EDS, TEM, and SAED measurements.

References

1. Arno MC, Inam M, Weems AC, Li Z, Binch ALA, Platt CI, Richardson SM, Hoyland JA, Dove AP, O'Reilly RK (2020) Exploiting the role of nanoparticle shape in enhancing hydrogel adhesive and mechanical properties. Nat Commun 11:1420
2. Hua Y, Chandra K, Dam DHM, Wiederrecht GP, Odom TW (2015) Shape-Dependent nonlinear optical properties of anisotropic gold nanoparticles. J Phys Chem Lett 6:4904–4908
3. Singh AK, Srivastava ON, Singh K (2017) Shape and Size-Dependent magnetic properties of Fe_3O_4 nanoparticles synthesized using piperidine. Nanoscale Res Lett 12:298
4. Essajai R, Benhouria Y, Rachadi A, Qjani M, Mzerda A, Hassanain N (2019) Shape-dependent structural and magnetic properties of Fe nanoparticles studied through simulation methods. RSC Adv 9:22057–22063
5. Huang MH, Mao S, Feick H, Yan H, Wu Y, Kind H, Weber E, Russo R, Yang P (2001) Room-temperature ultraviolet nanowire nanolasers. Science 292:1897–1899
6. Wang ZL, Song J (2006) Piezoelectric nanogenerators based on zinc oxide nanowire arrays. Science 312:242–246
7. Choi MY, Choi D, Jin MJ, Kim I, Kim SH, Choi JY, Lee SY, Kim JM, Kim SW (2009) Mechanically powered transparent flexible charge-generating nano devices with Piezoelectric ZnO Nanorods. Adv Materials 21:2185
8. Lee YJ, Ruby DS, Peters DW, McKenzie BB, Hsu JWP (2008) ZnO nanostructures as efficient antireflection layers in solar cells. Nano Lett 8:1501
9. Willander M, Nur O, Zhao QX, Yang LL, Lorenz M, Cao BQ, Perez JZ, Czekalla C, Zimmermann G, Grundmann M, Bakin A, Behrends A, Suleiman MA, Shaer AE, Mofor AC, Postels B, Waag A, Boukos N, Travlos A, Kwack HS, Guinard J, Dang DLS (2009) Zinc oxide nanorod based photonic devices: recent progress in growth, light emitting diodes and lasers. Nanotechnol 20:332001
10. Wang JX, Sun XW, Yang Y, Huang H, Lee YC, Tan OK, Vayssieres L (2006) Hydrothermally grown oriented ZnO nanorod arrays for gas sensing applications. Nanotechnol 17:4995
11. Jagiello K, Chomicz B, Avramopoulos A, Gajewicz A, Mikolajczyk A, Bonifassi P, Papadopoulos MG, Leszczynski J, Puzyn T (2017) Size-dependent electronic properties of nanomaterials. Struct Chem 28:635–643
12. Rodriguez-Sanchez L, Blanco MC, Lopez-Quintela MA (2000) Electrochemical synthesis of silver nanoparticles. J Phys Chem B 104:9683–9688
13. Therese GHA, Kamath V (2000) Electrochemical synthesis of metal Oxides and Hydroxides. Chem Mater 12:1195–1204
14. Morsy SMI (2014) Role of surfactants in nanotechnology and their applications. Int J Curr Microbiol App Sci 3:237–260
15. Niu Z, Li Y (2014) Removal and utilization of capping agents in nanocatalysis. Chem Mater 26:72–83
16. Burungale VV, Satale VV, Teli AM, Kamble AS, Kim JH, Patil PS (2016) Surfactant free single step synthesis of TiO_2 3-D microflowers by hydrothermal route and its photoelectrochemical characterizations. J Alloys Compd 656:491–499
17. Jeun YE, Baek B, Lee MW, Ahn HS (2018) Surfactant-free electrochemical synthesis of metallic nanoparticles via stochastic collisions of aqueous nanodroplet reactors. Chem Commun 54:10052–10055
18. Sinisterra JV, Fuentes A, Marinas JM (1987) $Ba(OH)_2$ as catalyst in organic reactions. 17. Interfacial solid-liquid wittig-horner reaction under sonochemical conditions, J Org Chem, 52, 3875–3879
19. Climent MS, Marinas JM, Mouloungui Z, Le Bigot Y, Delmas M, Gaset A, Sinisterra JV (1987) Ba(OH)2 as Catalyst in Organic Reactions. 20. Structure-Catalytic activity relationship in the wittig reaction, J Org Chem, 54, 3695–3701
20. Zhao W, Ye L, Zhang S, Yao H, Sun M, Hou J (2015) An easily accessible cathode buffer layer for achieving multiple high performance polymer photovoltaic cells. J Phys Chem C 119:27322–27329

21. Cui K, Liu L, Sun M (2017) Study on improving the heat storage property of Ba(OH)2·8H2O with paraffin. Mater Res Express 4:125502
22. Guisbiers G, Mejía-Rosales S, Deepak FL (2012) Nanomaterial properties: Size and shape dependencies. J Nanomater 2012:180976
23. Giorgi R, Ambrosi M, Toccafondi N, Baglioni P (2010) Nanoparticles for cultural heritage conservation: calcium and barium hydroxide nanoparticles for wall painting consolidation. Chem Eur J 16:9374–9382
24. Chelazzi D, Poggi G, Jaidar Y, Toccafondi N, Giorgi R, Baglioni P (2013) Hydroxide nanoparticles for cultural heritage: Consolidation and protection of wall paintings and carbonate materials. J Collidal Interface Sci 292:42–49
25. Saoud KM, Ibala I, El Ladki D, Ezzeldeen O, Saeed S (2014) Microwave assisted preparation of calcium hydroxide and barium hydroxide nanoparticles and their application for conservation of cultural heritage. In Proceedings Euro-Mediterranian Conference, 342–352
26. Kiazadeh A, Gomes HL, Rosa da Costa AM, Moreira JA, de Leuw M, Meskers SCJ (2012) Intrinsic and extrinsic resistive switching in a planar diode based on silver oxide nanoparticles. Thin Solid Films 522:407–411
27. Chen X, Guo Z, Xu WH, Yao HB, Li MQ, Liu JH (2011) Templating synthesis of SnO2 nanotubes loaded with Ag2O nanoparticles and their enhanced gas sensing properties. Adv Funct Mater 21:2049–2056
28. Agarwal RA, Gupta NK, Singh R, Nigam S, Ateeq B (2017) Ag/AgO nanoparticles grown via time dependent double mechanism in a 2D layered Ni-PCP and their antibacterial efficacy. Sci Rep 7:44852
29. Iqbal S, Fakhar-e-Alam M, Akbar F, Shafiq M, Atif M, Amin N, Ismail M, Hanif A, Farooq WA (2019) Application of silver oxide nanoparticles for the treatment of cancer. J Mol Struct 1189:203–209
30. Persson I, Sandström M, Yokoyama H (1995) Structure of the solvated strontium and barium ions in aqueous. Dimethyl Sulfoxide and Pyridine Solution, and Crystal Structure of Strontium and Barium Hydroxide Octahydrate, Zeitschrift für Naturforschung A 50:21–37
31. Irfan I, Graber S, So F, Gao Y (2012) Interplay of cleaning and de-doping in oxygen plasma treated high work function indium tin oxide (ITO). Org. Electronics 13:2028–2034
32. Kittel C (2010) Introduction to solid state physics, Chapter 1, 7th Edition, John Wiley and Sons, 2002
33. Manohar H, Ramashesan S (2010) The crystal structure of barium hydroxide octahydrate. Z Kristallogr 119:357–374
34. Vaufrey D, Khalifa MB, Tardy J, Ghica C, Blanchin MG, Sandu C, Roger JA (2003) ITO-on-top organic light-emitting devices: a correlated study of opto-electronic and structural characteristics. Semicond Sci Technol 18:253–260
35. Bochkova RI, Grishin IA, Kuzmin EA, Belov NV (1980) Refinement of the crystal structure of barium chloride dihydrate Ba Cl_2 2(H_2 O). Kristallografiya 25:1064–1065
36. Padmanabhan VM, Busing WR, Levy HA (1978) Barium chloride dihydrate by neutron diffraction. Acta Crystallographica B 34:2290–2292
37. Lutz HD, Kellersohn T (1990) Hydrogen bonding in barium hydroxide trihydrate by neutron diffraction. Acta Crystallogr C 46:361–363
38. Haram N, Ahmad N (2014) Formation of gold and silver nanochains and nanonetworks by liquid assisted laser ablation at elevated temperature, J Clust Sci, 3. 731
39. Salema MA, Bakra EA, El-Atta HG (2018) Pt@Ag and Pd@Ag core/shell nanoparticles for catalytic degradation ofCongo red in aqueous solution, Spectrochimica Acta A: Mol and Biomol Spectroscopy, 188, 155–163
40. Zhao J, Zeng Y, Liu C, Li Y (2010) Substrate temperature dependence of ZnTe epilayers grown on GaAs(0 0 1) by molecular beam epitaxy. J Crys Growth 312:1491
41. Lipponer MA, Armbrust N, Dürr M, Höfer U (2012) Adsorption dynamics of ethylene on Si(001). J Chemical Phys 136:144703
42. Kulkarni AM, Zukoski CF (2002) Nanoparticle crystal nucleation: influence of solution conditions. Langmuir 18:3090

Structure and Stability of Modern Electrolytes in Nanoscale Confinements from Molecular Dynamics Perspective

Harender S. Dhattarwal and Hemant K. Kashyap

Abstract Recent studies show that ionic liquids and high concentration salt solutions are promising alternatives to conventional electrolytes for high-performance batteries. The intercalation of electrolytes in nanoscale electrode confinements is a vital phenomenon governing the performance of batteries. A fundamental understanding of the electrolyte structure and stability inside electrode confinements helps explore the full potential of modern electrolytes for electrochemical devices. Factors such as the confinement shape, size, and flexibility govern the stability of electrolytes in nanoscale confinements. Enhanced molecular dynamics simulation can help delineate the free energy underlying the process of electrolyte evaporation or deintercalation from confinements. However, such studies in this direction are limited to few electrolytes only. This chapter highlights recent computational studies carried out in our group exploring the stability and structure of ionic liquids and water-in-salt electrolytes in nanoscale confinements, and provides a plausible mechanism for their intercalation and deintercalation behaviour.

Keywords Electrolytes · Intercalation · Water-in-salt · Ionic liquids · Umbrella sampling · Molecular dynamics

1 Introduction

The field of energy storage has been dominated by lithium ion batteries (LIBs) for the past few decades [1–3]. Electrolytes in conventional energy storage devices usually contain Li salt with volatile and flammable organic solvents, resulting in safety concerns while operating at higher temperatures and larger scale [1, 3–6]. The demand for clean and sustainable electrolytes is increasing for their use in energy

H. S. Dhattarwal · H. K. Kashyap (✉)
Department of Chemistry, Indian Institute of Technology Delhi, Hauz Khas, New Delhi 110016, India
e-mail: hkashyap@chemistry.iitd.ac.in
URL: https://web.iitd.ac.in/ hkashyap/

I. Uddin and I. Ahmad (eds.), *Synthesis and Applications of Nanomaterials and Nanocomposites*, Composites Science and Technology ,
https://doi.org/10.1007/978-981-99-1350-3_5

storage applications in electric vehicles and grid storage. Ionic liquids (ILs) and highlyconcentrated water-in-salt (WIS) electrolytes show great potential to replace conventional electrolytes in next-generation high-capacity batteries [7–9]. ILs are complex salts that are liquid below 100 °C and do not contain any neutral molecular species such as water or organic solvents [10]. ILs exhibit unique properties such as a wide electrochemical window, nonvolatility, high chemical and thermal stability, and high ionic conductivity, which make them suitable candidate for energy storage [11–20]. Several studies have demonstrated the capability of ILs to improve the performance of supercapacitors [21–26]. WIS electrolytes are another class of electrolytes that have emerged as novel electrolytes for high performance batteries. WIS electrolytes have a salt to solvent ratio greater than 1, i.e., they have more amount of salt than water in mass and volume [9]. WIS electrolytes belong to a wider range of "solvent-in-salt" electrolytes which are also termed as highly concentrated electrolytes [27]. These electrolytes consist of large amount of salts dissolved in different solvents [28–30]. Like ILs, WIS electrolytes also possess enhanced redox stability and a wide electrochemical window [9, 29, 31–34]. Highly concentrated or superconcentrated solutions have been frequently explored as electrolytes in batteries [27, 30, 32, 35–38].

In an electrochemical system, electron transfer occurs at the electrode-electrolyte interface. The distribution of electrolyte species at the interface governs the interfacial reactions occurring within the electrochemical devices. Understanding the structure and distribution of the electrolytes at the model electrode surfaces helps determine their capability to be used in batteries and capacitors [39]. Many studies on the nanodroplets of ILs on electrode surface have revealed that the structure of ionic liquids at the interface and the extent of wetting are dependent on the extent of hydrophobic or hydrophilic nature of the surface [40–48]. The structure of ILs has been studied extensively near different electrode surfaces at varying potentials [49–54]. Similar but limited number of studies are also available on the WIS electrolytes under confinement or on electrode surfaces [55–57]. However, most simulation studies explored the electrodes having smooth surfaces and only the van der Waal and electrostatic interactions between the electrode and electrolyte molecules have been considered. Although, the findings in the above mentioned endeavours were very insightful, an ideal electrode in energy storage devices possesses rough surfaces with pores of different shapes and sizes [58–60]. The actual performance of the device also depends on the wetting of these nanopores by the electrolyte molecules or ions. Additionally, in LIBs the process of charging and discharging involve intercalation and deintercalation of electrolyte species at the anode and cathode [3, 61]. The computational studies focusing on the structure of the electrolytes in the nanoscopic confinements and intercalation of electrolytes in electrodes are scarce in the literature. In this chapter, we made an effort to compile our recent studies on the structure and thermodynamics stability of confined ILs and WIS electrolytes using molecular simulations.

The interaction of a solvent/solution confined between sufficiently separated hydrophobic sheets induces the transition from liquid to vapor phase (or evaporation) inside the nanoscale confinement (capillary). The process of capillary evaporation of water from solvophobic confinements has been explored through number of theoretical and computer simulations studies [62–66]. Huang et al. showed that water can evaporate from a hydrophobic confinement when the separation between two confining plates is less than a critical separation d_c. Also, the vapor phase of water inside the confinement stays in vapor phase below d_c. However, near the d_c, the liquid and vapor phases of water are found to be in equilibrium [62]. Remsing et al. showed that the process of capillary evaporation of water from solvophobic confinement starts as vapor bubble nucleation on the confining surface. The vapor bubble then spans across the confinement to form a vapor tube, which further grows in size, leading to transition from liquid phase to vapor phase of water between the confining plates [64]. Altabet et al. further verified the vapor tube formation as a critical point in the process of water evaporation and showed that the process is significantly affected by the confinement flexibility [65]. Davoodabadi and Ghasemi have recently summarized the studies on the evaporation of water from nanoscale confinements [67]. The literature on the structure and thermodynamics of water inside different confinements is very extensive. However, number of such studies on ILs or WIS electrolytes is limited. Similar to capillary evaporation, the movement of electrolyte species in (intercalation) and out (deintercalation) of electrode vacant sites is also influenced by the electrode properties [61, 68–73]. The deintercalation of electrolytes from the electrode can be modelled by mimicking the process of capillary evaporation of water from solvophobic confinements [74–76].

In this chapter, we have reviewed the work done by our group on the stability of electrolytes in nanoscale carbon confinements. In the following sections, we highlight different key factors that govern the process of electrolyte intercalation/deintercalation and their stability inside solvophobic nanopores. Finally, the structure of electrolyte between the confining sheets is discussed. We limit the discussion in this chapter to the computer simulation studies of ionic liquids and water-in-salt electrolytes in nanoscopic graphene/graphite like hydrophobic confinements.

2 Stability of Electrolytes

Stability of electrolytes in confinement is usually discussed in terms of their phase behavior. When the electrolyte is stable inside the confinement, it usually tends to be in condensed phase, i.e., the density and structure of the electrolyte inside the confinement is comparable to that in the bulk [76]. However, when the electrolyte is unstable between the confining surfaces, it usually exists in vapor phase or does not prefer to stay in the confinement at all. In this case, the density of electrolyte within the confining region is significantly lower than the bulk density.

2.1 Process of Deintercalation

The process of change of electrolyte phase from its liquid to vapor phase inside a confinement is called deintercalation of the electrolyte, or sometimes termed as capillary evaporation [66, 76, 77]. The mechanism and thermodynamics of electrolyte intercalation-deintercalation have been studied by modelling electrolyte inside nanoscale confinement created by two parallel flat surfaces separated by a distance d [75, 76]. The interaction between the electrolyte and the confining sheets can be tuned by tweaking the parameters of the corresponding potential energy functions. The confining surfaces is considered solvophobic when the strength of the interaction, usually defined through the Lennard-Jones potential of the surface atoms, is weak and the contact angle formed by the electrolyte is more than 90° [48]. For example, the contact angles formed by 1-ethyl-3-methylimidazolium tetrafluoroborate ([EMIM][BF_4]) ionic liquid and Li bis(trifluoromethane)sulfonimide ([Li][TFSI]) WIS electrolyte on the model surfaces discussed here are around 140° and 130°, respectively [74, 76].

The electrolyte density fluctuations inside the confinement are probed using the indirect umbrella sampling (INDUS) technique [78, 79]. The free energy of electrolyte phase transition inside the confinement is estimated through probability ($P_v(N)$) of finding the N atoms or species of the electrolyte between the confining surfaces, separated by a distance d, with a confinement volume v. Shrivastav et al. used total number of heavy atoms of the ions of [EMIM][BF_4]as the order parameter of enhanced sampling whereas Dhattarwal et al. probed total number of atoms or species of [Li][TFSI] WIS electrolyte inside the confinement as the order parameter [74, 75]. A harmonic biasing potential is applied to maintain number of species inside the confinement to the target value N^* which typically varies in the range from vapor phase to liquid phase,

$$U(\widetilde{N}) = \frac{\kappa}{2}(\widetilde{N} - \widetilde{N}^*)^2. \tag{1}$$

Here, the biasing potential is applied on the continuous and coarse-grained variant of order parameter, $\widetilde{N}$ [78, 79]. κ is the spring force constant which guarantees the sampling of the phase space close to the target value of the order parameter. Figure 1 represents the simulation snapshots of a WIS electrolyte inside the nanoscale solvophobic sheets depicting a phase transition from vapor phase (VP) to liquid phase (LP). The free energy of phase transition is estimated from the probability distribution obtained from the enhanced sampling simulations using the weighted histogram analysis method (WHAM) as:

$$\Delta G(N) = -\beta^{-1} \ln P_v(N), \tag{2}$$

where $\Delta G(N)$ is free energy of the phase corresponding to N species inside the confinement, $\beta = 1/k_B T$, and T and k_B are the temperature and the Boltzmann's

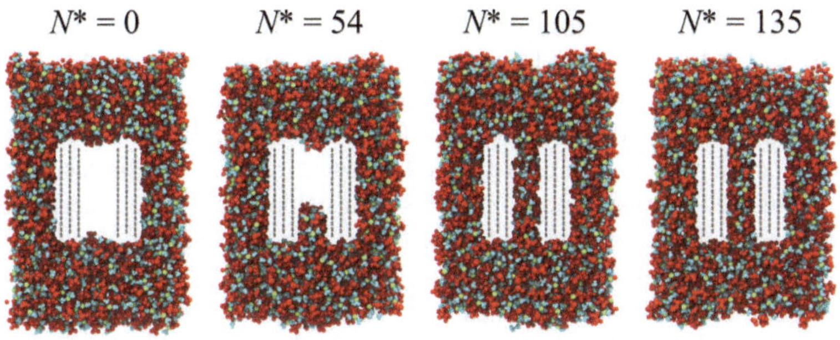

Fig. 1 Simulation snapshots of [Li][TFSI] based WIS electrolyte inside nanoscale solvophobic confinements for different number of target species between the confining sheets illustrating the transition between vapor and liquid phases of the electrolyte. The confining solvophobic carbon sheets are separated by 1.5 nm. Green, red, cyan, and grey beads represent $[Li]^+$, $[TFSI]^-$, water molecules, and carbon atoms of the sheets, respectively

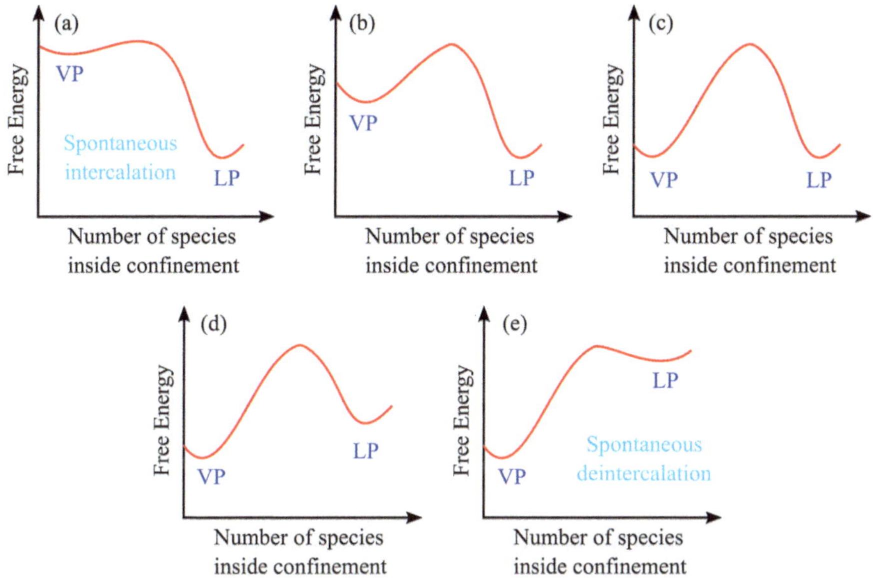

Fig. 2 Schematic illustration of different possible free energy profiles and stability of liquid/condensed (LP) and vapor phase (VP) of an electrolyte inside nanoscale confinements

constant, respectively [80]. The studies reviewed in this chapter discuss the simulations performed at 310 K temperature.

The relative stability of liquid and vapor phases, and the process of electrolyte deintercalation from a nanoscale solvophobic or ionophobic confinement can be described in five different scenarios summarized through the illustration shown in Fig. 2.

(a) In the first case, the free energy of LP is significantly lower than the VP, and there is no well-defined VP (see Fig. 2a). In this case, only the LP is stable inside the confinement and intercalation of electrolyte into the confinement is spontaneous.
(b) In the second case, the free energy of LP is lower than that of VP and both the phases are separated by a appreciable free energy barrier. In this case, the LP is stable inside the confinement and VP is metastable (see Fig. 2b). Here, the LP is more favourable inside the confinement, however, VP requires some amount of energy in order to make transition to LP or vice versa.
(c) In the third case, both VP and LP have same free energy but are separated by a significant free energy barrier (see Fig. 2c). Here, both the phases are stable inside the confinement and phase transition between them is energy intensive in both directions.
(d) In the fourth case, free energy of VP is lower than the LP and a free energy barrier separates both the phases. Here, the VP is more stable inside the confinement than the LP, which is metastable (see Fig. 2d). However, the free energy barrier needs to be crossed for the transition from LP to VP or vice versa.
(e) In the fifth case, the free energy of VP is significantly lower than the LP and there is no free energy barrier for LP to VP transition (see Fig. 2e). Here the process of deintercalation of the liquid or electrolyte is spontaneous, and their constituent species do not stay in the confinement.

There are multiple factors that govern the relative stability of vapor and liquid phases of electrolytes inside the confinement. In the following sections, we have discussed recent studies exploring the effect of these factors on the process of intercalation and deintercalation of electrolyte species.

2.2 *Dependence on the Surface Area and Separation of Confining Sheets*

Along with the surface area, the separation between the nanoscale sheets confining the electrolyte is one of the major factors that governs the relative stability of liquid and vapor states of electrolytes inside the confinement [74–76].

Figure 3 shows the simulated free energy plotted as a function of number (N) of [EMIM][BF_4] heavy atoms inside the solvophobic carbon confinements at different interplate separations [74]. Here, all the free energy profiles are shifted such that the free energy of the LP of [EMIM][BF_4] is at zero. Basins or minima with smaller N values on the x-axis correspond to VP of [EMIM][BF_4] inside the confinement. Similarly, the basins formed at larger N values correspond to the LP of [EMIM][BF_4]. The free energy profiles have a parabolic distribution at the LP basins, which deviates to a fat tail as the N value decreases (see inset of Fig. 3). At the larger d values, e.g. 1.6 nm, there is no well-defined basin for the VP, and the LP is the only stable state of [EMIM][BF_4] inside the confinement. As the inter-plate separation is decreased

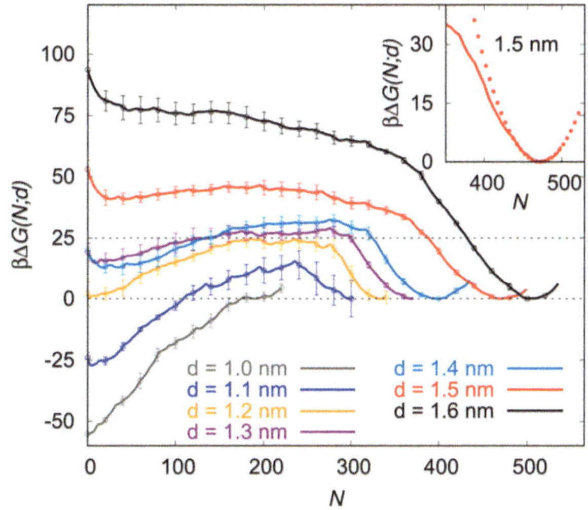

Fig. 3 Simulated free energy profiles showing variation in the free energy ($\beta \Delta G(N; d)$) for $[EMIM][BF_4]$ as a function of number of heavy atoms inside the solvophobic confinements at different separations. Inset highlights that in the region corresponding to the condensed or liquid phase of $[EMIM][BF_4]$ (higher N values), the free energy profiles follow a inverted Gaussian or parabolic distribution, and deviations from the inverted Gaussian distribution are observed while moving towards the vapor phase region. Reprinted with permission from Ref. [74], Copyright 2018 AIP

to 1.5 nm, a minimum corresponding to the VP starts to emerges in the smaller N region of the free energy profile.

However, the free energy of this basin is still more than the LP basin, and the VP is metastable. With further decrease in the inter-plate separation, the free energy of vapor basin decreases. When the inter-plate separation is 1.2 nm, the free energy of vapor and liquid phases is same, reflecting that both the states are stable at this separation. However, the free energy barrier separating the two states is significantly high. The separation at which the LP and VP are having same free energy is called as critical separation, d_c [62]. At this separation, the free energy cost of phase transition between vapor and liquid phases is very high. Moreover, below the critical separation, the VP tends to be more stable and LP become metastable. If the inter-plate separation is further decreased, the energy barrier between the liquid and vapor phases significantly decreases and the free energy cost for the liquid to vapor phase transition decreases. Below or at 1 nm separation between the confining surfaces, the free energy barrier vanishes, and the transition from LP to VP becomes a spontaneous process.

The dependence of stability of WIS electrolyte on the separation between confinement sheets is similar to ILs and water [76]. Figure 4 shows the simulated free energy plotted as a function of number of WIS species inside the solvophobic carbon confinements at different interplate separations. The figure reveals that the electrolyte is

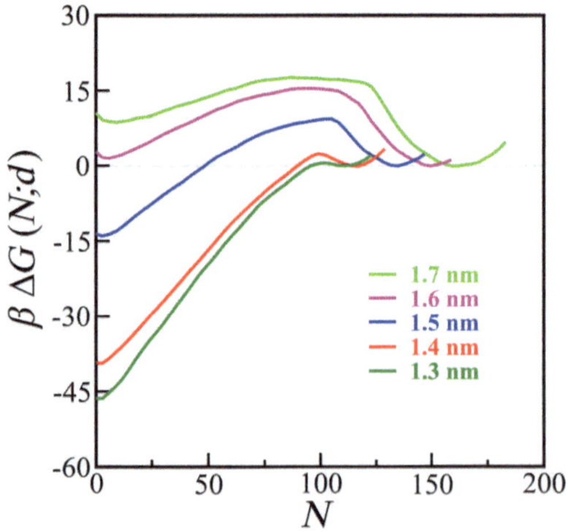

Fig. 4 Simulated free energy profiles ($\beta\Delta G(N;\,d)$) for [Li][TFSI]/water WIS electrolyte as a function of number of electrolyte species inside the solvophobic confinements at different separations. The height of the barrier separating the liquid and vapor phases of the WIS electrolyte within the confinement decreases with decrease in the separation between the confining surfaces. Reprinted with permission from Ref. [76], Copyright 2021 RSC

more stable inside nanoscale confinements at higher interplate separations, and at significantly lower separation, deintercalation of electrolyte is spontaneous. However, the critical separation, d_c, at which the liquid and vapor phases are equally stable is at around 1.6 nm, as compared to 1.2 nm in case of [EMIM][BF_4][74]. Similarly, the spontaneous evaporation of electrolyte from the confinement is observed at around 1.3 nm, which is significantly higher than 1 nm observed in case of [EMIM][BF_4]. These observations show that although with an increase in the confinement separation, the height of the free energy barrier separating the liquid and vapor phases increases for different electrolytes and water, the extent by which the barrier height increases depends on the type of electrolyte.

2.3 *Dependence on the Confinement Flexibility*

Most simulation studies on confined electrolytes have considered rigid sheets to model the confinement, by freezing the motion of their atoms. However, significant effect of confinement flexibility has been shown on the stability of VP and LP of water as well as ILs between solvophobic sheets [65, 75]. Figure 5 shows the simulated free energy profiles as function of number of [EMIM][BF_4] species inside rigid and flexible solvophobic carbon confinements at three different interplate separations. At all the three separations, introduction of flexibility resulted in slight broadening of the liquid basin. The effect of confinement flexibility is more prominent on the

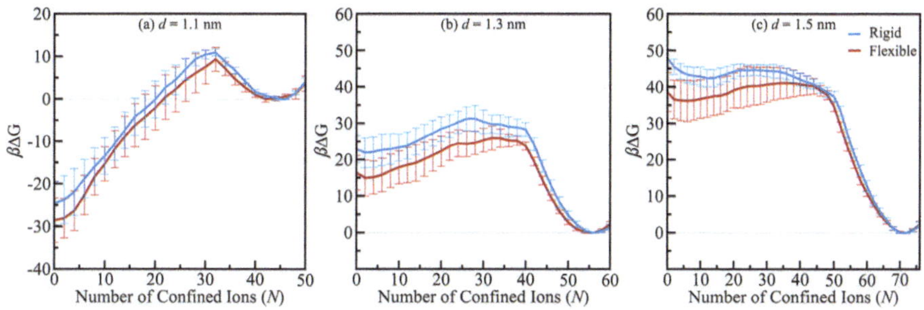

Fig. 5 Comparison between the simulated free energy profiles ($\beta\Delta G(N; d)$) for [EMIM][BF_4] as a function of number of ions inside the rigid (blue) and flexible (red) solvophobic confinements at three different separations. At all the three interplate separations studied, the free energy of [EMIM][BF_4] deintercalation from the flexible confinements is lower than the rigid confinements. Reprinted with permission from Ref. [75], Copyright 2020 ACS

vapor basin. The stability of vapor phase of [EMIM][BF_4] is significantly increased between the flexible solvophobic carbon sheets. Theoretical predictions attribute this enhanced stability to decrease in effective volume inside the confinement at lower N values [81]. When the electrolyte is in VP, a pressure difference builds up at the confinement sheets due to low density of electrolyte inside the confinement and high density of electrolyte outside the confinement. This pressure difference along with the interfacial tension in the flexible sheets lead to surface deformations, resulting in lowered confining volume [75, 81]. However, the effect of the pressure difference on the sheets is negligibly small when the electrolyte inside the confinement is in liquid state, hence, the stability of liquid state of electrolyte is not affected much by the confinement flexibility.

2.4 Dependence on the Electrolyte Concentration

Along with confinement characteristics like separation, size, and flexibility, salt concentration also governs the deintercalation process in WIS electrolytes. Figure 6 shows the free energy profiles for the WIS electrolyte as a function of number of species inside the confinement at three different concentrations for a fixed interplate separation of 1.7 nm. For 20 m (molal) salt solution, the LP of the electrolyte is more stable and VP is metastable inside the confinement. However, when the concentration is lowered to 10 m, the VP becomes more stable and LP is metastable. Just like critical separation, there is a critical concentration at which the liquid and vapor phases have similar free energies. For [Li][TFSI]-water system, the critical concentration lies between 10–20 m. With further lowering the concentration to 5 m, the VP is even more stable and the free energy barrier of liquid to vapor transition is lowered. It is expected that at even lower concentrations of the salt, the free energy barrier would further decrease and the process of electrolyte deintercalation will be spontaneous.

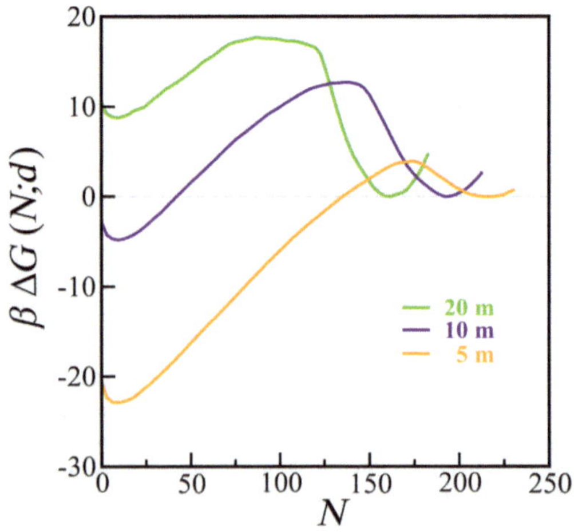

Fig. 6 Simulated free energy profiles ($\beta\Delta G(N; d)$) for three different concentration of [Li][TFSI]-water WIS electrolyte as a function of number of electrolyte species inside the solvophobic confinements at an interplate separation of 1.7 nm. Lowering the concentration of [Li][TFSI] in water resulted in significant increase in the stability of VP of electrolyte inside the confinement. Reprinted with permission from Ref. [76], Copyright 2021 RSC

3 Mechanism of Electrolyte Deintercalation

The mechanism of capillary evaporation or deintercalation of water from nanoscale solvophobic confinements involves the formation of a vapor bubble on the surface of the confinement that further grows to complete removal of water from the confinement [64, 65]. Similarly, the mechanism of deintercalation of WIS electrolyte was predicted by studying the growth of vapor bubble inside solvophobic confinement. Figure 7 shows the vapor bubble formed between the carbon surfaces (d = 1.5 nm), represented by the interface formed by the liquid (or solution) and vapor phases. When the electrolyte inside the confinement is in LP, there is no vapor bubble between the surfaces. As the number of species inside the confinement decreases, the vapor bubbles begin to appear on the opposite sides of the confinement. At an interplate separation of 1.5 nm, this process of vapor bubble formation is energy intensive. Decreasing the number of species inside the confinement requires further energy cost which leads to a gap-spanning vapor bubble formation across the confining sheets. The vapor bubble then further grows in size to eventually cover up the whole space between the surfaces, and the electrolyte is completely deintercalated from the confinement.

The process of vapor bubble formation starting from the LP follows a linear equation and growth of vapor tube can be described using macroscopic theory, [64, 76] i.e., $\Delta G_{\rm p}(r; d) = \Delta G_{\rm th}(r; d) - 2k_{\rm B}T \ln(1 - 2r/L)$, where r is the radius of vapor tube forming inside the confinement, $L = 3$ nm is the side length of the confining surface, and the translational entropy of the tube is represented by the logarithmic

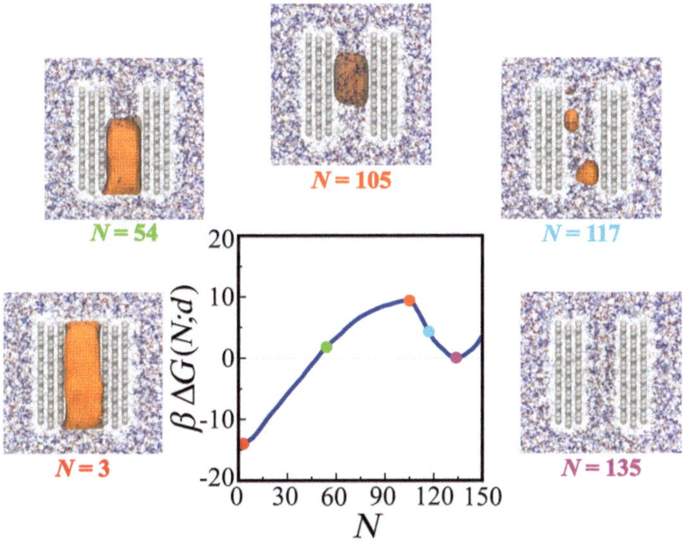

Fig. 7 Illustration of liquid-vapor interface formed inside the solvophobic carbon confinements rendered at different points along the free energy profile. Here, the concentration of the salt in the WIS electrolyte is 20 m and the separation between confining sheets is 1.5 nm. $N = 135$ corresponds to LP of WIS electrolyte, at $N = 117$ vapor bubbles begin to form, at $N = 105$ the bubbles join to form gap-spanning vapor tube, and $N = 54$ to $N = 3$ represent the growth of vapor tube inside the confinement. Reprinted with permission from Ref. [76], Copyright 2021 RSC

term. Also,

$$\Delta G_{\text{th}}(r; d) = 2\pi \gamma_{vl} \left[r^2 \cos\theta + r \left(\tilde{d} + \frac{2\lambda}{\gamma_{vl}} \right) \right], \tag{3}$$

where γ_{vl} and λ are the surface and line tensions obtained from the fitting of free energy profiles, respectively. Figure 8 shows the simulated free energy profiles for the process of electrolyte deintercalation from the confinement of nanoscale sheets at three different separations and two different concentrations of electrolyte. The free energy corresponding to the process of vapor bubble formation is fitted using linear equation and the growth of gap-spanning vapor tube is fitted using macroscopic theory [76]. For WIS system at smaller d value (1.4 nm), single free energy fitting is sufficient to describe the growth of vapor tube inside the confinement (see Fig. 8a). At this separation, the vapor tube once formed between the confining sheets is already of supercritical size and requires no further energy for complete deintercalation of electrolyte. Remsing et al. reported a similar kind of supercritical point in the capillary evaporation of water from solvophobic nanoscale carbon confinements [64]. However, at larger interplate separations ($d = 1.6$ and 1.7 nm) the free energy of vapor tube growth is best described by two separate fits. At larger d values, the vapor tube initially formed between the sheets is of subcritical size and requires more energy to reach the supercritical point after which the deintercalation of electrolyte

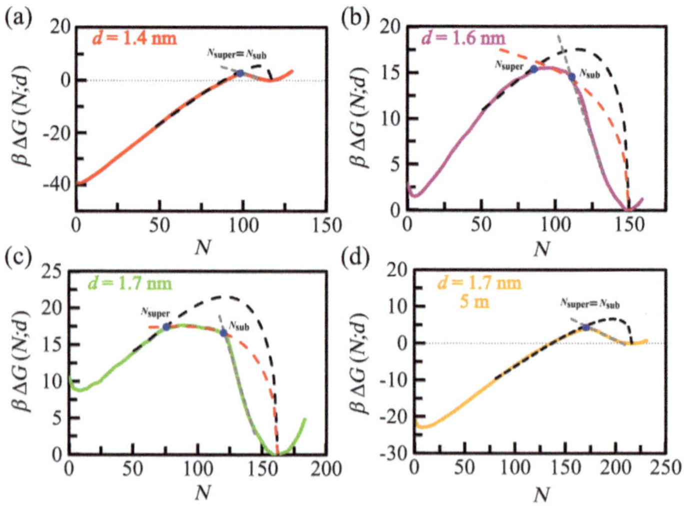

Fig. 8 The free energy profiles ($\beta\Delta G(N; d)$) for (**a–c**) 20 m and (**d**) 5 m concentration of WIS electrolyte inside carbon confinements as function of number of electrolyte species between the confining sheets. These profiles are fitted for different processes of electrolyte deintercalation using macroscopic theory as per Eq. 3. Grey dashed lines represent the formation of vapor bubble inside the confinement, red dashed lines represent the growth of subcritical vapor tube to its supercritical size, and black dashed lines represent the growth of supercritical vapor tube to vapor phase of the electrolyte. Reprinted with permission from Ref. [76], Copyright 2021 RSC

is spontaneous. Based on this analysis, Dhattarwal et al. described the electrolyte deintercalation from solvophobic confinements as a three step process [76].

1. The deintercalation is initiated by the formation of vapor bubbles on the opposite walls of the confinement surface. These vapor bubbles then join to form a gap-spanning vapor tube.
2. At the larger separations, the vapor tube formed is of subcritical size and grows further in size to reach at the supercritical size via an energy intensive process.
3. Once the vapor tube attain supercritical size, its size spontaneously increases, resulting in complete deintercalation of the electrolyte from the confinement.

4 Structure of Electrolytes Within the Confinement

Recent studies have highlighted the importance of local density fluctuations in governing the stability of water and electrolytes inside the nanoscale hydrophobic confinements [65, 75, 76]. The confining surface area and intersurface separation have been found to influence the distribution of electrolyte species within the confinement. The structural features of electrolytes between the confining sheets are usually stud-

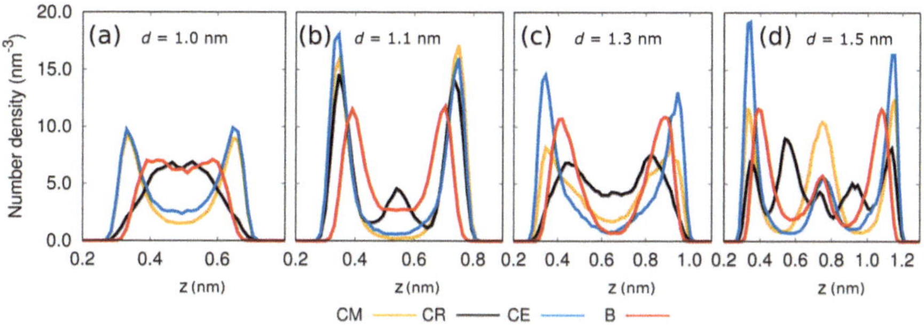

Fig. 9 Number density profiles corresponding to the terminal methyl and ethyl carbons (CM and CE) and ring carbon (CR) of $[EMIM]^+$ and boron atom (B) of $[BF_4]^-$. The profiles show highly structured and layered distribution of IL species inside the microscopic confinements. Reprinted with permission from Ref. [74], Copyright 2018 AIP

ied through number density distributions of the constituent species corresponding to the LP of confined electrolyte [74, 82].

Figure 9 depicts the number density distribution of terminal (CM and CE) and ring carbon (CR) of $[EMIM]^+$ cation and the boron atom (B) of $[BF_4]^-$ anion between two solvophobic carbon sheets at different separations [74]. Figure 9 shows the [EMIM][BF_4] species inside the confinement are highly structured. At small separation (d = 1.0 nm), the non-overlapping peaks corresponding to ring and terminal carbons suggest that the cations inside the confinement are distributed in a single layer and are perpendicular to the surface of the confining sheets. Note that the end-to-end expansion of $[EMIM]^+$ and $[BF_4]^-$ along the longest axis is around 0.75 nm and 0.19 nm, respectively. At interplate separation of 1.1 nm, the peaks for the terminal carbons (methyl and ethyl ends of $[EMIM]^+$) and ring carbon overlap near the confinement surface, suggesting two parallel layers of the ions. However, few cations also attain perpendicular orientation as reflected by the central peak for CR. At d = 1.3 nm, again two layers of the ions are present, however, their orientation is not parallel to the surface of the sheets. For intersurface separation of 1.5 nm, there are five peaks corresponding to the cation ring carbon. Three of these peaks overlap with the terminal carbon near the surface and at the center of the confinement, reflecting three layer distribution of the cation in parallel orientation. Some cations also attain orientations perpendicular to the confining surfaces between the parallel layers. Dhattarwal et al. also reported the layered distribution of [EMIM][BF_4] inside the nanoscale confinement and confirmed the orientational preferences of ions at different separations using orientational order parameter.

Similar to [EMIM][BF_4], highly structured distribution of electrolyte species inside nanoscale solvophobic confinement was observed in case of WIS solutions [75]. Figure 10 depicts the distribution of species of [Li][TFSI]-water electrolyte within the solvophobic confinement. Note that the end-to-end expansion of $[TFSI]^-$ along the longest axis is around 0.76 nm. At lower intersurface separation (d = 1.3 nm), the [TFSI]- anions prefer to align perpendicular to the confinement surface

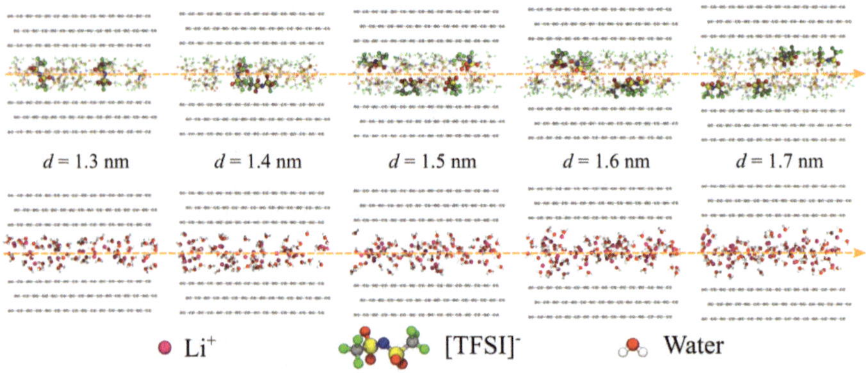

Fig. 10 Simulation snapshots representing the distribution of WIS electrolyte species between the solvophobic carbon sheets. The upper row shows the distribution of $[Li]^+$ and $[TFSI]^-$ ions, whereas, the lower row represents the distribution of water molecules inside the confinement. Magenta, blue, green, red, yellow, and white beads represent the lithium, nitrogen, fluorine, oxygen, sulphur, and hydrogen atoms, respectively. Reprinted with permission from Ref. [76], Copyright 2021 RSC

and water molecules are distributed throughout the confinement. As the separation between the sheets increases, some of the anions prefer to align parallel to the surface. Even at higher separations, most anions prefer parallel orientation and the separation between two parallel layers of anions increases with increase in interplate separation. The additional space results in more water molecules inside the confinement at higher separations, which is also reflected in the number density profiles. This also results in lower anion to water ratio between the confinement at higher interplate separations. It has also been shown that at the supercritical point, the height of the free energy barrier separating liquid and vapor phases of the electrolyte is directly correlated with the amount of salt ion pairs present within the confinement when the electrolyte is in the liquid phase [76].

5 Structural Changes During Electrolyte Deintercalation

In the previous section, we observed that the electrolytes within the nanoscale confinements are highly structured in their liquid phases. In this section, we shall examine the changes in the structure and orientation of electrolyte species during the process of electrolyte deintercalation (capillary evaporation). Figure 11 shows the joint density distribution ($\rho(z, \theta)$), which represents the number density of $[EMIM]^+$ ring vectors between the rigid and flexible confining sheets along z direction forming an angle θ with the axis perpendicular to confinement surfaces. When $[EMIM][BF_4]$ inside the confinement is in LP ($N^* = 56$), the cations prefer to align parallely near the surface ($\theta = 90°$) and perpendicularly at the center of the confinement ($\theta = 180°$). The plots

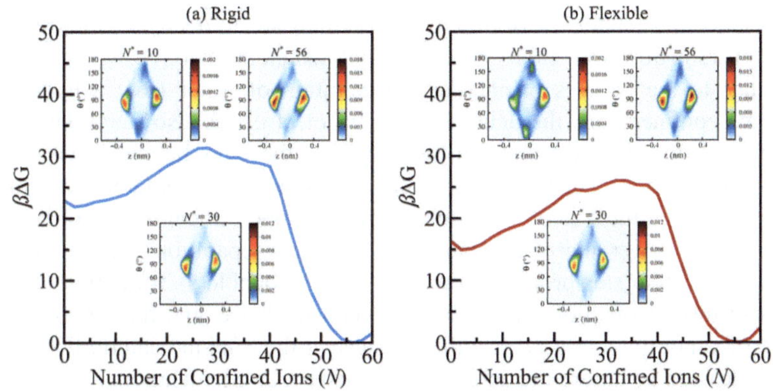

Fig. 11 The joint density distribution, $\rho(z, \theta)$, plotted along the free energy profile for $[EMIM][BF_4]$ representing the changes in orientation of $[EMIM]^+$ ions during the deintercalation (high to low N values) of electrolyte from the solvophobic confinement. Reprinted with permission from Ref. [75], Copyright 2020 ACS

show that for flexible sheets the number density of perpendicular cations at the center of the confinement are relatively higher. As the number of $[EMIM][BF_4]$ species inside the confinement is decreased ($N^* = 30$), the enhanced preference of perpendicular cations between flexible sheets facilitates the formation of vapor tube at the confinement surface. The difference in the preference of perpendicular orientation of cations is more prominent when the $[EMIM][BF_4]$ reaches towards the vapor region ($N^* = 10$). Here, the ease of vapor tube formation between flexible sheets results in lower energy barrier and relatively more stable vapor phase of $[EMIM][BF_4]$.

6 Summary and Future Directions

This chapter was focused on highlighting the recent advances in computer simulation studies on understanding the effect of confinement size, separation, flexibility, and electrolyte concentration on the process of intercalation-deintercalation of ions of neoteric electrolytes, such as ILs and WIS electrolytes, through nanoscopic confinements. These studies have clearly demonstrated that the stability of VP and LP of an IL or WIS electrolyte inside a hydrophobic confinement depends on the distance between the confining sheets. In the case of WIS electrolyte, stabilization of VP and LP also depends on the concentration of the salt. In LP, the structure of the electrolytes inside the confinement can be distinct from the bulk depending the interplate separation. It has also been demonstrated that above a critical separation between the sheets, LP is more stable for a given concentration of electrolyte. Similarly, for a given separation between the confining sheets, there exists a critical concentration of electrolyte above which the LP of the electrolyte is stable inside the confinement.

Additionally, introducing flexibility to the confining sheets decreases the free energy barrier of transition from LP to VP of the confined electrolyte.

Similar to the process of capillary evaporation of water from a hydrophobic nanoscale confinements, the deintercalation of electrolytes also starts with a vapor bubble formation at the confinement surface. Then the vapor bubble grows to form a gap-spanning vapor tube, which eventually covers the whole confinement. However, contrary to water, the size of vapor tube formed is supercritical in case of WIS electrolytes only at lower interplate separations. At higher separations, deintercalation of electrolyte is a four step process; formation of vapor bubble(s), formation of subcritical vapor tube, growth of vapor tube to supercritical size, and further spontaneous growth of vapor tube. It is evident that because of their unique nature, ILs and WIS electrolytes can be tailored to achieve higher extent of intercalation.

The distribution of ILs and WIS electrolytes within the microscopic confinements were found to be highly structured in their liquid phases. The structure of electrolyte between the confining sheets plays an important role in governing the process of electrolyte intercalation and deintercalation. Preferential deintercalation (dewetting) of ILs from flexible surfaces makes the process of electrolyte deintercalation more feasible from flexible confinements.

Although we have discussed some detailed studies on the stability of modern electrolytes inside nanoscale confinements but there are few important aspects that still remain unaddressed. Work needs to be done with a wider range of electrolytes along with utilization of confinement materials other than graphene or graphite. Effect of surface charge or potential on the stability of electrolytes inside microscopic confinements is also not fully understood. We believe that future studies in this direction would provide valuable insights in the field.

Acknowledgements HSD thanks UGC India for fellowship. This work was financially supported by Science and Engineering Research Board (SERB), Department of Science and Technology, India (Grant No. CRG/2022/007119); and Council of Scientific and Industrial Research (CSIR), Ministry of Science and Technology, India (Project No. 01 (3039)/21/EMR-II) awarded to HKK.

References

1. Winter M, Barnett B, Xu K (2018) Before li ion batteries. Chem Rev 118:11433–11456
2. Kim T, Song W, Son D-Y, Ono LK, Qi Y (2019) Lithium-ion batteries: outlook on present, future, and hybridized technologies. J Mater Chem A 7:2942–2964
3. Goodenough JB, Park K-S (2013) The li-ion rechargeable battery: a perspective. J Am Chem Soc 135:1167–1176
4. Goodenough JB, Kim Y (2010) Challenges for rechargeable li batteries. Chem Mater 22:587–603
5. Peled E, Menkin S (2017) Review-sei: past, present and future. J Electrochem Soc 164:A1703
6. Armand M, Tarascon J-M (2008) Building better batteries. Nature 451:652–657
7. Watanabe M, Thomas ML, Zhang S, Ueno K, Yasuda T, Dokko K (2017) Application of ionic liquids to energy storage and conversion materials and devices. Chem Rev 117:7190–7239
8. Matsumoto H (2014) Review: ionic liquids as a potential electrolyte for energy devices. In: 2014 IEEE International Nanoelectronics Conference (INEC), pp 1–4

9. Suo L, Borodin O, Gao T, Olguin M, Ho J, Fan X, Luo C, Wang C, Xu K (2015) "water-in-salt" electrolyte enables high-voltage aqueous lithium-ion chemistries. Science 350:938
10. Welton T (1999) Room-temperature ionic liquids solvents for synthesis and catalysis. Chem Rev 99:2071–2084
11. Lei Z, Chen B, Koo Y-M, MacFarlane DR (2017) Introduction: ionic liquids. Chem Rev 117(10):6633–6635
12. Seddon KR (2003) A taste of the future. Nat Mater 2:363
13. Davis JH Jr (2004) Task-specific ionic liquids. Chem Lett 33:1072–1077
14. Simon P, Gogotsi Y (2008) Materials for electrochemical capacitors. Nat Mater 7:845
15. Armand M, Endres F, MacFarlane DR, Ohno H, Scrosati B (2009) Ionic-liquid materials for the electrochemical challenges of the future. Nat Mater 8:621
16. Lu W, Fadeev AG, Qi B, Smela E, Mattes BR, Ding J, Spinks GM, Mazurkiewicz J, Zhou D, Wallace GG, MacFarlane DR, Forsyth SA, Forsyth M (2002) Use of ionic liquids for pi-conjugated polymer electrochemical devices. Science 297:983
17. MacFarlane DR, Forsyth M, Howlett PC, Pringle JM, Sun J, Annat G, Neil W, Izgorodina EI (2007) Ionic liquids in electrochemical devices and processes: managing interfacial electrochemistry. Acc Chem Res 40:1165–1173
18. Smiglak M, Metlen A, Rogers RD (2007) The second evolution of ionic liquids: From solvents and separations to advanced materials-energetic examples from the ionic liquid cookbook. Acc Chem Res 40:1182–1192
19. Greaves TL, Kennedy DF, Mudie ST, Drummond CJ (2010) Diversity observed in the nanostructure of protic ionic liquids. J Phys Chem B 114:10022–10031
20. Zein El Abedin S, Endres F (2007) Ionic liquids: The link to high-temperature molten salts? Acc Chem Res 40:1106–1113
21. Balducci A, Dugas R, Taberna PL, Simon P, Plée D, Mastragostino M, Passerini S (2007) High temperature carbon-carbon supercapacitor using ionic liquid as electrolyte. J Power Sources 165:922–927
22. Kim TY, Lee HW, Stoller M, Dreyer DR, Bielawski CW, Ruoff RS, Suh KS (2011) High-performance supercapacitors based on poly(ionic liquid)-modified graphene electrodes. ACS Nano 5:436–442
23. Lin R, Taberna P-L, Fantini S, Presser V, Pérez CR, Malbosc F, Rupesinghe NL, Teo KBK, Gogotsi Y, Simon P (2011) Capacitive energy storage from −50 to 100°c using an ionic liquid electrolyte. J Phys Chem Lett 2:2396–2401
24. Choi BG, Yang M, Jung SC, Lee KG, Kim J-G, Park H, Park TJ, Lee SB, Han Y-K, Huh YS (2013) Enhanced pseudocapacitance of ionic liquid/cobalt hydroxide nanohybrids. ACS Nano 7:2453–2460
25. Brandt A, Pohlmann S, Varzi A, Balducci A, Passerini S (2013) Ionic liquids in supercapacitors. MRS Bull 38:554–559
26. Wang H, Xu Z, Kohandehghan A, Li Z, Cui K, Tan X, Stephenson TJ, King'ondu CK, Holt CMB, Olsen BC, Tak JK, Harfield D, Anyia AO, Mitlin D (2013) Interconnected carbon nanosheets derived from hemp for ultrafast supercapacitors with high energy. ACS Nano 7:5131–5141
27. Suo L, Hu Y-S, Li H, Armand M, Chen L (2013) A new class of solvent-in-salt electrolyte for high-energy rechargeable metallic lithium batteries. Nature Commun 4:1481
28. Wang J, Yamada Y, Sodeyama K, Chiang CH, Tateyama Y, Yamada A (2016) Superconcentrated electrolytes for a high-voltage lithium-ion battery. Nature Commun 7:12032
29. Yamada Y, Usui K, Sodeyama K, Ko S, Tateyama Y, Yamada A (2016) Hydrate-melt electrolytes for high-energy-density aqueous batteries. Nat Energy 1:16129
30. Yamada Y, Yaegashi M, Abe T, Yamada A (2013) A superconcentrated ether electrolyte for fast-charging li-ion batteries. Chem Commun 49:11194–11196
31. Suo L, Borodin O, Sun W, Fan X, Yang C, Wang F, Gao T, Ma Z, Schroeder M, von Cresce A, Russell SM, Armand M, Angell A, Xu K, Wang C (2016) Advanced high-voltage aqueous lithium-ion battery enabled by "water-in-bisalt" electrolyte. Angew Chem Int Ed 55:7136–7141

32. Yamada Y, Yamada A (2015) Review-superconcentrated electrolytes for lithium batteries. J Electrochem Soc 162(14):A2406–A2423
33. Yang C, Chen J, Qing T, Fan X, Sun W, von Cresce A, Ding MS, Borodin O, Vatamanu J, Schroeder MA, Eidson N, Wang C, Xu K (2017) 4.0 v aqueous li-ion batteries. Joule 1:122–132
34. Kühnel R-S, Reber D, Battaglia C (2020) Perspective-electrochemical stability of water-in-salt electrolytes. J Electrochem Soc 167:070544
35. Yamada Y, Furukawa K, Sodeyama K, Kikuchi K, Yaegashi M, Tateyama Y, Yamada A (2014) Unusual stability of acetonitrile-based superconcentrated electrolytes for fast-charging lithium-ion batteries. J Am Chem Soc 136:5039–5046
36. Okuoka S-I, Ogasawara Y, Suga Y, Hibino M, Kudo T, Ono H, Yonehara K, Sumida Y, Yamada Y, Yamada A, Oshima M, Tochigi E, Shibata N, Ikuhara Y, Mizuno N (2014) A new sealed lithium-peroxide battery with a co-doped li2o cathode in a superconcentrated lithium bis(fluorosulfonyl)amide electrolyte. Sci Rep 4:5684
37. Dokko K, Watanabe D, Ugata Y, Thomas ML, Tsuzuki S, Shinoda W, Hashimoto K, Ueno K, Umebayashi Y, Watanabe M (2018) Direct evidence for li ion hopping conduction in highly concentrated sulfolane-based liquid electrolytes. J Phys Chem B 122:10736–10745
38. Qian J, Henderson WA, Xu W, Bhattacharya P, Engelhard M, Borodin O, Zhang J-G (2015) High rate and stable cycling of lithium metal anode. Nat Commun 6:6362
39. Yang J, Ding Y, Lian C, Ying S, Liu H (2020) Theoretical insights into the structures and capacitive performances of confined ionic liquids. Polymers 12(3):722
40. Herrera C, García G, Atilhan M, Aparicio S (2015) Nanowetting of graphene by ionic liquid droplets. J Phys Chem C 119(43):24529–24537
41. Burt R, Birkett G, Salanne M, Zhao XS (2016) Molecular dynamics simulations of the influence of drop size and surface potential on the contact angle of ionic-liquid droplets. J Phys Chem C 120(28):15244–15250
42. Li H, Sedev R, Ralston J (2011) Dynamic wetting of a fluoropolymer surface by ionic liquids. Phys Chem Chem Phys 13(9):3952–3959
43. Taherian F, Leroy F, Heim L-O, Bonaccurso E, van der Vegt NFA (2016) Mechanism for asymmetric nanoscale electrowetting of an ionic liquid on graphene. Langmuir 32(1):140–150
44. Malali S, Foroutan M (2017) Study of wetting behavior of bmim+/pf6- ionic liquid on tio2 (110) surface by molecular dynamics simulation. J Phys Chem C 121(21):11226–11233
45. Dong D, Vatamanu JP, Wei X, Bedrov D (2018) The 1-ethyl-3-methylimidazolium bis(trifluoromethylsulfonyl)-imide ionic liquid nanodroplets on solid surfaces and in electric field: a molecular dynamics simulation study. J Chem Phys 148:193833
46. Bordes E, Douce L, Quitevis EL, Padua AAH, Costa Gomes M (2018) Ionic liquids at the surface of graphite: wettability and structure. J Chem Phys 148:193840
47. Liu Z, Cui T, Li G, Endres F (2017) Interfacial nanostructure and asymmetric electrowetting of ionic liquids. Langmuir 33(38):9539–9547
48. Dhattarwal HS, Kashyap HK (2019) Molecular dynamics investigation of wetting-dewetting behavior of model carbon material by 1-butyl-3-methylimidazolium acetate ionic liquid nanodroplet. J Chem Phys 151:244705
49. Wang S, Li S, Cao Z, Yan T (2010) Molecular dynamic simulations of ionic liquids at graphite surface. J Phys Chem C 114:990–995
50. Feng G, Jiang X, Qiao R, Kornyshev AA (2014) Water in ionic liquids at electrified interfaces: the anatomy of electrosorption. ACS Nano 8:11685–11694
51. Sharma S, Kashyap HK (2015) Structure of quaternary ammonium ionic liquids at interfaces: effects of cation tail modification with isoelectronic groups. J Phys Chem C 119:23955–23967
52. Sharma S, Kashyap HK (2017) Interfacial structure of pyrrolidinium cation based ionic liquids at charged carbon electrodes: the role of linear and nonlinear alkyl tails. J Phys Chem C 121:13202–13210
53. Sharma S, Dhattarwal HS, Kashyap HK (2019) Molecular dynamics investigation of electrostatic properties of pyrrolidinium cation based ionic liquids near electrified carbon electrodes. J Mol Liq 291:111269

54. Gupta A, Dhattarwal HS, Kashyap HK (2021) Structure of cholinium glycinate biocompatible ionic liquid at graphite electrode interface. J Chem Phys 154:184702
55. Vatamanu J, Borodin O (2017) Ramifications of water-in-salt interfacial structure at charged electrodes for electrolyte electrochemical stability. J Phys Chem Lett 8:4362–4367
56. Li Z, Jeanmairet G, Méndez-Morales T, Rotenberg B, Salanne M (2018) Capacitive performance of water-in-salt electrolytes in supercapacitors: a simulation study. J Phys Chem C 122:23917–23924
57. McEldrew M, Goodwin ZAH, Kornyshev AA, Bazant MZ (2018) Theory of the double layer in water-in-salt electrolytes. J Phys Chem Lett 9:5840–5846
58. Ye J, Baumgaertel AC, Wang YM, Biener J, Biener MM (2015) Structural optimization of 3d porous electrodes for high-rate performance lithium ion batteries. ACS Nano 9:2194–2202
59. Vatamanu J, Vatamanu M, Borodin O, Bedrov D (2016) A comparative study of room temperature ionic liquids and their organic solvent mixtures near charged electrodes. J Phys: Condens Matter 28:464002
60. Vatamanu J, Bedrov D, Borodin O (2017) On the application of constant electrode potential simulation techniques in atomistic modelling of electric double layers. null 43:838–849
61. Zhang L, Wang H (2020) Anion intercalation into a graphite electrode from trimethyl phosphate. ACS Appl Mater Interfaces 12:47647–47654
62. Huang X, Margulis CJ, Berne BJ (2003) Dewetting-induced collapse of hydrophobic particles. Proc Natl Acad Sci 100:11953
63. Xu L, Molinero V (2010) Liquid-vapor oscillations of water nanoconfined between hydrophobic disks: thermodynamics and kinetics. J Phys Chem B 114:7320–7328
64. Remsing RC, Xi E, Vembanur S, Sharma S, Debenedetti PG, Garde S, Patel AJ (2015) Pathways to dewetting in hydrophobic confinement. Proc Natl Acad Sci 112(27):8181–8186
65. Altabet YE, Haji-Akbari A, Debenedetti PG (2017) Effect of material flexibility on the thermodynamics and kinetics of hydrophobically induced evaporation of water. Proc Natl Acad Sci 114(13):E2548–E2555
66. Cerdeiriña CA, Debenedetti PG, Rossky PJ, Giovambattista N (2011) Evaporation length scales of confined water and some common organic liquids. J Phys Chem Lett 2:1000–1003
67. Davoodabadi A, Ghasemi H (2021) Evaporation in nano/molecular materials. Adv Colloid Interface Sci 290:102385
68. Massé RC, Liu C, Li Y, Mai L, Cao G (2017) Energy storage through intercalation reactions: electrodes for rechargeable batteries. Natl Sci Rev 4:26–53
69. Zhu D, Fan H, Wang H (2021) Pf6- intercalation into graphite electrode from propylene carbonate. ACS Appl Energy Mater 4:2181–2189
70. Kondo Y, Miyahara Y, Fukutsuka T, Miyazaki K, Abe T (2019) Electrochemical intercalation of bis(fluorosulfonyl)amide anions into graphite from aqueous solutions. Electrochem Commun 100:26–29
71. Xia J, Wang J, Chao D, Chen Z, Liu Z, Kuo J-L, Yan J, Shen ZX (2017) Phase evolution of lithium intercalation dynamics in 2h-mos2. Nanoscale 9(22):7533–7540
72. Abbas G, Sonia FJ, Zafar ZA, Knížek K, Houdková J, Jiříček P, Bouša M, Plšek J, Kalbáč M, Červenka J, Frank O (2022) Influence of structural properties on (de-)intercalation of clo4- anion in graphite from concentrated aqueous electrolyte. Carbon 186:612–623
73. Rothermel S, Meister P, Schmuelling G, Fromm O, Meyer H-W, Nowak S, Winter M, Placke T (2014) Dual-graphite cells based on the reversible intercalation of bis(trifluoromethanesulfonyl)imide anions from an ionic liquid electrolyte. Energy Environ Sci 7(10):3412–3423
74. Shrivastav G, Remsing RC, Kashyap HK (2018) Capillary evaporation of the ionic liquid [EMIM][BF4] in nanoscale solvophobic confinement. J Chem Phys 148(19):193810
75. Dhattarwal HS, Remsing RC, Kashyap HK (2020) How flexibility of the nanoscale solvophobic confining material promotes capillary evaporation of ionic liquids. J Phys Chem C 124:4899–4906
76. Dhattarwal HS, Remsing RC, Kashyap HK (2021) Intercalation-deintercalation of water-in-salt electrolytes in nanoscale hydrophobic confinement. Nanoscale 13(7):4195–4205

77. Altabet YE, Debenedetti PG (2017) Communication: relationship between local structure and the stability of water in hydrophobic confinement. J Chem Phys 147:241102
78. Patel AJ, Varilly P, Chandler D (2010) Fluctuations of water near extended hydrophobic and hydrophilic surfaces. J Phys Chem B 114:1632–1637
79. Patel AJ, Varilly P, Chandler D, Garde S (2011) Quantifying density fluctuations in volumes of all shapes and sizes using indirect umbrella sampling. J Stat Phys 145:265–275
80. Kumar S, Rosenberg JM, Bouzida D, Swendsen RH, Kollman PA (1992) The weighted histogram analysis method for free-energy calculations on biomolecules. i. the method. J Comput Chem 13:1011–1021
81. Altabet YE, Debenedetti PG (2014) The role of material flexibility on the drying transition of water between hydrophobic objects: a thermodynamic analysis. J Chem Phys 141:18C531
82. Alibalazadeh M, Foroutan M (2015) Specific distributions of anions and cations of an ionic liquid through confinement between graphene sheets. J Mol Model 21:168

Applications: Biological

Synthesis and Biomedical Application of Coinage-Metal Nanoparticle and Their Composite

Piyali Sabui, Sadhucharan Mallick, and Adhish Jaiswal

Abstract Coinage metal nanoparticles including gold, silver, and copper are absorbed due to their size and shape-dependent distinct optoelectronic and chemical properties, in addition to their effective use in health-related applications. Among these nanoparticles, Because of their ease of fabrication, characterization, as well as surface modification, Au NPs have triggered a lot of interest in crucial biological applications. Coinage metal nanoparticles provide a robust platform for solving health-related problems because of their outstanding physical and chemical properties. Stable and biocompatible coinage metals NPs have been employed with targeted drug delivery and killing cancerous cells, diagnosing several types of cancers pharmacological applications, i.e., sensing probes, therapeutic agents, and drug delivery systems.

Keywords Coinage-metal nanoparticles · Photodynamic therapy · Sensing · Drug delivery · Multidrug resistant

1 Introduction

Nanotechnology has multiple applications in various fields of research. Nanoparticles (NPs) are indeed the foundational components of nanotechnology. From a historical

---

P. Sabui
Department of Zoology, The University of Burdwan Golapbag, Bardhaman, India

Department of Environmental Science, Indira Gandhi National Tribal University (Central University), Amarkantak, Madhya Pradesh 484886, India

S. Mallick (✉) · A. Jaiswal
Department of Chemistry, Indira Gandhi National Tribal University (Central University), Amarkantak, Madhya Pradesh 484886, India
e-mail: sadhucharan@igntu.ac.in

A. Jaiswal (✉)
Department of Chemistry, University of Lucknow, Hasanganj, Lucknow, Uttar Pradesh 226007, India
e-mail: adhish.jaiswal@igntu.ac.in

I. Uddin and I. Ahmad (eds.), *Synthesis and Applications of Nanomaterials and Nanocomposites*, Composites Science and Technology,
https://doi.org/10.1007/978-981-99-1350-3_6

perspective, the Coinage metals gold, silver, and copper are best known for their economic, ornamental jewelry pieces, and metallurgy values. In contrast, their metal nanostructures' remarkable properties and applications may be considered products of modern science, as nanotechnology's potential has only recently been realized [45, 66, 102, 106]. Metal NPs are nanometre-scale objects (1–100 nm) consisting of pure metals or metal compounds. Metal NPs have sparked a lot of attention because of their small size, making them preferable to their bulk counterparts in various applications [45, 66, 102, 106]. The nanoscale materials' surface area to volume ratio increases due to the unique features from their bulk counterparts. For over a century, metallic NPs have fascinated scientists and are not profoundly employed in biomedical sciences and engineering. NPs' features, including hardness, catalytically active surface, chemical reactivity, and biological action, can differ significantly from micron-sized particulates. It has been suggested that the biocidal effectiveness of metallic NPs is due to their size and high surface-to-volume ratio [26, 39, 60, 61, 106]. Group 11 coinage metallic nanostructures (i.e., Cu, Ag, and Au) have extraordinary size- and shape-dependent catalytic, optical, electrical, chemical, and other properties [7, 65, 68, 70]. The brilliant colours of colloidal solutions of coinage metal NPs have long intrigued scientists. When compared to other metals in the UV area, Au, Ag, and Cu nanoparticles have strong surface plasmon resonance (SPR) bands in the visible region [2–4, 7, 70, 79]. The surface plasmon band (SPB) is a prominent as well as broad absorption band, observed in the Ultra violet-visible spectra for metallic NPs greater than two nm. This originates from numerous interesting physical and related applications such as SPR spectroscopy and surface-enhanced Raman scattering (SERS), which have been widely applied in chemical analysis and biomedical detection. Due to their sole optical properties arising since the excitation of surface plasmon resonance, plasmonic coinage-metal NPs, namely Cu, Ag, and Au, have attracted a lot of attention for their potential applications in chemical and biochemical sensing, electrochemical sensing, SERS-based detection, electronics, optics, catalysis, medical diagnostics, therapeutics, and biological imaging [21, 22, 53, 68, 93].

The above properties have motivated in-depth study into the creation of coinage metal nanostructures and their potential applications. The modern scientific evaluation of metallic NPs did not start until Michael Faraday established the preparation of colloidal ruby colour gold (Au) solution by reducing the aqueous solution of chloroaurate ($AuCl_4^-$) salt with phosphorus in water [28]. Michael Faraday, with enthusiasm, has identified these finely divided gold particles as the origin of the ruby red to amethyst colour of colloidal solutions. Later in 1951, Turkevich et al. prepared Au NPs with an average diameter of about 20 nm with a very narrow size distribution and good reproducibility by reducing an aqueous solution of chloroaurate ($AuCl_4^-$) using sodium citrate [27]. In 1908, Gustav Mie provided the first justification for why Au NPs are red. He has given a theoretical explanation for spherical colloidal particle scattering and absorption. Gustav Mie developed a way to resolve Maxwell's equation for an electromagnetic light wave interacting with small spherical Au NPs with the same macroscopic frequency-dependent material dielectric constant as the bulk metal [73]. A high surface area to volume ratio raises the fraction of surface atoms increasing cases the material's reactivity. They, therefore, hold promise for ubiquitous commercial uses in consumer electronic, environmental, and energy technologies, data storage, therapeutic devices, and more due to their unique shape and size-dependent characteristics. It is now acknowledged that nano diagnostics and drug delivery using nanoparticles are complemental new technologies [18, 19, 34]. The popular term "theranostic" is used to describe the combination that can bring about unprecedented advances in medicine [12, 53]. Coinage-metal NPs' effective light-to-heat conversion enables the targeted thermal ablation of diseased or infected tissues. They can be used to increase CT scan contrast or improve cancer chemotherapy treatment due to their capacity to absorb large amounts of X-ray electromagnetic radiation (computed tomography). To detect microbial organisms and their metabolites, bio-imaging of tumor cells, detection of receptors on their surface, and investigation of Endocytosis, Au NPs are utilized in resonance scattering dark–field microscopy [12].

2 Synthetic Routes for Coinage-Metal Nanoparticle

Coinage-metal NPs are made in various ways, with top-down or physical procedures and bottom-up or chemical ones being the most common [72, 82, 86, 100]. A bulk solid is reduced to nanosized dimensions employing different physical and chemical strategies such as cutting, grinding, and etching techniques in top-down tactics [72]. In the bottom-up approach, the structure of NPs is generated by atoms, molecules, or clusters using chemical and biological procedures [68, 100]. Self-assembly refers to assembling atoms, molecules, or clusters to form NPs. For wet-chemical synthesis processes like chemical, electrochemical, polyol reduction, and Sonochemical, the bottom-up (reduction of salt precursors, nucleation, and growth) approach is widely used [12]. The most popular and cost-effective bottom-up technique utilizes chemical reduction of metallic ions in solutions. In theory, the metal salt is reduced using various reducing agents in the presence of a capping agent, stabilizing the nanostructures against aggregation under the proper reaction circumstances [21, 25, 54]. Many other reducing agents can be employed, including sodium citrate, $LiAlH_4$, $NaBH_4$, hydrazine, hydrogen, and ascorbic acid. The Lee–Meisel method uses $NaBH_4$, sodium citrate, and hydrogen to reduce nitrate and sulfate salts of Ag and Au [54]. The dimensions and shape of the particles are significantly influenced by the medium's pH. Both rod and spherical NPs were commonly observed at high pH due to rapid reduction. On the other hand, the slower reaction resulted in a lesser pH (5–6), polyhedral, and triangular forms. The size of the nanoparticles can be adjusted by adjusting the proportion of reducing and stabilizing agents as well as the pH of the system. Coinage-metal nanomaterials have been synthesized using a variety of physical and chemical processes. In 1982, Lee-Meisel prepared Au NPs from an aqueous hydrogen tetrachlorocuprate ($HAuCl_4$) brought to boiling for one hour in the presence of 1% sodium citrate solution [54]. Wine red Au NPs sols prepared by this method had absorption maxima at approximately 530 nm [54]. The Lee-Meisel process, the most common way for synthesizing Ag NPs, reduces aqueous $AgNO_3$

solution brought to boiling for one hour in the presence of 1% sodium citrate [54]. Ag sols prepared by procedures were yellow and had an absorption maximum of 420 nm. Sonit et al. had synthesized Ag NP using $NaBH_4$ in the Luria–Bertani (LB) bacterial growth medium [37]. They reported that Ag NPs were uniformly distributed and exhibited no agglomeration, with dimensions varying from 2 to 5 nm. Sanpui et al. made a Chitosan–AgNPs composite by mixing 50 ml of 0.2% (w/v) chitosan solution with 2 ml of freshly produced 10 mM $AgNO_3$ solution and 100 μl of 0.3 M sodium hydroxide (NaOH) solution at 95 °C with continual stirring [91]. They reported that the yellow-colored composite had only one, narrow, and strong absorption band at about 410 nm, which is the typical nature of Ag NP SPR. Rao and Paria reported synthesizing and stabilizing spherical size (60 nm) Ag NPs in aqueous leaf extract using *A. marmelos* at room temperature [44]. Sharma et al. demonstrated the preparation of Au NPs using Black Tea leaves in an aqueous medium at room temperature [94]. Biogenic synthesis are advantageous because they have a lower impact on the environment than some physicochemical production techniques and may create large numbers of contamination-free NPs with well-defined size and shapes [44]. Cheirmadurai et al. used aqueous Lawsonia inermis leaf extract to make reddish-brown colour Cu NPs soln with a 570 nm absorption maximum. They claimed that the diameter of as-synthesized Cu NPs (83 nm) was bigger than that of calcined Cu NPs (43 nm) [11]. Angajala et al. investigated the in vitro anti-inflammatory and mosquito larvicidal activity of spherical-shaped Cu NPs fabricated from *Aegle marmelos Correa* with greater particle sizes ranging 50–100 nm and 100–200 nm [5]. Nagar and Devra [76] generated crystalline, cubical Cu NPs using *Azadirachta indica* leaf broth and cupric chloride dihydrate as a precursor salt [76]. Cu^{2+} ions are reduced to Cu NPs by the biological molecules in *A. indica* leaves broth, which also acts as a capping and stabilizing agent.

3 Biomedical Applications

Coinage-metals are unique benign agents with many biomedical uses, including antibacterial agents, chemical and biochemical sensing, ultrasensitive diagnostic methods, thermal ablation as radiation enhancers, and drug and gene delivery vehicles [24, 66, 103]. Because of their unique properties of excellent penetration and traceability within the body, nanomaterials systems have become a powerful technique in diagnostics and therapeutics, allowing for more efficient treatment with a lower risk of toxicity than conventional treatments [10, 42, 97]. Man's valiant efforts to defeat sickness have resulted in revolutionary methods, detection, therapies, and new healthcare systems as a result of the fight between man and disease. Metals have been an essential aspect of human history for millennia. Nano silver, copper, and gold have all received much interest because of their unique characteristics and activity spectrum [14, 15]. However, combining them with recent advances in nanotechnology and material sciences has opened up new research opportunities in academia and industry, resulting in a diverse range of innovative biomedical applications such as antimicrobial and antiviral agents, diagnostic tools, controlled drug delivery, and imaging probes [10, 42]. The noble characteristics of these Coinage NPs, as opposed to other NPs like transition metal NPs like iron, nickel, or cobalt and their lower cytotoxicity, make them appealing for environmental and biological applications [50, 51, 56]. Coinage metal NPs have also been proven to be capable of inactivating viruses such as SARS, H1N1 flu, and H5N1 bird flu new wide-ranging nanomaterials (5–60 nm) can reduce viral loads by 80–100% through direct or indirect contact (New wide-ranging nanomaterials (5–60 nm) can reduce viral loads by 80–100% through direct or indirect contact [10, 42]. Chemical and biological sensing, biomedical such as detecting, targeted drug administration, imaging, photothermal and photodynamic treatment, and the regulation of two or three applications are all potential advantages of Au NPs, Ag NPs, and Cu NPs [1, 33, 62, 89, 104]. Nanotechnology, and more precisely nanomedicine, could advance quickly, allowing new and more efficient technologies to replace established treatment methods in the fight against the disease [30, 45, 49]. Nanotechnology has many applications in treating chronic human diseases, such as delivering precise medicines to specific locations [8, 9, 47, 99]. There have recently been numerous notable implementations of nanomedicine (chemotherapeutic, immunotherapeutic agents, and biological materials, so on) in the treatment of many diseases. Optical scattering has been employed in cancer cell diagnosis and is helpful in imaging approaches for detecting connected biosystems [2, 38, 40, 45, 107, 108]. On the other side, research into using nanoparticles as drug delivery methods for cancer treatment. Combining diagnosis and therapy features in one piece of equipment also has excellent potential.

3.1 Coinage-Metal Nanoparticles to Fight Against Multidrug-Resistant Bacteria

Infectious diseases continue to be a major cause of mortality and illness around the world. The World Health Organization (WHO) and the Centers for Disease Control and Prevention (CDC) have expressed great concern about the rise in multidrug resistance microorganisms [103]. As a result, one of the most critical challenges in global health security is antibiotic resistance [13, 29, 48, 59, 75]. Many multidrug-resistant bacteria can colonize and attach to surfaces via hydrophobic interactions by generating a biofilm, a matrix of extracellular polysaccharide molecules that acts as a barrier to antibiotic diffusion when treated [39]. It has highlighted nanotechnology-based medicines to treat the related diseases to tackle resistance to antibiotics, new infections, and biofilm formation. Recent studies have confirmed that antimicrobial formulations comprising coinage metal NPs could be effective bactericidal ingredients. Ag and Cu have been employed most extensively among inorganic antibacterial agents because of their effective biocidal ability at low concentrations and relatively less toxicity to mammalian cells [6, 7, 48, 55, 101]. Small dimensions and a high surface-to-volume ratio distinguish Ag and Cu-based NPs for improving interactions with biomolecules inside cells and on cell surfaces and cell permeability [7]. An inverse relationship between nanoparticle size and antimicrobial activity has been demonstrated, where nanoparticles in the size range of 1–10 nm have been shown to have the most significant biocidal activity against bacteria [75]. According to a previous investigation by Ahearn et al., pure silver, silver oxide, and silver chloride materials suppressed Candida albicans, Pseudomonas aeruginosa, Serratia marcescens, and Staphylococcus epidermidis growth of bacteria by increasing adhesion to an ion-beam-assisted-deposited silver surface and then losing viability [1]. Ag ions and Ag-based compounds have had bacteriostatic and bactericidal properties for centuries. In recent years, new bactericidal agents based on coinage metal nanomaterials have become essential due to the development of bacterial strains resistant to the most effective antibiotic preparations. Sonit et al. had reported the synthesis of Ag NPs in the LB medium, which had an antibacterial effect of Ag NPs against green fluorescent expressing (GFP) expressing *Escherichia coli* (*E. coli*) through perforation on the cell wall [37]. Sanpui et al. reported the antibacterial effect of chitosan–Ag NPs composite against GFP expressing *E. coli* where Ag NPs concentrations in the composite were only 2.2 and 2.6 μg/ml in the corresponding minimum inhibitory concentration (MIC) and minimum bactericidal concentration (MBC) of the composite [91]. Optical density (OD) at 595 nm of the sample is usually determined using a UV–Visible spectrophotometer at distinct intervals to verify the growth of the bacteria. The MIC value of the NPs is the minimum concentration of NPs at which there is no visible turbidity. In addition, cultures lacking visible turbidity are re-inoculated in fresh media. MBC is defined as the Nanoparticles concentration that killed at least 99.9% of the initial inoculation. Recently, Banerjee et al. synthesized ternary brown-colored iodinated chitosan-Ag NP composite and investigated antibacterial activity against GFP expressing *E. coli*. GFP expressing *E. coli* simplifies the monitoring of the cell population changes during the action of the iodinated chitosan-Ag NP composite at various time points [7]. The Minimum Inhibitory Concentration (MIC) of the iodinated chitosan-silver nanoparticle (CS-Ag NP) composite (87.20 μg/mL) comprised 38.40 and 48.80 μg/mL of chitosan-Ag NP and iodine, respectively, whereas the corresponding MBC (119.00 μg/mL) consisted of 51.70 and 67.30 μg/mL, respectively. In these above MIC and Concentrations, the quantity of Ag NP was only 0.82 and 1.10 μg/mL, respectively [7]. The positively charged chitosan might help attach the composite to the negatively charged bacterial cell wall. Ag NPs embedded in the composite induce pore formation on the cell wall, making bacteria unviable. The CS-Ag NP composite and generated iodine atom blend exert reactive oxygen species (ROS) that produce oxidative stress in the cytoplasm, causing more damage to the cell which is illustrated in Fig. 1.

In a nutrient broth and saline, Ghosh et al. assessed the antibacterial effect of Ag/agar film on *E. coli*, *Staphylococcus aureus* (*S. aureus*), and *Candida albicans* (*C. Albicans*) [35]. Further, Ag NPs have also been proposed for anticancer therapy [78, 108]. For example, Ag NPs have been widely used in household products such

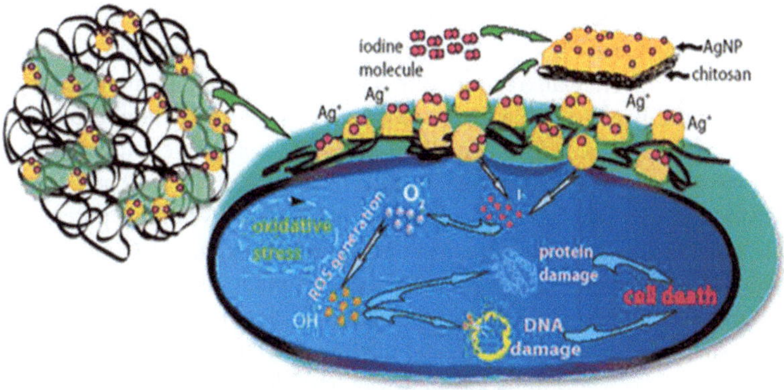

Fig. 1 Schematic representation of the proposed mechanism of antibacterial activity of the iodinated chitosan-Ag NP composite. Reprinted with permission from [7]. Copyright © 2010 American Chemical Society

as textiles, medical equipment disinfecting, domestic appliances, and their potential water treatment capacity. As a result of the dramatic rise in Ag NPs manufacturing, use, and customer needs, it is unavoidable that large quantities of Ag NPs discharged into the environment will disrupt typical sewage behavior. The World Health Organization has established an upper limit of 0.1 mg/L for the amount of silver that is permitted in drinkable water. Recent studies show that combining Ag NPs with polymer improves Ag NPs' antibacterial activity at reduced Ag concentrations. Numerous categories of Ag-containing antibacterial agents using diverse carriers, such as zeolite TiO_2 material, mesoporous silica, chitosan, polyacrylonitrile, siloxane, carbon aerogels, montmorillonite sol, and epoxy resin, have been observed [7, 66]. Combining Ag NPs with these polymers increases their antimicrobial activity efficiency by exploiting the individual substances' active components. These composites may find use in a variety of industries, including those that deal in food processing, hospital instruments, water purification, and surgical equipment and devices. However, Ag NPs have recently found wide applications in water and air purification systems. Furthermore, adding Ag NPs or Au NPs to existing antibiotics such as salinomycin, penicillin G, kanamycin, amoxicillin, erythromycin, clindamycin, chloramphenicol, and ampicillin are bactericidal vancomycin toward Staphylococcus aureus was enhanced [74, 81, 88]. Cu has long been recognized as an effective antibacterial agent with low toxicity. Cu has already been employed as a powerful fungicidal agent for many decades. Since a period of time ago, Cu's germicidal activity has been known and reported for both free and complex species. Since the nineteenth century, the paint industry has used Cu or Cu oxide particle dispersions in organic matrixes as antifouling coatings, mainly for maritime applications. In 2006, Kim et al. deposited metallic Cu NPs with sizes ranging from 3.5 to 11 nm on the spherical silica matrix, i.e., the hybrid structure of $Cu\text{-}SiO_2$ nanocomposite formed, which exhibited a remarkable inhibitory effect on various bacteria [48]. In 2008, Anyaogu et al. reported the synthesis of acrylic functionalized, with 2-phenoxyethyl acrylate polymer composites, showing antimicrobial activity and claiming that Cu ions leach from chemically bound acrylated Cu NPs is the biocidal species [6]. They also reported that Cu NPs chemically modified with positively charged thiols were stable in aqueous solution for many months and exhibited outstanding antifungal activities. Cu NPs and Cu NP-polymer composites are attractive alternatives to common antibiotics as they have substantial antibacterial and antifungal properties [13, 70]. These stable nanocomposites have a high potential for use in antibacterial or antifouling coatings in the marine environment. Copper-containing NPs have been shown effective against animal and plant pathogens impeding MDR biofilms formation when CuO NPs enter bacteria; metabolic functions are affected by active transport, electron transfer, and nitrogen metabolism [55]. It is involved in iron homeostasis's oxygen transport pathway and electron transport chain. At high concentrations in free ionic form, it can start generating ROS by reducing O_2 and producing superoxide anion O^{2-}, which come into contact with the cell membrane and produce free radicals that enter the cell and disrupt the internal cellular contents and biochemical processes, as well as preventing DNA and amino acid synthesis. Cu^{2+} ions generated by nanostructured materials can cause ROS through a broad range of chemical reactions, including breaks in DNA strands that affect the expression of genes. Furthermore, Cu^{2+}, Cu^{+} ions, as well as relatively small Cu NPs have been shown to stimulate the production of intracellular reactive oxygen species in bacterial cells. Recently, Mallick et al. prepared the chitosan-Cu NP composite from the reaction of alkaline $CuSO_4$ and hydrazine in the presence of chitosan in an aqueous medium [70]. It was taken out of the reaction mixture and dispersed in water with dilute acetic acid. The spherical Cu NPs (9.8 ± 4.0 nm) were stable after synthesizing if molecular iodine was added to the medium, as depicted in Fig. 2. Mallick et al. investigated the bactericidal action of iodinated chitosan-Cu NP composite on gram-negative GFP expressing *E. coli* and Gram-positive *Bacillus cereus* (*B. cereus*) bacteria which is shown in Fig. 3. The MIC of the Iodine-stabilized Cu nanoparticle chitosan composite on *E. coli* was 130.80 μg/mL, which consisted of 21.50 μg/mL Cu NPs.

Further, Mallick et al. found that chitosan-Cu@Ag NPs composites were stable in solution for more than two weeks under ambient atmospheric conditions and exhibited bactericidal activity using a lower concentration of Cu and Ag against *E. coli* and *B. cereus* bacteria [69]. They reported that when chitosan-supported copper nanoparticles were treated with silver nitrate ($AgNO_3$) solution, Ag shells formed on

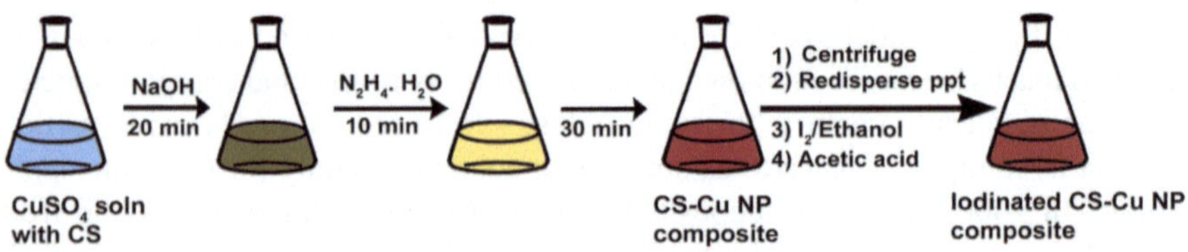

Fig. 2 The procedure for making iodinated CS–Cu NP composite is illustrated graphically. Reprinted with permission from [70]. Copyright © 2012 American Chemical Society

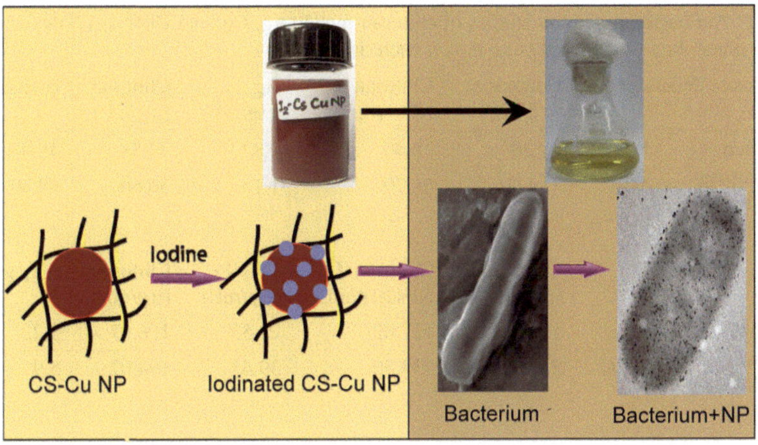

Fig. 3 A schematic illustration of iodinated CS–Cu NP composite synthesis and bactericidal application. Reprinted with permission from [70]. Copyright © 2012 American Chemical Society

the Cu NP cores in the composite. Mallick et al. ensure that through their research work over the preceding few years [7, 69, 70] a contribution has been made to the progress of nanoscience and nanotechnology, which are summarized in Table 1, which would be relevant for the improvement of humankind. Cu and Ag NPs can exert their antibacterial activity via a multitude of mechanisms, such as (a) Direct interaction of NPs through the bacterial cell wall leading to perforation of the cell wall, (b) Biofilm formation inhibition; (c) Activation both of essential and adaptive host immune function; (d) Generation of ROS including oxidative damage to proteins and DNA; and (e) Intracellular effects e.g., interactions with proteins and DNA have been induced. In terms of the molecular pathways of Ag (+) inhibiting action on microorganisms, it has been observed that DNA loses its ability to replicate and the expression of ribosomal subunit proteins and other cellular proteins and enzymes required for ATP production is inactivated [29].

3.2 *Photothermal Therapy*

Photothermal therapy (PTT) is a minimally invasive cancer treatment known as thermal ablation or optical hyperthermia [2, 59]. In recent years, it has received much attention, compared to traditional therapeutic methods, PTT is a highly efficient and distinct method for treating diseases because it exhibits spatiotemporal selectivity, with the benefits of high responsiveness, lesser side effects, non-invasiveness, quick and effective therapeutic interventions, and low cost [20, 23, 84, 95, 98]. Due to their unique features, with high photothermal activity and small diameters, Au NPs are generally utilized for PTT as photothermal agents, allowing tumor penetration greater than normal cells [77, 104]. The optical absorption of Au NPs, particularly

Table 1 Antibacterial activity studies of chitosan-metal (Ag, Cu, and Cu@Ag) NPs composites against green fluorescent expressing *E. coli* bacteria

Composite details		Composite (μg/mL)	Chitosan-Ag NPs (μg/mL)	Ag (μg/mL)	Chitosan (μg/mL)	I_2 (μg/mL)
Iodinated chitosan-Ag NP composite [7]	MIC	87.20	38.40	0.82	37.58	48.8
	MBC	119.00	51.70	1.1	50.60	67.30
		Composite (μg/mL)	Chitosan–Cu NPs (μg/mL)	Cu NPs (μg/mL)	Chitosan (μg/mL)	I_2 (μg/mL)
Iodinated chitosan-Cu NP composite [67]	MIC	130.84	127.62	21.55	106.07	3.22
	MBC	239.4	233.50	39.43	194.07	5.9
		Composite (μg/mL)	Cu (μg/mL)	Ag (μg/mL)	Chitosan (μg/mL)	I_2 (μg/mL)
Chitosan Cu@Ag NP composite [69]	MIC	63.40	3.03	1.47	58.90	0.0
	MBC	92.99	4.44	2.13	86.42	0.0

gold nanorods (Au NRs), is dominating, and the absorbed light energy is generally dissipated thermally, resulting in localized heat, which is the foundation of healthcare medical strategies such as photothermal therapy [46]. Au NPs have an extremely good binding affinity, which makes them ideal for attaching ligands to improve biomolecular interactions. Au NPs also exhibit an intense color in the visible range and strongly contrast with electron microscopy imaging [58]. Tunable Au NR surface Plasmon and photothermal effects and Au NRs' widely accepted biocompatibility and functional diversity are used in various biomedical sensing, drug transport thermo-chemotherapeutic, and PTT applications [80]. The Au NPs-enhanced PTT method can more accurately target tumor cells than conventional chemotherapeutics. Photothermal therapy is confirmed by Au NPs, which can absorb specific wavelengths of light, resulting in photoacoustic and photothermal characteristics, making them highly beneficial for hyperthermic treating cancer. The smaller Au NRs, among other nanostructures, are much more attractive for in vivo PTT [80]. In early clinical trials, PTT was investigated with different forms of Au nanostructures. For example, the Food and Drug Administration approved pilot clinical studies with Au–Silica (AuroShell) Nanoshells around 155 nm in diameter, with a layer of 5000 molecular weight polyethylene glycol (PEG). Au–Silica is a multifunctional theranostic platform based on photosensitizer-loaded plasmonic vesicular assemblies of Au NPs developed for practical cancer imaging and primary and metastatic lung cancer tumors or head and neck cancer treatment [30]. In vitro and in vivo, Au NRs serve as the most appropriate photothermal contrast agents. Mackey et al. [64] synthesized three different PEG-AuNRs (median size 38 × 11, 28 × 8, and 17 × 5 nm) and examined (in vitro) their photothermal contrast ability for plasmonic photothermal therapy application on human oral squamous cell carcinoma

(HSC-3 Cancer cells) by irradiation of continuous near-infrared (NIR) laser with a wavelength of 808 nm. Using experimental and discrete dipole approximation (DDA) theoretical technique, Mackey et al. had established that Au NR (dimensions 28 × 8 nm) exhibits the ideal size for the application as a photothermal contrast agent [64]. Plasmonic photothermal therapy (PPTT) is a cancer therapy in which AuNRs are injected at the tumor site before near-infrared light is transiently applied to the tumor, causing localized cell death [2, 3]. Moustafa et al. reported effectiveness, molecular mechanism, toxic effects, and pharmacokinetics of PPTT in cancer cells in vitro and in vivo using PEGylated AuNRs-PPTT as implemented in natural cancerous cells in the mammary glands (in vivo) of animal clinical cases (cats and dogs), which more plausibly demonstrate their human equivalents at the molecular scale, and cancer MCF-7 cells (in vitro) by NIR laser irradiating the tumors for 2 or 5 min. Generally, cell death follows one of two distinct pathways: apoptosis or necrosis [2, 3]. Two minutes of laser irradiation generated a moderate alteration (apoptosis) within the tumor, but five mins of laser irradiation caused severe burning (necrosis). Their findings imply that giving AuNRs to canines and felines with natural tumors under extremely low photothermal temperatures of 42–44 °C for two mins causes apoptosis, which is better for tumor treatment than necrosis. Prostate cancer is major cancer in males and the second most significant cause of death from cancer. Rastinehad et al. [84] demonstrate that targeting prostate cancers with gold–silica nanoshells (GSN) guided laser excitation and ablation is a safe and feasible approach. GSNs, which comprise a silica core and a gold exterior with an overall diameter of 150 nm range, are intended to absorb near-infrared light as efficiently as possible and convert it to heat. Photothermal heating of the tumor causes selective hyperthermic cell death without heating the adjacent non-tumorous tissue [84]. Tao and colleagues developed a cancer therapeutic platform based on immunomodulatory cytosine-guanosine (CpG) oligodeoxynucleotides (ODNs) that are attached with NIR-responsive Au NRs and doxorubicin (Dox) [95]. In 2014, Tao et al. established a novel DOX-loaded Au NRs (AuNRs -CpG-DOX) for a combination of chemotherapy and PTT in the mouse hepatocellular carcinoma cell (H22) lines (in vitro) and cancer-induced BALB/c mice (in vivo). The viability decreased from 88 to 52 percent (%) when cells were incubated with Dox, depending on the Dox concentration. Furthermore, after near-infrared irradiation, the viability of cells treated with AuNRs-CpG conjugates was reduced to 58 percent, indicating that AuNRs may function as a photothermal agent to efficiently kill cancerous cells (irradiated by the 808 nm laser). Cell viability was drastically reduced to 12% when the two treatments were combined under single irradiation, as expected ($P < 0.005$) is the significance level. Based on the significant reduction cell viability, AuNRs-CpG-Dox conjugates (ACD) could be used for photothermal ablation of cancerous cells as well as a drug delivery vehicle for cancer chemotherapy. Notably, combining chemotherapeutic and photothermal therapy outperformed solo treatments in terms of cancer treatment. These findings demonstrated that ACD was an effective thermo-chemotherapy agent for cancer cells [95]. During NIR laser irradiation, the AuNRs may efficiently increase the thermal death of cancer cells. Furthermore, the CpG motif-mediated synthesis of soluble molecules like antitumor cytokines prevented H22 cells from proliferating. Furthermore, the enhanced chemotherapeutic activity was due to dox molecules liberated from ACD after NIR irradiation. These effects had a powerful anticancer impact and resulted in long-term tumor-specific immunity. This triple combination of chemotherapeutics, thermotherapy, and immunotherapeutic has the potential to synergize antitumor effects and maximise the therapeutic benefits of cancer treatment. These findings suggested that ACD could be a promising tool for inhibiting tumour cell growth by delivering a proinflammatory signal as well as an anticancer agent to tumour cells. Dong et al. had synthesized biocompatible and multifunctional Cu-Ag_2S/PVP NPs PTT agents with high photothermal conversion efficiency. Under 808 nm laser illumination, the new Cu-Ag_2S/PVP NPs (average diameter of 8.6 nm) have a high photothermal conversion efficiency (58.2%), which is significantly higher than most of the described PTT drugs versus tumor-bearing mice (in vivo) previously. Furthermore, Cu-Ag_2S/PVP NPs might be excreted by renal clearance and hepaticclearable excretion pathway within a few days signifying a high level of biosafety for further potential clinical translation [20]. Gallic acid functional silver nanoparticles (GA-Ag NP hydrogel) (average diameter of 43.1 nm) work as a photothermal agent that efficiently and controllably alters 808 nm near-infrared ray light into heat to kill bacteria against Gram-positive *Staphylococcus aureus* (*S. aureus*) and Gram-negative *E. coli* [59]. The enhanced local temperature caused by NIR laser light damaged the bacterial cell membrane and caused protein denaturation. The photothermal properties of GA-Ag NPs and the innate germicidal activity of Ag^+ can disrupt cell membrane integrity and induce the death of bacteria [59]. Although PTT by itself is sufficient for medical assistance, synergistic therapies that combine PTT with other treatment methods like as photodynamic therapy (PDT), radiation, immunotherapeutic, and chemotherapy have recently been investigated to improve PTT's treatment effectiveness. Vijayaraghavan et al. reported the multi-branched gold nanoechinus (AuNEs) exhibiting photodynamic and photothermal therapeutic effects in both the NIR-I and NIR-II windows [98]. They demonstrated the in vivo biomedical application of Au NEs (average crystallite size of 350 $\pm$ 50 nm) as dual-modal PDT and PTT constituents for solid tumour obliteration using Light energy in the first and second biological windows.

Photo-irradiation of Au NEs internalized HeLa cells with 550, and 940 nm LED arrays results in cell death. To create a new type of NP for cancer treatment, Au NRs are encapsulated with chitosan (CS) derivatives to combine chemical and photothermal activities. [23] described that thiol-modified CS derivatives are chemically conjugated to DOX and produce nanocarriers DOX-CS-Au NRs [23]. The biostability of Au NRs is enhanced by CS derivatives. Through the chemical effect of DOX and the photothermal effect of Au NRs, Au NRs coated with CS derivatives can be employed for cancer therapy. The photothermolysis impact was enhanced by increasing the Au NR-loaded nanoparticle concentration (depending on the Au NR quantity) for CS-Au NR and DOX-CS-Au NR nanocarriers. Because of the anticancer drug, the clinical efficacy of DOX-CS-Au NR nanocarriers was greater than that of CS-Au NR nanoparticles at varied concentrations. It is worth noting that a combined effect of hyperthermic treatment and traditional chemotherapy can greatly enhance the therapeutic potential of anticancer drugs. In 2014, Vankayala et al. demonstrated through In vitro experiments that the PDT effect via irradiation of lipid bilayer-coated Au NRs (an average length 37.3 ± 2.4 nm and diameters of 11 ± 1.2 nm) by 940 nm could destroy the Hela cell tumor far more effectively than the PTT effect (via irradiation of Au NRs by 550 nm light). The amount of cellular deaths induced by 940 nm light (PDT effect), however, is about tenfold higher than that by 550 nm light (PTT effect) at Au NRs concentration of 25 μg/mL [96]. Experimental studies in mice demonstrated that the PDT effect (via irradiation of Au NRs by 915 nm light) could destroy the B16F0 melanoma tumour much more effectively than the clinically used anticancer drug doxorubicin, as well as the PTT effect (via irradiation of Au NRs by 780 nm light).

3.3 Photodynamic Therapy

Photodynamic therapy (PDT) offers the promise to become incorporated into the mainstream of cancer therapy [16, 31, 96]. Cancer seems to be the best example of a disease where both its diagnosis and therapy have profited from nonmedical technologies. Cancer is the leading cause of mortality in developed countries and the second most significant cause of death in developing countries, according to the WHO [103]. The most common cancer treatment modalities that are used effectively in cancer therapy are surgical procedure, chemo, and radiation therapy. PDT promises to hold whenever surgery is impossible [16, 17, 32, 41, 43, 49]. In contrast to radiotherapy, the low-cost and repeatable PDT aids in long-term cancer management. The PDT was approved by the United States Food and Drug Administration as the first drug-device combination for cancer treatment. PDT is efficacious both against malignant and non-malignant diseases. In the future, the abnormal structure of the tumor vasculature may allow the design of NPs that preferentially accumulate at the tumor site and, if loaded with a drug, to deliver their cargo directly to the tumor. PDT uses light and photosensitizers (PS) to create reactive oxygen species to treat cancer and infectious disease. PDT is a minimally invasive and nontoxic approach for treating solid malignant tumours that have been clinically validated. PDT is a site-specific treatment that causes cell death by altering normal tissue oxygen ($^3O_2$) into harmful ROS such as singlet oxygen ($^1O_2$) and other free radicals. This ROS can directly damage the tumour or destroy cancer cells by triggering apoptosis and necrosis, causing damage to the tumour vasculature or increasing immunological effects [36]. The three factors (Photosensitizer, Light, and ROS) effectively treat tumors when combined [87]. Organic photosensitizers employed in PDT have a number of disadvantages, including high toxicity, tumour non-selectivity, and low light absorption. As a result, interest in harmless inorganic photosensitizers such as noble metal NPs has grown. NPs replace organic dyes in bioscience since they have photostability and non-toxicity [16, 17]. The key features of Au NPs used in photodynamic treatment include strong fluorescence quenching and SPR absorbance. Au NPs demonstrate PDT is considered a crucial therapy for oncological disorders, and photosensitizers as light-sensitizing chemicals and a laser are used to treat specific skin or contagious diseases. The heightened performance of nanohybrids in PDT compared to the free sensitizers was also shown for other systems, including Au NPs functionalized with zinc phthalocyanine and a lactose derivative employing human breast adenocarcinoma cells (SK-BR-3) breast cancer cells [32]. García et al. [32] had described how functionalizing phthalocyanine-AuNPs with lactose produced water-dispersible NPs that can create singlet oxygen and cause cell death when irradiated. Haimov et al. had reported the PDT agent, meso-tetrahydroxy-phenylchlorin (mTHPC) photosensitizer, to AuNPs that serve as carriers for the drug [32]. Haimov et al. [41] demonstrated higher SH-SY5Y human neuroblastoma cell death rates when incubated with the stable, soluble AuNP-mTHPC complex than with the free photosensitizer drug. The Meso-tetrahydroxy-phenylchlorin (mTHPC) photosensitizer, also known as "Temoporfin" (brand name "Foscan®"), has been approved for the treatment of nonmelanoma skin cancers by the EMA in 2001 [41, 92]. mTHPC is a second-generation photosensitizer, a 650 nm laser beam activates it, and in the triplet state, the mTHPC has a reasonably extended lifetime. mTHPC also possesses the unusual trait of accumulating significantly more strongly in tumours than in normal tissues. As a result, for effective PDT, comparatively lower dosages of mTHPC are needed. The XTT assay has been used to evaluate cell death quantitatively. In this case, mTHPC causes singlet oxygen generation, which impacts cell survival. Unlike free mTHPC, which induces non-specific cytotoxicity at greater concentrations, the AuNP-mTHPC complex only kills cells when exposed to radiation at concentrations ranging as 1.2 M. Therefore, cells treated with the AuNP-mTHPC combination died faster in the investigation than cells cultured with the free drug. In the presence of endogenous photosensitizer Riboflavin, Hela cells were exposed to pectin-coated silver nanoparticles (Pec@AgNP) (2.3 ± 0.7 nm). Riboflavin (vitamin B2, Rf) is a component of proteins that binds flavin mononucleotide and flavin adenine dinucleotide as cofactors in biological systems. The Annexin V test was positive in irradiated HeLa cells pre-incubated with 50 μM Rf, while the Propidium Iodide assay was negative, demonstrating early apoptosis. Annexin V and PI, on the other hand, showed positive results when Pec@AgNP was added to the culture medium, confirming late apoptosis or necrosis. Rivas et al. [87] had described that when Pec@AgNP and Rf are present, photodamage increases. The MTT test also indicated decreased cell viability when the nanoparticles were added along with Rf to the incubation medium [87]. Aggregation is one of the significant drawbacks with the Au and Ag NPs of diameter less than 200 nm, consequential in decreased reach to the tumor sites. Also, the stability decreases by inhibiting its use in long-term applications; however, this has been overcome by using a wide range of materials such as TiO_2, SiO_2, and ZrO_2 to support Ag and Au NPs [9, 102]. Dhanalekshmi et al. synthesized core–shell Au@SiO_2 NPs are very stable with high photodynamic efficacy under visible light illumination. Dhanalekshmi et al. [16] proved that core–shell type Ag@SiO_2 NPs retains their potential to kill HeLa cells after being exposed to light. These results show that the SiO_2 shell prevents aggregation and enhances the photodynamic activity of Ag NPs [16]. Cell viability of the highly stable Au@SiO_2, core–shell NPs against the human cervical cancer line (HeLa cell lines) were studied by [71] using the MTT assay method [71]. The cell viability studies using HeLa cell a line in the occurrence of a light source show that the cell viability depends on the concentration of the NPs, time, and light dose [71].

3.4 Coinage-Metal Nanoparticle-Enabled Bioimaging

The versatility of metal NPs has been beneficial in various studies related to disease diagnostics, early detection studies, and well contrast agents for better-quality imaging techniques. In disease diagnostics and signalling pathway elucidation, the discovery of molecular imaging agents for fluorescence imaging of cells is of tremendous importance. Au NPs and Ag NPs have emerged as exciting tools for cell imaging applications when coupled to target molecules and fluorophores [53, 62, 109]. Their more excellent brightness and shorter lives have recently been highlighted as advantages of their utilization as cell imaging agents. Fluorescence-based single-molecule detection in cells needs the use of powerful laser beams that injure cells. Ag NPs are becoming increasingly popular. Various research teams have started looking into alternate methods for creating optical sensors and imaging labels in light of the extraordinary optical characteristics of these metal NPs. Silver Sulfide (Ag_2S) has recently been found appealing for biological imaging due to its near-infrared fluorescence [63].

3.5 Sensing

Several sensors based on the plasmonic properties of coinage-metal NPs, such as very high molar extinction coefficients and resonant Rayleigh scattering on the one hand, and enhanced local electromagnetic fields near the Nanoparticle's other surface on the other, have been developed over the last decade. Chemical and biological sensing is one of the most important uses of Plasmonic metal nanoparticles (Au NPs, Ag NPs, Cu NPs). Plasmonic metal nanoparticles have been utilized as efficient sensors for the detection of various analytes such as metal ions, anions, and biomolecules such as saccharides, nucleotides, and proteins [105] and toxins by employing their inherent characteristics [58]. Au NPs sensors can be colorimetric, surface Plasmon resonance, fluorescence-based, electrical and electrochemical, SERS, quartz crystal microbalance-based, and Bio-Barcode assay sensor, depending on the sensing approach [57, 85, 90]. The basic principles of colorimetric sensing are based on visible colour changes caused by Au NP agglomeration. This is an essential feature for naked-eye colorimetric sensing applications because changes in their surface charge are translated into visible colour changes. Using a copper oxide/gold nanocomposite, Rajpurohit et al. recently reported the fabrication of an electrochemical sensor for the simultaneous detection of methylglyoxal (MGO) and glyoxalase (GLO) as possible diabetic indicators [83] The Rajpurohit et al. used the adsorptive stripping differential pulse voltammetry (AdSDPV) method for the specific and reliable detection of MGO levels in human plasma and urine samples from healthy people and diabetes patients. The excellent conducting capabilities of nanoplate CuO and anchored Au nanoparticles, as well as their synergistic effect, made MGO and GLO determination extremely simple and sensitive [83]. Ag NP- and Au NP-based assays have been used to detect metal ions, small molecules, proteins, deoxyribonucleic acid DNA, and enzymes. According to Lee et al. [52], sialic acid-stabilized Au NPs ($d = 20.1 \pm 1.8$ nm) can be utilized to detect influenza viruses quickly by attaching to the influenza virus envelope protein hemagglutinin. Amino acids, for example, arginine (Arg), histidine (His), and lysine (Lys) have been detected using quercetin-AuNPs as a colorimetric probe (Lys). Indeed, quercetin-Au NPs agglomeration produced by amino acids leads to a color alteration from red to blue [52].

3.6 Drug Delivery

Nanomedicine and nano-delivery systems are a comparatively novel but fast-growing field. Coinage metals-based nanomaterials are used as disease diagnosis tools, treatment, or to deliver therapeutic drugs to precisely targeted areas in a controlled manner [2, 38, 40, 45, 79, 107, 108]. According to the WHO, lung cancer is the most prevalent cause of cancer-related death, followed by colorectal, stomach, liver, and breast cancers. Using NPs as drug delivery vehicles can change how cancer drugs are delivered to tumors. Liposomes (lipid-based NPs) are already being employed as drug delivery systems to prevent drugs from degradation, improve pharmacokinetics, and deliver a significant amount of drugs [67]. These drugs can be better by using drug-delivery vehicles that passively or actively target cancers. Depending on the physical and chemical properties of the pharmaceuticals, nanoparticles are chosen for drug delivery. Nanoparticle-delivery systems are typically between 10 and 100 nm in size. While many antibiotics' efficacy is limited by poor membrane transport, drug-loaded NPs vehicles can penetrate host cells via Endocytosis, making intracellular entrance easier [99]. Au NPs are an appropriate platform for developing effective delivery systems as Au NPs have unique optical and physicochemical properties, functional flexibility, tuneable monolayer, controlled disparity, and biocompatibility. Au NPs have a large surface area to volume ratio, which permits more drug and prodrug loading capacities, remarkably reducing the minimum effective dosages compared to free drug molecules.

As a result, Au NPs can transfer several drugs such as plasmid DNAs (pDNAs), small interfering RNAs (siRNAs), peptides, proteins, and chemotherapeutic agents. Verigene (FDA-approved) and Aurimmune (Phase-II) are gold-based nanomaterials for the application of therapeutics. Tamoxifen, a breast cancer treatment medicine, has been commonly used for three decades. Dreaden et al. utilized Au NPs conjugated to thiol-PEGylated tamoxifen derivatives to select MCF-7 breast cancer cells. Tamoxifen-gold nanoparticle conjugates have been found to specifically target estrogen receptor alpha in human breast cancer cells in vitro, with up to 2.7-fold increased therapeutic potency [22]. The vital chemotherapy components are the platinum-based anticancer medicines cisplatin, carboplatin, and oxaliplatin. Although oxaliplatin has shown promise in treating various cancers in the past decade, it is presently used to treat colorectal tumors. In the colon cancer cell lines (HCT116, HCT15, HT29), the platinum-tethered Au NPs are up to 5.6-fold more cytotoxic than or at least as active as oxaliplatin.

In contrast, the platinum-tethered Au NPs had an IC_{50} of 0.495 nM in the human lung cancer (A549) cell line (which is nearly sixfold more active than oxaliplatin (IC_{50}) 0.775 μM). Because Au NPs are nontoxic and nonimmunogenic, they are also an excellent drug delivery scaffold. Wang et al. developed a drug delivery system by conjugating DOX onto the AuNPs surface with a poly (ethylene glycol) spacer via an acid-labile linkage (DOX-Hyd@AuNPs) which can remarkably inhibit the growth of multidrug-resistant MCF-7/ADR cancer cells due to the high efficiency of cellular uptake by Endocytosis and following acid-responsive release in cells [99]. Yuan et al. reported that combining Ag NPs with gemcitabine increased cytotoxicity and apoptosis in Human Ovarian Cancer A2780 cells as compared to using the gemcitabine alone. According to their findings, Ag NPs can boost ovarian cancer cells' responsiveness to therapeutic drugs, and Ag NPs can be employed as chemosensitizing agents in ovarian cancer therapy [107].

4 Conclusion

We have discussed the synthesis and applications of coinage metal NPs in the medical domain.

According to preceding research, physical, chemical, and biological approaches have been used to synthesize Coinage metals NPs and their nanocomposites of specific sizes and morphologies. The antimicrobial and anticancer characteristics of these coinage metal NPs (Au NPs, Ag NPs, Cu NPs) and their composites are highly encouraging. When the size of Nanostructures is reduced, the percentage of surface area will increase; this means that biological activity, such as antifungal, antibacterial, antiviral, and anticancer properties, is enhanced compared to larger particles. Coinage metals NPs and their nanocomposites have involved extreme care in the medical domain to develop the advanced generation of targeted drug-delivery systems, diagnosis of diseases, and treatments of various life-threatening diseases. One of the primary difficulties in nanotechnology and nanobiotechnology is to improve the therapeutic efficacy of NPs while minimizing any inherent side effects. With the rapid advancement of nanomaterials, novel treatment procedures are being investigated that may solve current problems employing coinage metal NPs. The science of Coinage metals-based nanomedicine and nano-delivery is presently among the fascinating research areas. In the previous two decades, extensive research in this area has resulted in the filing of 1500 patents and the achievement of dozens of clinical trials [79]. Colloids of Au, Ag, Pt, and Pd have been generally used and recognized in medical trials to treat cancer. The commencement of nanotechnology has enlarged the therapeutic standards and taken the technology to advanced stages. The idea of manipulating nanosized objects, which was once a dream, has become authentic and has a lot of futuristic prospects. Because of the SPR effect, Coinage metals NPs have thermally ablated tumour tissue in in vivo and in vitro and have quickly benefited from PDT anticancer therapy.

Such NPs are specifically designed to attract diseased cells, allowing for direct treatment of those cells, continuing to improve efficacy, reducing adverse effects, and overall improving public health. The application of these metals, including Au and Ag, both in diagnosis and therapy, is an area of research that could potentially lead to a wider application of nanomedicine in the future [79]. One source of significant interest in this direction is Au NPs, which appear to be efficaciously absorbed in soft tumour tissues and make the tumour susceptible to radiation (e.g., in the near-infrared region) based on thermal ablation for selective elimination. Due to a favorable safety profile that allows for repeated dosing with no total dose restrictions, PDT treatment of malignant tumours has many benefits over other treatment approaches such as chemotherapy or radiotherapy. NPs combined with therapeutic agents address the issues associated with traditional therapy; however, a few issues, such as side effects and toxicity, remain unresolved and must be properly examined before their use in biological systems.

Ag^+ ions have now been probed for a variety of biomedical usages including use as an antimicrobial constituent in dental resin composites. The antibacterial and antiviral properties of elemental Ag, Ag^+ ions, as well as Ag compounds have been thoroughly studied.

The use of Ag NPs rather than elemental Ag or complex Ag compounds to prevent biofilm formation on biomedical and other surfaces has been explored. The optical, electronic, and physicochemical SPR properties of Au spherical, as well as nanorods, have obtained them a special place. MTT assays, ROS measurement, RT-PCR, and western blotting techniques are commonly used to evaluate Ag NP's capability to prevent cellular growth and mediate cell death. This property has led to the development of medical uses such as diagnostics and therapy. The combined effect of Ag NPs and other compounds (natural and synthetic antibiotics, anticancer or antiviral compounds) has a synergistic effect that could aid in the development of new therapeutic approaches, overcoming the current limitations in microbial resistance that have become more prominent in recent years. PVP-coated Ag NPs and Ag^+ ions were also found to be effective against human lung cancer. Although both Ag NPs and Ag^+ had a similar toxic effect on mitochondrial function in an alveolar cell line, A549, it should be noted that Ag NPs produce more ROS than Ag^+ (leading further to DNA damage). The findings demonstrated that Ag NPs might also preferentially inhibit HeLa's cellular mechanisms of DNA damage as well as caspase-mediated cell death. Despite substantial advancements in the laboratory-based implementations of coinage metallic NPs in the medical sciences for disease detection and drug administration, clinical trials and real-world applications are still in their early stages. As a result, there is plenty of room for future studies to bring this technology to ordinary people.

References

1. Ahearn DG, May LL, Gabriel MM (1995) Adherence of organisms to silver-coated surfaces. J Ind Microbiol 15(4):372–376
2. Ali MR, Ibrahim IM, Ali HR, Selim SA, El-Sayed MA (2016) Treatment of natural mammary gland tumors in canines and felines using gold nanorods-assisted plasmonic photothermal therapy to induce tumor apoptosis. Int J Nanomed 11:4849–4863. https://doi.org/10.2147/IJN.S109470
3. Ali MRK et al (2019) Gold-nanoparticle-assisted plasmonic photothermal therapy advances toward clinical application. J Phys Chem C 123(25):15375–15393. https://doi.org/10.1021/acs.jpcc.9b01961
4. Allaker RP, Ren G (2008) Potential impact of nanotechnology on the control of infectious diseases. Trans R Soc Trop Med Hyg 102(1):1–2
5. Angajala G et al (2014) One-step biofabrication of copper nanoparticles from *Aegle marmelos Correa* aqueous leaf extract and evaluation of its anti-inflammatory and mosquito larvicidal efficacy. RSC Adv 4(93):51459–51470
6. Anyaogu KC, Fedorov AV, Neckers DC (2008) Synthesis, characterization, and antifouling potential of functionalized copper nanoparticles. Langmuir 24(8):4340–4346
7. Banerjee M, Mallick S, Paul A, Chattopadhyay A, Ghosh SS (2010) Heightened reactive oxygen species generation in the antimicrobial activity of a three-component iodinated chitosan-silver nanoparticle composite. Langmuir ACS J Surf Colloids 26(8):5901–5908. https://doi.org/10.1021/la9038528
8. Brown SD et al (2010) Gold nanoparticles for the improved anticancer drug delivery of the active component of oxaliplatin. J Am Chem Soc 132(13):4678–4684. https://doi.org/10.1021/ja908117a
9. Chaloupka K, Malam Y, Seifalian AM (2010) Nanosilver as a new generation of nanoproduct in biomedical applications. Trends Biotechnol 28(11):580–588. https://doi.org/10.1016/j.tibtech.2010.07.006
10. Chang SY, Huang KY, Chao TL, Kao HC, Pang YH, Lu L, Chiu CL, Huang HC, Cheng TR, Fang JM, Yang PC (2021) Nanoparticle composite TPNT1 is effective against SARS-CoV-2 and influenza viruses. Sci Rep 11(1):8692. https://doi.org/10.1038/s41598-021-87254-3
11. Cheirmadurai K, Biswas S, Murali R, Thanikaivelan P (2014) Green synthesis of copper nanoparticles and conducting nanobiocomposites using plant and animal sources. RSC Adv 4(37):19507–19511. https://doi.org/10.1039/C4RA01414F
12. Chen G, Roy I, Yang C, Prasad PN (2016) Nanochemistry and nanomedicine for nanoparticle-based diagnostics and therapy. Chem Rev 116(5):2826–2885. https://doi.org/10.1021/acs.chemrev.5b00148
13. Cioffi N et al (2005) Copper nanoparticle/polymer composites with antifungal and bacteriostatic properties. Chem Mater 17(21):5255–5262
14. Cui Y, Zhao Y, Tian Y, Zhang W, Lu X, Jiang X (2012) The molecular mechanism of action of bactericidal gold nanoparticles on *Escherichia coli*. Biomaterials 33:2327–2333. https://doi.org/10.1016/j.biomaterials.2011.11.057
15. Dakal TC, Kumar A, Majumdar RS, Yadav V (2016) Mechanistic basis of antimicrobial actions of silver nanoparticles. Front Microbiol 7:1831. https://doi.org/10.3389/fmicb.2016.01831
16. Dhanalekshmi KI, Magesan P, Sangeetha K, Zhang X, Jayamoorthy K, Srinivasan N (2019) Preparation and characterization of core-shell type Ag@SiO_2 nanoparticles for photodynamic cancer therapy. Photodiagn Photodyn Ther 28:324–329. https://doi.org/10.1016/j.pdpdt.2019.10.006
17. Dhanalekshmi KI, Sangeetha K, Magesan P, Johnson J, Zhang X, Jayamoorthy K (2020) Photodynamic cancer therapy: role of Ag- and Au-based hybrid nano-photosensitizers. J Biomol Struct Dyn 1–8. https://doi.org/10.1080/07391102.2020.1858965

18. Dhar S, Reddy EM, Shiras A, Pokharkar V, Prasad BLV (2008) Natural gum reduced/stabilized gold nanoparticles for drug delivery formulations. Chem A Eur J 14(33):10244–10250. https://doi.org/10.1002/chem.200801093
19. Ding Y, Jiang Z, Saha K, Kim CS, Kim ST, Landis RF, Rotello VM (2014) Gold nanoparticles for nucleic acid delivery. Mol Ther J Am Soc Gene Ther 22(6):1075–1083. https://doi.org/10.1038/mt.2014.30
20. Dong L, Ji G, Liu Y, Xu X, Lei P, Du K, Song S, Feng J, Zhang H (2018) Multifunctional Cu-Ag_2S nanoparticles with high photothermal conversion efficiency for photoacoustic imaging-guided photothermal therapy in vivo. Nanoscale 10(2):825–831. https://doi.org/10.1039/c7nr07263e
21. Dreaden EC, Alkilany AM, Huang X, Murphy CJ, El-Sayed MA (2012) The golden age: gold nanoparticles for biomedicine. Chem Soc Rev 41(7):2740–2779. https://doi.org/10.1039/c1cs15237h
22. Dreaden EC, Mwakwari SC, Sodji QH, Oyelere AK, El-Sayed MA (2009) Tamoxifen-poly (ethylene glycol)-thiol gold nanoparticle conjugates: enhanced potency and selective delivery for breast cancer treatment. Bioconjug Chem 20(12):2247–2253. https://doi.org/10.1021/bc9002212
23. Duan R, Zhou Z, Su G, Liu L, Guan M, Du B, Zhang Q (2014) Chitosan-coated gold nanorods for cancer therapy combining chemical and photothermal effects. Macromol Biosci 14(8):1160–1169. https://doi.org/10.1002/mabi.201300563
24. Durán N, Durán M, de Jesus MB, Seabra AB, Fávaro WJ, Nakazato G (2016) Silver nanoparticles: a new view on mechanistic aspects on antimicrobial activity. Nanomed Nanotechnol Biol Med 12:789–799. https://doi.org/10.1016/j.nano.2015.11.016
25. Elahi N, Mehdi KM, Baghersad MH (2018) Recent biomedical applications of gold nanoparticles: a review. Talanta 184:537–556
26. El-Sayed MA (2001) Some interesting properties of metals confined in time and nanometer space of different shapes. Acc Chem Res 34(4):257–264
27. Enustun BV, Turkevich J (1963) Coagulation of colloidal gold. J Am Chem Soc 85:3317–3328
28. Faraday M (1857) Experimental relations of gold (and other metals) to light. Philos Trans 147:145–181
29. Feng QL, Wu J, Chen GQ, Cui FZ, Kim TN, Kim JO (2000) A mechanistic study of the antibacterial effect of silver ions on *Escherichia coli* and *Staphylococcus aureus*. J Biomed Mater Res 52(4):662–668
30. Gad SC, Sharp KL, Montgomery C, Payne JD, Goodrich GP (2012) Evaluation of the toxicity of intravenous delivery of auroshell particles (gold–silica nanoshells). Int J Toxicol 31(6):584–594. https://doi.org/10.1177/1091581812465969
31. Gamaleia NF, Shishko ED, Dolinsky GA, Shcherbakov AB, Usatenko AV, Kholin VV (2010) Photodynamic activity of hematoporphyrin conjugates with gold nanoparticles: experiments in vitro. Exp Oncol 32(1):44–47
32. García Calavia P, Chambrier I, Cook MJ, Haines AH, Field RA, Russell DA (2018) Targeted photodynamic therapy of breast cancer cells using lactose-phthalocyanine functionalized gold nanoparticles. J Colloid Interface Sci 512:249–259. https://doi.org/10.1016/j.jcis.2017.10.030
33. Gharatape A, Davaran S, Salehi R, Hamishehkar H (2016) Engineered gold nanoparticles for photothermal cancer therapy and bacteria killing. RSC Adv 6(112):111482–111516. https://doi.org/10.1039/c6ra18760a
34. Ghiuță I, Cristea D (2020) Silver nanoparticles for delivery purposes. In: Nanoengineered biomaterials for advanced drug delivery, pp 347–371. https://doi.org/10.1016/B978-0-08-102985-5.00015-2
35. Ghosh S, Kaushik R, Nagalakshmi K, Hoti SL, Menezes GA, Harish BN, Vasan HN (2010) Antimicrobial activity of highly stable silver nanoparticles embedded in agar-agar matrix as a thin film. Carbohyd Res 345(15):2220–2227
36. Girotti AW (2001) Photosensitized oxidation of membrane lipids: reaction pathways, cytotoxic effects, and cytoprotective mechanisms. J Photochem Photobiol B 63(1):103–113

37. Gogoi SK, Gopinath P, Paul A, Ramesh A, Ghosh SS, Chattopadhyay A (2006) Green fluorescent protein-expressing Escherichia coli as a model system for investigating the antimicrobial activities of silver nanoparticles. Langmuir ACS J Surf Colloids 22(22):9322–9328
38. Gonçalves ASC, Rodrigues CF, Moreira AF, Correia IJ (2020) Strategies to improve the photothermal capacity of gold-based nanomedicines. Acta Biomater. https://doi.org/10.1016/j.actbio.2020.09.008
39. Gupta A, Mumtaz S, Li CH, Hussain I, Rotello VM (2019) Combatting antibiotic-resistant bacteria using nanomaterials. Chem Soc Rev 48(2):415–427
40. Gurunathan S, Raman J, Abd Malek SN, John PA, Vikineswary S (2013) Green synthesis of silver nanoparticles using Ganoderma neo-japonicum Imazeki: a potential cytotoxic agent against breast cancer cells. Int J Nanomed 8:4399–4413
41. Haimov E, Weitman H, Polani S, Schori H, Zitoun D, Shefi O (2018) meso-Tetrahydroxyphenylchlorin-conjugated gold nanoparticles as a tool to improve photodynamic therapy. ACS Appl Mater Interfaces 10(3):2319–2327. https://doi.org/10.1021/acsami.7b16455
42. Horie M, Ogawa H, Yoshida Y et al (2008) Inactivation and morphological changes of avian influenza virus by copper ions. Adv Virol 153(8):1467–1472
43. Horlings RK, Terra JB, Witjes MJ (2015) mTHPC mediated, systemic photodynamic therapy (PDT) for nonmelanoma skin cancers: case and literature review. Lasers Surg Med 47(10):779–787. https://doi.org/10.1002/lsm.22429
44. Jagajjanani Rao K, Paria S (2013) Green synthesis of silver nanoparticles from aqueous Aegle marmelos leaf extract. Mater Res Bull 48(2):628–634
45. Jahangirian H, Lemraski EG, Webster TJ, Rafiee-Moghaddam R, Abdollahi Y (2017) A review of drug delivery systems based on nanotechnology and green chemistry: green nanomedicine. Int J Nanomed 12:2957–2978. https://doi.org/10.2147/IJN.S127683
46. Kah JC, Chen J, Zubieta A, Hamad-Schifferli K (2012) Exploiting the protein corona around gold nanorods for loading and triggered release. ACS Nano 6(8):6730–6740. https://doi.org/10.1021/nn301389c
47. Kim C, Ghosh P, Rotello VM (2009) Multimodal drug delivery using gold nanoparticles. Nanoscale 1(1):61–67
48. Kim YH et al (2006) Preparation and characterization of the antibacterial Cu nanoparticle formed on the surface of SiO_2 nanoparticles. J Phys Chem B 110(49):24923–24928
49. Konan YN, Gurny R, Allemann E (2002) State of the art in the delivery of photosensitizers for photodynamic therapy. J Photochem Photobiol B 66(2):89–106
50. Kumari K, Singh P, Bauddh K, Sweta, Mallick S, Chandra R (2019) Implications of metal nanoparticles on aquatic fauna: a review. Nanosci Nanotechnol Asia 9(1):30–43
51. Lai HZ, Chen WY, Wu CY, Chen YC (2015) Potent antibacterial nanoparticles for pathogenic bacteria. ACS Appl Mater Interfaces 7:2046–2054. https://doi.org/10.1021/am507919m
52. Lee C, Gaston MA, Weiss AA, Zhang P (2013) Colorimetric viral detection based on sialic acid stabilized gold nanoparticles. Biosens Bioelectron 42:236–241. https://doi.org/10.1016/j.bios.2012.10.067
53. Lee K-S, El-Sayed MA (2006) Gold and silver nanoparticles in sensing and imaging: sensitivity of plasmon response to size, shape, and metal composition. J Phys Chem B 110(39):19220–19225. https://doi.org/10.1021/jp062536y
54. Lee PC, Meisel D (1982) Adsorption and surface-enhanced Raman of dyes on silver and gold sols. J Phys Chem 86(17):3391–3395
55. LewisOscar F, Mubarak Ali D, Nithya C, Priyanka R, Gopinath V, Alharbi NS et al (2015) One-pot synthesis and anti-biofilm potential of copper nanoparticles (CuNPs) against clinical strains of Pseudomonas aeruginosa. Biofouling 31:379–391. https://doi.org/10.1080/08927014.2015.1048686
56. Li Q, Mahendra S, Lyon DY, Brunet L, Liga MV, Li D, Alvarez PJJ (2008) Antimicrobial nanomaterials for water disinfection and microbial control: potential applications and implications. Water Res 42(18):4591–4602

57. Lin C-Y, Yu C-J, Lin Y-H, Tseng W-L (2010) Colorimetric sensing of silver(I) and mercury (II) ions based on an assembly of tween 20-stabilized gold nanoparticles. Anal Chem 82(16):6830–6837. https://doi.org/10.1021/ac1007909
58. Lin J et al (2013) Photosensitizer-loaded gold vesicles with strong plasmonic coupling effect for imaging-guided photothermal/photodynamic therapy. ACS Nano 7(6):5320–5329
59. Liu Y, Li F, Guo Z, Xiao Y et al (2019) Silver nanoparticle-embedded hydrogel as a photothermal platform for combating bacterial infections. Chem Eng J 122990. https://doi.org/10.1016/j.cej.2019.122990
60. Liz-Marzán LM (2004) Nanometals: formation and color. Mater Today 7(2):26–31. https://doi.org/10.1016/s1369-7021(04)00080-x
61. Liz-Marzán LM (2006) Tailoring surface plasmons through the morphology and assembly of metal nanoparticles. Langmuir 22(1):32–41
62. Lu HD, Wang LZ, Wilson BK et al (2018) Copper loading of preformed nanoparticles for PET-imaging applications. ACS Appl Mater Interfaces 10(4):3191–3199. https://doi.org/10.1021/acsami.7b07242
63. Ma L, Li L, Li X, Deng L, Zheng H et al (2016) Silver sulfide nanoparticles as photothermal transducing agents for cancer treatment. J Nanomater Mol Nanotechnol 5:2. https://doi.org/10.4172/2324-8777.1000182
64. Mackey MA, Ali MRK, Austin LA, Near RD, El-Sayed MA (2014) The most effective gold nanorod size for plasmonic photothermal therapy: theory and in vitro experiments. J Phys Chem B 118(5):1319–1326. https://doi.org/10.1021/jp409298f
65. Majumdar KC, Sinha B (2014) Coinage metals (Cu, Ag and Au) in the synthesis of natural products. RSC Adv 4(16):8085–8120
66. Mallick S, Sabui P (2021) Green synthesis of copper and copper-based nanoparticles for their use in medicine. In: Rai M, Patel M, Patel R (eds) Nanotechnology in medicine, pp 174–194
67. Mallick S et al (2021) Bionanomaterials utility for therapeutic applications. In: Bionanomaterials. IOP Publishing: 7-1-7-29
68. Mallick S, Mukhi P, Kumari P, Mahato KR, Verma SK, Das D (2019) Synthesis, characterization and catalytic application of starch supported cuprous iodide nanoparticles. Catal Lett 149(12):3501–3507
69. Mallick S, Sanpui P, Ghosh SS et al (2015) Synthesis, characterization and enhanced bactericidal action of chitosan supported core-shell copper-silver nanoparticle composite. RSC Adv 5(16):12268–12276
70. Mallick S, Sharma S, Banerjee M, Ghosh SS, Chattopadhyay A, Paul A (2012) Iodine-stabilized Cu nanoparticle chitosan composite for antibacterial applications. ACS Appl Mater Interfaces 4(3):1313–1323
71. Meena KS, Dhanalekshmi KI, Jayamoorthy K (2016) Study of photodynamic activity of Au@SiO_2 core-shell nanoparticles in vitro. Mater Sci Eng C 63:317–322
72. Merkel TJ et al (2010) Scalable, shape-specific, top-down fabrication methods for the synthesis of engineered colloidal particles. Langmuir 26(16):13086–13096
73. Mie G (1908) Contributions to the optics of turbid media, particularly of colloidal metal solutions. Ann Phys 25:377–445. https://doi.org/10.1002/andp.19083300302
74. Mohammed Fayaz A, Girilal M, Mahdy SA, Somsundar SS, Venkatesan R, Kalaichelvan PT (2011) Vancomycin bound biogenic gold nanoparticles: a different perspective for development of anti VRSA agents. Process Biochem 46:636–641. https://doi.org/10.1016/j.procbio.2010.11.001
75. Morones JR, Elechiguerra JL, Camacho A, Holt K, Kouri JB, Ramírez JT, Yacaman MJ (2005) The bactericidal effect of silver nanoparticles. Nanotechnology 16(10):2346–2353
76. Nagar N, Devra V (2018) Green synthesis and characterization of copper nanoparticles using Azadirachta indica leaves. Mater Chem Phys 213:44–51
77. Nagy-Simon T, Potara M, Craciun A-M, Licarete E, Astilean S (2018) IR780-dye loaded gold nanoparticles as new near infrared activatable nanotheranostic agents for simultaneous photodynamic and photothermal therapy and intracellular tracking by surface enhanced resonant Raman scattering imaging. J Colloid Interface Sci 517:239–250. https://doi.org/10.1016/j.jcis.2018.02.007

78. Park MV, Neigh AM, Vermeulen JP, de la Fonteyne LJ, Verharen HW, Briedé JJ, van Loveren H, de Jong WH (2011) The effect of particle size on the cytotoxicity, inflammation, developmental toxicity and genotoxicity of silver nanoparticles. Biomaterials 32(36):9810–9817. https://doi.org/10.1016/j.biomaterials.2011.08.085
79. Patra JK, Das G, Fraceto LF et al (2018) Nano based drug delivery systems: recent developments and future prospects. J Nanobiotechnol 16(1):71. https://doi.org/10.1186/s12951-018-0392-8
80. Pattani VP, Shah J, Atalis A, Sharma A, Tunnell JW (2015) Role of apoptosis and necrosis in cell death induced by nanoparticle-mediated photothermal therapy. J Nanopart Res 17(1):20
81. Payne JN, Waghwani HK, Connor MG, Hamilton W, Tockstein S, Moolani H et al (2016) Novel synthesis of kanamycin conjugated gold nanoparticles with potent antibacterial activity. Front Microbiol 7:607. https://doi.org/10.3389/fmicb.2016.00607
82. Prajapati JP, Das D, Katlakunta S, Maramu N, Ranjan V, Mallick S (2021) Synthesis and characterization of ultrasmall Cu_2O nanoparticles on silica nanoparticles surface. Inorg Chim Acta 515:120069
83. Rajpurohit AS, Punde NS, Srivastava AK (2019) An electrochemical sensor for the simultaneous detection of potential diabetic biomarkers: methylglyoxal and its detoxification enzyme Glyoxalase employing copper oxide/gold nanoparticles modified electrode. New J Chem 43:16572–16582. https://doi.org/10.1039/c9nj03553b
84. Rastinehad AR et al (2019) Gold nanoshell-localized photothermal ablation of prostate tumors in a clinical pilot device study. Proc Natl Acad Sci 116(37):18590–18596
85. Ravi SS, Christena LR, Sai Subramanian N, Anthony SP (2013) Green synthesized silver nanoparticles for selective colorimetric sensing of Hg^{2+} in aqueous solution at wide pH range. Analyst 138(15):4370. https://doi.org/10.1039/c3an00320e
86. Raj SI, Jaiswal A, Uddin I (2020) Ultrasmall aqueous starch-capped CuS quantum dots with tunable localized surface plasmon resonance and composition for the selective and sensitive detection of mercury(ii) ions. RSC Adv 10(24):14050–14059
87. Rivas Aiello MB, Castrogiovanni D, Parisi J et al (2018) Photodynamic therapy in HeLa cells incubated with riboflavin and pectin-coated silver nanoparticles. Photochem Photobiol 94(6):1159–1166. https://doi.org/10.1111/php.12974
88. Roshmi T, Soumya KR, Jyothis M, Radhakrishnan EK (2015) Effect of biofabricated gold nanoparticle-based antibiotic conjugates on minimum inhibitory concentration of bacterial isolates of clinical origin. Gold Bull 48:63–71. https://doi.org/10.1007/s13404-015-0162-4
89. Russell AD, Hugo WB (1994) Antimicrobial activity and action of silver. Prog Med Chem 31:351–370
90. Sabela M, Balme S, Bechelany M, Janot J-M, Bisetty K (2017) A Review of gold and silver nanoparticle-based colorimetric sensing assays. Adv Eng Mater 19(12):1700270. https://doi.org/10.1002/adem.201700270
91. Sanpui P et al (2008) The antibacterial properties of a novel chitosan–Ag-nanoparticle composite. Int J Food Microbiol 124(2):142–146
92. Senge MO, Brandt JC (2011) Temoporfin (Foscan®, 5,10,15,20-tetra(m-hydroxyphenyl) chlorin)—a second-generation photosensitizer. Photochem Photobiol 87(6):1240–1296. https://doi.org/10.1111/j.1751-1097.2011.00986.x
93. Sharma VK, Yngard RA, Lin Y (2009) Silver nanoparticles: green synthesis and their antimicrobial activities. Adv Coll Interface Sci 145(1):83–96
94. Sharma RK et al (2012) Preparation of gold nanoparticles using tea: a green chemistry experiment. J Chem Educ 89(10):1316–1318
95. Tao Y, Ju E, Liu Z, Dong K, Ren J, Qu X (2014) Engineered, self-assembled near-infrared photothermal agents for combined tumor immunotherapy and chemo-photothermal therapy. Biomaterials 35(24):6646–6656. https://doi.org/10.1016/j.biomaterials.2014.04.073
96. Vankayala R, Huang Y-K, Kalluru P, Chiang C-S, Hwang KC (2014) First demonstration of gold nanorods-mediated photodynamic therapeutic destruction of tumors via near infra-red-light activation. Small 10:1612–1622

97. Vargas A, Pegaz B, Debefve E, Konan-Kouakou Y, Lange N, Ballini JP, van den Bergh H, Gurny R, Delie F (2004) Improved photodynamic activity of porphyrin loaded into nanoparticles: an in vivo evaluation using chick embryos. Int J Pharmaceut 286(1–2):131–145. https://doi.org/10.1016/j.ijpharm.2004.07.029
98. Vijayaraghavan P, Liu C-H, Vankayala R, Chiang C-S, Hwang KC (2014) Designing multibranched gold nanoechinus for NIR light activated dual modal photodynamic and photothermal therapy in the second biological window. Adv Mater 26(39):6689–6695. https://doi.org/10.1002/adma.201400703
99. Wang F, Wang Y-C, Dou S, Xiong M-H, Sun T-M, Wang J (2011) Doxorubicin-tethered responsive gold nanoparticles facilitate intracellular drug delivery for overcoming multidrug resistance in cancer cells. ACS Nano 5(5):3679–3692. https://doi.org/10.1021/nn200007z
100. Wang X, Zhuang J, Peng Q, Li Y (2005) A general strategy for nanocrystal synthesis. Nature 437(7055):121–124. https://doi.org/10.1038/nature03968
101. Wei Y et al (2010) Synthesis of stable, low-dispersity copper nanoparticles and nanorods and their antifungal and catalytic properties. J Phys Chem C 114(37):15612–15616
102. Whitesides GM (2005) Nanoscience, nanotechnology, and chemistry. Small 1(2):172–179. https://doi.org/10.1002/smll.200400130
103. World Health Organization (2008) The global burden of disease: 2004 update. World Health Organization. https://apps.who.int/iris/handle/10665/43942
104. Yang W et al (2019) Gold nanoparticle based photothermal therapy: development and application for effective cancer treatment. Sustain Mater Technol 22:e00109
105. Ye X, Gao Y, Chen J, Reifsnyder DC, Zheng C, Murray CB (2013) Seeded growth of monodisperse gold nanorods using bromide-free surfactant mixtures. Nano Lett 13(5):2163–2171. https://doi.org/10.1021/nl400653s
106. Yeh YC, Creran B, Rotello VM (2012) Gold nanoparticles: preparation, properties, and applications in bionanotechnology. Nanoscale 4(6):1871–1880
107. Yuan YG, Peng QL, Gurunathan S (2017) Silver nanoparticles enhance the apoptotic potential of gemcitabine in human ovarian cancer cells: combination therapy for effective cancer treatment. Int J Nanomed 12:6487–6502
108. Zhang X, Gurunathan S (2016) Combination of salinomycin and silver nanoparticles enhances apoptosis and autophagy in human ovarian cancer cells: an effective anticancer therapy. Int J Nanomed 11:3655–3675
109. Zhang Z et al (2014) Near-infrared laser-induced targeted cancer therapy using thermoresponsive polymer encapsulated gold nanorods. J Am Chem Soc 136(20):7317–7326

Role of Inorganic Nanocomposite Materials in Drug Delivery Systems

Hira Ateeq, Afaf Zia, Qayyum Husain, and Mohd Sajid Khan

Abstract Nanotechnology comprehends the study of various attributes of nanostructured materials at the molecular and sub-molecular levels. Nanoparticles (NPs) have garnered attention recently as a potential tool for targeted therapy in the clinical setting. Distinct benefits of nanotechnology-based drug-delivery systems to overcome the pharmacokinetic limitations of conventional treatments are due to their small size and thus promise a good course of action in all aspects of life. Small biological targets such as RNA, DNA and proteins can be easily manipulated by nano drug delivery vehicles. The ultimate goal of research interest in nanotechnology is to create therapeutically relevant nanoparticles (NPs) with suitable dimensions, chemical composition, and surface properties that can encapsulate the appropriate dosage of a targeted drug or molecule with improved kinetics and dynamics in a biological system. To increase safety and effectiveness, NPs facilitate transport across membranes, enhance the stability and solubility of encapsulated cargos, and improve circulation time. Inorganic NPs are used in various drug delivery and imaging applications and are manufactured to have different sizes, topologies and geometries. These factors have led to substantial NPs research that has produced positive outcomes, with inorganic NPs serving as the main focus for the delivery of the drug.

Keywords Drug delivery · Inorganic nanoparticles · Targeted therapy · Endocytosis · Oncology · Skin

H. Ateeq · Q. Husain · M. S. Khan (✉)
Department of Biochemistry, Faculty of Life Sciences, Aligarh Muslim University, Aligarh 202002, India
e-mail: researchsajid@gmail.com

A. Zia
Dr. Ziauddin Ahmad Dental College, Aligarh Muslim University, Aligarh 202002, India

I. Uddin and I. Ahmad (eds.), *Synthesis and Applications of Nanomaterials and Nanocomposites*, Composites Science and Technology,
https://doi.org/10.1007/978-981-99-1350-3_7

Abbreviations

NPs	Nanoparticles
P-gp	P-glycoprotein
MDR	Multidrug resistance
ROS	Reactive oxygen species
AuNPs	Gold nanoparticles
NIR	Near-infrared
AgNPs	Silver nanoparticles
EPR	Enhanced permeability and retention
RES	Reticuloendothelial system
SiNPs	Silica nanoparticles
IONPs	Iron oxide nanoparticles

1 Introduction

Nanotechnology has widened domains of unexplored science in nature. Nanoparticles (NPs) have revolutionised engineering, industry and nanomedicines sectors. The enhanced structural integrity and unique chemical, optical, mechanical and magnetic properties owing to the characteristic size, distribution and structure promise more comprehensive applications of NPs in all aspects of life. NPs are smaller than cellular organelles, alike in dimensions to macromolecules present biologically, thus providing new elements to translational research.

Research efforts worldwide are invested simultaneously to develop nanoproducts to improve health care and advance medical research. Various biomedical applications such as drug delivery have garnered increased nanotechnology use due to their ability to alter the drug pharmacokinetics. The traditional drug delivery affects normal body cells and has limited specificity, uncontrolled biodistribution and risks of intracellular trafficking while targeting specific cells on the conjugation of the drug with NPs [44]. On administration into the body, orally or through injections, the interactions between NPs and biomolecules are inescapable before the NPs reach the effective sites via the bloodstream. The biodistribution, therapeutic efficacy, and biocompatibility of NPs are affected by the formation of a protein layer known as protein corona due to their interaction with serum opsonin proteins. As a result, there is a need to design a nano-drug delivery system where NPs serve as a drug delivery vehicle for drugs to reach target cells such as cancer cells. With the advancement of nanotechnology, more potent drug delivery systems with improved drug efficacy and lower side effects have evolved to overcome these challenges of traditional drug delivery [58]. Regarding its characteristics and transport, a drug carrier at the nanoscale functions as a single entity. These nanoclusters have a confined size range and at least one dimension between one and 100 nm (nm). It is anticipated that revolutionary pharmaceutical therapies that target the disease site and help lower the cost and toxicity of the active component in healthcare would be facilitated by nanotechnology-enabled drug delivery vehicles. The targeting moiety of the functionalised NPs recognises and preferentially binds to the receptors expressed by the target cells once they are close to them [6]. The active drug may occasionally be released inside the target cells, including the nucleus, after the NPs have entered the cells through receptor-mediated internalisation processes [17, 25]. Nano encapsulated cargos have increased stability and solubility, easy transport across membranes, and enhanced safety and efficacy with prolonged circulation time [32]. Enormous quantity of cargo, such as chemotherapeutic medicines, plasmid DNA and small inhibitory (siRNA) can be loaded onto NPs owing to their high surface-area-to-volume ratios and easy surface functionalisation chemistries. NPs can simultaneously deliver more than one drug [25]. Nanomedicine arose from the widespread use of nanotechnology in treating and detecting diseases [9].

Biological applications of NPs are determined by their method of synthesis and properties. The colloidal NPs have been proved to be highly stable and aqueous suspension cater optimised physico-chemical properties including size, shape, surface charge and chemistry, stability, and optical properties. Biodistribution of these NPs is dependent on their size and various surface characteristics following their dispension in the target cell. The respiratory, gastric, transdermal, intravenous, and intramuscular systems can all be used to administer NPs. NPs eventually enter the bloodstream and lymphatics through most of these pathways, collect in the organs and are cleared [11].

2 Inorganic Nanoparticles

Since inorganic NPs possess highly advanced chemical properties, they find major applications as drug carriers. Ascribed to their unique material characteristics and size-dependent physicochemical characteristics, inorganic NPs have recently attracted much attention. NPs generated from metal, carbon-based materials, ceramic, etc., are known as inorganic NPs. They have large surface-to-volume ratios and their surface functionalisation strongly affects their behaviour. Various physical, optical, and magnetic properties, in addition to ease of functionalisation, stability and inertness, make inorganic NPs an attractive alternative to organic NPs. Heavy metals such as gold (Au) and silver (Ag) are being used to produce standard inorganic NPs. Metals such as nickel (Ni), cobalt (Co), iron (Fe), magnetite (Fe_30_4) and iron-platinum (FePt) are used to design magnetic NPs particles with supra magnetic properties in a magnetic field. Fluorescent NPs include quantum dots, silicon dioxide (SiO_2), etc. [47].

Inorganic NPs have received a great deal of attention as possible disease modifiers and drug distribution channels [8]. The stability and resistance of inorganic NPs to microbial attack can be imparted to shallow toxicity profile, biocompatibility and hydrophilic nature. Inorganic NPs are preferred over organic NPs in biomedical science to develop a drug delivery system because they can internalize into the cells without interacting with P-glycoprotein (P-gp) and potentiating controlled release of drugs [52]. P-gp is multidrug resistance (MDR) protein belonging to the ATP-binding cassette (ABC) family of efflux transporters that has a prominent role in cancer research and drug delivery. MDR can be referred to as cross-resistance acquired by cancerous cells towards major chemotherapeutic drugs, thus vitally contributing to the therapeutic collapse. P-gp is normally expressed in the breast, brain, kidney, liver, and colon but overexpressed in cancerous cells. P-gp promotes tumor progression and compromises treatment by reducing therapeutic efficacy. P-gp is the primary therapeutic target to combat MDR as its overexpression is highly noticeable in tumors [33]. Over the years, the designed P-gp inhibitors have shown dismaying results due to unexpectedly low selectivity, poor inhibitory efficacy and accumulated toxic plasma concentrations causing side effects in patients. Other drug delivery systems, such as NPs, liposomes, etc., have depicted promising ability to evade P-gp-mediated drug efflux and improve intracellular concentration without altering the pharmacokinetics of the co-administered anticancer drug and limited toxicity. Nanocarrier systems characteristically evade the drug efflux transporters by efficiently escaping MDR [53]. Nanomaterials derived from Au, Fe, Si and lanthanides have shown promising carrier platforms for therapeutics attributed to enhanced stability and optical responsiveness [41]. High cellular absorption capacity, lack of an immune response, and minimal toxicity are further benefits of using inorganic NPs as drug or gene delivery vehicles. Their physical, chemical and biological characteristics are markedly different from those of their bulk counterparts. The most widely Food and Drug Administration (FDA) approved inorganic NPs are Iron oxide (Fe_2O_3) NPs, including several imaging agents, and iron (Fe) carboxy maltose for anemia. However, in Phase II clinical trials, AuNPs against multiple sclerosis, amyotrophic lateral sclerosis, and Parkinson's Disease [8]. Other characteristic properties of inorganic NPs are electron-rich surfaces due to high surface area to volume ratios. High surface energy possessed by inorganic NPs facilitates the easy exchange of outer electrons and surface vacancies with electron acceptors and donors in a biological system. Inorganic NPs increase reactive oxygen species (ROS) production brought on by metabolic processing of nanoparticles through enzyme- or metal-catalyzed chemical reactions in a compromised cellular environment, such as a tumor. ROS overproduction further induces oxidative stress, disturbing normal physiological redox-regulated functions, damaging DNA, disrupting cell signalling cascade, changing cell motility and enhancing cytotoxicity and apoptosis [19, 29]. Several inorganic phosphate NPs such as Ca, Sr, Mg, and Mn phosphates, CNTs, Si, Au, QDs, and double hydroxides are used for different applications to deliver DNA. Modifying the surface of inorganic NPs helps in the targeted drug delivery and keeps track of drug release [42]. Inorganic nanoparticles offer excellent biocompatibility, low immunogenicity, low cost, highly scalable production, and strong loading capacity in addition to having highly favourable chemical, thermal, and mechanical properties. Large proteins or oligonucleotide strands and a variety of small drugs can be transported owing to the configurable diameter of porous inorganic NPs. Drug conjugated inorganic NPs provide a versatile platform for image diagnostic and therapeutic applications. Metallic NPs and inorganic nanomaterials such as graphene,

Si and silicon are rigid and resistant. Still, they can be easily modified chemically and mechanically, felicitating more accessible transport into tumor cells but encountering hampered penetration due to limited flexibility. Organic nanocarriers such as liposomes can easily penetrate tumors due to their soft nature but present premature drug release.

Inorganic NPs have revolutionised the biomedical field- nano biosensors have aided in cellular tracking, imaging and disease diagnosis. Luminescent nanodevices have contributed substantially to nano diagnosis in identifying disease at the cellular or molecular level. Nanotherapy involving nanocarrier systems to target the disease site with increased accuracy and enhanced drug efficacy and minimum side effects has garnered tremendous success. Another important branch of biomedicine that has received attention is regenerative medicine-nanotechnology. Tools and techniques such as gene and cell therapy remove or replace damaged tissues and organs. Dosage therapy to promote healing includes replacing or restoring damaged tissues or organs using nanotechnology methods, such as bio regenerative tissue engineering.

Apart from various advantages, naked or bare inorganic NPs cannot be used since they have a harmful effect on biological living cells. A biological system lacks a mechanism to eliminate these NPs, which penetrate membranes and accumulate in the body, causing toxicity. Induction of oxidative stress, which causes damage to biological macromolecules, genotoxicity and immunotoxicity or inflammation, is the most typical mechanism of NPs toxicity. Owing to their distinctive surface properties and interactions with various biomolecules, unbound NPs directly produce radical and non-radical ROS. Therefore the inorganic NPs are encapsulated with a nontoxic, biocompatible layer. Encapsulation with another metal forms core/shell NPs and compliments properties such as quantum yield and photoluminescence. The non-radiative combination sites are passivated, providing air stability and reducing surface defects [12]. Encapsulated NPs are transported between cells by either endocytosis or exocytosis. Endocytosis mediated transport is caveolin or clathrin proteins dependent or independent and plays a crucial role in the cellular internalisation of NPs [5, 30, 45].

The size of NPs is relatable to their deposition in the respiratory tract; smaller particles tend to accumulate less. Inhaled NPs are phagocytosed by the macrophages or cleared via a mucociliary escalator. The NPs may retain, aggregate, or metabolise on passage to the lungs. The particles further move to different tissues via the bloodstream. It has been shown that mediocre-sized NPs (hydrodynamic diameter <34 nm) pass from the lungs to mediastinal lymph nodes very fast, whereas small-sized particles (hydrodynamic diameter <6 nm) diffuse into the bloodstream and are eventually cleared from the body via urine. After entering the bloodstream, NPs disseminated all over the body, with the liver being the most common target, though other organs can also be targeted [4, 35].

Chemical and biological processes are used to synthesize NPs using two directional approaches. The bottom-up approach manipulates atoms, whereas the top-down method relies on breaking larger structures into nanostructures using different chemical, physical, or biological techniques [16]. The environmentally friendly production of metallic NPs using biological components, i.e., green synthesis, has gained much attention. Biomass, plant extracts, microbes, and a range of additional reductants are used in green synthesis [27].

2.1 Classes of Inorganic Nanoparticles as Drug Delivery Vehicles

2.1.1 Gold Nanoparticles (AuNPs)

The valuable properties of Au have been explored in many applications. Due to their distinctive combination of physical, chemical, electrical, and optical features, AuNPs, which first appeared in the late 1990s, have garnered substantial interest in targeted drug delivery, disease diagnostics, and monitoring surgical procedures. Au can be functionalised to generate monodisperse nanostructures with high specificity and chemical inertness with a wide range of ligand species and other chemical moieties. The primary purpose of nanomedicine is to achieve targeted delivery. AuNPs play a significant role in achieving this goal as they can be easily synthesised, functionalised and are biocompatible. AuNPs have emerged as a promising candidate for delivering a variety of payloads to the desired location. AuNPs have a large surface to volume ratio, which allows them to transport multiple drug molecules. Small drug molecules and large macromolecules like DNA, RNA, and proteins can be easily transported. Without modifying an AuNPs monolayer for administration, some pharmaceutical drugs can be directly coupled with AuNPs through physical absorption and ionic or covalent interaction [23]. The optical behaviour of AuNPs less than 2 nm is based on plasmonic characteristics, which allows AuNPs to absorb and scatter light with incredible efficiency. Plasmonic characteristics can be tailored to absorb specific wavelengths based on form and aspect ratio. AuNPs have far higher absorption and scattering intensity than most absorbing and scattering organic molecular dyes, making them effective contrast agents in imaging. When AuNPs are exposed to near-infrared (NIR) light (650–900 nm), electron–phonon and phonon–phonon interactions generate heat. Due to negligible absorption by water and blood in this area of the wavelength spectrum, NIR light penetrates human tissue from half a mm to a few centimetres (cm). Au nanoshells, nanocages, and nanorods prepared using various chemical and electrochemical synthesis methods absorb light in the near-infrared range and transfer heat to the surrounding tissue. They have been widely used and investigated for tumor imaging and ablation. AuNPs with a core size of mid-to-large differ from ultrasmall counterparts in various respects, the most significant of which is efficient light-to-heat conversion. This feature of AuNPs was first employed in photothermal treatment to eliminate malignant tissue by converting NIR irradiation into heat. Photothermal therapy with AuNPs can be combined with other therapeutic techniques, allowing for multimodal cancer treatment. However, this method has problems such as high irradiation power density and low selectivity.

Photothermal properties of AuNPs nanoshells are also being investigated for prostate cancer treatment [48].

Another parameter essential for the coherent delivery of therapeutic agents via AuNPs is the efficient release of the drug. Internal stimuli such as glutathione, enzymes and pH and external stimuli such as light affect the release of play loads from AuNPs. Covalent attachment of cargo to the NPs presents challenges such as early detachment before reaching the target site; therefore, non-covalent encapsulation is a good alternative. AuNPs are coated with polyethylene glycol (PEG)- (PEGylation NPs), oligonucleotides or conjugated with amino acid and peptides to deliver other payloads. PEGlyations alter the size of NPs and protect them from recognition by enzymes and antibodies. Thus coated NPs evade rapid degradation and secretion from the body. The coating does not, however, stop macrophages or other immune cells from fully recognising the NPs. When formed in higher concentrations in response to PEGlyation NPs, the anti-PEG antibodies can rapidly remove NPs from the body [56]. Hydrophobic pockets of PEGylated AuNPs provide enhanced loading capacity. The ligand monolayer on AuNPs also allows for drug encapsulation and delivery. For delivery to cancer cells, a hydrophobic small molecular anticancer drug can be put into a monolayer of 2 nm core AuNPs. AuNPs are conjugated with PEG alone or in the presence of another molecule during the PEGylation process to improve cellular absorption of AuNPs. Biotin, peptides, and oligonucleotides are examples of these molecules. Because of their capacity to bind to cell membranes, these functionalised AuNPs are employed for targeted drug administration. The production of AuNPs functionalised with lectin, lactose, biotin, and PEG has been studied. Such functionalised AuNPs can be used for inter and intracellular internalisation. Additionally, PEGylated AuNPs can be coupled to thiols that have the fluorescent dye coumarin bound to one side. These fluorescent (fAuNP) moieties operate as hetero-bifunctional components. The dye molecule attached at one end allows them to enter the cells and be tracked at the same time. There has been a study that involves internalising such fAuNPs for use in tumor ablation in mice. The findings of this study demonstrated that physiochemical characteristics, ligand conjugation, and AuNP size all impacted how well AuNPs were internalised by cells [2].

2.1.2 Silica (Si) Nanoparticles

Si-based nanocarriers have a high loading capacity and are simple to functionalise for a wide range of applications. Si NPs find tremendous attraction in biological applications due to their unique qualities such as size, shape, structure, and morphology, all of which can be easily manipulated. The made-to-order nature of porous inorganic NPs has also offered tremendous advantages in drug delivery applications due to their high drug loading capacity and a high exterior surface area ranging from 500 to 10,000 m^2g^{-1}. These particles can easily cross cell membranes and deliver wide-ranging biomolecules, from macromolecules to tiny organic molecules. For their potential to solve critical therapeutic concerns such as drug solubility of drug and prolonged release of drug, Si nanomaterials, particularly porous NPs, have picked interest as delivery platforms. Mesoporous silica NPs (MSiNPs) exhibit hollow channels and honeycomb-like structures and are employed as silica-based delivery vehicles. Due of their extended cargo loading, heat resistance, chemical stability, and adjustable structure, MSi NPs have become popular delivery vehicles. They have shown to be promising in delivering a variety of treatments and anticancer drugs for various malignancies. MSiNPs are typically synthesised using a cationic surfactant as a template for the base-catalysed sol–gel method. The method and template selection are highly adjustable, which allows for controlling the particle size, pore quantity, and pore radius of the final MSi NPs according to applications. The cargos are encapsulated via non-covalent binding inside the pores of MSiNPs; however, the controlled cargo release at targets can be benefited from rationally designed covalent surface changes. The surface functionalization and pore sizes influence drug loading capacity, and the adjustable release rate of MSiNPs is influenced by pore size and surface functionalization. Cargo molecules are often loaded into MSi NPs by weak non-covalent interactions such as hydrogen bonding, physical adsorption, electrostatic contact or aromatic stacking. Charge density and steric effects within the pore can be significantly influenced by changing pore parameters, altering weak electrostatic interactions, and affecting cargo release [26].

Porous silicon (PsiNPs) have attracted a lot of interest due to their tailorable properties, making them a flexible and multipurpose substance in nanomedicine for drug administration, diagnostics and therapeutic applications interest. Psi is composed of crystalline Si traversed by pores with nanometer-scale ranging from 2–50 nm for drug delivery applications. The pore structure depends on manufacturing conditions that offer exact control of physical properties such as pore diameter, shape and NPs size to be altered with ease to accommodate various qualities. Because it decomposes into orthosilicic acid, which is safe for cells, it has in vivo biocompatibility and biodegradability. PSi also has a programmable pore nanostructure, customised surface features, and a large surface area and pore volume, allowing it to load a therapeutic cargo of varying sizes, ranging from massive macromolecules to tiny chemotherapeutic chemicals. The optoelectronic properties of Psi also allow them to function as an imaging agent, photosensitiser, and theranostic platform, releasing therapeutic drugs while providing information about the diseased tissue. Immersion or chemical conjugation can be used to achieve drug loading. Thus, the payload will be held within the Si network by relatively gentle electrostatic interactions in the first situation. Still, in the second case, it will be retained by robust covalent connections. The chosen technique will determine the drug withdrawal process. When drugs are retained electrostatically, they usually tend to leak out from the nanocarrier through simple diffusion. In some circumstances, such as the slightly acidic conditions present in tumor tissue, this process speeds up. As opposed to covalent grafting, which allows the payload to be released in response to specific stimuli via matrix breakdown or controlled release. The drug release rate is influenced by PSi particle pore size and surface chemistry. In many therapeutic situations, a sustained drug release over an extended period of time is necessary, and the drug must be delivered to particular tissues or cell populations. To gain fine control over the drug release process in these circumstances, the cargo should be covalently linked to the PSi pore walls.

As a result, the drug leak out of the nanocarrier via regulated covalent bond rupture or matrix degradation accelerated by precise conditions in the target tissue or cell. Additionally, stimuli-responsive gatekeepers can be used to keep the drug confined within the silicon matrix until the presence of a specific trigger removes the pore blocker [28].

2.1.3 Iron Oxide Nanoparticles

Iron oxide nanoparticles (IO NPs) have excellent therapeutic application potential due to their inherent magnetic characteristics and variable size and activity [31]. Maghemite (γ-Fe_2O_3) and magnetite are the two main types of IONPs employed for biological applications because of their relative colloidal stability (Fe_3O_4). Due to their inherent magnetic properties (superparamagnetism), IONPs have received an extensive review, making them useful in a variety of scientific domains, including electronics or the environment. In addition to their exceptional magnetic properties, IO NPs are the perfect platform for biological applications due to their biocompatibility, stability, and environmental friendliness. In recent years, research on the delivery of therapeutics using IONPs as carriers has attracted great interest. Loading drugs onto IO nanocarriers has shown to be a successful method to increase the therapeutic efficacy of various drugs by using the biological and magnetic features of these nanocarriers. Most medications can be conjugated to IONPs to overcome undesirable features such as nonspecific delivery, low solubility, high toxicity and short circulation time. In the realm of oncology, the formation of leaky blood vessels characterises the growth of malignant tumors, permitting the diffusion and storage of IONPs within the tumors. IONPs, can be deliberately targeted to specific sites using either an external magnetic field known as magnetic targeting or by functionalising their surface with vectors that interact with specific biomarkers. In drug delivery, IONPs can be employed as individual NPs or as magnetic nano assemblies, i.e. NPs enclosed in macromolecular matrices. Despite the recent discovery of several ferrites, the inorganic core is usually made of Fe_3O_4 or γ-Fe_2O_3 which is covered by an inorganic shell in both cases. The characteristics of IONPs are greatly influenced by their size and shape in any of these applications and these factors are critical. Other methods for increasing drug release at specific sites include the use of stimuli-responsive coatings or functional groups. Drugs loaded onto IONPs via temperature-responsive polymers cause hyperthermia and boost drug release rate. In recent years, pH-responsive drug delivery systems based on IONPs have received a lot of attention. Tumors have a considerably lower extracellular pH than healthy tissues around them. pH-sensitive bonding or pH-sensitive coatings such as polymers or liposomes can be employed to initiate the delivery of drugs grafted onto nanocarriers. Fe_3O_4NPs have attracted much interest in biomedical applications like targeted delivery, localised hyperthermia therapy, and magnetic resonance imaging contrast agents (MRI). Fe_3O_4NPs have been shown to upregulate ROS, including H_2O_2, which is useful in biomedical applications [18]. The influence of anticancer drug-loaded Fe_3O_4NPs on elicitation of ROS-dependent (Cyt-c mediated) and independent (NADH oxidase) cell apoptosis needs to be fully explored. Fe_3O_4NPs can not be used as a drug carrier independently because they can't survive as an independent entity in aqueous conditions. Therefore, surface functionalization with natural (liposomes, Lipids, proteins, carbohydrates) or synthetic biomolecules is used as a coating material to counteract this disadvantage.

Ayyanaar et al., synthesised chitosan (CS) modified oleic acid (OA) based Fe_3O_4NP. OA acts as a surfactant to stabilise magnetic NPs by forming a bond between the carboxylic group and IONPs. Drug loaded aggregates of Fe_3O_4@OA-CS-5-FLU-NPs were used as a combined drug carrier for pH-responsive 5-flourouracil (5-FLU) administration in A549 and HeLa S3 cell cultures. Results showed that Fe_3O_4@OA-CS-5-FLU-NPs released the drug 5-Flu in response to acidic pH. The nanoformulation was toxic to the cell cultures. It upregulated ROS production, thus potentiating its role as anticancer drug carrier with promising magnetic properties advantageous for magnetic targeted and pH-responsive release of drugs [3]. Independent.

2.1.4 Cadmium Oxide Nanoparticles

Cadmium oxide (CdO) has attracted a lot of interest because of its intriguing and beneficial properties. Brown CdONPs are formed when Cd is burned in the air. It collects CO_2 from the atmosphere and converts it to conducting oxides. CdO nanostructures are polycrystalline wurtzite structured n-type semiconductors with a direct wide bandgap (4.05 eV). CdONPs find applications in gas sensors, solar cells, lasers, spintronics, infrared reflectors, and medicine. CdONP is of great importance in biomedical and pharmaceutical as it possesses antibacterial and anti-cancerous properties and is also employed as a nano-drug carrier and in image diagnostics such as magnetic resonance imaging (MRI). However, CdONPs are small in size, thus limiting the amount of drug carried to the target cell. Therefore millions or billions of these NPs are used to deliver the drug at the exact location [50]. The formation of glycated end products and oxidative stress can be linked to the progression of various diseases such as diabetes, Alzhimer's disease, and certain cancers. NPs have been reported to inhibit glycation. Zahera et al., explored the anti glycation properties of CdONPs. Saos-2 human osteosarcoma cell line and the HeLa cervical cancer cell line were used to assess the cytotoxicity of manufactured glucose encapsulated CdONPs (G-CdONPs). The synthesised CdONPs were passively coated with glucose to prevent aggregation and strengthen biocompatibility. The generated G-CdONPs were cytotoxic to HeLa and Saos-2 cells in vitro due to ROS production, induced apoptosis, and reportedly contained glycation [57].

Amongst other semiconductors, CdONPs have low toxicity. The anticancer nature of CdONPs is due to their unique optical and chemical properties, such as high resolution, fluorescence, two-photon emission, second harmonic generation, and electrical and photoelectrochemical properties. CdONPs destroy cancerous cells by damaging cell wall, DNA and proteins. $CdO/CdCO_3$ nanocomposite induces apoptosis of cancerous cells by generating free radicals, which damage the DNA, inhibit DNA repair, damage mitochondria and disrupt intracellular calcium signaling. Lefojane et al., synthesised $CdO/CdCO_3$ nanocomposites using plant extracts of *Syndenium cupulare* to determine the effectiveness and selectivity of the nanocomposite against breast cancer. *S cupulare*, also known as dead man's tree, is found in southern African and Mozambique regions and possesses antimicrobial, anti-cancerous, anti-inflammatory and antioxidant properties. Plant metabolites play an essential role in the synthesis and the hydroxyl and carboxyl functional groups found in secondary metabolites act as reducing agents. Alkaloids functioned as capping agents and stabilised the reduced NPs. The synthesised $CdO/CdCO_3$ nanocomposites showed inhibitory activity against MF-7, MDA and MB-231 breast cancer cell lines. It was seen that the cytotoxic activity of $CdO/CdCO_3$ nanocomposite was attributed to CdONPs, which induced apoptosis via increased concentration of p53 mRNA and proteins along with the formation of ROS. It was also found that the inhibitory activity was irrespective of binding to receptors as the MDA, MB-231 cell lines lacked receptors [37].

2.1.5 Silver Nanoparticles

Silver (Ag) is a low-cost and abundant natural resource with intriguing material characteristics. The shift of focus to AgNPs can be attributed to their documented antibacterial properties. The unique physical, chemical and biological properties of AgNPs have been intensively studied. Anticancer therapies, diagnostic, optoelectronics, water disinfection and other biomedical and pharmaceutical uses have all benefited from the appealing features of AgNPs. Ag-based dressings and Ag covered medicinal products such as nano lotions and nano gels have shown optimal results. AgNPs have been found to be effective as an alternative to conventional drugs again MDR gram-positive and negative bacteria, viruses, protozoans and other pathogenic organisms.

In comparison to relatively stable AuNPs, AgNPs offer untapped potential. The physicochemical, optical and catalytic properties of AgNPs are significantly impacted by their size, distribution, morphological shape and surface properties, which can be altered using various synthesis processes, reducing agents and stabilisers. AgNPs can be customised to fit a given application; for example, drug delivery AgNPs are often larger than 100 nm to accommodate the amount of drug to be supplied. AgNPs can be shaped into various shapes with varying surface properties, including rods, triangles, rounds, octahedral and polyhedral. Optical properties of AgNPs surface plasma resonance (SPR) are exploited in drug delivery, diagnostic, imaging and biosensors. Analytical sensors used in Surface-Enhanced Raman Spectroscopy (SERS), immune sensors in biological markers and probes and Metal Enhanced Fluorescence (MEF) find applications of AgNPs. Due to the photocatalytic properties of AgNPs, they are used as a catalyst in oxidation reactions and hydrogenation of organic compounds [43].

Mechanism of Antibacterial Action of AgNPs

AgNPs penetrate the outer membrane and accumulate in the inner membrane. Adhesion of AgNPs increases membrane permeability, damage the cell by leaking intracellular components and finally kills the cell. AgNPs also interact with sulphur-containing proteins in the cell wall of bacteria causing structural damage and rupturing the cell. AgNPs also interact with sulphur and phosphorous-containing moieties of DNA and proteins and alter their structure and functioning. Interaction of AgNPs with the thiol group of enzymes disrupts the respiratory chain by damaging the intracellular machinery by generating free radicals and ROS, activating the apoptotic pathway. Silver ions released from the AgNPs disrupt the membrane and genetic material and alter the metabolic pathways. The antibacterial activity of AgNPs is also due to their small size, which increases the diffusion rate in different media and releases silver ions.

The acquisition of resistance by microorganisms has increased the burden of infectious diseases. Existing antimicrobial therapies have limited diversity and show antagonistic reactions. Nano antibiotics have emerged as an attribute to prevent the growth of MDR bacteria. Synergistic antimicrobial therapy of NPs-antibiotics have shown promising results. A combination of AgNPs and antibiotics amoxicillin, chloramphenicol, and ampicillin have reportedly limited the growth of MDR bacteria. Munoz et al. [55] evaluated the antimicrobial activity of AgNPs with a combination of different antibiotics chloramphenicol, aztreonam, biapenem, kanamycin and ampicillin, against various cellular targets through diverse mechanisms. A synergistic effect was seen by a combination of AgNPs-kanamycin against *S typhimurium*, *E coli*, and *S aureus*, whereas AgNPs- chloramphenicol showed less significant results. It can be speculated that the synergy between AgNPs and antibiotics is due to the binding of the sulphur group of antibiotics and AgNPs, which requires more conclusive studies. AgNPs were found to increase membrane permeability by altering the integrity of the membrane. It was proposed that the destabilisation of the membrane by AgNPs complimented the entry of antibiotics and enhanced antimicrobial activity by promoting access to antibiotic targets such as ribosomes for kanamycin and chloramphenicol. On the contrary, beta-lactam antibiotic ampicillin did not show any synergistic effect as the integrity of bacterial cell wall need to be disrupted for their course of action. Therefore it was concluded that the enhancement of antimicrobial activity did not involve direct interaction of AgNPs. It was aided due to the disruption of membrane structure by the NPs [10, 55].

3 Characteristics of Nanoparticles and Drug Delivery Parameters

The drug delivery system employs safe and effective technologies to deliver therapeutic substances into a biological system to achieve desired results while minimising adverse effects. The physicochemical properties of a drug determine the delivery media, drug release kinetics, and mechanism of action. An in vitro drug release experiment is followed by an in vivo pharmacokinetics investigation in animals and humans to establish the suitability of a formulation for clinical use. The quantity of the drug loaded and the duration of the presence of a drug in the body are the two most common variables used to evaluate how efficient the drug delivery system is. As a result, the chemical formulation of the drug, its route of administration, dosage amount, and usage of auxiliary medical equipment are considered when designing drug delivery systems. Nanoparticle-based drug delivery systems provide a platform for transporting large and diverse cargos to tumor cells, including hydrophobic and amphiphilic drugs and genes. The strategy of the targeted delivery system is that the directed drug is released in a regulated manner over a period of time and is only active in the targeted location of the body. The potential of a nanoparticle-based nano-drug delivery system is determined by several parameters, including the production method, surface features, and administration route. These nanoparticle-based methods improve therapeutic efficacy by increasing targeted distributions and lowering off-target systemic toxicity [7].

The systemic transport of NPs to solid tumors after intravenous (i.v.) administration is never simple; it involves multiple major biological obstacles at all levels, from organs to tissues to cells. From the injection site to the site of action, NPs face five biological barriers: (1) serum opsonin proteins adsorption on the surface of NPs, (2) immune system interactions, (3) selective extravasation of NPs at tumor sites, (4) penetration of NPs into solid tumors and (5) internalisation of NPs by tumor cells. Aside from size, shape, and surface chemistry, the mechanical property of NPs has only lately been recognised as being critical in regulating their biological function. This is partly motivated by the fact that many cells, including viruses, can change their mechanical characteristics to perform certain biological roles. The nanometer size range can significantly modify the physical, chemical, and biological aspects of a substance. Various compounds, including polysaccharides, proteins, and polymers, can be used to change the surface properties of NPs. For certain medications, the size and surface qualities of NPs can be tuned. This opportunity permits many different therapies to be reformulated into new pharmaceutical products with greater activity and lower toxicity. Various functional and chemical structures of NPs can be used to encapsulate active substances. This enables the creation of a wide range of NP-based delivery systems with specific features tailored to their intended purpose. Encapsulation or bioconjugation of the drug(s) with NPs facilitates their precise localization and delivery. The specific delivery for a specific role can be manipulated via surface charge, size, and shape of particles.

3.1 Particle Size

Because core/shell NPs have size-dependent distinctive features, particle size and size distribution are fundamental properties of NPs. The size of the delivery vehicles determines the biodistribution of the desired therapeutic molecule. For many NPs, size characteristics are expected to have a key influence in regulating cell contact and adhesion. The size of NPs has a big impact on their cellular absorption. In the degradation and removal of NPs, size can also play an important role. Colloidal NPs smaller than 10 nm are quickly excreted by the kidneys, while bigger NPs are removed via the reticuloendothelial system (RES) of the liver. As a result, it is critical to remember that NP size has been linked to clearance. The rate of clearance increases as the size of the NPs grows [36]. Toxicity is caused by NPs greater than 300 mn accumulating in the liver. However, no single geometry is ideal for every stage of delivery or transportation. There are two types of drug release mechanisms: active targeting and passive targeting. Active targeting is the method by which cells take up NPs via the endocytosis mechanisms. NPs are coated with specific ligands in this scenario. The uptake process of cells in passive targeting is EPR. In this type of targeting, neither the receptor nor the targeted ligands are present. The EPR effect is considered to cause NPs to concentrate in cancerous cells. The primary purpose of some NP systems is to avoid the RES, which is responsible for destroying foreign bodies. Bioavailability and total blood circulation time and are increased by avoiding this process. NPs easily evade the mononuclear phagocytic system (MPS) with hydrophilic surfaces and particle sizes less than 100 nm [39]. MPS is a crucial part of physiological methods for eliminating external pollutants from the body. Opsonin proteins expressed in blood serum can effectively tag larger NPs for MPS degradation. The total blood circulation period, opsonization, and MPS degradation can be avoided or exploited using small-sized and hydrophilic NPs [14].

3.2 Particle Charge

The activity and efficiency of NPs distribution to and through cellular membranes are both influenced by the charge of the NPs. The notch of surface charge on NPs facilitates an NP system's stability. The repulsion between like-charged particles is substantially stronger in a highly charged system. This net repulsive force acts to keep NPs aggregation at bay. It has been demonstrated that NPs with more pronounced surface charges stabilise NPs suspension and prevent particle aggregation [21]. Highly positively charged NPs create significant mucoadhesion, which causes entrapment of NPs inside the mucus layer; surface charge features can impact the degree of absorption. Negatively charged membranes are found in a large number of biological structures. Poly (lactic-co-glycolic acid) (PLGA) and polyvinyl alcohol formulated NPs have resulted in NPs with substantially negative surface charges. Net negative surface charge will be higher in NPs made with known anionic polymers or surfactants. When the NP comes into contact with cell membranes, its increased negative surface charge will cause it to repel them. Cellular absorption is hampered and cellular adhesion is diminished as a result of this repulsive force. The effects of positively charged NPs are polarised. The cationic NP promotes membrane attraction and adhesion, providing favourable conditions for cellular uptake via endocytosis or other processes [13].

3.3 Particle Shape

Particle shape has been discovered to have significant implications on the biological characteristics of NPs in recent years. Shorter polymer micelles have been reported to have a longer circulation time when injected intravenously. Shorter spheres have a higher degree of cell absorption and can efficiently deliver the drug paclitaxel to tumor cells as compared to longer micelles. The length of the NPs is said to have an inverse effect on cellular adherence. Previous research has shown that as particle length increases, subsequent NP binding reduces, implying cellular length affects attachment and adhesion. These findings suggest that NP shape significantly impacts therapeutic results in drug formulation and delivery. The characteristics and design of NPs must be considered more than just particle size and surface charge. The effect of shape on targeted NP outcomes must be examined in tandem [40].

3.4 Drug Loading

The main characteristics to contemplate during developing nano-drug delivery systems are drug loading and encapsulation efficiency. Different methodologies, including hydrophobic, electrostatic, and covalent interactions, are practiced to conjugate varieties of drug molecules on the surface of NPs. The methods of creation of nano-drug carriers dictate the pharmacological treatment of diseases. The biological, chemical, and physical aspects of the developing nanoconjugates heavily rely on the composition, size, shape, surface charge, stability, dispersity, solubility, and bioavailability of nanoconjugates. The synthesis technique has an impact on the release characteristics of drug carriers. A successful nano-drug carrier must have a higher loading capacity so that base matrix material should be required minimum for administration. The drug attachment on the nanocarrier surface can be achieved via physical absorbance, where a concentrated drug is incubated with a nanocarrier to form either hydrophobic or electrostatic bonds. However, in the chemical method, the covalent bonds are established by coupling agents. The direct physical methods are not considered very accurate compared to chemical coupling methods because dissociation of drug molecules can be observed non-specifically (at an unintended location) due to variation in the chemical environment of the cellular compartments.

The drug release rate is also affected by the degradation of the nanocarrier and, eventually, breaking down the bond between the carrier and the drug molecule. As a result, drugs are leaked during transportation, and the efficiency at the target site is affected due to a lower effective concentration. Several different approaches and crosslinkers can be used to create covalent bonds. The loading capacity of inorganic NPs is high due to the large surface area. The physical state of the NPs, the molecular weight, and the stability of the drug molecule in the synthesis solvent, along with the bond strength between the drug and NPs, can affect the loading and encapsulation efficiency of the drug. The encapsulation efficiency percentage and loading capacity of the loaded drug over nanostructure are measured using UV–Visible/Photoluminescence spectrometer or High-Performance Liquid Chromatography (HPLC) [20].

3.5 Drug Release

The overall functionality of NP based drug delivery systems is critically determined by their ability to release the entrapped drug, which requires a thorough understanding of drug release properties to create drug delivery vehicles for biomedical applications. Researchers can appropriately study new NP system behaviour and evaluate their potential efficacy in clinical application by measuring NP release kinetics. Surface-bound drug desorption, polymer matrix erosion and drug diffusion are three ways for NPs to release drugs. Matrix diffusion or erosion are primary routes for the release of a drug, while in many NP formulations, rapid burst release is seen for the desorption of the surface bound drug, which is either unbound or weakly bounded. The overall content of drug release in cells or animal models is evaluated using chromatographic techniques. So far, no standard method for determining drug release qualities has been developed. Researchers have developed many ways, including continuous flow, dialysis membrane, sample and separation and novel techniques like turbidimetry and voltammetry [15]. Every approach has its own set of benefits, drawbacks, and obstacles for setup and sampling. The type of drug, its solubility, the amount of drug utilised, particle size, and crystallinity can all affect the release kinetics. Controlled release of the drug includes extended and pulsatile release. Based on nanotechnology, a pH-dependent, temperature-sensitive or photosensitive drug delivery mechanism can release the drug at the target site. In the case of nanospheres, the drug usually is uniformly deposited throughout the surface, and the drug is released through a diffusion process. The burst release mechanism releases unbound or weakly bound drugs. The nanoparticle and drug molecule conjugation approach impacts the release profile. When a drug is combined with a strong connection, it releases slowly, referred to as sustained release. If drug coupled NPs are enclosed with biodegradable polymer drug is released through the polymer matrix. The dialysis membrane technique is commonly used to analyse NP release, which measures the released drug as a function of continuous diffusion across the membrane [14].

3.6 Nanoparticle Uptake and Transport

Cellular uptake of NPs is highly dependent on various intrinsic and extrinsic properties. Different intrinsic, physiochemical properties of NPs such as size, surface charge and chemistry influence their uptake and interaction in a biological system. NPs with positive surface charge have shown higher cellular uptake but higher toxicity than NPs with a negative surface charge which are more efficiently transported. NPs can interact with cellular plasma membranes and enter the cell by passive diffusion or endocytosis when positioned outside the cell.

3.6.1 Endocytosis Mediated Transport

It can be further classified into two categories-Pinocytosis can be receptor-mediated or macropinocytosis: receptors are engulfed into coated pits in conjugation with the associated ligand in receptor-mediated endocytosis. The classical form of receptor-mediated endocytosis is represented by clatherin protein-coated pits responsible for the uptake of nutrients such as iron and cholesterol. Caveolae mediated endocytosis engulfs molecules bound to caveola surfaces. Structural caveolin proteins function as integral membrane proteins and are abundantly found in adipocytes, epithelial cells and muscles. Larger solute molecules measuring up to 5 μm are engulfed by macropinocytosis. The clathrin pathway plays a crucial role in the uptake and transport of positively charged NPs. Micropinocytosis participates in the uptake of both positively and negatively charged NP. Phagocytosis-uptake of large macroparticles such as microorganisms and dead cells occurs through phagocytosis and the ingested particles are transported via phagosomes. Phagocytosis occurs only in mammalian cells such as macrophages which recognise and eliminate foreign substances.

3.6.2 Passive Diffusion

Lipophilic NPs cross the lipid membrane via passive diffusion. The charge on NPs highly influences the diffusion of NPs across the membrane; cholesterol-containing lipid bilayer facilitates a 3-fivefold increase in uptake of NPs.

Negatively charged extracellular plasma membrane proteins impart negative surface charge to epithelial cells. Electrostatic interaction between positively charged NPs and negatively charged plasma membrane promote the association of NPs with the cell and cellular uptake by endocytosis [15].

4 Applications of Inorganic Nanoparticles

4.1 Inorganic Nanoparticles in Oncology

One of the fatal diseases and the major cause of death worldwide is cancer. The most frequent causes of cancer are damage or mutations in the genetic makeup of the cells brought on by hereditary or environmental factors. Cancer thrives in acidic environments and cannot survive in alkaline conditions. Chemotherapeutic drugs can destroy both healthy and unhealthy cells because they are non-selective. Some of the adverse consequences brought on by the non-selectivity of chemotherapy medications include immunosuppression, mucositis, and baldness. These drugs must be administered at significant quantities with a high frequency, which causes toxicity because the bioaccessibility of these drugs to malignant cells is restricted. Maeda and Matsumura, two Japanese researchers, made an important discovery in 1986: NPs tend to concentrate in solid tumors due to a process known as EPR [51]. The unique blood artery architecture found in solid tumors causes EPR. Solid tumors necessitate a lot of nutrients to keep their cells growing at a fast rate. Therefore they force the formation of blood vessels to irrigate the neoplastic tissue. The EPR effect has long been thought to be the main mechanism that allows NPs to accumulate preferentially in tumor tissues over normal tissues. These new vessels are produced quickly and have a number of flaws, including high tortuosity and wide inter endothelial junctions, which result in huge fenestrations with sizes of a few hundred nanometers. Various NP-based drug delivery systems have been developed to precisely deliver therapeutic drugs to solid tumors to improve anticancer efficacy while limiting systemic toxicity. NPs can travel through these gaps and accumulate within solid tumors, but they can not traverse the walls of healthy blood vessels, which is why they are only seen in diseased tissues. This offers a novel technique to deliver chemotherapeutic drugs selectively to tumoral lesions by simply loading them onto nanometric carriers.

Preclinical development of inorganic NPs as possible diagnostic and therapeutic systems in oncology for various applications, including tumor imaging, tumor medication delivery, and radiation augmentation, has gained a lot of attention. Because of the EPR effect, NPs between 10 and 100 nm in size preferentially concentrate in tumor sites in cancer disease. NPs have the potential to dramatically improve cancer diagnostics for early tumor identification and act as configurable therapeutic devices for integrated imaging and therapy, i.e. theranostics, of a variety of cancer disorders [46]. In vivo biodistribution is aided by the nanoscale size, huge surface-to-volume ratios and surface functionalisation capacity. Furthermore, multifunctional nanocarrier systems can be used to deliver chemotherapeutic agents to specific locations. The NPs are functionalised and targeted to the malignant cell through active processes. Attaching biomolecules or a ligand as a receptor to NPs might sometimes improve the delivery of the NPs to malignant cells while causing no harm to normal cells.

In children and adolescents, osteosarcoma is one of the most prevalent primary malignant tumors. In children, it is the sixth most frequent cancer. Osteosarcoma is characterised by the uncontrolled growth of osteoid-producing mesenchymal cells.

Compared to surgery alone, neoadjuvant and adjuvant chemotherapy and surgery have raised the 5-year survival rate for localised illness by more than 60%. On the other hand, patients with metastases and a poor prognosis had a 5-year survival rate of less than 30%. The current successful chemotherapy regimens for osteosarcoma also have harmful effects on normal cells, resulting in life-threatening consequences. Combinations of high dosages of cisplatin, methotrexate, ifosfamide, and doxorubicin have shown a considerable improvement in the survival rate in the chemotherapeutic therapy of osteosarcoma. However, due to nonspecific uptakes, these anticancer medications have various adverse effects, including the need for high dosages, the insufficient blood supply in the case of osteosarcoma, secondary cancers, and drug resistance phenotype [49]. Lung cancer is the most significant cause of cancer-related death, accounting for 23% of fatalities globally, more than the total deaths from colon, breast, and prostate cancer. Lung cancer is fatal due to a lack of efficient chemotherapeutic treatments and diagnostic procedures for early detection. Cisplatin is a highly effective cancer therapy, including lung cancer and osteosarcoma, that is now widely available, although its use is restricted due to its numerous side effects [38]. In their scientific report by Iram et al., used bromelain, a pineapple-based cysteine protease, as a reducing and surface functionalising agent to manufacture monodispersed AuNPs. Cisplatin was conjugated with AuNPs in this study to distribute it selectively on the directed site delivery with high patient compliance. Due to their enormous size, AuNPs preferentially aggregate at tumor sites and in inflamed tissues. Enzyme mediated synthesis of NPs has been reported to enhance the adequacy of the medication; bromelain acts as a capping agent to improve therapeutic efficacy and overcome constraints. Bromelain is hydrophilic in nature and has numerous polar functional groups on its surface, and NPs with a hydrophilic surface can resist identification by RES [25].

Furthermore, because of the EPR effect, NPs will have a longer circulation duration in the circulatory compartment and will be able to accumulate in solid tumors. Because these groups will be engaged in the covalent bonding for bioconjugation between the drug and NPs, bioconjugation will also block the functional groups responsible for nonspecific binding. Anticancer action of bromelain is mainly related to its protease component. Even at lower concentrations, the synthesised Cisplatin bioconjugated bromelain encapsulated AuNPs were potent against osteosarcoma and lung cancer which can be due to the synergy of bromelain. The encapsulation of NPs rendered the loss in secondary structure, but the tertiary structure was not altered and this improved anticancer potential, as reported earlier [25].

4.2 Inorganic Nanoparticles in Skin Regeneration Treatment

Numerous nano-drug delivery methods have been devised for application in skin regeneration and allied fields due to nanotechnology's state-of-the-art and striking development. Nano-encapsulated drugs are harmless and compatible with the milieu of skin, ensuring a moist environment for faster wound healing. The nanoformulations can also pass across the cytoplasmic gaps, triggering the drug retention mechanism. By avoiding destruction by proteases found in skin wounds, the therapeutic efficiency of nano conjugated bioactive compounds is increased. Sustained drug release maintains the drug's prolonged effective concentration, reduces administration frequency, and lowers costs while improving compliance [41]. By their very nature, inorganic NPs have similar benefits in bacterial infection and wound healing; such as, AgNPs are frequently used as antimicrobial agents. In order to create a synergistic stimulating impact of both materials and medications, researchers choose to use a combination of inorganic NPs.

Ali et al., used the co-precipitation approach to design and produce ZnO_2 NPs. These NPs, due to their effective antibacterial inorganic substances, showed promising inhibition against the microbes of Pseudomonas aeruginosa and Aspergillus species isolated from burn patient's infected wound tissues. Histopathological results showed ZnO_2 NPs accelerated the healing of skin lesions in animal models in vivo [1].

To improve efficacy, extended studies have been performed to include the combination of different types of inorganic NPs. Zn is essential in the body to regulate the immune system, synthesise DNA and proteins and cell division. CuONPs keep the wounds clean by controlling inflammation and also promote collagen formation by fibroblasts. Thanusha et al., produced a hydrogel for second burn wound healing that was co-encapsulated with Asiatic acid, ZnO, and CuONPs. Asiatic acid is used in skincare products as a skin toning, antiageing and suntan controlling agent. Porous shape, substantial water uptake, incomparable tensile strength, and prominent antibacterial potential were all found in physicochemical investigations of the formulation. The in vivo efficacy of hydrogel revealed that the total protein, DNA, hydroxyproline, and hexosamine, re-epithelialization, collagen fibers arrangement, and angiogenesis in wounds were significantly better than commercially available therapeutics. Thus hydrogels containing inorganic NPs as components have the potential to treat skin wounds and other skin related disorders [54].

4.3 Inorganic Nanoparticles Against MDR Bacteria

Persistent bacterial infections with high morbidity and mortality rates and antibiotic resistance have presented a dire situation. Nosocomial infections count up to 14% of hospital admissions. Bacterial resistance, or the ability of bacteria to survive even after exposure to antimicrobials, is the defining attribute of this problem. Bacterial biofilm makes them resistant to antibiotics and provides stability. Other mechanisms to antibiotic resistance are alteration of drug binding sites, reduced permeability to drugs, enzymatic inactivation and production of flow pumps. Expansion of MDR microbes is due to inaccessibility to appropriate antibacterial therapeutics. The emergence of MDR strains and the high cost of their treatment raises the need to find novel therapeutic strategies to control highly mutating MDR bacteria. Numerous biological processes, including as bioremediation, microbial corrosion, bioleaching, and biomineralization, depend largely on the interaction between bacteria and metal. Due of their many uses, inorganic NPs are currently being utilised on a global scale. The various metal and metal oxide NPs are used as antibacterial agents such as AuNPs, AgNPs, ZnONPs and titanium dioxide NPs. Baker et al., synthesised levofloxacin (Levo) and balofloxacin (Balo) bioconjugated AuNPs to amplify the synergistic effect of antibiotics. AuNPs were capped with antibacterial compound sericin against *E. coli* and *S. aureus*.

A multi-targeting approach showed the synergistic effect of sericin capped (SrAuNPs) bioconjugated with drugs. The EDC-mediated reaction facilitates covalent bridging between capping agents on the surface of NPs and drugs, where carboxylate forms peptide bonds with amino groups and ester bonds with hydroxyl groups, respectively. According to the findings, functionalised NPs had lower intrinsic toxicity, improved antibacterial effectiveness, and a lower likelihood of resistance across a larger spectrum of action. AuNPs are regarded safe up to very high concentrations. After bioconjugation with SrAuNPs, the nonspecific interactions were minimized, and the subsequent efficacy of Levo and Balo (quinolones) was greatly augmented. Both the drugs interact with the same sequence of amino acids (QRDR) of topoisomerase IV and DNA gyrase to destroy bacteria. The high surface energy (intrinsic properties) of inorganic nanoparticles channel the aggressive use of capping agents and bioconjugated drugs, thus enhancing the antibacterial activity of the functionalized NPs [7].

Another work used *C. roseus* and *A. Indica* aqueous leaf extracts mediated synthesized economical biogenic AgNPs and checked they're antibacterial (against MDR strains of *E. coli*, *K. pneumoniae*, *S. aureus*, and *P. aeruginosa*), anti-inflammatory and wound healing potential on animal models. These outcomes strengthened the use of biogenic AgNPs as therapeutics in microbial infection against MDR and wound healing [34].

5 Conclusion

A well-coordinated series of biological events in synergism with physicochemical properties of NPs form the basis of NP based drug delivery to overcome biological barriers at the target delivery site. Despite extensive research, inconsistency between mechanical and physicochemical properties such as shape, size and surface modification has contributed to contradictory results in most NP based drug delivery studies. This necessitates thorough clarification of linked investigations on regulating mechanical and other nano-bio interaction factors. To achieve similar results, NP systems with a wide range of mechanical properties should be employed and their mechanical characteristics should be evaluated using standardised techniques. Additional consideration should be given to other crucial elements, such as the role of NP mechanical properties in controlling their extravasation from tumor vasculature and retention at tumor tissues by taking advantage of the EPR effect, as well as their interactions with the extracellular matrix, tumor-associated cells, and tumor stromal cells in the tumor microenvironment. The impact of NPs mechanical qualities on the development of protein corona and on cell signalling pathways utilising the newly developed omics approach, which have an impact on NP cellular uptake, are two additional crucial issues that should be thoroughly researched. The diverse fields of nanotechnology, biophysics, cell biology, and mechanical engineering would need to work together to achieve this [24]. In order to deliver the majority of the drug to the target site, significant progress has been achieved in altering the characteristics of NPs that can be delivered directly to the blood. The success of gene therapy and the treatment of tumors, among others, completely rests on the ability of drug delivery systems to reach their intended targets. Despite substantial primacy, the availability of nanomedicines to patients is much behind forecasts due to the translational gap between animal and human studies. This gap results from a lack of knowledge about the physiology and pathological variations between human beings and animal model species, mainly how these differences affect the behaviour and functionality of nanomedicines in the body. Because the growth, structure, and physiology of diseased tissue modify NP distribution and functionality, NP efficacy is altered. This is partly owing to the understudied heterogeneity both in the molecular foundations of diseases and among patients [22]. In small animal models, all nanoparticle drug delivery systems reduced tumor size more effectively than the control; however, none of the nanoparticle formulations have been successfully applied clinically. The confluence of size, physical characteristics and structure of NPs makes them essential contributors to developing novel pharmaceutical and imaging agents. The successes and explicit promises illustrated in the contributions mentioned above suggest that these nanoscale platforms will advance our knowledge of diseases and improve our capacity to treat them, even though several obstacles still need to be addressed and overcome before implementing these platforms in the medicinal field. In terms of biomedicine, nanostructured materials have a lot to offer.

Acknowledgements The authors are thankful to the Chairman Department of Biochemistry, Faculty of Life Science, Aligarh Muslim University, for kind support.

Conflicts of Interest The authors declare no conflict of interest.

References

1. Ali SS, Morsy R, El-Zawawy NA, Fareed MF, Bedaiwy MY (2017) Synthesised zinc peroxide nanoparticles (ZnO_2-NPs): a novel antimicrobial, anti-elastase, anti-keratinase, and anti-inflammatory approach toward polymicrobial burn wounds. Int J Nanomed 12:6059–6073
2. Amina SJ, Guo B (2020) A review on the synthesis and functionalisation of gold nanoparticles as a drug delivery vehicle. Int J Nanomed 15:9823–9857

3. Ayyanaar S, Balachandran C, Bhaskar RC, Kesavan MP, Aoki S, Raja RP et al (2020) ROS-responsive chitosan coated magnetic iron oxide nanoparticles as potential vehicles for targeted drug delivery in cancer therapy. Int J Nanomed 15:3333–3346
4. Bagga P, Ansari TM, Siddiqui HH, Syed A, Bahkali AH, Rahman MA, Khan MS (2016) Bromelain capped gold nanoparticles as the novel drug delivery carriers to aggrandise effect of the antibiotic levofloxacin. Excli J 15:772–780
5. Bagga P, Siddiqui HH, Akhtar J, Mahmood T, Zahera M, Khan MS (2017) Gold nanoparticles conjugated levofloxacin: for improved antibacterial activity over levofloxacin alone. Curr Drug Deliv 14(8):1114–1119
6. Bahrami B, Hojjat-Farsangi M, Mohammadi H, Anvari E, Ghalamfarsa G, Yousefi M et al (2017) Nanoparticles and targeted drug delivery in cancer therapy. Immunol Lett 190:64–83
7. Baker A, Syed A, Alyousef AA, Arshad M, Alqasim A, Khalid M, Khan MS (2020) Sericin-functionalized GNPs potentiate the synergistic effect of levofloxacin and balofloxacin against MDR bacteria. Microb Pathog 148:104467
8. Baskin J, Jeon JE, Lewis S (2021) Nanoparticles for drug delivery in Parkinson's disease. J Neurol 268(5):1981–1994
9. Blanco E, Shen H, Ferrari M (2015) Principles of nanoparticle design for overcoming biological barriers to drug delivery. Nat Biotechnol 33(9):941–951
10. Bruna T, Maldonado-Bravo F, Jara P, Caro N (2021) Silver nanoparticles and their antibacterial applications. Int J Mol Sci 22(13):7202
11. Cardoso S, Silva CFD, Severino P, Silva AM, Souto SB, Zielinska A et al (2021) Genotoxicity assessment of metal-based nanocomposites applied in drug delivery. Materials 14(21):6551
12. Chatterjee K, Sarkar S, Jagajjanani Rao K, Paria S (2014) Core/shell nanoparticles in biomedical applications. Adv Colloid Interface Sci 209:8–39
13. Cooley M, Sarode A, Hoore M, Fedosov DA, Mitragotri S, Sen Gupta A (2018) Influence of particle size and shape on their margination and wall-adhesion: implications in drug delivery vehicle design across nano-to-micro scale. Nanoscale 10(32):15350–15364
14. Cooper DL, Conder CM, Harirforoosh S (2014) Nanoparticles in drug delivery: Mechanism of action, formulation and clinical application towards reduction in drug-associated nephrotoxicity. Expert Opin Drug Deliv 11(10):1661–1680
15. D'Souza S (2014) A review of in vitro drug release test methods for nano-sized dosage forms. Adv Pharmaceutics
16. Das RK, Pachapur VL, Lonappan L, Nghdi M, Pulicharla R, Maiti S et al (2017) Biological synthesis of metallic nanoparticles: plants, animals and microbial aspects. Nanotechnol Environ Eng 2:18
17. Degli Esposti L, Carella F, Adamiano A, Tampieri A, Iafisco M (2018) Calcium phosphate-based nanosystems for advanced targeted nanomedicine. Drug Dev Ind Pharm 44(8):1223–1238
18. Estelrich J, Escribano E, Queralt J, Busquets MA (2015) Iron oxide nanoparticles for magnetically-guided and magnetically-responsive drug delivery. Int J Mol Sci 16(4):8070–8101
19. Fan J, Yin JJ, Ning B, Wu X, Hu Y, Ferrari M et al (2011) Direct evidence for catalase and peroxidase activities of ferritin-platinum nanoparticles. Biomaterials 32(6):1611–1618
20. Fang C, Kievit FM, Veiseh O, Stephen ZR, Wang T, Lee D, Ellenbogen RG, Zhang M (2012) Fabrication of magnetic nanoparticles with controllable drug loading and release through a simple assembly approach. J Control Release 162(1):233–241
21. Guerrini L, Alvarez-Puebla RA, Pazos-Perez N (2018) Surface modifications of nanoparticles for stability in biological fluids. Materials (Basel, Switzerland) 11(7):1154
22. Hua S, de Matos M, Metselaar JM, Storm G (2018) Current trends and challenges in the clinical translation of nanoparticulate nanomedicines: pathways for translational development and commercialization. Front Pharmacol 9:790
23. Huang HC, Barua S, Sharma G, Dey SK, Rege K (2011) Inorganic nanoparticles for cancer imaging and therapy. J Control Release 155(3):344–357
24. Hui Y, Yi X, Hou F, Wibowo D, Zhang F, Zhao D et al (2019) Role of nanoparticle mechanical properties in cancer drug delivery. ACS Nano 13(7):7410–7424

25. Iram S, Zahera M, Wahid I, Baker A, Raish M, Khan A (2019) Cisplatin bioconjugated enzymatic GNPs amplify the effect of cisplatin with acquiescence. Sci Rep 9:13826
26. Jafari S, Derakhshankhah H, Alaei L, Fattahi A, Varnamkhasti BS, Saboury AA (2019) Mesoporous silica nanoparticles for therapeutic/diagnostic applications. Biomed Pharmacother 109:1100–1111
27. Jalil SU, Zahera M, Khan MS, Ansari MI (2019) Biochemical synthesis of gold nanoparticles from leaf protein of *Nicotianatabacum* L. cv. xanthi and their physiological, developmental, and ROS scavenging responses on tobacco plant under stress conditions. IET Nanobiotechnol 13(1):23–29
28. Kaasalainen M, Aseyev V, von Haartman E, Karaman DS, Makila E, Tenhu H et al (2017) Size, stability, and porosity of mesoporous nanoparticles characterised with light scattering. Nanoscale Res Lett 12(1):74
29. Khan S, Danish S, Ahmad V, Baig MH, Kamal M, Ahmad S et al (2015) Magnetic nanoparticles: properties, synthesis and biomedical applications. Curr Drug Metab 8(12):16
30. Khan S, Haseeb M, Baig MH, Bagga PS, Siddiqui HH, Kamal MA, Khan MS (2015) Improved efficiency and stability of secnidazole: an ideal delivery system. Saudi J Biol Sci 22(1):42–49
31. Khan S, Rizvi SMD, Avaish M, Arshad M, Bagga P, Khan MS (2015) A novel process for size controlled biosynthesis of gold nanoparticles using bromelain. Mater Lett 159:373–376
32. Kou L, Bhutia YD, Yao Q, He Z, Sun J, Ganapathy V (2018) Transporter-guided delivery of nanoparticles to improve drug permeation across cellular barriers and drug exposure to selective cell types. Front Pharmacol 9:27
33. Kou L, Sun R, Bhutia YD, Yao Q, Chen R (2018) Emerging advances in P-glycoprotein inhibitory nanomaterials for drug delivery. Expert Opin Drug Deliv 15(9):869–879
34. Lakkim V, Reddy MC, Pallavali RR, Reddy KR, Reddy CV, Inamuddin B et al (2020) Green synthesis of silver nanoparticles and evaluation of their antibacterial activity against multidrug-resistant bacteria and wound healing efficacy using a murine model. Antibiotics 9(12):902
35. Lebedova J, Blahova L, Vecera Z, Mikuska P, Docekal B, Buchtova M et al (2016) Impact of acute and chronic inhalation exposure to CdO nanoparticles on mice. Environ Sci Pollut Res Int 23(23):24047–24060
36. Lee S-Y, Cheng J-X (2013) Clearance of nanoparticles during circulation. In: Gad SC (ed) Pharmaceutical sciences encyclopedia
37. Lefojane RP, Sone BT, Matinise N, Saleh K, Direko P, Mfengwana P, Mashele S, Maaza M, Sekhoacha MP (2021) CdO/$CdCO_3$ nanocomposite physical properties and cytotoxicity against selected breast cancer cell lines. Sci Rep 11(1):30
38. Liu Q, Wang K (2019) The induction of ferroptosis by impairing STAT3/Nrf2/GPx4 signaling enhances the sensitivity of osteosarcoma cells to cisplatin. Cell Biol Int 43(11):1245–1256
39. Liu Y, Hardie J, Zhang X, Rotello VM (2017) Effects of engineered nanoparticles on the innate immune system. Semin Immunol 34:25–32
40. Liu Y, Tan J, Thomas A, Ou-Yang D, Muzykantov VR (2012) The shape of things to come: importance of design in nanotechnology for drug delivery. Ther Deliv 3(2):181–194
41. Luther DC, Huang R, Jeon T, Zhang X, Lee YW, Nagaraj H et al (2020) Delivery of drugs, proteins, and nucleic acids using inorganic nanoparticles. Adv Drug Deliv Rev 156:188–213
42. Meir R, Popovtzer R (2018) Cell tracking using gold nanoparticles and computed tomography imaging. Wiley Interdiscip Rev Nanomed Nanobiotechnol 10(2). https://doi.org/10.1002/wnan.1480
43. Mukherjee S, Chowdhury D, Kotcherlakota R, Patra S, Vinothkumar B, Bhadra MP et al (2014) Potential theranostics application of bio-synthesised silver nanoparticles (4-in-1 system). Theranostics 4(3):316–335
44. Nunez C, Estevez SV, Del PilarChantada M (2018) Inorganic nanoparticles in diagnosis and treatment of breast cancer. J BiolInorg Chem 23(3):331–345
45. Oh N, Park JH (2014) Endocytosis and exocytosis of nanoparticles in mammalian cells. Int J Nanomed 9:51–63
46. Parra-Nieto J, Del Cid M, de Carcer IA, Baeza A (2021) Inorganic porous nanoparticles for drug delivery in antitumoral therapy. Biotechnol J 16(2):e2000150

47. Paul W, Sharma CP (2019) Inorganic nanoparticles for targeted drug delivery. Biointegr Med Implant Mater 333–373
48. Pissuwan D (2017) Monitoring and tracking metallic nanobiomaterials in vivo. In: Monitoring and evaluation of biomaterials and their performance in vivo. Woodhead Publishing, pp 135–149
49. Roche B, Vanden-Bossche A, Normand M, Malaval L, Vico L, Lafage-Proust MH (2013) Validated laser doppler protocol for measurement of mouse bone blood perfusion—response to age or ovariectomy differs with genetic background. Bone 55(2):418–426
50. Sadhukhan S, Ghosh TK, Roy I, Rana D, Bhattacharyya A, Saha R et al (2019) Green synthesis of cadmium oxide decorated reduced graphene oxide nanocomposites and its electrical and antibacterial properties. Mater Sci Eng C 99:696–709
51. Shi Y, van der Meel R, Chen X, Lammers T (2020) The EPR effect and beyond: Strategies to improve tumor targeting and cancer nanomedicine treatment efficacy. Theranostics 10(17):7921–7924
52. Shityakov S, Forster C (2013) Multidrug resistance protein P-gp interaction with nanoparticles (fullerenes and carbon nanotube) to assess their drug delivery potential: a theoretical molecular docking study. Int J Comput Biol Drug Des 6(4):343–357
53. Singh MS, Lamprecht A (2016) P-glycoprotein inhibition of drug resistant cell lines by nanoparticles. Drug Dev Ind Pharm 42(2):325–331
54. Thanisha AV, Dinda AK, Koul V (2018) Evaluation of nano hydrogel composite based on gelatin/HA/CS suffused with Asiatic acid/ZnO and CuO nanoparticles for second degree burns. Mater Sci Eng C Mater Biol Appl 89:378–386
55. Vazquez-Munoz R, Meza-Villezcas A, Fournier P, Soria-Castro E, Juarez-Moreno K, Gallego-Hernandez AL et al (2019) Enhancement of antibiotics antimicrobial activity due to the silver nanoparticles impact on the cell membrane. PLoS ONE 14(11):e0224904
56. Yang Q, Jacobs TM, McCallen JD, Moore DT, Huckaby JT, Edelstein JN et al (2016) Analysis of pre-existing IgG and IgM antibodies against polyethylene glycol (PEG) in the general population. Anal Chem 88(23):11804–11812
57. Zahera M, Khan SA, Khan IA, Sharma RK, Sinha N, Al-Shwaiman HA et al (2020) Cadmium oxide nanoparticles: an attractive candidate for novel therapeutic approaches. Colloids Surf, A 585:124017
58. Zhang K, Cheng Y, Ren W, Sun L, Liu C, Wang D et al (2018) Coordination-responsive longitudinal relaxation tuning as a versatile MRI sensing protocol for malignancy targets. Adv Sci 5(9):1800021

Bio-nanocomposites: A Next Generation Food Packaging Materials

Arati Dubey, S. Irudhayaraj, and Adhish Jaiswal

Abstract Bio-nanocomposites are next generation food packaging materials that promote better food quality. Bio-nanocomposites are the replacement for current non-biodegradable and non-renewable materials which are applied as food packaging materials. These nanocomposites are promising materials for human health, food storage and eco-system. Advantages of incorporating nanomaterials into the packaging materials include better physico-chemical, mechanical, antibacterial and antimicrobial properties. Several kinds of nanomaterials including metal, metal oxide and other inorganic and organic nanostructures have proven to be effective to increase the shelf life and reduce the spoilage of food by different mechanism of action. Incorporation of nanomaterials in biopolymer makes the production and application of food packaging material.

Keywords Food spoilage · Packaging materials · Antimicrobial properties · Polymer · Biodegradation

1 Introduction

Now a days, the fear of rising and spreading CORONA virus among the people increases which leads to the consumption of ready to eat food products globally. Recently, fresh packed food's sales in market have been raised more than 60% than the previous year. Specially in metro and tyre I city demands and sales of frozen foods like meat and fish products, fries, patties and products such as batter, paste, curries, and desserts continuously rising which shows that there will be a massive blow in ready to eat meals in the market. Due to COVID-19 outbreak, online shopping are also growing among the people which again accelerate this business. The

A. Dubey · S. Irudhayaraj
Department of Chemistry, Indira Gandhi National Tribal University, Amarkantak, Madhya Pradesh 484887, India

A. Jaiswal (✉)
Department of Chemistry, University of Lucknow, Lucknow, Uttar Pradesh 226007, India
e-mail: adhish.jaiswal@igntu.ac.in

I. Uddin and I. Ahmad (eds.), *Synthesis and Applications of Nanomaterials and Nanocomposites*, Composites Science and Technology,
https://doi.org/10.1007/978-981-99-1350-3_8

current challenge in this sector is that ready to eat food is a pre-cooked food so it is important that packaging of these food should be such that it can protect food from surrounding contamination, dust, odors, temperature, physical damage during transport, light, microorganism and humidity. The factors which can deteriorate food are microbial contamination, chemical and physical reactions which affect the taste, odor and appearance of food. Here, Food packaging plays a important role to: ensure the safety, increase health consciousness, and shelf life of the food. In the modern age, the need for food packaging development is changing due to men's lifestyle. Over an extended period, people ate whatever they could collect from their local surroundings. When people changed their peripatetic lifestyle to staying in a sheltered area, they needed to store food items. There has been little sophistication in packaging materials until the 1800s. The people used various things found in nature like leaves, shells, and gourds to hold food items and bamboos, grasses, and wood used for the weave baskets and also for shaping into food containers, people used some materials such as paper, glass, and pottery. Egyptians formed the glass and pottery near about 7000 B.C., which was the first evidence; however, industrialization was not seen by Egyptians as long as around 1500 B.C. Figure 1 shows more details about the evolution of packaging materials growing demand for the ease of food products due to change in lifestyle [11, 50]. After early development, there was rising attention on food and food quality. There were several modifications for improving the quality of food. Plastics is one of the significant fields that has seen a considerable improvement in materials and their properties. Plastics have as food packaging materials in the last 50–60 years for a very long time. Presently, a wide range of plastic materials are employed for food packaging applications. These packaging application generate waste materials, such as metals, glass, and plastic, which causes pollution. In a record, 42% plastics produced enter into the world primarily as a packaging material. Global plastic waste per year is 275 million tones and world use 500 billion plastics every year. Furthermore, limitation of petroleum resources in our country pushed researchers to focus on different ecofriendly, safe and non-toxic packaging material without compromising the main features of the packaging such as strong mechanical properties, moisture barrier properties and extended product shelf life. Nanotechnology is considered as an essential tool for ensuring good quality and safety of food and food products during storage and transportation. It also help to extend the food product shelf life by avoiding the issues related to chemical contaminants, oxygen, damage microorganisms, light, moisture barriers, etc., as a food packaging materials [14]. Nanomaterials plays an vital role in the food packaging industry as fillers, nutrient carriers, and antimicrobial agents. Nano-food packaging is described as a novel food packaging because it has various properties like antimicrobial, antioxidant, optoelectronic that could assess the food quality and improving their barrier properties and mechanical strength. The main nanomaterials used in food packaging are silver nanoparticles, Zinc oxide, titanium dioxide, and silicon dioxide to develop strong light and heat resistance plastic with gas barrier properties to prevent food spoilage and improved the shelf-life of the packaging product [19]. Additionally, some active agents could use as functional food packaging in the food industry, shown in Fig. 2 [64].

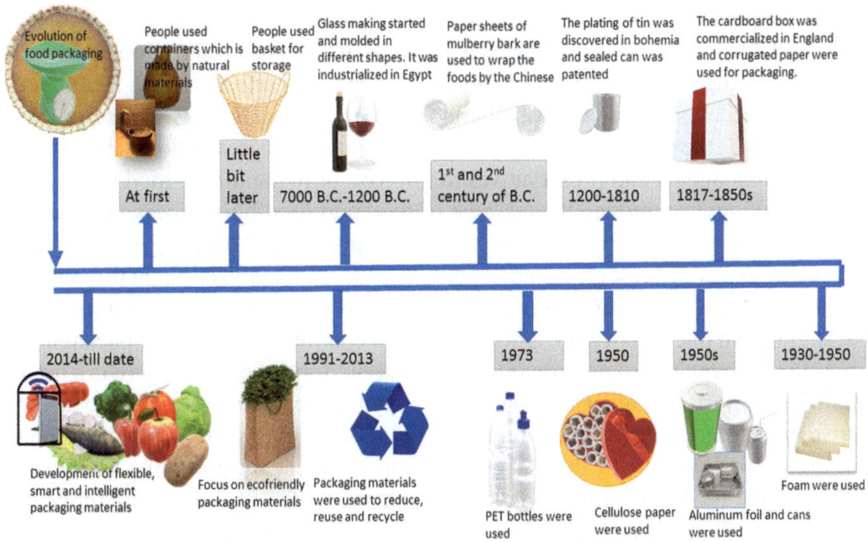

Fig. 1 Evolution of food packaging

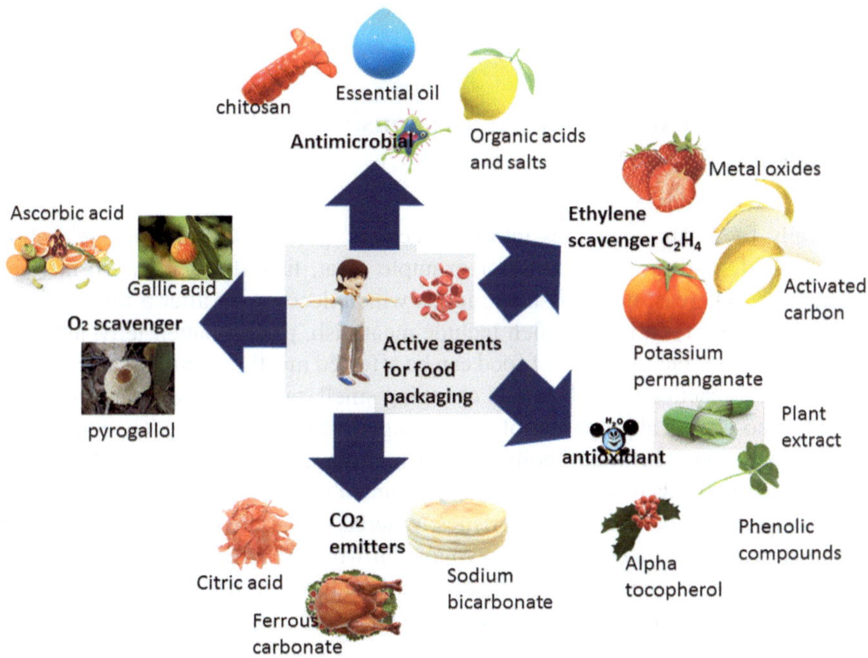

Fig. 2 Active agents for food packaging

Nanotechnology has recently focused on monitoring the spoilage and increasing the shelf life of food products. The incorporation of nanomaterials in food packaging makes it intelligent and active packaging. Food spoilage affects human health, has a tremendous economic impact globally, and raises medical care costs. The active packaging involves the coating of materials and sachet that releases antioxidants, flavors, antimicrobial agents and preservatives to improve or maintain the food quality. The new technology can reduce the wastage of food and food born illness via real-time food quality monitoring throughout the food supply chain. There is growing interest in specific types of biological and chemical sensors to analyze the quality of food. Active packaging can utilize as increasing demand for standard and traditional food packaging. An intelligent packaging system is a continuation of active packaging by improving the quality to provide information about food spoilage. There is a limited number of reports that are using nanotechnology for such kinds of applications. Innovative packaging aims to develop a new type of packaging system that can monitor packaged food quality to the surrounding environment by providing information related to the freshness of food via various signals. Different food quality indicators and devices are used in commercial places to detect food spoilage in intelligent packaging systems [68]. This chapter mainly cover the factors that are responsible for spoilage of food products and the various type of food packaging materials including biodegradable polymers and nanoparticles to control the barrier properties.

2 Type of Food and Factors Responsible for Spoilage of Food

Generally foods are categories into two sections: perishable food, having a short period of life and quickly spoils for example meat, fish, dairy, bread, fruits, and vegetables. Whereas, non-perishable foods which decompose after a long time for example canned food items which include meat, fish, peanut butter, jelly, tea bags, dry soups, stews, etc. All sorts of food can be affected mostly by microbial, chemical and physical reactions which results changing smell, taste and appearance of food. The physico-chemical properties of food changed after spoilage and it becomes unfit for consumption. The food spoilage from microorganism involves production of an enzyme that leads to the formation of unwanted products in the food which react together to change the food characteristics. Whereas in the chemical spoilage of food, chemical reaction e.g., oxidation, enzymatic browning, and non-enzymatic reaction occur in the food to change the smell or color of the food. In case of Physical spoilage of food, occurs when foods are physically damaged during harvesting, processing, packaging or distribution.

3 Microbial Food Spoilage

The microbial spoilage of food has a great concern and has always been problematic for human being. Microbial spoilage caused by large number of microorganisms like bacteria and fungi (molds, yeasts). The microorganisms that are responsible for the food spoilage are shown in Fig. 3.

3.1 Bacteria

Generally, the bacteria are considered as significant organisms for the spoilage of food. There are some non-pathogenic bacteria also that are responsible for food spoilage but they do not pose any threat to human health. However, some pathogenic bacteria may cause serious concern to human health. Proteinaceous food like milk, dairy product, meat, fish, shellfish and poultry are affected by the pathogenic species of microorganisms for example Bacillus cereus, have been found in a fluid milk product, responsible for spoilage and human illness. Clostridium perfringens are found in meat and poultry products that are stored under anaerobic conditions. Additionally, some Enterobacteriaceae include pathogenic species such as Escherichia coli and Salmonella, responsible for the spoilage of food aerobically and anaerobically. In the seafood products, obtained from hot atmospheric conditions, the Vibrio. Spp., species are dominant for the spoilage of marine food products at temperature above 20 °C.

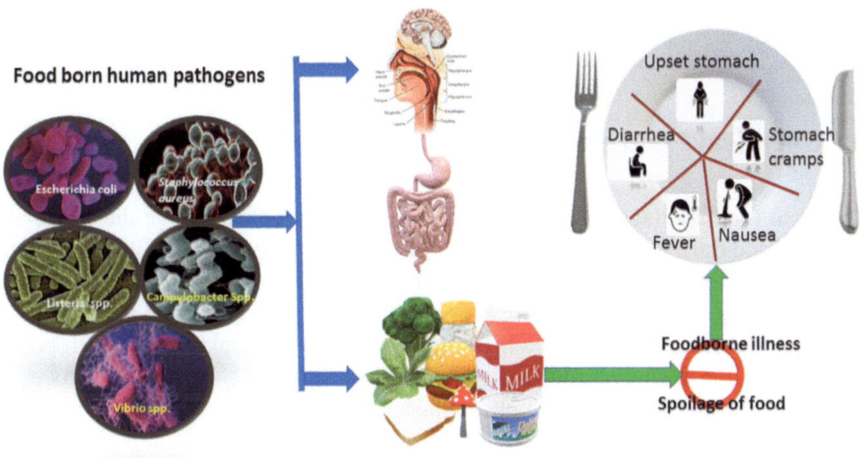

Fig. 3 Microorganisms for food spoilage

Furthermore, in fish and meat, the amine is produced by living organisms via several members of this group. The remaining species that are Obesumbacterium have disagreeable colors in beer and orders.

3.2 *Yeast*

The spoilage of Yeast generally occurs at pH 5.5 or lower. Generally, the products with high sugar and salt content, spoilage occur by the genera related to the Zygosaccharomyces species. Few species like Z. baillii, Z. bisporus, and Z. rouxii are responsible for the spoilage of foods and beverages. Candida species is one of the significant contributors to the spoilage of foods [45].

3.3 *Molds*

Molds are filamentous fungi that are responsible for food spoilage. Molds are found in wide varieties and number of genera. Usually, the Zygomycetes, termed as the "pin molds," are frequently considered fast producers. These species spoiled various kinds of food by a "hit and run" approach. Generally, at higher temperatures and lower water activities, Aspergilli and Penicillia, known as common molds, are responsible for the spoilage. Molds are responsible for the genesis of unwanted mycotoxins grow on numerous food stuff. Some fungi also produces small secondary metabolites belonging mainly to the genera Aspergillus, Penicillium, Fusarium, and Alternaria [45].

4 Chemical Food Spoilage

Chemical food spoilage refers to the chemical reactions that change the flavor and color of food during processing and storage. After harvesting the fruits and vegetables or slaughtering animals, chemical changes begin naturally within the foods and start degrading the quality. The breakdown of fats turns into rancid, and naturally—occurring enzymes develop which results extensive chemical changes in foods as they age [10]. Enzymatic and non-enzymatic browning is two different processes which are responsible for chemical spoilage.

4.1 Enzymatic Browning

Enzymatic browning is a naturally process that occurs in many fruits, vegetables, and seafood. Generally, the negative impact of enzymatic browning in fruits and vegetables are change in color, flavor, taste and nutritional value. Enzymatic browning occurs when they are cut, injured, peeled and diseased or exposed to any unusual conditions. On exposure to air the fruits and vegetables develop brown colored melanin pigment because of the oxidation of phenolic compounds. It became functional with the oxidation of phenols by polyphenol oxidase into quinones. It has been reported that polyphenol oxidase present in fruits, vegetables and some seafood is highly responsible for enzymatic browning [11].

4.2 Non-enzymatic Browning

Non-enzymatic browning is a strong chemical reaction responsible for the characteristic change in color during the heating or long-time storage of fruits and vegetables. It is attributed to a Maillard chemical reaction, where carbonyl group of reducing sugars reacts non-enzymatically with free amino groups of amino acids and proteins. In this chemical reaction, the brown coloration of food occurs without the activity of enzymes. Non-enzymatic browning can produce fluorescent, brown, and highly cross-linked pigments such as chromophores and melanoidins. It is commonly apparent that L-ascorbic acid degradation products and sugars, e.g., furfural, 5-(hydroxymethyl) furfural, or other carbonyl compounds are responsible for browning in juice and further, polymerize with each other or react with amino acids to yield browning materials [45].

5 Physical Food Spoilage

Physical spoilage of food occurs during harvesting, processing, or distribution due to physical damage. Due to the physical deterioration of food, there is an increase in the chance of chemical or microbial spoilage and contamination because the outer covering of the food is injured or busted, and microorganisms can penetrate the foodstuff more easily [45].

6 Condition Required for Spoilage of Food

6.1 Nature of Food

Generally, liquid food like fruit juices, water, butter, sugar, soup broth, etc., spoil faster since the organisms can lay out all over the food through their movability or by convection currents. Some semi-solid foods like soup, tinned fruits, and meat stews can spoil more quickly than liquid foods. The solid food items to be more prone to tarnish from their outward appearance.

The growth of the most appropriate microorganisms can be controlled by food that is rich in nutrients. Proteinaceous foods like meat, eggs, and fish are more prone to be attacked by the organisms that can break down the proteins. On the other hand, various foods like bread, pasta, jams, and syrups rich in carbohydrates are more prone to be attacked by fermentative organisms. In contrast, fats are more prone to be attacked by lipolytic organisms. Enzymes can degrade the foods with lipids to produce free fatty acids, which have a putrid and annoying odor, and it is crucial in olive oil, meat, and dairy systems. Several foods have found naturally occurring antimicrobial and inhibitory substances. These substances can slow down the growth of microorganisms [42, 45].

Several yeasts grow very well in very high sugar or salt concentrations. Notably, Debaryomyces hansenii is tolerant to very high NaCl concentration and some strains are resistant even up to 24% (*w/v*) NaCl concentration. Furthermore, molds are tolerant to high concentrations of salt or sugar.

6.2 Activity of Water

The water activity (WA) of food is the ratio of vapor pressure of the food to the vapor pressure of pure water. WA has a dominant role in the growth of the microorganisms. The highest WA for most microorganisms is in the range 0.995–0.980. Since an aqueous phase is a principal requirement for the metabolism therefore any process like drying, curing by addition of sugar or salt will decrease the WA value. Decreasing the value of WA will slow down the microbiological growth which results longer shelf life of food. Various microorganisms have different requirements concerning WA; the composition of the microflora is impacted by the WA.

Bacteria need a high WA and therefore do not cause the spoilage of dry foods in dry condition. Usually, yeasts grow in a WA value above 0.87–0.94, however Osmotoler species grow in a low WA value of 0.60. Next, molds can grow in foods with the most insufficient WA of 0.70–0.80; notably Xeromyces can grow in WA value of 0.60 [45].

6.3 pH

The growth of different microorganisms requires different pH values of the food. Most natural habitat of microorganisms evolve around neutral pH, which is excellent for bacterial enzyme activities. Yet, few bacteria like Lactobacillus and Acetobacter are allow to grow at lower pH ranges (3.0–4.4).

Generally, Yeasts can be grown in the range of pH 4.5–7.0. Some yeasts such as P. membranifaciens and Z. bailii are more permissive to grow at low acidic pH value while, others such as S. cerevisiae and D. hansenii are more permissive towards higher pH values within 5.0–7.0 [45].

The stress response mechanisms by microorganisms enables them to sustain in certain pH conditions. There are two distinct processes with Salmonella Typhimurium are resistance response to more acidic pH and tolerance response between pH 4.5 to 6. A few proteolytic bacteria can soar at a higher pH because they produce amines to buffer the high pH at the time of soaring. Oxidative yeasts oxidize organic acids, sugar, and alcohol which raise the pH during the growing period; so, they favor to grow on the surface of liquors forming a film.

6.4 Temperature

Temperature conditions play an essential role in the spoilage process during processing, transportation and storage of foods. When the temperature increases, the lag phase decreases which increases the growth rate. Temperature also control enzyme activity, protein synthesis, solute uptake, and shelf life length. Various microorganisms can grow within a wide range of temperatures, but maximum growth occur at the optimum temperature.

Organisms are divided into mesophiles, psychrophiles, and thermophiles based on their temperature preferences. Several human and animal pathogens in the mesophiles category have food spoilage microorganisms. The minimum growth range of mesophiles occur at 5–15 °C, optimum 30–45 °C, and maximum 35–47 °C temperature range. Escherichia coli, Salmonella, Clostridium botulinum, and Staphylococcus aureus are the examples of these pathogens. Molds and yeasts grow at and below room temperature, therefore they are highly responsible for spoilage of food at cool temperatures [45].

6.5 Gaseous Conditions

The oxidation–reduction potential (ORP) of the food also have impact on the types of organisms that will grow. The spoilage occurs at the surfaces of foods by the aerobic organisms; maximum animal foods and fresh plants have a small ORP throughout.

Facultative microorganisms grow on the surface of the foods and within them, e.g. spoilage of canned foods by the members of the genus Bacillus. Anaerobic organisms can grow in foods stored in the absence of oxygen. Spoilage of food stored in vaccum packs occur due to the fermentation process of bacteria and yeasts. Yeasts grow under both aerobic and anaerobic condition. Generally, molds are aerobic, therefore spoilage occurs at the surface, but mycelium penetrates deep into the food [45].

6.6 Interaction Phenomena

All of the Physico-chemical parameters have a definite impact, but it is essential to study their mixed interaction concerning the growth of microorganisms. Depending on the microorganisms, the interactions may be positive or negative, and the permissible levels of food additives may support the growth of microorganisms if it is present in food. The bacteria will become dominant first; but at a later stage, if the food conditions permits then mold or yeast spoilage may occur. The waste products produced by the dominant organisms may act as stimulant or depressant in the growth of another organism's. For example, few molds of the Penicillium species may produce antibiotics during their development, blocking the growth of other microorganisms [45].

7 Basic Properties Required for Food Packaging Materials

7.1 Barrier properties

7.1.1 Mechanical Properties

The polymer architecture of food packaging material plays a crucial role in several polymer preparation process like injection molding, film forming, sheet extrusion, blow molding, etc., and further polymeric structure tailors the mechanical properties of the end product. Additionally, numerous packaging materials are employed for storage at below room temperature, so it is crucial to examine the mechanical performance under these storage conditions [13, 58].

The tensile strength, percent elongation and the elastic modulus are determined by performing the tensile test analyses. These values provide mechanical information of the biopolymer materials.

7.1.2 Chemical Resistance Properties

Chemical resistance is an essential property required for food packaging because it exhibits its behavior towards acid or base. Consequently, it is required to estimate the performance and the suitability of food packaging material towards acidic and basic condition as a function of time. Generally, the tensile stress elongation at break and modulus of elasticity of sample immersed in weak and strong acid solutions as a function of time, simulating actual conditions, at ambient temperature [58].

7.1.3 Biodegradation/Antimicrobial Properties

Novel bio-nano composites show unique properties such as biodegradability which should not be lost during practical applications. Physical properties of biodegradable polymers need to to improved to replace the current petroleum-based materials. Exploitation of the antimicrobial properties of metal nanocomposites has received great attention. The packaging films with metal nanocomposites are found to control the growth of harmful pathogenic and spoilage microorganisms. The nanocomposite films with antimicrobial activity are especially advantageous due to their admissible architectural integrity and barrier properties impregnated within [6].

8 Food Packaging Materials

Materials used for food packing include paper, paperboards, glass and metals (aluminum, foils, and tin-free steel). In addition, a wide range of plastics in both flexible and rigid forms are employed [13]. These days, food packaging materials combine various materials with peculiar functional or aesthetic properties.

In the food industry, packaging is an essential component at each stage. Its permeability is the fundamental weakness therefore there is a need for innovative, cost-effective, environmentally safe, and intelligent food packaging material. Consequently, various major factors handle the continuous discovery of intelligent food packaging materials that promote transportation, handling, and storage. So far, the glass, plastics, and metals are employed in packaging applications but they face the problem of non-biodegradability. Bio-nanocomposite materials are promising candidate for food packaging applications. Green packaging materials, including biodegradable, edible materials, plant extracts, and nanocomposite materials characteristics, can reduce the negative environmental impacts.

8.1 Paper as Packaging Materials

The reedy plant papyrus was used by Egyptians to produce the world's first writing material. Later bamboo and mulberry barks were used in the development of paper-making process. Nowadays, paper are manufactured even from cottonseed hair, sunflower stalk, agricultural waste and leaves [52].

Based on grade the paper can be classified as processed and recycled paper. Further, paper for food packaging can be classified into two broad categories: (1) based on pulp or paper treatment (2) based on shape and combination of various materials. Wood pulp treatment effects the paper properties and its use significantly [15] (Table 1).

Advantage of using paper as a packaging material are cost-effectiveness, lightweight, printability, easy availability, and strong mechanical properties. But its major drawback is humidity and moisture absorption. The paper products were progressively combined with a layer of biopolymer coating to improve their barrier properties, functionality and hydrophobicity by the surface modification with the mixing of biopolymers [13].

Table 1 Different types, properties and uses of paper in food packaging application

Types of paper	Preparation and properties	Use in food packaging
Kraft paper	Unbleached pulp and bleached pulp Soft, expensive and White color	Flour, sugar, and fruits and vegetables
Tissues	Wrapping tissue is of two types Ordinary tissue and neutral tissue	Bakery products, and tea and coffee bags
Grease proof paper	Prolonged heating of wood pulp Translucent and impermeable to fats and oils	Oil/fat rich food, milk products and meat
Glassine paper	Extreme hydrationof grease proof paper High density, transparency, smooth and glassy surface	Liner for baked goods, biscuits and cooking fat rich food
Vegetable parchment paper	Acid treatment of pulp High wet-strength with poor gas barrier property	Oil/fat rich food and pastries,
Waxed paper	Wax coated on paper base Liquid and gas barrier properties	Fruit juices, milk, baked food and pastries
Sulfite paper	Light weight, glazed to improve appearance	Bakery and confectionary products
Paper board	Prepared from 100% virgin pulp or 100% bleached virgin pulp or 100% recycled paper	Milk and bakery products, fruit juices and dry fruits
Paper bags	Prepared from various types of paper and available in different forms	'Carry home' and 'grocery carry use' bags

Recycling of paper refers to reuse of recovered paper after proper processing in form of new paper or other paper based products [21]. But, recycled paper can never match the quality of virgin paper, however recycling of waste paper conserves raw materials such as wood, forest and biomass [65].

8.2 Glass and Metals Based Packaging Materials

Both glass and metals are totally resistant to vapors and gases, they act as an effective barrier between the external atmosphere and air inside the material. Glass containers are one of the oldest method and currently practiced method of packaging wide variety of materials (pharmaceuticals, nutraceutical, dietary supplements and several other food products) [59]. Glass is manufactured from silica precursor, in presence of sodium carbonate, limestone or calcium carbonate and alumina. For storing the pharmaceutical and biopharmaceutical specially designed glass of Type I, Type II, Type III and Type IV are employed, among these Type I and II are preferred for medical formulation [55].

Metal-based packaging is one of the most versatile and widely utilized method for packaging of different products. This packaging provides physical protection and recyclability [71]. Metals, such as aluminum and steel are durable for food products. Tin-plated steel are durable properties, such as recyclable, eco-friendly, provide physical protection, and thermal and chemical resistance (Figs. 4 and 5).

Bio-based packaging materials exhibit better barrier properties, including permeability of vapor and gases, while hydrophobic surface treatments should boost resistance to humidity. The barrier properties of bio-based packaging materials can have improved by coatings using deposition of atomic layers, while hydrophobic surface modifications should increase resistance to humidity [13].

Fig. 4 Global production of petroleum-based polymer

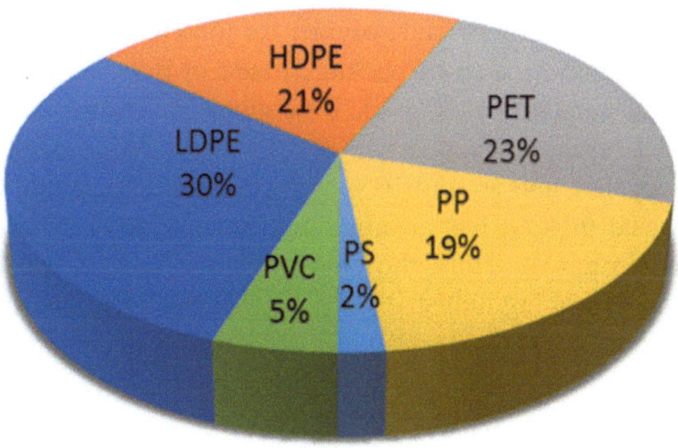

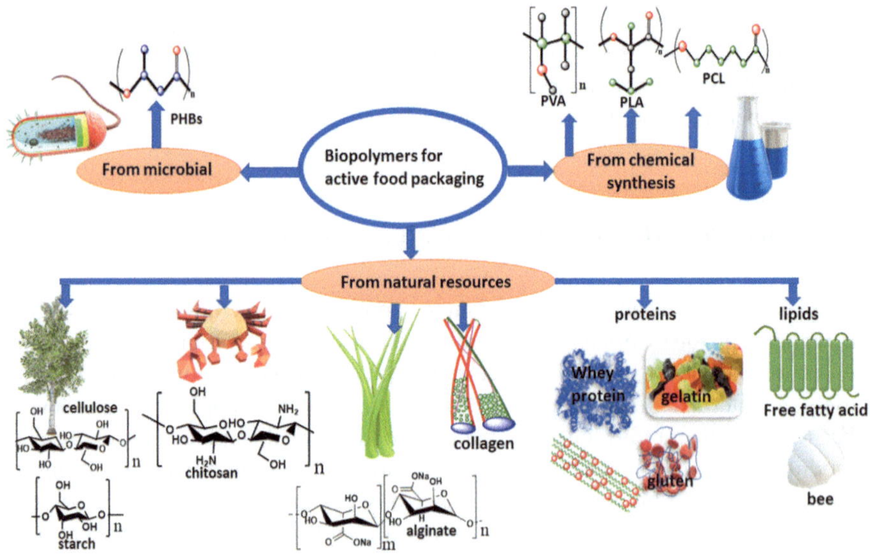

Fig. 5 Biopolymers for food packaging

9 Petroleum Polymers Based Packaging Material and Its Draw Back

In the middle of the twentieth century, petroleum-based polymer materials were widely used by human beings. These materials are advantageous in terms of cost-effectiveness, usability, aesthetic quality and physicochemical properties. Generally, plastics are employed as packaging materials, notably they are employed as food packaging materials. Major drawback of these plastic packages are non-biodegradability and use of petroleum polymer as source. Percentage of more often used petroleum-based polymer for food packaging are shown in Fig. 6. Blackberries packaged in snap-fit package made of poly(lactic acid) and poly(styrene) met the "US standard No 1" grade for commercialization for more than 12 days at 3 °C [33].

Consequently, nearly 80% of plastics used for landfills, less than 10% of plastics recycled, and around 10% incinerated. So far, above 6 billion metric tons of plastics waste in total has accumulated worldwide, and it is a matter of concern for the environment. Thus, the biodegradable packaging materials give more attention to the packaging industry [47, 73]. Few petroleum based polymers has been discussed here.

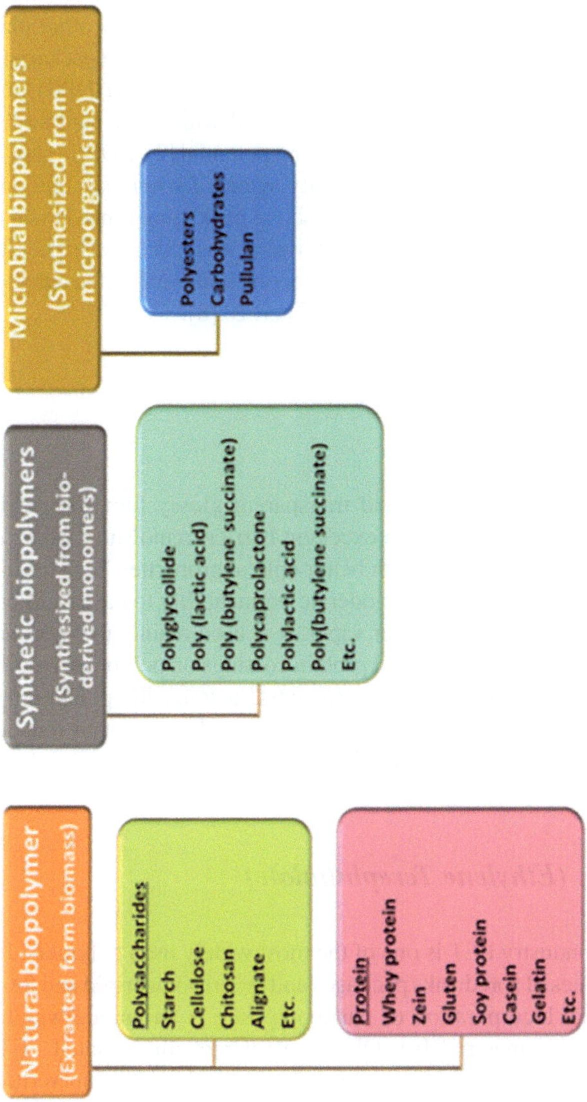

Fig. 6 Different types of biopolymers

9.1 Polyethylene (PE)

PE is one of the most common plastics produced in the world and has a wide range of physical properties. PE are rigid, hard or soft and malleable. The large varieties of products stored and packaged made by soft and malleable films in the packaging industry. Polyethylene has a low cost in comparison with many other plastics. The lower softening point favors lower processing energy costs. The packaging industry used two types of polyethylene: high-density polyethylene (HDPE), low-density polyethylene (LDPE) [6]. LDPE film incorporated with gallic acid and potassium chloride exhibited excellent oxygen-scavenging potential [2]. LDPE/(clay/carvacrol) films exhibited excellent antibacterial activity against *Escherichia coli* and *Listeria innocua* [57]. LDPE/ethylene vinyl acetate active packaging films shown controlled release of ferulic acid as natural antioxidant, which was 635 times higher than that of LDPE film [26]. HDPE/copper nanofiber nanocomposites exhibited enhanced tensile strength, oxygen barrier and antibacterial properties [8].

9.2 Polypropylene (PP)

Polypropylene is a rigid, rugged, and transparent glossy film with high strength and puncture resistance. It acts as an excellent barrier to moisture, odors and gases. Oriented PP is a clear glossy film with better optical properties, high tensile strength and puncture resistance. It exhibits moderate permeability to gases and odors while higher barrier property towards water vapor. It is used mainly to pack snack foods, biscuits, and dry foods [6]. Plasticized protein coatings on polypropylene film exhibited significant bacterial growth inhibition against Lactobacillus plantarum [35]. Corn zein nanocomposite coating of polypropylene films demonstrated reduced oxygen permeability and water permeablility [35].

9.3 PET Poly (Ethylene Terephthalate)

In the packaging industry PET is one of the most widely used polymers. It increases the barrier properties of soft drinks packages and several other rigid and flexible packaging applications. Incorporation of nanocomposite is an unique way to improve the properties of this material [6]. Olive leaf extract impregnated PET/PP films exhibited significant lipid oxidation of sunflower seeds [12]. Supercritical carbon dioxide impregnated PET/PP exhibited adsorption of α-tocopherol (TOC), a natural antioxidant [24].

9.4 PVC (Polyvinyl Chloride)

It is a thermoplastic material produced via free radical polymerization of vinyl chloride. It withstands high temperatures and shows excellent resistance to oil and fat. PVC are used in the form of blow-molded bottles. PVC based film is tough, high elongation, relatively low tensile and tear strength. It is appropriate for the packaging of mineral water, fruit drinks, and fruit juice in bottles. Generally, in the food packaging application, the use of PVC is decreasing because of its toxicity and environmental issues [6] (Table 2).

10 Biopolymers Based Packaging Material and Its Benefit

Polymers have played an essential role in packaging materials for a long time because they possess various desired features like softness, lightness, and transparency. Although, synthetic packaging films can cause some severe environmental problems because of their total non-biodegradability. The complete replacement of synthetic film with an eco-friendly packaging film is exclusively difficult for particular applications like food packaging. Biopolymers are natural polymer which has introduced as sustainable packaging materials. There is another possibility for non-petroleum-based polymers including polysaccharides, proteins, lipids, poly hydroxyl butyrate (PHB), polylactic acid (PLA), polyvinyl alcohol (PVA), poly butylene succinate (PBS), poly caprolactone (PCL) and their biopolymer blends. Biopolymers possess different properties such as non-toxicity and biodegradability that can be boosting their application in food packaging.

In the packaging industry, several natural and synthetic biodegradable polymers are used. The polymers can classify as natural, artificial, and modified natural polymers based on their origin and an environmental point of view Fig. 6. These polymers with excellent physicochemical properties are promising candidate for food packaging [6].

Table 2 Different petroleum based product and their respective packaging applications

Petroleum based product based	Packaging applications
LDPE	Carrier bags, bin bags
HDPE	Milk and fruit juice bottles
PP	Drink bottle, juice bottles
PET	Carbonated drink bottles and other transparent drink bottles
PVC	Chemical bottles, trays and cups

10.1 Natural Polymers

Natural polymers play an essential function in the food packaging business. They have excellent properties, nutritional food value, antioxidant and inclusive antimicrobial properties, cost-effective, renewable origin, spreading, and the fact that are environment friendly. Different types of biodegradable polymers occur in nature. A brief explanation of few of them described below.

10.2 Starch

Starch is a natural polymeric carbohydrate present in many green plants (potato, rice, corn, wheat, barley, soybean and oat). The molecules of starch are composed of two types of polymers of d-glucose that is amylopectin (70–80%) and amylose (20–30%). Additionally, it is considered one of the promising alternative biopolymer materials. The advantages include nontoxicity, biodegradability, easy availability, and renewability. Starch-based films showed excellent barrier to oxygen at relatively low humidity. The brittleness nature of the starch film can be improved by adding few common plasticizers viz. glycerol, xylitol and sorbitol in the food packaging materials. It can also be plasticized using materials like poly (ε-caprolactone) (PCL), poly (vinyl alcohol) (PVA), and others. Starch blended with thermoplastic were used for cups, food wrapping, plates and other food containers. Yet, the blending technique do not show improve film properties, such as barrier properties and mechanical properties as compare to plain starch film. The sugar palm starch, combined with various plasticizer can be used for preparation of biodegradable films [53].

10.3 Cellulose

Cellulose is a biopolymers in found in all plant materials, algae, fungi, bacteria and made up of a chain of β-(1 → 4)-linked glucose residues. Due to their poor solubility, hydrophilic nature and a highly crystalline structure faces some difficulties in food packaging.

Cellulose has the potential to conserve water loss from dry areas and it can absorb undesirable liquids. These properties sharpen the treatment of deep ulcers and as an antimicrobial agents in wound dressings. Surface improved wood cellulose fibers, like taurine cellulose, α-hydroxysulfonic acid cellulose (HSAC) are used to prepare environmentally friendly film. The film made by HSAC showed excellent mechanical properties. Improved cellulose fiber blended with PVA played vital role in improving the tensile strength of the composites and can be used as a packaging material [53].

10.4 Chitosan

Chitosan (CS) is a nontoxic linear polycationic polysaccharide composed of randomly distributed β-(1 → 4)-linked 2-amino-2-deoxy-d-glucose (d-glucosamine) and 2-acetamido-2-deoxy-d-glucose (N-acetyl-d-glucosamine) units. Commercially it is produced by chemical N-deacetylation of chitin shells of shrimp and other crustaceans. The molar mass of the polymer and the extent of deacetylation is one of its most important chemical characteristics which dictate its use. Additionally, it showed different properties, biodegradability and biological roles, depending on the relative proportions of d-glucosamine and N-acetyl-d-glucosamine residues. CS showed many valuable features as a packaging material because of partially or well enough solubility in the water, films forming nature without the use of other additive and heat resistance. Further, CS has excellent barrier properties towards oxygen and carbon dioxide and shown excellent antimicrobial activity. The membrane of chitosan is used for coating fresh fruit, in particularly, strawberries, berries, and grapes. The addition of antioxidant and preservative like garlic extract, lysozyme, nisin allow for longer storage of products [53].

10.5 Protein-Whey Protein

Whey protein (WP) is a combination of alpha lactalbumin, beta-lactoglobulin and immunoglobulin. They are adequate of forming flexible films, and they are used as raw materials because they have good oxygen barriers, average moisture permeability, and good biodegradability. WP are one of the cost-effective raw materials used in the manufacute of edible films. The WP with Plastic films is used as biodegradable and showed high barrier to oxygen. The commercial biodegradable Bio-Flex, blended with WP to form coatings films which is used as biodegradable packaging material [53].

10.6 Zein

Zein is a mixture of proteins from corn and is classified as a prolamin, which contains 44%-79% of the endosperm protein. It has both hydrophilic and hydrophobic properties. Although is shown poor mechanical properties, it can be employed as food packaging material, mainly as an impermeable protective coating because of its good gas barrier, biocompatibility, and biodegradation properties. Zein-based films with glycerol, nanocarbonate, oleic acid and nanocarbonate show the possibility to pack food products [53].

10.7 Gluten

Wheat gluten (WG) is a main protein of wheat. It is a water-insoluble protein consisting of more than 60 different polymeric polypeptides. It has a molar mass of 30,000–100,000 g/mol. WG has been widely modified chemically or enzymatically because of its low solubility. The laminated film of WG with glycerol as a plasticizer and polylactic acid (PLA) as a reinforcing component has higher strength than the gluten films and it shows good barrier properties [53].

10.8 Microbial Polymer

Microbial polymers, also indicated to as biopolymers and have been employed in food, pharmaceutical, cosmetics and other industries applications. The controlled microbial fermentation has admitted the use of these significant components as coating, packaging, stabilizing, thickening or gelling ingredients in the food industry. As an additive, these polymers can act as an antioxidant, antimicrobial, sweetener preservative, antioxidant material and can be used to boost properties of modern functional food. In the last decades, these natural polymers have garnered boost interest due to their biodegradable, nontoxic, eco-friendly, nontoxic, and modifiable features, along with decreased production costs [53].

10.9 Polyhydroxyalkanoates (PHA)

PHA are naturally occurring biodegradable polymers which are produced by bacterial fermentation of lipid and sugars. Depending on the monomer it can be thermoplastic or elastomeric materials. These polymers produce excellent packaging films either alone or in combination with synthetic plastic or starch. Though these polymers are currently too much expensive than petrochemically based plastics. Few modifications of PHA have been developed to expand the range of its packaging applications. For the food packaging applications polymer must illustrate a high purity level. The obtained material involves minor impurities like lipids and proteins during the microbial production of PHA. The small amount of impurities may cause a significant odor problem when using PHA as a package [53].

10.10 Synthetic/Artificial Polymers

Man-made biodegradable polymers and polymers from renewable resources have found applications in the packaging industry. These polymers are found to be advantageous over natural polymers because they can be tailored to provide a wider range of properties for specific applications [53].

10.11 Polyglycolide

Polyglycolide (PGA) or polyglycolic acid is the linear, simple, rigid, aliphatic hydrolyzable polyester. It exhibits high crystallinity and barrier properties against CO_2 and O_2. PGA is use as protective layer in multilayer packaging systems, for carbonated soft drinks and beer bottle for packaging applications [53].

10.12 Polylactic Acid (PLA)

Polylactic acid (PLA) derived from biodegradable and renewable resources like corn, fermentation of sugar feedstock, etc., which has excellent physical and chemical properties. PLA is recyclable, biocompatible, biodegradable and compostable, non-toxic, renewable, hydrophilic and highly transparent. Generally, commercial PLA is a copolymer synthesized form poly (D-lactic acid) and poly (L-lactic acid) monomers. The properties of PLA can vary from semi crystalline to amorphous nature depending on the D-lactide/L-lactide enantiomers ratio. It also has a high heat distortion temperature, good gas barrier properties and mechanical properties. Concerning the optical, physical, and mechanical performance of the oriented PLA polymer (OPLA) in food application, the comparative study has achieved two of the commonly used materials used for new food packaging applications, which are polyethylene terephthalate (PET) and oriented polystyrene (OPS) [53].

10.13 Polybutylene Succinate (PBS)

PBS is a commercially available aliphatic polyester having succinic acid as one of the monomers. It is principally manufactured through polycondensation of 1, 4-butanediol and succinic acid. These polymers give a wide range of environmentally friendly thermoplastics. When PBS mixes with other polymers such as poly (ethylene succinate) (PES) and thermoplastic starch, the materials' cost reduces [53].

10.14 Polycaprolactone (PCL)

Poly- ε-caprolactone is a biodegradable polyester obtained by the ring-opening polymerization of ε-caprolactone. It is soluble in a broad range of solvents. It has a low glass transition temperature and exist as semi-rigid material at room temperature. It is readily biodegradable by Enzymes and fungi. Various copolymers with lactide or glycoside developed to improve the degradation rate. It uses for packaging materials with composite films [53] (Table 3).

11 Nanomaterials Used in Food Packaging

In recent years nanomaterials have drawn much attention due to their extraordinary properties and mainly used to impart antimicrobial function and improve the gas barrier, mechanical and thermal properties to extend the shelf life and freshness of packaged food items. The different types of nanomaterials used in food packaging are discussed below-

11.1 Silver NPs (AgNPs)

Generally, AgNPs is widely used as an antimicrobial agent for food and beverage safety. AgNPs possess a larger surface area at a nano-scale as compared to micro-scale or bulk. In addition, AgNPs interact with the surface of the cell via different mechanism. AgNPs is incorporated into the plastic polymers by different methods for food packaging. It has been reported that AgNPs and ZnO-NPs containing LDPE can protect and elongate the shelf life of orange juice. The active-nanocomposite of this material has proven to be highly efficient for antimicrobial nanomaterial in combination with heat treatment [13] (Table 4).

11.2 Nanoclay

Nano clay was the most demanding material in the food packaging industry because it was the first emerging material in the market among the other polymer nanocomposites. It is used for enhancing the physical of plastic and barrier properties of packaging material. Montmorillonite (MMT) is an example of nano clay. These polymer composites is extensively used because of their magnificent cation exchange efficiency, good swelling behavior, and wide surface area. Naturally, it is readily available because it derived from volcanic ash and rocks. Depending on the degree of nanoparticles distribution in the polymer matrix, the nano clay composites have

Table 3 Literature report of some examples of biopolymers used for food packaging

Polymer	Matrix	Properties	Applications	References
Starch	Cassava starch based foams incorporated with grape stalks	Biodegradable	Suitable for storage of food with low moisture content	[20]
	cassava starch based films incorporated with cinnamon essential oil and sodium bentonite clay nanoparticles	Biodegradable	Packaging of meat balls	[32]
Cellulose	Carboxymethyl cellulose (CMC) based films containing Chinese chives root extract (CRE)	Improved moisture content, water-solubility, swelling degree and thickness by the addition of CRE	Good antioxidant and antimicrobial activity	[48]
	Lysozyme and lactoferrin were incorporated into paper containing CMC	CMC improves the protein payload of paper	Lysozyme was most effective in preventing growth of microbiota	[7, 46]
Chitosan	Grapefruit seed (GFSE) extract incorporated chitosan film	GFSE made the films more amorphous and decrease the tensile strength	Inhibited the proliferation of fungal growth	[61]
	CRE incorporated chitosan based film	Optical properties were improved	Good antioxidant and antimicrobial activity	[49]
Whey protein (WP)	Lysozyme with Polyacrylic acid incorporated WP isolate films	Incorporating lysozyme into the film in complexed form extended its release time	Good anti-bacterial film	[43]
	WP incorporated with *Fucus vesiculosus* extract	Strengthened the mechanical properties	Inhibited the chicken breasts lipid oxidation	[4]
Zein protein (ZP)	Chilto fruit extract incorporated ZP	Zein fibers improved coating integrity upon water contact	Zein fibers delayed the release of phenolic compounds, hence suitable for food packaging	[35]
	Glycerol and polyethers added zein films	Excellent elongation and UV barrier properties	Promising for food packaging	[31]

(continued)

Table 3 (continued)

Polymer	Matrix	Properties	Applications	References
Soy protein	Bilayer of soy protein and poly(lactic acid)	Improved mechanical properties	Inhibition of mold, yeast and two strains of bacteria	[28]
	Soy protein isolate: bees wax edible coating	High O2 modified atmosphere did not extend shelf life	Antioxidant capacity of cut artichoke was maintained	[25]
Microbial	PLA films coated with a cellulose derivative/cocoa butter	Coatings decreased the water vapour permeability of PLA	Reduced the number of bacterial strains	[40]

Table 4 Ag NPs based composite food packaging materials reported in literature

Polymer matrix	Tested food	Tested microorganism	Shelf life	References
Ag/LDPE nanocomposite film	Chicken breast fillets	Psychotropic bacteria	8 days	[5]
Ag NPs/regenerated cellulose film	Cherry tomatoes	E.coli and S.aureus	9 days	[30]
Ag NPs/Alignate	Cheese	B.cereus, S.aureus, E.coli and S. typhi	14 days	[41]
Ag NPs/PE packages	Olivier salad	Coliform, mold and yeast	15 days	[63]
Silver and clay nanocomposite	Shrimp	V.parahaemolyticus, S. aureus and E.coli	6 days	[44]

either an intercalated or exfoliated pattern. It helps in the dispersion of the nanoparticles in the matrix of the polymer. In general, nano clay acknowledges as a vital filler for polymer, which is bio-based reinforcements. Therefore, the reinforced polymeric material reduces the barrier properties and gives mechanical strength to the biopolymer. It is used for packaging, carbonated drinks, beer bottles, and thermoformed containers for industrial purposes. Reportedly, the nano clay in plastic bottles keeps the juice fresh and prolongs the shelf life up to 30 weeks. The use of nano clay-based hybrid materials will provide sustainability and reduce environmental risks associated with dumping synthetic polymer-based packaging materials [38].

11.3 ZnO

It is extensively used in various applications like medical devices, cosmetics, medical devices, medication, atmospheric processed (MAP) packaging, and delivery. ZnO-NPs is specifically more attractive for packaging applications than AgNPs because

Table 5 ZnO NPs based composite food packaging materials reported in the literature

Polymer matrix	Tested food	Tested microorganism	Shelf life	References
ZnO NPs/PVA/Spathodea campanulata bud fluid matix	Black grapes	E.coli, P. aeruginosa and E.aerogenosa	7 days	[29]
Chitosan/potato/protein/linseed oil/ZnO NPs	Raw meat	Total bacterial count reduced	7 days	[66]
Betanin nanoliposomes incorporated gelatin/chitosan nanofiber/ZnO NPs	Fresh beef	E. coli and S.aureus	16 days	[3]
ZnO NPs/Chitosan	Poultry meat	Total aerobic mesophilic and psychotrophic, and Enterobacteriaceae	11 days	[60]
ZnO NPs/ Pullulan/chitosan	Meat	E.coli and L. monocytogenes	15 days	[51]

its cost-effectiveness and less toxicity. Also, ZnO can produce a large amount of hydrogen peroxide under UV irradiation that can cause oxidative stress in bacteria cells. Zinc ions play a crucial role in blocking the growth of bacteria. ZnO-NPs can oxidize ethylene into carbon dioxide and water under UV irradiation and decrease the accumulation of malondialdehyde (MDA) and pyrogallol peroxidase (POD) activity. PVC–ZnO-nano-composite films can use this mechanism to extend the shelf life of food products [13] (Table 5).

11.4 Titanium NPs (TiO_2-NPs)

TiO_2-NPs consider as most crucial metal oxide nanomaterials with inertia and thermo-stability which can improve the properties of biodegradable films. US-FDA can approve it in 1996 as a food additive. In the food packaging industry, it is frequently used as a whitener, photocatalyst, air and water purification, self-cleaning structures, water filtering, and antimicrobial. Many studies carried out on the antimicrobial effects of TiO_2, which indicated that, under ultraviolet light or sunlight, it could produce reactive oxygen species which directly destroy microbial cell walls. The particles of white TiO_2 efficiently scatter the visible light, therefore giving the coated object power, brightness, and whiteness. The TiO_2 in the form of thin coated solid is non-flammable, non-volatile, totally inert, and completely insoluble in all foods and food materials [13] (Table 6).

Table 6 TiO_2 NPs based composite food packaging materials reported in the literature

Polymer matrix	Tested food	Tested microorganism	Experimental findings	References
k-carrageenan/konjac glucomnann/TiO_2 film	Strawberry	Penicilliumm viridicatum	Irradiated for 6 h	[17]
Polylactic acid/TiO_2/GO film	Green peppers	E.coli and S.aureus	High antibacterial activity after 24 h of UV irradiation	[16]
CMC/Arabic gum/gelatin/garlic extract/TiO_2 NPs	Fresh Nile tilapia fish fillets	S. aureus, B.cereus, L. monoctogenes, S. typhmium, P. aeruginosa, E. coli and yeast C. albicans	Delayed bacteriological development, lost less water during cold storage for 21 days	[70]
Chitosan/whey protein/TiO2 NPs and Zataria multiflora essential oil	Cheese	Coliforms, spore-forming bacteria and lacotose-fermenting yeasts	Inhibit the growth of Listeria monocytogenes on the surface of cheese	[27]
SiO, ZnO and CuO/4A zeolite	Shrimp	Staphylococcus aureus, Listeria monocytogenes, Escherichia coli, Pseudomonas fluorescens, Vibrio parahaemolyticus *and* Aeromonas caviae	Reduced the total viable, Enterobacteriaceae, *Pseudomonas* spp. and *Shewanella putrefaciens*, in raw shrimp and *S. aureus, L. monocytogenes*, and *E. coli* count in inoculated shrimp	[56]

Table 7 Metal/metal oxide NPs based composite food packaging materials reported in the literature

Polymer matrix	Tested food	Tested microorganism	Shelf life	References
LDPE/Cu NPs	Peda (sweet)	E. coli and S.aureus	8 days	[37]
(Ag–Cu) NPs/polylactide films	Chicken samples	S. typhi, C. jejuni and L. monocytogenes	21 days	[1]
CuO/ZnO NPs	Guava fruit	E. coli and S.aureus	7 days	[34]
CMC/PVA/CuO	Processed cheese	Total bacteria count, coliforms, moulds and yeasts, and pscychrotropics	28 days	[69]
SiO, ZnO and CuO/4A zeolite	Shrimp	Staphylococcus aureus, Listeria monocytogenes, Escherichia coli, Pseudomonas fluorescens, Vibrio parahaemolyticus *and* Aeromonas caviae		[56]

11.5 Copper and Copper Oxide (Cu/CuO)

Copper nanoparticles have relatively Cu^0/Cu^{2+} low reduction potential hence it is quickly oxidized. It is used as an antimicrobial agent because it can reduce the growth of few microorganisms such as bacteria, fungi, and viruses. CuO nanoparticles increase the shelf life of food packaging materials [19] (Table 7).

11.6 Magnetic Nanoparticles (Fe_3O_4)

Magnetic nanoparticles may come into different shapes, sizes, and crystalline forms that might alter their toxicity. They were used in food packaging by utilizing a nanofiller. Several studies have found that the food packaging applications related to magnetic nanoparticles show better characteristics as a composites film [22].

11.7 Nano-starch

Starch is the cheapest, biodegradable, renewable, and non-toxic polysaccharide. It is used in food, pharmaceuticals, paper-making, plastic, rubber, and packaging materials because of its effectiveness, environmentally friendly, and ample supply. Generally, starch can be found in discrete and partially crystalline granules and mainly made up of two glycosidic macromolecules: branched amylopectin and linear amylose. It is vital to separate crystalline starch from the amorphous and crystalline complexes for the production of nano-starch. Flexible food packaging uses starch nanocrystals as a promising nanofiller. The starch nanocrystals improved barrier properties and mechanical properties since it is highly susceptible to hydration. The molecular structure of starch possesses hydrophilic functional groups; hence it is not applicable in a humid environment. The film of starch is appropriate as an antimicrobial packaging material [13].

11.8 Carbon Nanotubes (CNTs)

Carbon nanotubes are hollow tubes with a diameter of a nanometer in range. The two types of CNTs are single-wall CNTs and multiwall CNTs. It provides an outstanding high tensile strength and elastic modulus when combined with the polymer matrix. In recent times, CNTs have been composed of polymers and employed for packaging purposes and intelligent antimicrobial sensors. The carbon nanotube-based sensor could be producing a transparent, thin-film embedded with wireless chips which communicate to the market manager and customer about the spoilage of fruit and meal [13].

11.9 Nano-silica

The nano-silica mainly use at the time of hydrophobic coatings, especially for self-cleaning materials. Inside the container, the non-hydrophobic coating can allow food to be free-flowing material inside the containers or jars. There are various products such as wine, beer, and powdered soup that benefitted from this technology [13].

12 US and Indian Safety Guidelines in Food Packaging

The packaging and labeling of food is regulated by the U.S. food and drug administration (FDA). These regulations aim to aware the consumers about the food and increase the safety of food given out all over the United States. The center for food safety and applied nutrition (CFSAN) and the office of food additive safety can ascertain the protection from harmful food contact substances. The guidelines of FDAs also need that the labeling of the package involves the expiration, nutrition guidelines and best before used by dates, preparation instruction and handling, and the packaging company's contact information. The labeling of allergen has also been necessary since 2006. The labeling of food allergen and consumer act requires the disclosure when the product contains potential allergens like cereals, crustaceans, mollusks, eggs, fish, lupin and milk soybeans, peanuts. Also, a notification for the packaging of a product is made using any of the allergens mentioned earlier (U.S. FDA Administration) (Table 8).

13 Obstacles in Commercialization

There are several issues in the commercialization of nanopackaging films. First, the production and studies of nanopackaging materials are still in the laboratory level and is not yet scaled up to industrial level due to their high cost of production. Second, poor barrier and mechanical propertied as compared to that of the synthetic plastics. Third, leaching of NPs into the food materials. Fifth, lack of awareness. Sixth, cost of nanopackaging films are higher than currently available synthetic plastics packaging materials. Finally, it is essential for to emphasize manufacturers to label the nanopackaging packaging materials with requisite information or label.

14 Future Trends

Presently we focus the bio-based packaging materials. It is necessary to transformed bioplastic research to address industrial application via academic and industrial collaborations. The growth of the bioplastic market can reduce petroleum-based polymers. Currently, we are crossing active and intelligent packaging materials. Innovative packaging reduces the wasting of food to check the spoilage of food products in real-time. Bioplastics will become a core part of the bioplastics market. There are many types of research needed to improve the quality of bioplastic for food packaging applications. Still, much research is required to overcome the next generation of innovative and intelligent packaging that can improve the shelf life of food products and know the spoilage of food products by incorporating various materials and devices. The researcher can create an intelligent barcode by installing the sensors on the food packaging and scanned to clarify the freshness of packaged items.

Table 8 Different other nanocomposites with promising packaging applications

Nanocomposite	Method	Properties	References
Starch NPs/poly(dimethylsiloxane)	Super hydrophobic coating	Excellent water resistance, self-cleaning, and liquid-food residue reduction	[67]
Water chestnut starch composite films	Acid hydrolysis method	Starch NPs decrease moisture content, water vapour transmission rate and solubility, while increase thickness and burst strength	[18]
Polylactic acid/CNTs/chitosan	Electrospinning	Exhibited better antimicrobial activity against S. aureus than E.coli	[36]
Pectin-CNTs films	Physical mixing or chemical bonding	Improvement of properties of pectin-CNTs films by chemical bonding as compared to physical mixing	[23]
Nanoclay/cinnamon oil/ Cassava starch films	Green technology	Exhibited significant antibacterial potential against E.coli, S. typhi and S. aureus	[32]
LDPE/nano clay composite films	Twin-screw extrusion	Beef color maintained for up to 4 days and native micro flora was suppressed	[62]
Cellulose nanofibrils and nano clay	Spray technique	Higher clay content increased the barrier properties	[39]
Silica–carbon/Ag NPs	One-step ball milling process	Exhibited bacteriostatic effect	[9]
Nisin/nano silica/chitosan	Coating	Provided longer storage life for mushroom	[54]
Mesoporous silica NPs/potato starch films	Casting	Better antibacterial protection for mushrooms against M. circinelloids and mucur sp.	[72]

References

1. Ahmed J, Arfat YA, Bher A et al (2018) Active chicken meat packaging based on polylactide films and bimetallic Ag–Cu nanoparticles and essential oil. J Food Sci 83:1299–1310. https://doi.org/10.1111/1750-3841.14121
2. Ahn BJ, Gaikwad KK, Lee YS (2016) Characterization and properties of LDPE film with gallic-acid-based oxygen scavenging system useful as a functional packaging material. J Appl Polym Sci 133. https://doi.org/10.1002/app.44138
3. Amjadi S, Nazari M, Alizadeh SA, Hamishehkar H (2020) Multifunctional betanin nanoliposomes-incorporated gelatin/chitosan nanofiber/ZnO nanoparticles nanocomposite film for fresh beef preservation. Meat Sci 167:108161. https://doi.org/10.1016/j.meatsci.2020.108161
4. Andrade MA, Barbosa CH, Souza VGL et al (2021) Novel active food packaging films based on whey protein incorporated with seaweed extract: development, characterization, and application in fresh poultry meat. Coatings 11. https://doi.org/10.3390/coatings11020229
5. Azlin-Hasim S, Cruz-Romero MC, Morris MA et al (2015) Effects of a combination of antimicrobial silver low density polyethylene nanocomposite films and modified atmosphere packaging on the shelf life of chicken breast fillets. Food Packag Shelf Life 4:26–35. https://doi.org/10.1016/j.fpsl.2015.03.003
6. Balakrishnan P, Thomas MS, Pothen LA et al (2014) Polymer films for packaging. In: Kobayashi S, Müllen K (eds) Encyclopedia of polymeric nanomaterials. Springer, Berlin, Heidelberg, pp 1–8
7. Barbiroli A, Bonomi F, Capretti G et al (2012) Antimicrobial activity of lysozyme and lactoferrin incorporated in cellulose-based food packaging. Food Control 26:387–392. https://doi.org/10.1016/j.foodcont.2012.01.046
8. Bikiaris DN, Triantafyllidis KS (2013) HDPE/Cu-nanofiber nanocomposites with enhanced antibacterial and oxygen barrier properties appropriate for food packaging applications. Mater Lett 93:1–4. https://doi.org/10.1016/j.matlet.2012.10.128
9. Biswas MC, Tiimob BJ, Abdela W et al (2019) Nano silica-carbon-silver ternary hybrid induced antimicrobial composite films for food packaging application. Food Packag Shelf Life 19:104–113. https://doi.org/10.1016/j.fpsl.2018.12.003
10. Blackburn C de W (2006) Food spoilage microorganisms
11. Bopp AF (2019) The evolution of food preservation and packaging. In: Chemistry's role in food production and sustainability: past and present. American Chemical Society, pp 15–211
12. Cejudo Bastante C, Casas Cardoso L, Fernández Ponce MT et al (2018) Characterization of olive leaf extract polyphenols loaded by supercritical solvent impregnation into PET/PP food packaging films. J Supercrit Fluids 140:196–206. https://doi.org/10.1016/j.supflu.2018.06.008
13. Chaudhary P, Fatima F, Kumar A (2020) Relevance of Nanomaterials in food packaging and its advanced future prospects. J Inorg Organomet Polym Mater 1–13. https://doi.org/10.1007/s10904-020-01674-8
14. Chawla R, Sivakumar S, Kaur H (2021) Antimicrobial edible films in food packaging: current scenario and recent nanotechnological advancements—a review. Carbohydr Polym Technol Appl 2:100024. https://doi.org/10.1016/j.carpta.2020.100024
15. Deshwal GK, Panjagari NR, Alam T (2019) An overview of paper and paper based food packaging materials: health safety and environmental concerns. J Food Sci Technol 56:4391–4403. https://doi.org/10.1007/s13197-019-03950-z
16. Dong X, Liang X, Zhou Y et al (2021) Preparation of polylactic acid/TiO2/GO nano-fibrous films and their preservation effect on green peppers. Int J Biol Macromol 177:135–148. https://doi.org/10.1016/j.ijbiomac.2021.02.125
17. Duan N, Li Q, Meng X et al (2021) Preparation and characterization of k-carrageenan/konjac glucomannan/TiO2 nanocomposite film with efficient anti-fungal activity and its application in strawberry preservation. Food Chem 364:130441. https://doi.org/10.1016/j.foodchem.2021.130441

18. Dularia C, Sinhmar A, Thory R et al (2019) Development of starch nanoparticles based composite films from non-conventional source—water chestnut (Trapa bispinosa). Int J Biol Macromol 136:1161–1168. https://doi.org/10.1016/j.ijbiomac.2019.06.169
19. Emamhadi MA, Sarafraz M, Akbari M et al (2020) Nanomaterials for food packaging applications: a systematic review. Food Chem Toxicol 146:111825. https://doi.org/10.1016/j.fct.2020.111825
20. Engel JB, Ambrosi A, Tessaro IC (2019) Development of biodegradable starch-based foams incorporated with grape stalks for food packaging. Carbohydr Polym 225:115234. https://doi.org/10.1016/j.carbpol.2019.115234
21. Ervasti I, Miranda R, Kauranen I (2016) A global, comprehensive review of literature related to paper recycling: a pressing need for a uniform system of terms and definitions. Waste Manag 48:64–71. https://doi.org/10.1016/j.wasman.2015.11.020
22. Fahmy HM, Salah Eldin RE, Abu Serea ES et al (2020) Advances in nanotechnology and antibacterial properties of biodegradable food packaging materials. RSC Adv 10:20467–20484. https://doi.org/10.1039/D0RA02922J
23. Farahnaky A, Sharifi S, Imani B et al (2018) Physicochemical and mechanical properties of pectin-carbon nanotubes films produced by chemical bonding. Food Packag Shelf Life 16:8–14. https://doi.org/10.1016/j.fpsl.2018.01.004
24. Franco P, Incarnato L, De Marco I (2019) Supercritical CO2 impregnation of α-tocopherol into PET/PP films for active packaging applications. J CO2 Util 34:266–273. https://doi.org/10.1016/j.jcou.2019.06.012
25. Ghidelli C, Mateos M, Rojas-Argudo C, Pérez-Gago MB (2015) Novel approaches to control browning of fresh-cut artichoke: effect of a soy protein-based coating and modified atmosphere packaging. Postharvest Biol Technol 99:105–113. https://doi.org/10.1016/j.postharvbio.2014.08.008
26. GilakHakimabadi S, Ehsani M, Khonakdar HA et al (2019) Controlled-release of ferulic acid from active packaging based on LDPE/EVA blend: experimental and modeling. Food Packag Shelf Life 22:100392. https://doi.org/10.1016/j.fpsl.2019.100392
27. Gohargani M, Lashkari H, Shirazinejad A (2021) The effect of chitosan-whey protein based edible coating containing bionanocomposite material and Zataria multiflora essential oil on UF-Feta type cheese shelf life. Iran Food Sci Technol Res J 17:729–745. https://doi.org/10.22067/ifstrj.v17i5.88681
28. González A, Alvarez Igarzabal CI (2013) Soy protein—poly (lactic acid) bilayer films as biodegradable material for active food packaging. Food Hydrocoll 33:289–296. https://doi.org/10.1016/j.foodhyd.2013.03.010
29. Goudar N, Vanjeri VN, Kasai D et al (2021) ZnO NPs doped PVA/Spathodea campanulata thin films for food packaging. J Polym Environ 29:2797–2812. https://doi.org/10.1007/s10924-021-02070-0
30. Gu R, Yun H, Chen L et al (2020) Regenerated cellulose films with amino-terminated hyperbranched polyamic anchored nanosilver for active food packaging. ACS Appl Bio Mater 3:602–610. https://doi.org/10.1021/acsabm.9b00992
31. Huo W, Wei D, Zhu W et al (2018) High-elongation zein films for flexible packaging by synergistic plasticization: preparation, structure and properties. J Cereal Sci 79:354–361. https://doi.org/10.1016/j.jcs.2017.11.021
32. Iamareerat B, Singh M, Sadiq MB, Anal AK (2018) Reinforced cassava starch based edible film incorporated with essential oil and sodium bentonite nanoclay as food packaging material. J Food Sci Technol 55:1953–1959. https://doi.org/10.1007/s13197-018-3100-7
33. Joo M, Lewandowski N, Auras R et al (2011) Comparative shelf life study of blackberry fruit in bio-based and petroleum-based containers under retail storage conditions. Food Chem 126:1734–1740. https://doi.org/10.1016/j.foodchem.2010.12.071
34. Kalia A, Kaur M, Shami A et al (2021) Nettle-leaf extract derived ZnO/CuO nanoparticle-biopolymer-based antioxidant and antimicrobial nanocomposite packaging films and their impact on extending the post-harvest shelf life of Guava Fruit. Biomolecules 11. https://doi.org/10.3390/biom11020224

35. Lee J-W, Son S-M, Hong S-I (2008) Characterization of protein-coated polypropylene films as a novel composite structure for active food packaging application. J Food Eng 86:484–493. https://doi.org/10.1016/j.jfoodeng.2007.10.025
36. Liu Y, Wang S, Lan W, Qin W (2019) Fabrication of polylactic acid/carbon nanotubes/chitosan composite fibers by electrospinning for strawberry preservation. Int J Biol Macromol 121:1329–1336. https://doi.org/10.1016/j.ijbiomac.2018.09.042
37. Lomate GB, Dandi B, Mishra S (2018) Development of antimicrobial LDPE/Cu nanocomposite food packaging film for extended shelf life of peda. Food Packag Shelf Life 16:211–219. https://doi.org/10.1016/j.fpsl.2018.04.001
38. Majeed K, Jawaid M, Hassan A et al (2012) Potential materials for food packaging from nanoclay/natural fibres filled hybrid composites. Mater Des 46:391–410. https://doi.org/10.1016/j.matdes.2012.10.044
39. Mirmehdi S, Hein PRG, de Luca Sarantópoulos CIG et al (2018) Cellulose nanofibrils/nanoclay hybrid composite as a paper coating: effects of spray time, nanoclay content and corona discharge on barrier and mechanical properties of the coated papers. Food Packag Shelf Life 15:87–94. https://doi.org/10.1016/j.fpsl.2017.11.007
40. Mizielińska M, Kowalska U, Salachna P et al (2018) The influence of accelerated UV-A and Q-SUN irradiation on the antibacterial properties of hydrophobic coatings containing Eucomis comosa extract. Polymers (Basel) 10. https://doi.org/10.3390/polym10040421
41. Motelica L, Ficai D, Oprea O-C et al (2021) Antibacterial biodegradable films based on alginate with silver nanoparticles and lemongrass essential oil–innovative packaging for cheese. Nanomaterials 11. https://doi.org/10.3390/nano11092377
42. Odeyemi OA, Alegbeleye OO, Strateva M, Stratev D (2020) Understanding spoilage microbial community and spoilage mechanisms in foods of animal origin. Compr Rev Food Sci Food Saf 19:311–331. https://doi.org/10.1111/1541-4337.12526
43. Ozer BBP, Uz M, Oymaci P, Altinkaya SA (2016) Development of a novel strategy for controlled release of lysozyme from whey protein isolate based active food packaging films. Food Hydrocoll 61:877–886. https://doi.org/10.1016/j.foodhyd.2016.07.001
44. Paidari S, Ahari H (2021) The effects of nanosilver and nanoclay nanocomposites on shrimp (Penaeus semisulcatus) samples inoculated to food pathogens. J Food Meas Charact 15:3195–3206. https://doi.org/10.1007/s11694-021-00905-x
45. Petruzzi L, Corbo MR, Sinigaglia M, Bevilacqua A (2017) Chapter 1—microbial spoilage of foods: fundamentals. In: Bevilacqua A, Corbo MR, Sinigaglia M (eds) The microbiological quality of food. Woodhead Publishing, pp 1–21
46. Quadrifoglio F, Crescenzi V (1971) The interaction of methyl orange and other azo-dyes with polyelectrolytes and with colloidal electrolytes in dilute aqueous solution. J Colloid Interface Sci. https://doi.org/10.1016/0021-9797(71)90145-7
47. Rabnawaz M, Wyman I, Auras R, Cheng S (2017) A roadmap towards green packaging: the current status and future outlook for polyesters in the packaging industry. Green Chem 19:4737–4753. https://doi.org/10.1039/C7GC02521A
48. Riaz A, Lagnika C, Luo H et al (2020a) Effect of Chinese chives (Allium tuberosum) addition to carboxymethyl cellulose based food packaging films. Carbohydr Polym 235:115944. https://doi.org/10.1016/j.carbpol.2020.115944
49. Riaz A, Lagnika C, Luo H et al (2020b) Chitosan-based biodegradable active food packaging film containing Chinese chive (Allium tuberosum) root extract for food application. Int J Biol Macromol 150:595–604. https://doi.org/10.1016/j.ijbiomac.2020.02.078
50. Risch SJ (2009) Food packaging history and innovations. J Agric Food Chem 57:8089–8092. https://doi.org/10.1021/jf900040r
51. Roy S, Priyadarshi R, Rhim J-W (2021) Development of multifunctional pullulan/chitosan-based composite films reinforced with ZnO nanoparticles and propolis for meat packaging applications. Foods 10. https://doi.org/10.3390/foods10112789
52. Rudi H, Resalati H, Eshkiki RB, Kermanian H (2016) Sunflower stalk neutral sulfite semi-chemical pulp: an alternative fiber source for production of fluting paper. J Clean Prod 127:562–566. https://doi.org/10.1016/j.jclepro.2016.04.049

53. Rydz J, Musioł M, Zawidlak-Węgrzyńska B, Sikorska W (2018) Present and future of biodegradable polymers for food packaging applications. In: Biopolymers for food design, pp 431–467
54. Sami R, Elhakem A, Alharbi M et al (2021) The combined effect of coating treatments to nisin, nano-silica, and chitosan on oxidation processes of stored button mushrooms at 4 °C. Sci Rep 11:6031. https://doi.org/10.1038/s41598-021-85610-x
55. Shamblin SL, Tang X, Chang L et al (1999) Characterization of the time scales of molecular motion in pharmaceutically important glasses. J Phys Chem B 103:4113–4121. https://doi.org/10.1021/jp983964+
56. Shao J, Wang L, Wang X, Ma J (2021) Enhancing microbial management and shelf life of shrimp Penaeus vannamei by using nanoparticles of metallic oxides as an alternate active packaging tool to synthetic chemicals. Food Packag Shelf Life 28:100652. https://doi.org/10.1016/j.fpsl.2021.100652
57. Shemesh R, Krepker M, Goldman D et al (2015) Antibacterial and antifungal LDPE films for active packaging. Polym Adv Technol 26:110–116. https://doi.org/10.1002/pat.3434
58. Siracusa V, Rocculi P, Romani S, Rosa MD (2008) Biodegradable polymers for food packaging: a review. Trends Food Sci Technol 19:634–643. https://doi.org/10.1016/j.tifs.2008.07.003
59. Sorrentino A, Gorrasi G, Vittoria V (2007) Potential perspectives of bio-nanocomposites for food packaging applications. Trends Food Sci Technol 18:84–95. https://doi.org/10.1016/j.tifs.2006.09.004
60. Souza VGL, Rodrigues C, Valente S et al (2020) Eco-friendly ZnO/chitosan bionanocomposites films for packaging of fresh poultry meat. Coatings 10. https://doi.org/10.3390/coatings10020110
61. Tan YM, Lim SH, Tay BY et al (2015) Functional chitosan-based grapefruit seed extract composite films for applications in food packaging technology. Mater Res Bull 69:142–146. https://doi.org/10.1016/j.materresbull.2014.11.041
62. Tornuk F, Hancer M, Sagdic O, Yetim H (2015) LLDPE based food packaging incorporated with nanoclays grafted with bioactive compounds to extend shelf life of some meat products. LWT Food Sci Technol 64:540–546. https://doi.org/10.1016/j.lwt.2015.06.030
63. Valipour Motlagh N, Aghazamani J, Gholami R (2021) Investigating the effect of nano-silver contained packaging on the olivier salad shelf-life. Bionanoscience 11:838–847. https://doi.org/10.1007/s12668-021-00876-9
64. Vilela C, Kurek M, Hayouka Z et al (2018) A concise guide to active agents for active food packaging. Trends Food Sci Technol 80:212–222. https://doi.org/10.1016/j.tifs.2018.08.006
65. Villanueva A, Wenzel H (2007) Paper waste—recycling, incineration or landfilling? A review of existing life cycle assessments. Waste Manag 27:S29-46. https://doi.org/10.1016/j.wasman.2007.02.019
66. Wang C, Chang T, Dong S et al (2020) Biopolymer films based on chitosan/potato protein/linseed oil/ZnO NPs to maintain the storage quality of raw meat. Food Chem 332:127375. https://doi.org/10.1016/j.foodchem.2020.127375
67. Wang F, Chang R, Ma R et al (2021) Eco-friendly and pH-responsive nano-starch-based superhydrophobic coatings for liquid-food residue reduction and freshness monitoring. ACS Sustain Chem Eng 9:10142–10153. https://doi.org/10.1021/acssuschemeng.1c02090
68. Yousefi H, Su H-M, Imani SM et al (2019) Intelligent food packaging: a review of smart sensing technologies for monitoring food quality. ACS Sensors 4:808–821. https://doi.org/10.1021/acssensors.9b00440
69. Youssef AM, Assem FM, El-Sayed HS et al (2020) Synthesis and evaluation of eco-friendly carboxymethyl cellulose/polyvinyl alcohol/CuO bionanocomposites and their use in coating processed cheese. RSC Adv 10:37857–37870
70. Youssef AM, El-Sayed HS, El-Nagar I, El-Sayed SM (2021) Preparation and characterization of novel bionanocomposites based on garlic extract for preserving fresh Nile tilapia fish fillets. RSC Adv 11:22571–22584. https://doi.org/10.1039/d1ra03819b
71. Yu L (2001) Amorphous pharmaceutical solids: preparation, characterization and stabilization. Adv Drug Deliv Rev 48:27–42. https://doi.org/10.1016/S0169-409X(01)00098-9

72. Zhang R, Cheng M, Wang X, Wang J (2019) Bioactive mesoporous nano-silica/potato starch films against molds commonly found in post-harvest white mushrooms. Food Hydrocoll 95:517–525. https://doi.org/10.1016/j.foodhyd.2019.04.060
73. Zhao X, Cornish K, Vodovotz Y (2020) Narrowing the gap for bioplastic use in food packaging: an update. Environ Sci Technol 54:4712–4732. https://doi.org/10.1021/acs.est.9b03755

Applications: Miscellaneous

Synthesis of Silver and Copper Nanowires and Their Application for Transparent Conductors

Conghua Zhou and Yongli Gao

Abstract Silver and copper nanowires have been used in many areas, like sensors, catalysis, optoelectronics, etc. They are well-known by the excellent optoelectronic and chemical properties. Here in this chapter, some interesting aspects of these metal nanowires are reviewed, including: (i) the synthesis methods of the metal nanowires; (ii) application of these nanowires in transparent conductors; (iii) application of the transparent conductors in flexible optoelectronic devices, like solar cells, OLEDs, touch screens, transparent heaters, and so on. Besides the applications, this chapter also sheds light to some fundamental problems, including the coarsening dynamics of nanowires, corrosion and protection of nanowires, and also the conducting mechanism of the transparent conductors basing on nanowires. Finally, perspective is given.

Keywords AgNWs · CuNWs · Transparent conductor · TC · Transparent and conductive · Solar cells · Flexible

1 Introduction

Metal nanowires (NWs) based on gold [1], silver [2], and copper [3], have arisen as a new class of one-dimension (1D) materials that attract ever-increasing attentions due to the excellent optoelectronic and chemical properties. They have been applied widely in transparent conductor (TC, hereafter) [4, 5], conductive composite [6, 7], optoelectronic devices [8], heater [9, 10], catalysis [11–13], and so on. The

C. Zhou (✉)
Hunan Key Laboratory of Super-microstructure and Ultrafast Process, Hunan Key Laboratory of Nanophotonics and Device, Institute of Super-Microstructure and Ultrafast Process in Advanced Materials, School of Physics and Electronics, Central South University, Changsha, Hunan 410083, P.R. China
e-mail: chzhou@csu.edu.cn

Y. Gao
Department of Physics and Astronomy, University of Rochester, Rochester, NY 14627, USA

I. Uddin and I. Ahmad (eds.), *Synthesis and Applications of Nanomaterials and Nanocomposites*, Composites Science and Technology,
https://doi.org/10.1007/978-981-99-1350-3_9

mechanical flexibility of these metal nanowires renders it possible for the application in adaptive optoelectronic devices and wearable electronics [14–16].

The exploration of 1D nanomaterials could be ascribed to the successful synthesis of 0D nanomaterials or nanoparticles [17–20]. Originally in late 1990s, metal based nanowires was synthesized via templating methods, like anodic aluminum oxide (AAO) porous membranes or carbon nanotube (CNT) confined growth [21–23], or DNA chain basing assembly [24]. These methods could help to grow metal based nanostructures from 0 to 1D, though "large-scale" synthesis is somehow difficult to realize. This bottleneck was then broken in the early 2000s, owing to the pioneering works done by Xia et al. (2002). Using the so-called "polyol reaction", silver nanowires with aspect ratio up to 1000 could be obtained [25–27]. Following the success in synthesis, the silver nanowires were quickly observed to show conductivity of 0.8×10^5 S/cm (one eighth of the bulk conductivity of 6.3×10^5 S/cm) [25], which triggered the application in TCs that were usually dominated by oxide films like indium tin oxide (ITO) or F-doped tin oxide (FTO). From then on, researches towards the synthesis and application of silver nanowires have bloomed quickly and widely. Such TCs have been used in many optoelectronic devices, like solar cells, touch panels, heaters and so on. Recently, due to the development of wearable electronics, the nanowires are also applied to fabricate flexible and stretchable electrodes under the form of "conductive composite". Meanwhile, robustness of the nanowire based TCs and composite is discussed due to the long-term use concerns. Questions about the conducting mechanism of such nanowire-based TCs also arise, since the organization mechanism of these TC differs greatly from those oxide-based ones. During the past decade, the synthesis [2, 28, 29], application as TCs (including optoelectronic devices, heaters) [4, 5] or composite [9, 10, 30], and robustness [31] of metal-based NWs have been discussed in several separate reviewing works, though less has been done for the conducting mechanism. In current chapter, the authors will try to put these topics together under the main topic of "nanomaterial and application" designated for the book. In more, the chapter will mainly focus on two kinds of metal nanowires, or silver and copper nanowires (AgNWs and CuNWs). Such consideration is based on the practical availability and the application of them in TCs.

The rest of the chapter is organized as the following: in Sect. 2, the synthesis methods of metal nanowires and the related coarsening dynamics are presented. Section 3 contains the application of these nanowires as TCs, and the application of TCs in optoelectronic devices. Finally, perspective is given in Sect. 4.

2 Synthesis of Silver and Copper Nanowires

During the past two decades, several methods have been developed to synthesis metal nanowires, for example polyol method [25–27], solvothermal reaction (including hydrothermal reaction) [32–34], templating [21, 23, 24, 35], and so on, among which the polyol method and solvothermal reaction are the mostly applied ones. As a result, these two methods are mainly discussed.

2.1 Polyol Method of Silver Nanowires (AgNWs)

Polyol method was originally proposed by Fievet and coworkers in 1980s in synthesizing metal particles, like nickel, silver, cobalt, platinum and so on [17, 20]. Micro or submicron sized metal particles could be obtained by reactions using polyol (for example ethylene glycol, diethylene glycol) as solvent and reducing agent, and corresponding salt or metal oxide as metal source. The proposed reaction can be described by formulas (1) and (2) [17, 20]:

$$CH_2OH\text{-}CH_2OH \xrightarrow{\Delta} CH_3CHO + H_2O, \quad (1)$$

$$2CH_3CHO + M(II) \xrightarrow{\Delta} CH_3COCOCH_3 + M\downarrow + H_2O, \quad (2)$$

whereas M(II) represents related salt or oxide of the metal. Noting that, the reaction could take place even the oxide was less soluble [17].

The reaction process was then developed by Sun et al. (2002) in the synthesis of silver nanowires in the early 2000s. As shown in Fig. 1, with the assistance of the so-called "seeds" of platinum or silver particles, and the coordination effect of polyvinylpyrrolidone (PVP), silver nanowire (AgNWs) with width of about 40 nm and aspect ratio (defined as the ratio between length and width) up to 1000 were successfully obtained. During the reaction, silver nitrite ($AgNO_3$) and ethylene glycol (EG) were used as the metal source and reducing agent, respectively, following the reaction routines described by formulas (1) and (2) [25]. From the crystallographic study [XRD from Fig. 1d], one can see that these nanowires are of silver. In addition, effects of several parameters were examined with respect to the morphology and aspect ratio of the nanowires, including PVP concentration, reaction temperature, and seeding condition, which will be discussed later. Following that, they also demonstrated the possibility of AgNWs growth without the addition of exotic seeds of platinum particles. With the assistance of PVP and silver nanoparticles seeds (reduced from $AgNO_3$ by EG, also marked as the "self-seeding process"), AgNWs with width of ~20 nm (in average) and aspect ratio up to 1000 were harvested. PVP was suggested to act as the coordination reagent in the reaction [26]. It is worth noting that, "seeding" is somehow a rather complicated process. Two year later in 2004, Wiley et al. fingered out that hence chloride was present in the reagents of commercial obtained ethylene glycol (EG), depending on the vendor and the lot number. The concentration was from 0.0019 to 0.116 mM [36]. As a result, they suggested that, the previous AgNWs synthesis would have been possibly been contaminated by the hence chloride.

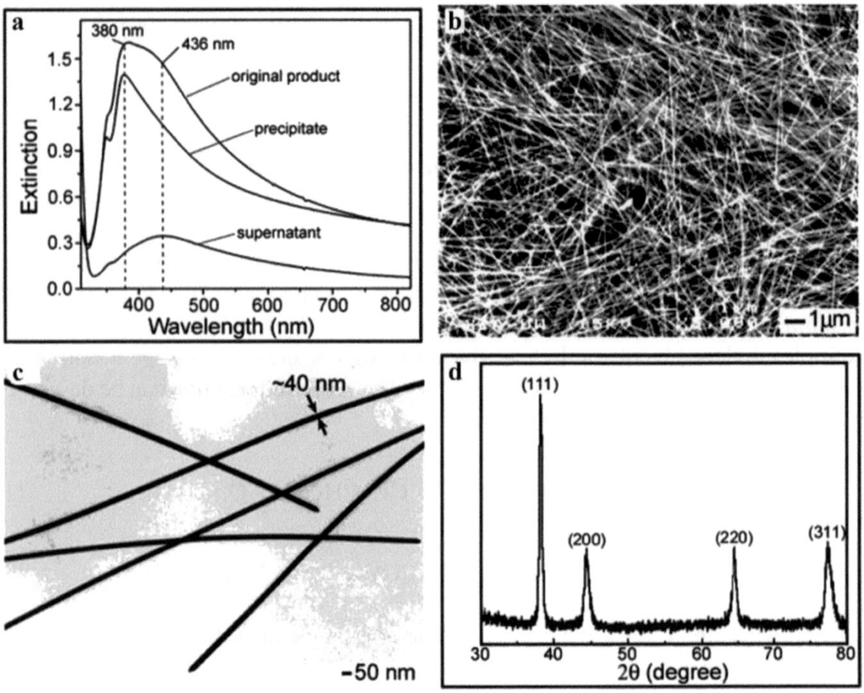

Fig. 1 **a** UV-visible extinction spectra of the synthesized product; **b** Scanning electron microscopy (SEM) image, **c** transmission electron microscopy (TEM) image, and **d** X-ray diffraction (XRD) pattern of the synthesized silver nanowire (AgNWs). Reproduced with permission [25]. Copyright © 2002, American Chemical Society

Following that, modifications were proposed to the polyol method. For example, in 2012, Lee et al. reported a so-called successive multistep growth (SMG) method in growing very long AgNWs. During the reaction, the obtained AgNWs were purified and used as the "seeds" for the next round of growth, while the reaction followed the similar polyol process described above. After 7 rounds of growth, they could obtain nanowires with length longer than 300 μm (width ~150 nm), and aspect ratio up to 1000–3000. The schematic of this strategy and typical length distribution is shown in Fig. 2 [37]. This study shows that AgNWs could also act as the seeds. Besides that, other methods were also proposed. In 2013, Lee et al. reported a high-pressure polyol method. By controlling the reaction pressure, AgNWs with width of 15–30 nm and length of ~20 μm were obtained. Increasing the reaction pressure could decrease the width of the nanowires. Typically, at pressure of 200 psi (1.38 $\times$ 10^6 Pa), AgNWs with width of 15–22 nm were harvested [32]. As a common synthesis routine of nanomaterials, solvothermal (similar to hydrothermal) reaction was also tried. Jiu et al. (2013) synthesized AgNWs with width of 70 nm and length of 5–20 μm using the solvothermal method, in which $AgNO_3$, PVP and EG were used as the metal source, coordination reagent and reducer, respectively [38]. Then

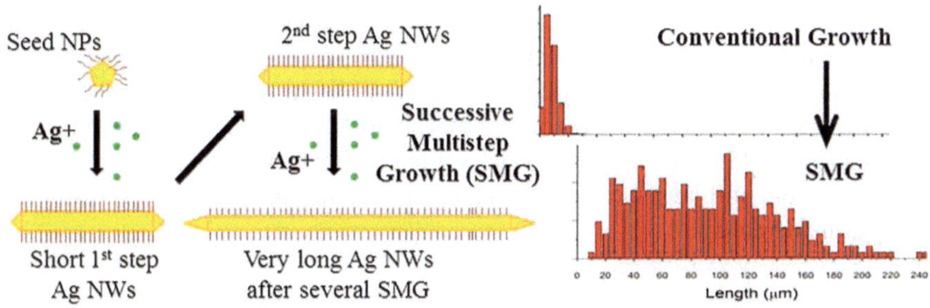

Fig. 2 Schematic for the successive multistep growth (SMG) and the length distribution of silver nanowires (AgNWs) synthesized from using conventional growth and SMG method. Reproduced with permission [37]. Copyright © 2012, American Chemical Society

in 2016, Xu et al. (2016) showed that, by importing small amount of HNO_3 in the solvothermal reaction, AgNWs with length longer than 400 μm and average aspect ratio of 821 could be obtained in a one-spot reaction [33]. In such "one-step" growth method the reaction process is greatly simplified, thus the method provides a direct strategy to grow ultra-long AgNWs.

Along with the development of synthesis methods, the crystallographic properties were also studied. With combination of high-resolution transmission electron microscopy (HRTEM) and the selected area electron diffraction (SAED) technique, Xia et al. in 2002 illustrated that, these nanowires were not "single crystalline", rather, a twinned structure was adopted [25, 27]. One year later in 2003, with the assistance of scanning electron microscopy (SEM) and selected area electron diffraction (SAED), Gao et al. proposed a fivefold twinned structure in the silver nanowire, which was suggested to be bounded by five {100} side planes and capped by five {111} end planes, with growth direction following [110], as is shown in Fig. 3a [39]. Similar results were depicted by Sun, Xia and coworkers in the same year [41]. Following that, in 2005, Gao et al. embedded AgNWs in epoxy and microtomed the matrix into slices with thickness less than 50 nm, by which the cross-section of these nanowires could be directly examined by HRTEM and SEAD. As is shown in Fig. 3b, pentagonal cross-section is clearly depicted. In more, they demonstrated that, the cross-section could hardly be separated into five uniform crystals. Since if that was true, there would be a small region left with angle of 7.5°. However, no such region was observed. Instead, many defects were observed in the boundaries between neighboring twin domains. Accordingly, "fivefold twinned structure" is named, though in fact many defects exist in the twinning boundaries or the "twin planes". This could also be clearly seen in Fig. 4b [40]. This feature is very interesting, showing that the growth of AgNWs is in fact accompanied by formation of defects. In addition, the growing direction is confirmed to be [110] of the FCC structure. It might be worthy to note that, AgNWs with single crystalline structure were reported to be synthesized by templating method in 2001 [35]. Though the width was as less as few silver atoms (or 0.4 nm). On the other hand, similar five-fold twinned structure was also believed to be present in CuNWs [11], and even Pd–Au–Pd segmental nanorod [42], pending detailed examination on the cross-section of these nanostructures. However, in 2016 in the atomic study of the Au nanorod, Yu et al. revealed a "necking and breaking" morphological property in the synthesized Au nanorod, and meanwhile many stacking fault defects were disclosed [43]. As a result, it would still be meaningful to verify the relationship between 1D structure formation and defect.

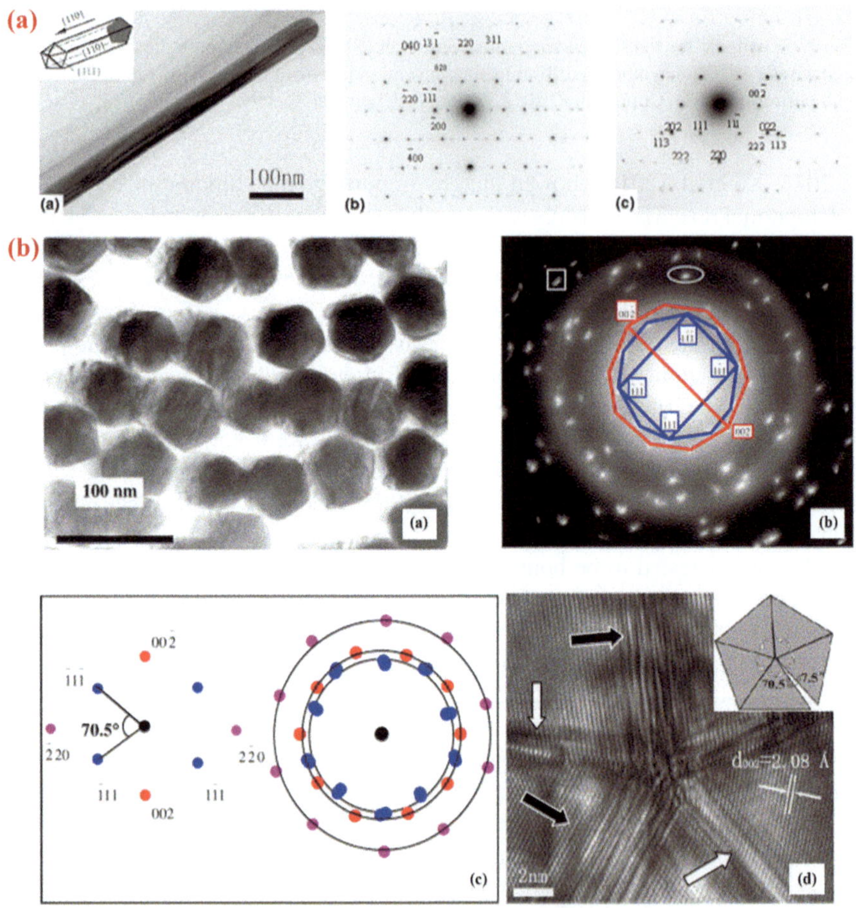

Fig. 3 **a** Fivefold twinned structure proposed in literature and corresponding selected area electron diffraction (SAED) patterns. Reproduced with permission [39]. Copyright © 2003, Elsevier. **b** Transmission electron microscopy (TEM) image of cross-section of AgNWs, corresponding SAED pattern, and high resolution TEM (HRTEM) image of the cross-section. Clearly, fivefold twinned structure is shown. Reproduced with permission [40]. Copyright © 2004, Elsevier

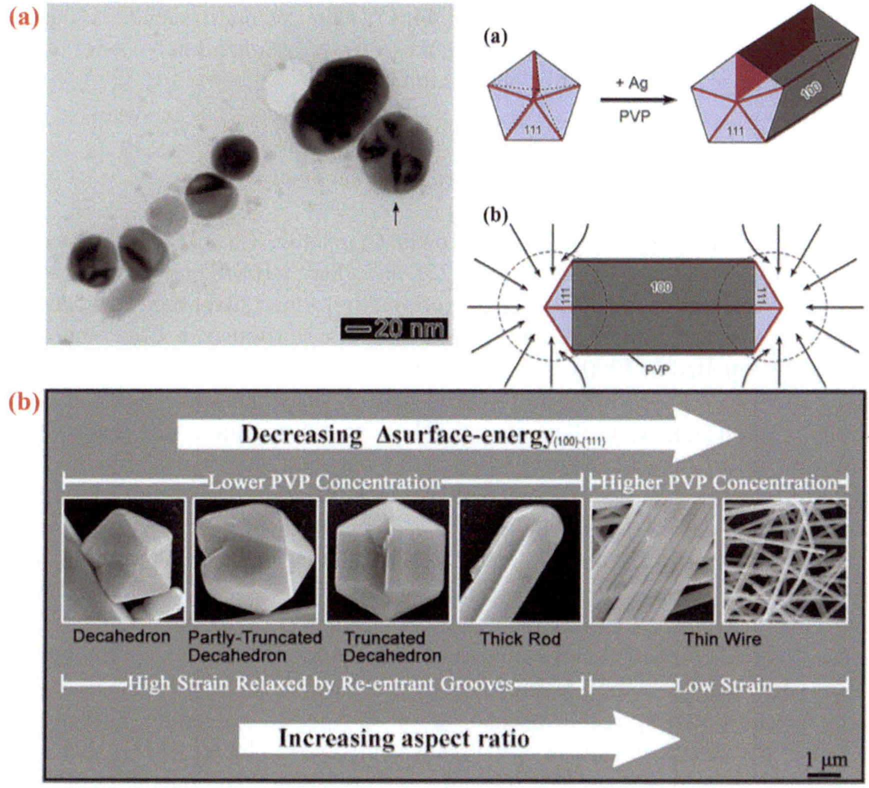

Fig. 4 **a** TEM image of fivefold twinned silver nanoparticles picked from the solution in polyol growth of AgNWs; and **b** Schematic illustration of the unidirectional growth of twinned silver nanoparticles into AgNWs. The nanowire owns five {100} facets on the side, and ten {111} facets at the end. Heavy capping of PVP molecule on {100} facets is marked by dark grey, while the relative weak capping at the end {111} facets is marked by light-blue color. Reproduced with permission [41]. Copyright © 2003, American Chemical Society. **c** Schematic presentation of the effect of synergy between strain and surface energy difference [$\Delta\varphi_{(100)-(111)}$] on morphology of the fivefold twinned silver structures. Clearly, when PVP concentration is low, lateral growth is favored; while at higher concentrations, unidirectional growth dominates. Reproduced with permission [51]. Copyright © 2008, American Chemical Society

2.2 Coarsening Dynamics of Metal Nanowire

Comparison between the synthesis of 1D metal nanowire and 0D metal particles shows that nanowire growth is different from 0D particles, even though similar reducer and metal source are used [17, 20, 25–27]. Nanowire growth is a unique and dynamic process that can be separated into two distinctive procedures: nucleation and growth. From the ever published literatures, one could find several parameters relating to these two processes, for example, seeds, coordination reagent, reaction temperature and pressure, and even processing techniques. Readers are encouraged to refer to several reviewing works [2, 28, 29, 44, 45]. Here the main issues are discussed: (i) Transformation of 1D nanowire from 0D seeds; (ii) oxidization behavior during the nanowire growth; (iii) role of precipitation anions.

2.2.1 Transformation of 1D Nanowire from 0D Seeds

Seed plays key role in the 1D nanowire growth. Originally, Xia et al. used platinum or silver nanoparticles as the exotic seeds [25], and then "self-seeding" was proposed, in which the seeds were provided by the originally formed silver nanoparticles [26, 27]. As discussed above, fivefold twinned structure is formed in silver nanowire. But works reported in 1990s usually showed the growth of single-crystalline silver nanoparticles [19]. According to previous studies, such single nanoparticles could hardly develop along typical direction that leads to 1D structure. As a result, before nanowire growth, it is crucial to grow seeds ("or nanoparticle") with similar fivefold-twinned structure like that discussed above. In fact, twinned noble metal particles were observed before the growth of metal nanowires [46–50].

In 2003, following the observation of AgNWs, the seeding process was studied by Sun, Xia et al. In the so-called polyol process, where PVP was used as the coordination molecule, they picked the solution in the starting stage of the reaction, and observed some "fivefold twinned" silver nanoparticles, or the seeds in the system (though not all the particle are twinned). As shown in Fig. 4a, "twin boundaries" are marked [41]. Then they proposed a possible mechanism for 1D AgNWs coarsening. They pointed out three possible stages. At first, the seeds were covered by ten {111} facets, thus decahedral shape was observed. Secondly, defects gathered at boundaries of seeds, make boundaries more attractive place for silver atoms to deposit (after being reduced by polyol reaction), and thus lead to the formation of "nanorods". Following that, the other five {100} facets appeared. At last, PVP molecules could interact with silver atoms at nanowire surfaces, but the interaction would be stronger at the side {111} facets than that at end facets. The difference between the interaction strength then leads to unidirectional growth of AgNWs. Such processes are clearly depicted in Fig. 5b, whereas the interaction strength is marked by different color. The difference in the interaction strength was confirmed by Zhang et al. [51] Using first-principle calculations basing on density function theory (DFT), the surface energy (eV/atom) was calculated for facets of both {100} and {111}. It was observed that, the surface energy was 0.40724 eV/atom [$\varphi_{(100)}$] and 0.32411 eV/atom [$\varphi_{(111)}$] for pure facets {100} and {111}, respectively; After absorption of α-pyrrolidone molecule (similar to the unit in PVP), it decreased to be 0.39056 and 0.31755 eV/atom, respectively. As such, the difference between the two, or marked by $\Delta\varphi_{(100)-(111)}$, was reduced from 0.08313 to 0.07301 eV/atom. Thus the difference was minimized by 12.2% [51]. However, from the calculated surface energy (though the practical physical chemistry is more complicated) of the two facets, one may expect lateral growth that leads to thick nanowires. Then why usually long and thin AgNWs could be obtained? The authors suggested another possible mechanism, or the confinement of strain. Indeed, from previous studies, it had been known that the nanowires were fivefold twinned.

But the nanowires could hardly be divided into 5 crystals (If not, there should be a small angle of 7.5°). To fill the gap (according to the small angle region), there in fact existed many defects at the grain boundaries as illustrated by Gao et al. [40]. Similarly, strain was suggested to exist in the nanowires, the thicker was the nanowires, the higher was the strain. In contract, increment in nanowire length would not add to the strain (given straight nanowire). As a result, a so-called synergy effect was proposed, which is shown in Fig. 5c. When PVP concentration was low, lateral growth dominated due to the relatively high surface energy of facets {100}. Meanwhile the strain of the 1D structure also rose. To release the strain, "re-entrant grooves" appeared. When PVP concentration became higher, passivation happens at the outside facets of {100}, the surface energy difference $[\Delta\varphi_{(100)-(111)}]$ was then decreased. Then the superiority of lateral growth was reduced, and longitudinal growth happened, since it would not increase the strain of the system. Indeed, the synergy effect provided clear explanation towards the role played by the coordination reagent.

2.2.2 Etching Behavior

Up to now, the mostly used metal source in AgNWs growth is $AgNO_3$. Based on the reaction formula (2), a typical formula (3) is expected, in which a byproduct of HNO_3 appears. Due to the pristine oxidization behavior formula (4) could be anticipated, though later could be seen, the oxidization behavior might be sensitive to the concentration of nitric acid. Such phenomenon was disclosed two years later after the successful growth of AgNWs, along with the two possible reaction formulas listed as following [36, 52]:

$$2CH_3CHO + 2AgNO_3 \longrightarrow CH_3COCOCH_3 + 2Ag \downarrow + 2HNO_3, \quad (3)$$

$$3Ag + 4HNO_3 \longrightarrow 3AgNO_3 + NO \uparrow + 2H_2O. \quad (4)$$

Accordingly, oxidization becomes an important issue in nanowire growth. Wiley et al. (2014) observed that oxygen played a key role in the coarsening dynamics of silver nanostructures. They added trace amount of NaCl (0.22 mM) in the typical polyol reaction, and observed two kinds of silver nanoparticles: single crystal and multi-twinned particles in the starting stage of the reaction. Different products were observed when oxygen was expelled or not from the reaction or not [36]. Concretely, in presence of oxygen, only silver cubes and tetrahedrons were obtained. After oxygen was excluded and replaced by argon (Ar) gas, thin and long AgNWs were harvested. Accordingly, they proposed a etching couple made of "O_2/Cl^-" and suggested that, the originally formed multi-twinned silver particles could be easily be etched by this couple, thus nanowires could be less productive; After oxygen is expelled, the etching effect disappeared, and the seeds could grow into nanowires. At the same time, they fingered out the chloride contamination problem in previous AgNWs synthesis work (as stated in previous section). Indeed, this work drew the attention to the effect of O_2, chloride to AgNWs formation at the first time, though the exact etching mechanism would be more complicated. One year later in 2005, Xia et al. added HCl in the polyol reaction (with HCl concentration of 0.125, 0.25 and 0.375 mM, respectively), and studied the effect of HCl concentration on the resultant silver nanostructures [52]. It was observed that, at concentration of 0.125 mM, both silver particles and nanowires were obtained. While silver nanocubes were obtained at concentration of 0.25 mM, and thick wires and irregular particles came out as the concentration increased to 0.375 mM. Similarly, they attributed the formation of cube (at case of 0.25 mM) to the etching effect brought by "O_2/Cl^-". From the above two studies one could know the important role of "etching" in the growing dynamics of silver nanostructures.

To verify the exact mechanism, more works were done after that. In 2008, Zhang, Tang and coworkers focused on the acidic etching behavior of the byproduct of HNO_3 [according to formula (3)]. To retard the etching behavior, they added steel plates into the polyol reaction, and obtained AgNWs in high concentration (0.5 M) [53]. In 2016, Xu and Zhou et al. studied the effect of HNO_3 in AgNW growth in solvothermal reaction. By adding small amount of HNO_3 in the reaction, they observed 50-fold increase in nanowire length, reaching >400 μm [33]. In order to study the role of HNO_3, they designed 16 batches of experiments, 8 for reaction with addition of HNO_3 but with varied reaction period, while the other 8 for references (W/O HNO_3). This designation eliminated the disruption that might be drawn by "picking" itself. Then, the samples were examined by both UV–Visible extinction spectra and SEM techniques, with photo-images taken at the same time.

As shown in Fig. 5a, b, when there was no HNO_3 added, three cyclic round color changes appeared in the solution. For example, it changed from dark red (1 h) to colorless (4 h) at the first cycle, then turned back to grey at 5 h, but clear again at 6 h. When HNO_3 was pre-added, different behavior was observed. The solution remained colorless for the first 5 h, and changed gradually to brown. Extinction behavior and morphological properties (SEM images could be referred to the literature [33]) showed that, in each cycle , AgNWs and nanoparticles appeared at first, and then disappeared. This implied that, dissolution happened. To explain the beneath mechanism, they carefully examined the reaction environment, and proposed that oxygen played a key role in the reaction. Indeed, there exists less O_2 in the upper space of the autoclave, calculated to be 2.7 mM given that O_2 could be all dissolved in the solution. This concentration was about 11% of [Ag]. Considering the byproduct of HNO_3, then following reactions could possibly take place, as represented by formulas (5) and (6):

$$4Ag + O_2 \longrightarrow 2Ag_2O, \tag{5}$$

$$Ag_2O + 2HNO_3 \longrightarrow 2AgNO_3 + H_2O. \tag{6}$$

As such, nanowires could grow steadily only after the consumption of O_2, which is quite similar to that stated by Xia et al. (2004), whereas O_2 was expelled and Ar

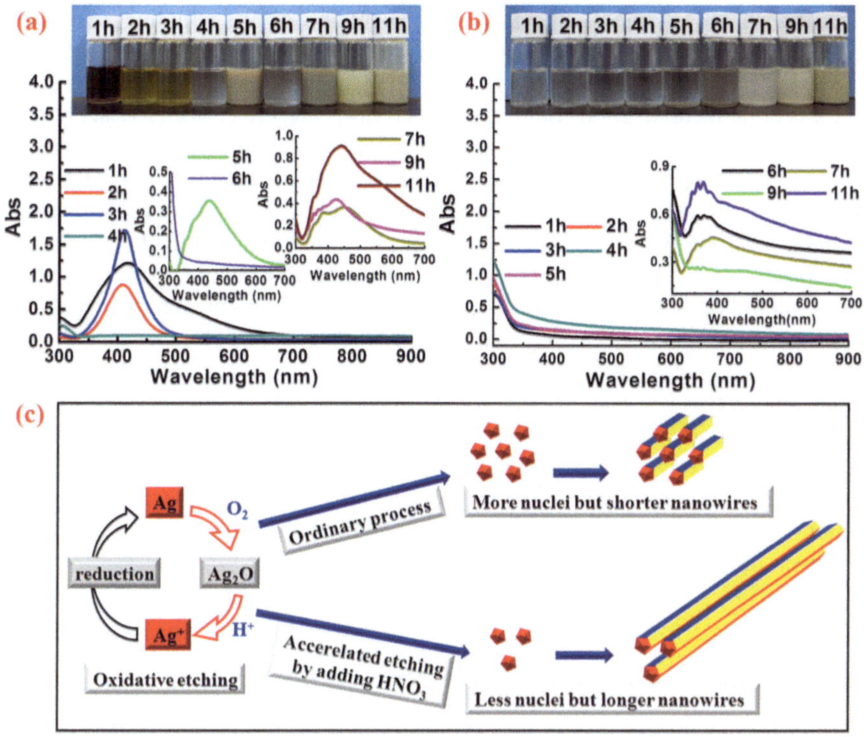

Fig. 5 UV–Visible spectra of solution obtained from solvothermal reaction at different reacting time: **a** W/O HNO_3, **b** with HNO_3 (1.44 mM). **c** Schematic of the oxidative etching behavior brought by the etching couple of "O_2/H^+", and the effect on long AgNWs growth [33]

was imported as the protection gas [36]. Moreover, it is worthy to note that, to verify the acidic behavior of HNO_3, specific experiments were designated (refer to literature [33]), showing that when concentration of HNO_3 was low, acidic role appeared, rather than the oxidization behavior. Accordingly, after HNO_3 was added, a screening effect appears. After addition of small amount of HNO_3, etching behavior brought by the couple of "O_2/H^+" could dissolve certain amount of multi-twinned silver nanoparticles or seeds. Then the existing seeds could find more opportunity to grow longer, as depicted schematically in Fig. 5c.

2.2.3 Role of Precipitating Anions

Besides PVP, O_2, HNO_3, AgNW growth is also affected by anions in the reaction system. AgNWs have been successfully synthesized in solution containing Cl^- [27], Br^- [54], S^{2-} [55, 56], etc. These anions were found to affect both nanowire width and aspect ratio. Usually, incorporation of Br^- was favorable for thin nanowire growth. For example, in the so-called high pressure polyol method, Lee et al. (2013) observed that, mixture of NaCl and KBr was helpful to obtain AgNWs with thickness low to 15–22 nm [32]. Then, Li et al. (2015) showed, by tuning reacting procedure and also the concentration of Br^- in polyol reaction, AgNWs with diameters of 20 nm and aspect ratios up to 2000 could be obtained [54]. The effect of Br^- concentration on the nanowire width is clearly depicted in Fig. 6. If only Cl^- exists, nanowire width is 72 ± 15 nm, after incorporating small portion of NaBr (2.2 mM), the width decreases to 20 ± 2 nm. However, increasing concentration up to 4.4 mM results in particles for the most. As such, anions should play important roles in those reactions. To explore the cause, the authors suggested that, the Br^- adsorbed on the AgNWs surface could reduce the coarsening rate perpendicular to nanowire length [32, 57]. However, the detailed mechanism was not clear yet. One possible cause might arise from the "precipitation behavior" brought by the halide anions. It is well-known that, halide anions like Cl^- and Br^- could react with Ag^+ and get precipitates; in more, AgBr is less dissolvable than AgCl. As such, when adsorbed on AgNW surface, Br^- could provide stronger protection for the nanowires. As such, it not only prevents the further Ag atom deposition on the surface (for transverse growth), but also makes it more robust when facing with corrosive and oxidizative reagents in the solution. However, more works are needed to clarify this assumption.

2.3 *Synthesis of Copper Nanowires*

Copper nanowires (CuNWs) have also been synthesized. Toimil Molares et al. (2013) demonstrated that CuNWs with width of 30–60 nm could be grown by electrochemical method [58]. Shi et al. (2015) synthesized CuNWs from hydrothermal reaction (120–180 °C, 48 h) containing $CuCl_2$, octadecylamine (ODA) [59]. CuNWs with width of 50–100 nm, and aspect ratio up to 10^5 were obtained by this one pot reaction. In more, reaction temperature was observed to affect nanowire growth. 120 °C reaction helped to obtain uniform distribution in nanowire width (30–50 nm), while 180 °C reaction produced wider distribution (or 50–1000 nm) and meanwhile reduced aspect ratio. They suggested that ODA played a key role in nanowire growth. It could serve as both of the reducing agent and the coordination reagent. The reduction behavior was easily to distinguish by judge the production from reaction. As for the coordination reagent, ODA molecules were suggested to coordinate with Cu (II) cations. They noticed that, since ODA molecules could be hardly dissolved by water, these molecules tended to form micelles with lamellar structure. Such micelle could serve as "micro-reactor", in which 1D CuNWs could grow separately. Similar "micelle-derived" growth mechanism was proposed later. For example, Zhang et al. (2015) proposed a growth model basing on "liquid-crystalline medium". During the reaction, HDA and cetyltriamoninum bromide (CTAB) were mixed at first. At elevated temperatures, this mixture will form a liquid-crystalline medium. Then copper source, or copper acetylacetonate [$Cu(acac)_2$] was added and reacted, with presence of Pt catalyst in the medium [60]. The growth dynamics are shown in Fig. 7a. They suggested that, due to the coordination behavior between Cu salt and Br^- (from

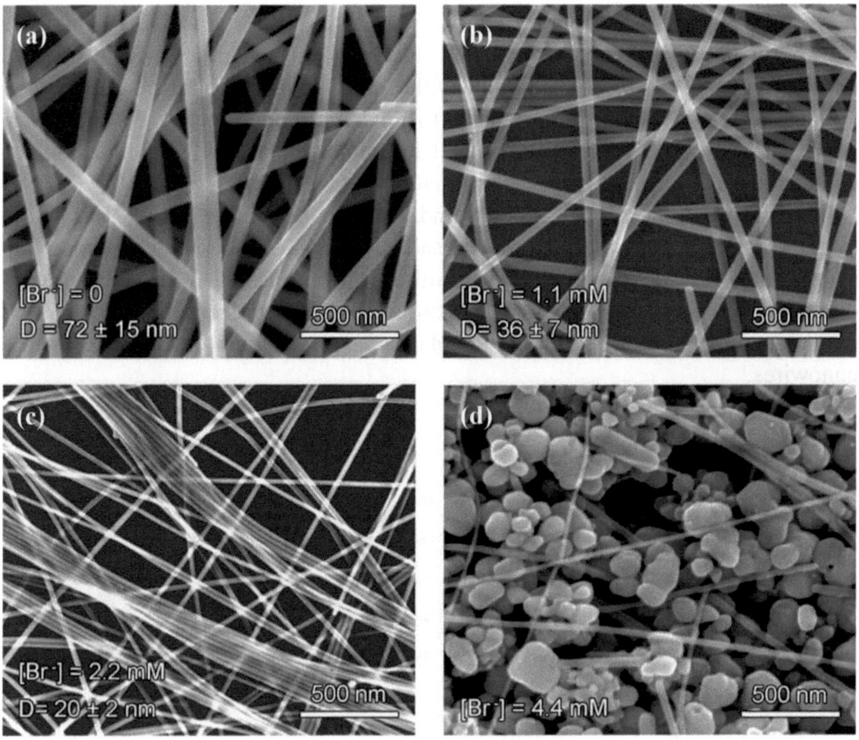

Fig. 6 Effect of concentration of NaBr on AgNWs width (mM): **a** 0, **b** 1.1, **c** 2.2 and **d** 4.4. Concentration of other reagents was kept the same, or 26.5, 50.5 and 4.2 mM, for $AgNO_3$, PVP and NaCl, respectively. Reproduced with permission [54]. Copyright © 2015, American Chemical Society

CTAB) and HDA, the added Cu precursor molecules could fill the tubular channels of the pre-formed liquid-crystallines. After reduced by HDA, the Cu atoms were mounted to nanowires. Due to the preferential adsorption of HAD and Br^- on {200} planes, unidirectional growth is favored at [110] direction. CTAB and Pt catalyst were found to play important role in nanowire formation. When CTAB concentration was relatively low, only polyhedrals were obtained. On the other hand, without addition of Pt, neither nanowires nor nanoparticles were observed even after 10 h reaction. It is worthy to note that, according to the authors, single crystalline-like CuNWs, with hexagonal cross-section were obtained. This is different from those AgNWs, releasing an open question. Other methods were also imported for CuNW growth. For example and also shown in Fig. 7b, copper nanowires (CuNWs) with width of 17.5 nm and aspect ratio of 1000 were reported by Cui et al. (2015) [61], using tris(trimethylsily)silane, oleylamine (OLA) and $CuCl_2$ as the reducing regent, coordinating ligand and metal source, respectively. The reaction routine was similar to that used in AgNWs growth, though with different reducing regent and coordination ligand. The coordination behavior between OLA and Cu(II) was suggested to play a key role in 1D structure formation [61]. Besides, Jin et al. (2015) synthesized CuNWs in oil bath (100 °C for 6 h), using glucose, hexadecylamine (HDA) as reducer and capping agent [also, to coordinate with Cu(II)], respectively [62]. CuNWs with width of 24 ± 4 nm, and length from tens to hundreds of micrometers were harvested. Moreover, these works also demonstrated the application of nanowires in transparent conductors (TCs), which will be discussed later [60, 61].

From above description, one could find that reacting routine of CuNW is similar to AgNW. Both of them contain metal salts, reducers, and capping (coordinating) agents, though the detailed materials are different from each other. In addition, coordination agent plays a key role in the reactions. It could coordinate with metal atoms, thus reduce the surface energy of side planes and prompt unidirectional growth of nanowires.

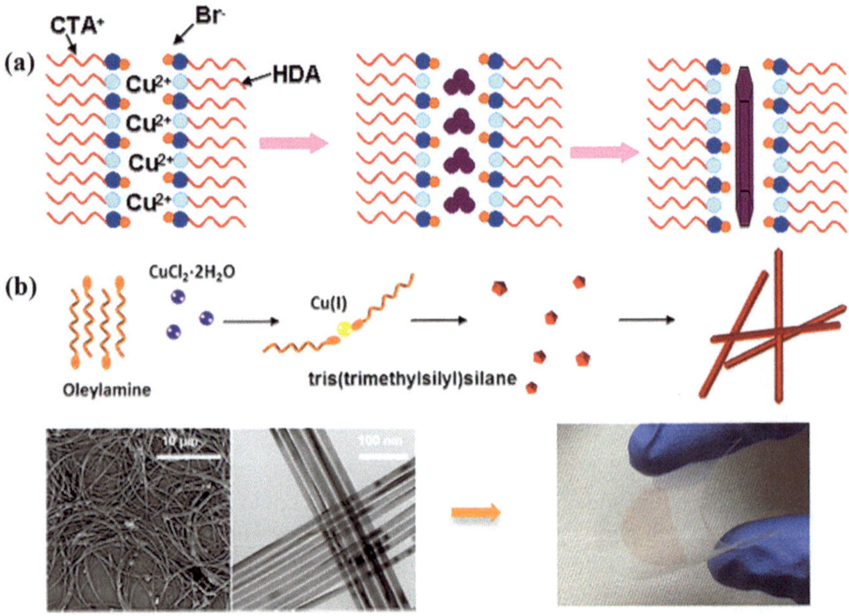

Fig. 7 **a** Formation scheme of CuNWs in liquid-crystalline medium. Reproduced with permission [60]. Copyright © 2012 American Chemical Society. **b** Schematic for the growth of ultrathin copper nanowires (CuNWs), and corresponding SEM image, TEM images of the synthesized CuNWs, and typical TC prepared from the nanowires. Reproduced with permission [61]. Copyright © 2015, American Chemical Society

2.4 Summary

In this section, synthesis routines of AgNWs and CuNWs are described, including the synthesis methods, coarsening mechanisms of nanowires, as well as the nanowire structural properties. Honestly, there still remain some interesting questions to answer: (i) What causes the formation of the five-fold seeds and then the nanowires? (ii) Could single crystalline AgNWs be realized? (iii) What is the exact role played by halide anions in the growth of AgNWs? More works are needed to clarify those mechanisms.

3 Application of Metal Nanowires in Transparent Conductors

One of the applications of these metal nanowires is the transparent & conductive film (TCs). Nowadays the commercialized TCs are mainly basing on oxide films, like ITO or FTO. They have shown excellent optoelectronic performance, though with relatively poorer flexibility. Metal nanowires is flexible, thus could solve this problem. Since the first attempt in TCs done by Lee et al. [63], researches on nanowire network basing TCs widespread, as well as the application in optoelectronic devices. In this section, these aspects will be all introduced, in addition, conducting mechanism of the networks, and protection method are also covered. Prior to detailed discussion on these aspects, it would be meaningful to list typical progress during the last decade in TCs made of nanowires, with respect to preparation methods, treatment and corresponding optoelectronic performances. Two tables are summarized as Tables 1 and 2, for AgNWs and CuNWs, respectively. It is worthy to note that, for both of these two kinds of nanowires, T% of 90%, and R_{sh} of 5–30 Ω/□ have been achieved on both rigid and flexible substrates, which is comparable to ITO based TCs.

3.1 Preparation of TC

As noted above in 2008, Lee et al. reported a kind of AgNW basing TCs, with transmittance ($T\%$ at 550 nm) of 84%, and sheet resistance (R_{sh}) of 10.3 Ω/sq or (Ω/□). Typical SEM images, effect of annealing period on sheet resistance of AgNWs films, and typical UV–Visible transmission spectra of AgNWs basing TCs are shown in Fig. 8a–c. When prepared on flexible substrate (Kapton tape), the film conductance only changed 1% after bending the substrate (along with the network deposited on) to radii of 4 mm. Also, effect of contact resistance between nanowires was noticed, and heating was performed to decrease the resistance. Effect of nanowire density on the film conductance was also studied [inset of Fig. 8c] [63]. Moreover, they showed power conversion efficiency (PCE) of 0.38% for organic photovoltaic solar

Table 1 Optoelectronic performance of AgNWs based TCs

	T (%)[a]	Rsh (Ω/□)	Haze (%)	Coating method	Substrate	Treatment	Year
Ag	84[b]	10.3	–	Spin coating	Glass	Heating[c]	2008 [63]
Ag	95	580	–	Spray coating	Saran wrap	Plasma	2012 [64]
Ag	90 85	19 47.78	–	Transfer	Glass Plast[c]	Heating	2012 [65]
Ag	88	40	≤1	Casting	PET	–	2013 [32]
Ag	89.4	14.9	–	Spin coating	Glass	Plasma/AgNPs	2014 [66]
Ag	92	5	7.26	Roll to roll	PET	Solution bath	2014 [67]
Ag	99.1 93[b]	130.0 25[b]	–	Meyer rod	Glass	Dry[c]	2015 [54]
Ag	90.2	12.5	–	Spin coating	PET	Bending	2015 [68]
Ag	92.6	35.0	–	Spin coating	PET	$NaBH_4$	2018 [69]
Ag	95	28	1.2	Transfer	Glass	$NaBH_4$	2018 [70]
Ag	94	30	–	Meyer	PET	Dry[d]	2020 [71]

[a]Measured at wavelength of 550 nm; [b]estimated from the figure of corresponding literatures [c]"heating" is used when temperature is higher than 150 °C in the reported document, otherwise, dry is noted; [d]According to the literature, the fabricated AgNWs TCs are rinsed by ethyl alcohol and DI respective

Table 2 Optoelectronic performance of CuNWs based TCs

	T (%)[a]	Rsh (Ω/□)	Haze (%)	Coating method	Substra-te	Treatment	Year
Cu	90	90	–	Transfer	Glass	Heating	2012 [60]
Cu	90	34.8	2–3	Transfer	Glass	Heating	2015 [61]
Cu	85.5	19.8	–	Spray	–	Heating	2016 [72]
Cu	83.0	24.5	–	Roll to roll	PC[b]	Pulse light	2016 [73]
Cu	89.3	28.2	~2	Transfer	Glass	Cu@r-GO	2016 [74]
Cu	89	35	<3	Transfer	Glass	Cu@Au	2018 [75]
Cu	70.2[c]	24	–	Transfer	PC[b]	Heating	2018 [76]
Cu	87	33	–	Transfer	Glass	Pulse laser	2020 [77]

[a]Measured at wavelength of 550 nm; [b]polycarbonate; [c] the T% was measured containing substrate

cells using such network as bottom electrode [inset of Fig. 7d]. Indeed, this work opened a door for the metal nanowires to participate in the fast development progress of optoelectronics. Following that, many state-of-art strategies have been developed, for the fabrication of nanowire basing TCs, the methods to enhance the optoelectronic performance of such TCs, and the application of the TCs in optoelectronic devices. For the sake of organization, the device applications will be discussed separately.

In laboratory, nanowire basing TCs were usually fabricated by spin-coating method, which it is applicable for small area preparation. Roll-to-roll method might be more suitable for mass production. In 2014, Lee et al. reported a "roll-to-roll" routine for AgNW TC [67]. As shown schematically in Fig. 9a, 3 distinct steps were involved in this strategy. At first (step 1), AgNWs were coated on flexible plastic substrate (like PET) by "Meyer rod" method, and transferred to a "solvent spray system", where several kinds of solvents could be used, like distilled water, methylene chloride and tetrahydrofuran. Then (step 2), the TC (along with substrate) were laminated by two rollers to weld the nanowires. At last, the compressed TCs were rinsed by NaCl and $FeCl_2$ solution, and finally by distilled water. Such fabrication routine holds four aspects of merits: (i) the method is suitable for scalable production due to fluent treating processes; (ii) Contact resistance between nanowires could be

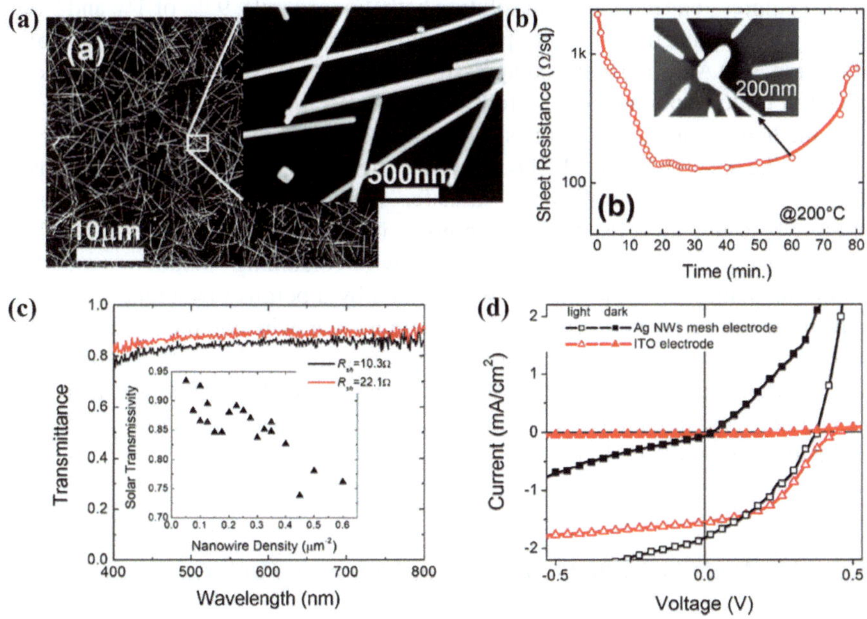

Fig. 8 **a** SEM images of AgNWs film on silicon substrates. **b** Effect of annealing period on sheet resistance of AgNWs films. **c** Typical UV—Visible transmission spectra of AgNWs TCs with two kinds of R_{sh} of 22.1 and 10.3 Ω/□. Inset shows the effect of nanowire density on T%. **d** Typical current density—voltage (J-V) curves of organic photovoltaic solar cells prepared on ITO and AgNWs TCs. Reproduced with permission [63]. Copyright © 2008, American Chemical Society

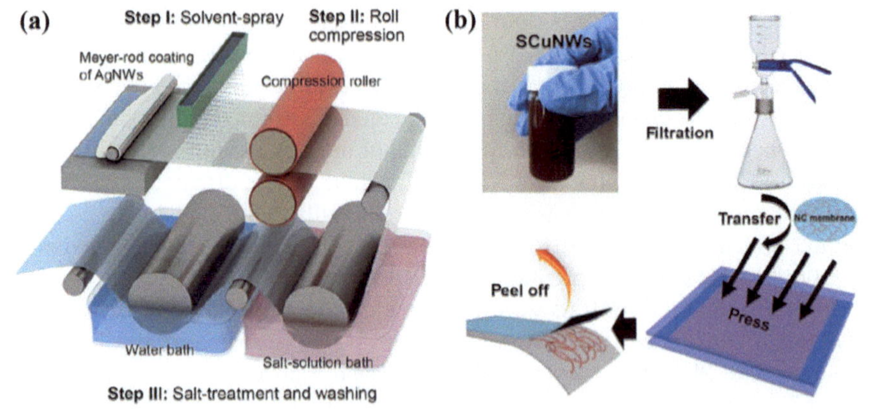

Fig. 9 **a** Schematic of roll-to-roll (R2R) coating method for the fabrication of AgNWs TCs. Reproduced with permission [67]. Copyright 2009, Royal Society of Chemistry. **b** Schematic of spray coating used in fabricating CuNWs TC. Reproduced with permission [78]. Copyright © 2018, Springer Nature

largely reduced and thus excellent optoelectronic performance could be obtained due to the compression technique and solution bath. For example, 92% of T% and 5 Ω/□ of R_{sh} were obtained on flexible plastic substrate, even without high temperature annealing. (iii) Also due to the compression technique, TC surface could be flattened which is quite important for the following optoelectronic device preparation. (iv) This method is compatible to the scalable fabrication of the flexible optoelectronic devices like solar cells, or OLEDs. Anyway, the haze could be decreased by using thin nanowires. Besides roll-to-roll, "transfer" method was also proposed. As shown in Fig. 9b, Zhang et al. used a 3-step transfer method to prepare CuNWs TCs [78]. The first step is filtration in which the nanowire containing "solution" (exactly, it should be named as turbid liquid) was filtrated by a porous membrane. After that, network could be formed between the remaining nanowires on the membrane. The second step is "transfer", or the formed networks are transferred to a substrate. At last, the transferred networks were annealed. After that, T% of 96.65%, and R_{sh} of 115.45 Ω/□ was obtained on PET substrate, while the haze was as low as 1.5% due to the ultrathin width of 15 nm.

3.2 Methods to Upgrade the Optoelectronic Performance

Normally, there exists a compromise between $T\%$ and R_{sh} of TCs. Given certain kind of conductive material, high $T\%$ usually leads to high R_{sh}. This is obeyed by nanowire based TCs. As a result, the enhancement in the optoelectronic performance always account on the reduction of R_{sh} at given T%. Such topic is directed to the conducting mechanism of such nanowire based networks. Usually, the film conductivity is relating the aspect ratio of nanowires, nanowire width and also the contact resistance between nanowires. As seen in literatures, and also the following discussions, longer nanowires are useful in providing more conducting paths, thus favoring the film conductivity. Thus one possible strategy is to synthesis high aspect ratio nanowires, which represents an important sub-topic in the field, and could be refer to several interesting literatures [33, 37, 65, 68, 70, 78–80]. Here we would like to pay more attention to the reduction on contact resistance.

Due to the loose stacking of nanowires, the contact between neighboring nanowires is hardly close enough; meanwhile, organic ligands may cap nanowires. As a result, large contact resistance exists. In 2016, Selzer et al. measured the contact resistance through the junction made by cross-linking nanowires, and showed that the resistance ranged from dozens to several hundreds of ohm for a junction, which is far too larger than the nanowire itself (for example, about 4.4 Ω/μm for AgNW with diameter of 90 nm) [82]. As a result, it is crucial to enhance the contact between nanowires.

In 2012, Garnett et al. showed that, silver nanowires could be welded by light induced plasmas [64]. As shown in Fig. 10a, they used a broadband tungsten-halogen lamp to irradiate the AgNWs network for 10–120 s, and then examined morphological properties near the junctions. Before illumination, loose stack could be seen. After illumination, welding behavior appeared at the junction region. Furthermore, they performed finite element method simulation on the optical heat generation between nanowires. As being demonstrated in Fig. 10a, heat is mainly generated in the bottom nanowires. Such melting behavior is appealing for junction resistance reduction, making the network a unit. Following that, similar welding techniques were developed, for example, laser [83], intense-pulsed light [84, 85], and pulsed UV laser irradiation [77].

Besides plasma welding, joule heat was also proposed. As shown in Fig. 10b, Song et al. (2014) demonstrated that, by applying suitable current across the AgNWs TC for short period (timescale of tens of seconds), R_{sh} could be cut down quickly. Careful examination showed that the contact region between nanowires could be welded [81], like that obtained by Garnett et al. (2012) [64]. They ascribed the welding effect to the synergy between thermal treatment and electro-migration. With assistance of simulation and conductivity measurement, they found that the contact resistance was quite larger than that of AgNW itself. In such situation, the joule heat mainly distributed at the contact region. Since noble atoms were originally found to diffuse easily (with surface diffusion activation energy of <1 eV [81, 86]), welding effect could be harvested under heating, between contact region. The study provides an efficient strategy to reduce contact resistance. However, joule heating was also noticed to cause break down in nanowire TCs, which is due to the unevenly distributed contact resistance, and the high surface energy of nanowire (especially for thinner nanowires) [87–89]. As such, protection is needed [90], as will be discussed later.

In order to reduce contact resistance, nanowire surface should be highlighted. In the original works towards AgNWs basing TCs, the surfactant of PVP remaining from the polyol method synthesis was noticed to hinder the network conductivity [63, 64].

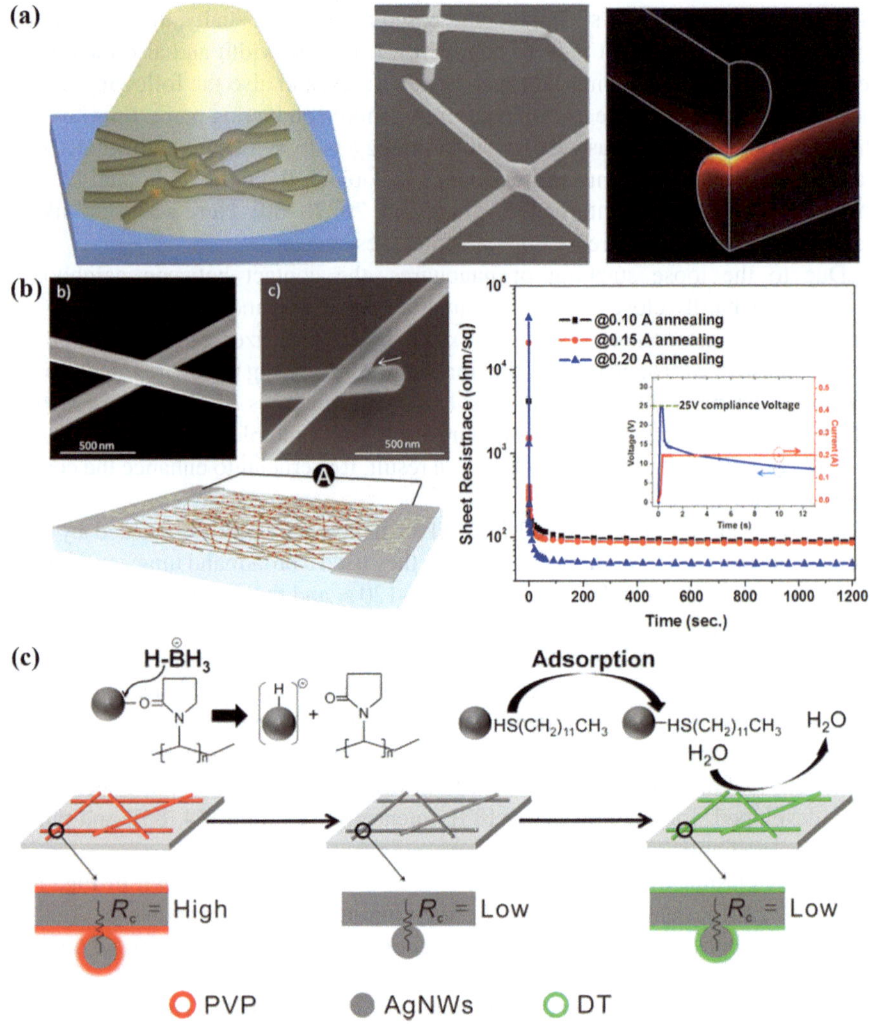

Fig. 10 **a** Plasma treatment on AgNWs networks: schematic of the treatment, enlarged contact region after the treatment, and finite-element method simulation the treatment are shown. One can see that heat is mainly generated from the bottom nanowire. Reproduced with permission [64]. Copyright © 2019, Springer Nature. **b** Welding AgNWs by current induced joule heating. Equipment for joule heating on AgNWs TC, SEM image of the contacting region, and the effect of treating period on R_{sh} of TC is shown. Reproduced with permission [81]. Copyright©2014, American Chemical Society. **c** Schematic for the "Room temperature welding" on AgNWs TCs: typically, PVP ligands could be removed from AgNWs by borohydride ($NaBH_4$). Reproduced with permission [69]. Copyright © 2018, American Chemical Society

Thus such ligands should be moved away. Indeed, this was usually done by multi-time rinsing process. In 2018, Ge et al. proposed an idea of "room temperature welding" between silver nanowires, which was realized through replacing PVP ligands by borohydride ($NaBH_4$) [69]. As shown in Fig. 10c, the replacement is dominated by the competition of binding energy between silver atom and different ligands. According to the authors, it is about 50.9 kcal/mol for "Ag–O" bond (O is from PVP), while 81.71 kcal/mol between Ag atom and hydride ion (which is caused by the decomposition of $NaBH_4$). Thus the insulating PVP ligands could be replaced by the hydride ions, which helps to form atom-level-clean interface between neighboring AgNWs, by which "room temperature welding" was obtained. Such welding effect helped to decrease R_{sh} from 81.5 to 35.0 Ω/□ (T% at 92.6%). Moreover, another capping layer of hydrophobic dodecanethiol [or DT, in Fig. 9c] was observed to improve the stability of the TCs. This work clearly indicates that: (i) PVP could be hardly completely washed off by common solvent due to the relatively high binding energy between "Ag–O"; (ii) Chemical replacement is effective to remove these insulating ligands, thus leading to lower contact resistance. Beyond these three strategies, other methods have also been tried, including mechanical rolling, [67, 91] or bending (which is similarly to compressing) [68], silver nanoparticle welding [66], etc. As for the bending treatment, it is usually used to determine the flexibility of TCs or devices, thus usually seen to deteriorate film conductance of electrode like ITO. However, Xia et al. (2015) found that, bending treatment could sharply increase the film conductance of AgNWs TCs [68]. In more, the improvement was found to be sensitive to bending direction. Inner bending (or nanowires are put to the inner side of the bended circle) produced better film conductance. Finally, with the assistance of ultra-long AgNWs, R_{sh} of 12.5 Ω/□ was obtained at T% of 90.2% (@550 nm) in 2015. The improvement is also due to the improved contact between nanowires.

3.3 Stability and Protection

Due to the high surface energy of both AgNW and CuNW, surface atoms are easily to diffusion, cause breakdown problem [81, 86–90], or Rayleigh like instability problem on one hand [33, 92, 93]; On the other hand, bare nanowire TCs could be easily oxidized or corroded in air or harsh electrochemical environment. As a result, protection should be performed. Usually, this was done by depositing another layer of inert material on the NW surface. Here below three methods are introduced. The first one is electroplating. In 2018, Lee and coworkers electroplated nickel layer on silver microgrid using the equipment shown in Fig. 11a [94]. Nickel electroplating helped to achieve 4.3 Ω/□ in R_{sh} and 96% in T%. In addition, Ni coated silver grid showed largely upgraded corrosion resistance. As shown in Fig. 11b, even being put in a quite harsh box filled by sulfur (heated by 150 °C) for 60 min, there was less change in the R_{sh} of the Ni-coated grid. As such, nickel coating is effective in enhance the anti-corrosion ability of Ag. In fact, similar methods have been applied in AgNWs or CuNWs basing TCs [95–99].

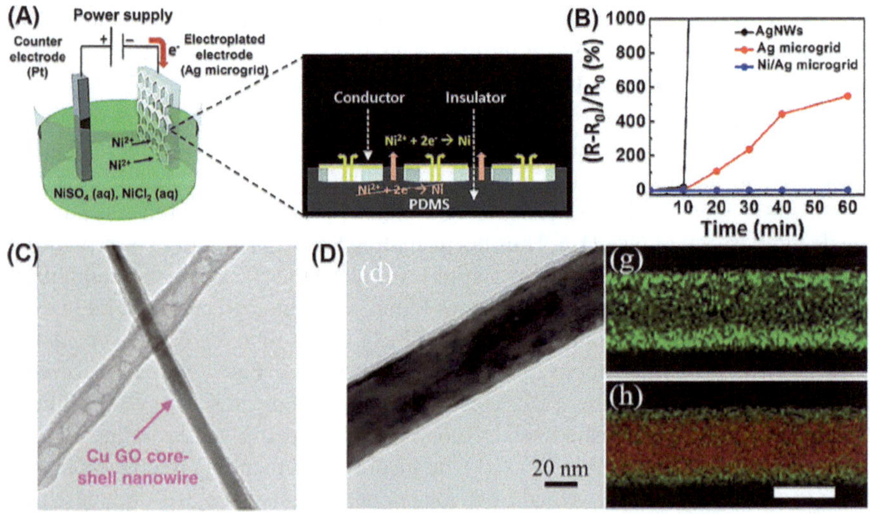

Fig. 11 **a** Schematic of the electroplating equipment by which silver microgrid electrode embedded in PDMS was coated by thin layer of nickel. **b** Stability test (chemical sulfurization test, 150 °C) on three kinds of electrode basing on AgNWs, Ag microgrid and Ag microgrid coated with nickel layer. Clearly, nickel coated electrode shows the strongest robustness. Reproduced with permission [94]. Copyright © 2013, Royal Society of Chemistry. **c** TEM image of CuNWs wrapped with r-GO layer [100] Reproduced with permission. Copyright 2016, American Chemical Society. **d** TEM image of a typical Ag-Au core—shell nanowire and corresponding EDX mapping (green-Au, red-Ag) [74]

In 2016, Dou et al. wrapped CuNWs by r-GO, which helped to increase both optoelectronic performance and stability of the TCs [100]. The wrapping process could be briefly described. At first, the CuNWs were immersed in graphene oxides (GO) suspension to get core–shell like NWs. Then the GO-wrapped CuNWs were filtrated and transferred to substrate. After that, the sample was annealed in "forming gas" (containing 10% H_2 in Ar) to reduce GO, which led to r-GO wrapped core–shell CuNWs, as shown in Fig. 11c. Such treatment could improve the optoelectronic performance to 89.3% (T%) and 28.2 Ω/□ (R_{sh}), along with low haze of ~2%. Moreover, storage-stability of 200 days was achieved.

In 2021, Huang et al. reported a kind of Ag-Au core–shell nanowire [74]. During experiments, AgNWs were added to Au precursor where $HAuCl_4$ was used as the Au source, and kept in room temperature for about 6 h. The reduced Au atoms were observed to deposit on the surface of AgNWs, with thickness of several nanometers. Typical TEM image of the resultant sample and corresponding EDX maps could clearly be seen in Fig. 11d. Such decoration was found to improve the robustness of the AgNWs against harsh electrochemical environment.

3.4 Conducting Mechanism of Networks Made of 1D Conductive Materials

Different from the conventional conductors basing on continuous films or bulks, above discussed TCs are consisting of randomly arranged nanowires. In fact, they are networks, rather than continuous films. From this point of view, distinctive conducting behavior should be anticipated. Indeed, understanding the conducting behavior of such disordered systems is not only a need for the application of TCs basing on nanowires, but also a classical task in statistical physics. For example, experiments showed that network conductance could be affected by aspect ratio (length/width) [33, 65], density [101, 102] and geometrical parameters [103, 104], and also junction resistance in addition [64, 67]. While on the other hand, from the theoretical aspect, the problem could be ascribed to "stick percolation", which belongs to the classic percolation systems [105–107]. Then two theories were applied to explain the conducting mechanism. Near percolation threshold, the network conductivity could be described by the scaling formula (7):

$$\sigma \sim (p - p_c)^t, \tag{7}$$

whereas p is the nanowire density, and p_c the percolation threshold; while t is dimension-relating parameter, equaling to 1.33 for 2D system [28, 106, 108]. For region away from threshold, or conducting phase has been stably built up, formula (7) is not applicable. Instead, large gap arises, as has been seen in previous report [109]. For this situation, Kirkpatrick's Effective Medium Theory (EMT, sometimes noted as effective medium approximation, or EMA) proposed in 1970s seems to give reliable description on the conducting behavior, as seen in formula (8) [110]:

$$g_m = 1 - \frac{1 - p}{1 - 2/z} = (2p - 1)|_{z=4}, \tag{8}$$

whereas g_m, p, z is normalized conductivity, existence probability of bonds, and maximum coordination number of nodes in 2D lattice. This theory was also used in AgNWs basing networks [111]. However, as being stated above, the nanowire basing TCs are disordered nanowire networks, rather than continuous films. The conductance should be relating closely to the "conducting paths" in the networks. As such, the formation dynamics and hence the effect of such "conducting paths" on the network conductivity should be considered. Honestly, the importance of conducting paths was noticed in literatures [112–115], though the detailed relationship between conducting path and network conductivity was unclear. This problem was then solved by He et al. (2018), with the assistance of simulation and experiment [116]. As shown in Fig. 12, an algorithm was developed to monitor the formation dynamics of conducting paths and network conductance simultaneously. Obviously, more conducting paths could be formed if area fraction of nanowire or nanowire length increases. What is more, two parameters were defined, or the length-ratio

(η_{cp}) of conducting-path to all nanowires, and normalized network conductivity (σ). According to the numerical simulation, an equation was obtained, as shown in formula (9):

$$\sigma = k(\eta_{cp} - b). \tag{9}$$

When contact resistance is not considered, k, b equals to 2 and 0.5 respectively, for 2D networks with maximum coordination of 4 for a node. As shown in Fig. 12f, a line could be drawn according to formula (9), which coincides well with the simulated results. When contact resistance is considered, linear behavior could also be seen between network conductivity (σ) and length ratio (η_{cp}), though with smaller slope (k). Moreover, such linear relationship was verified by experiments basing on AgNWs networks [116]. As a result, parameter of "η_{cp}" could be used as a basic topological parameter in describing the conducting behavior of disordered systems [117–119]. In fact, this theory could be named as "Effective Path Theory" (EPT) [120]. This theory differs to previous EMT theory in three aspects: (i) EPT only concerns the

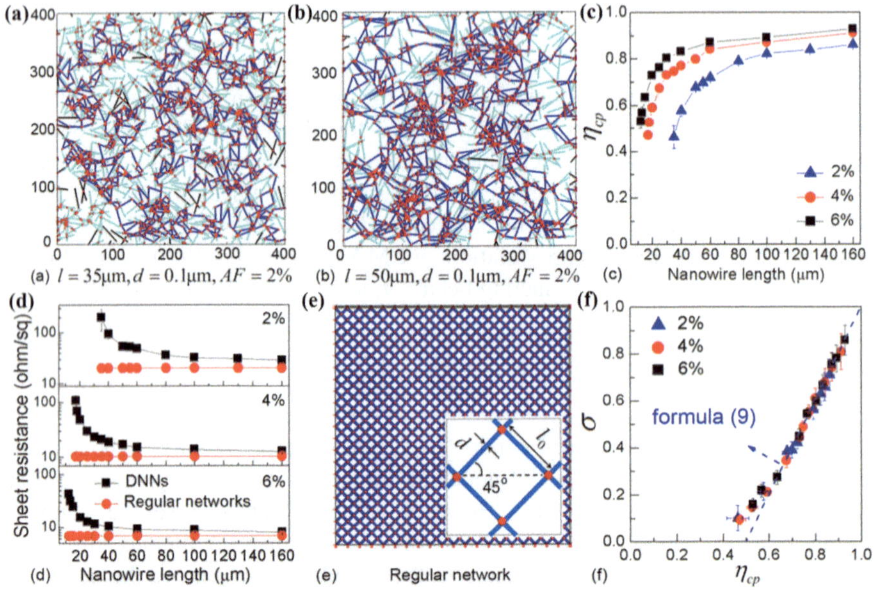

Fig. 12 Illustration for the suggestion of "Effect-Path Theory" (EPT). Typical 2D disordered nanowire networks (DNNs) simulated from two nanowire lengths: **a** 35 μm and **b** 120 μm. The area fraction (AF) of the nanowires in the network was fixed at 2%. For comparison, the conducting paths are marked by dark blue. Effect of nanowire length on: **c** length-ratio (η_{cp}) between conducting-path and all nanowires, and **d** R_{sh} of TC (R_{sh} of regular networks is plotted for reference). **e** Schematic of regular 2D network. **f** Plotting normalized network conductivity (σ) against length-ratio (η_{cp}). Dashed line in (F) shows fitted results according to formula (9) using $k = 2$ and $b = 0.5$. Coincidence could be clearly seen when η_{cp} is larger than 0.6. Reproduced with permission [116]. Copyright 2018, AIP Publishing

formation of conducting paths, which are the collective of the segments in nanowires that really carry current, other than those "dead ends" or branches; while in EMT theory, this could not be distinguished, since the existing probability p formula (8) contains all of the existing bonds. (ii) The normalization process of EPT is basing on the conductance of a regular network with nanowire width and area fraction equaling to that of disordered nanowire networks; while for EMT, the normalization is basing on regular network with highest area fraction (originally, the normalization is basing on resistor network with existing probability $p = 1$. Then if a resistor network is transferred into bond network, "$p = 1$" implies the highest area faction [110]); (iii) Local conductance or the conductance between neighboring nodes is allowed to vary one by one in EPT, but not so in EMT [110]. However, it is interesting that the two theories share the same mathematic form when comparing the formulas of (8) and (9). Thus further examination is worthy to follow. According to the EPT theory, more attention should be paid to the formation of conducting-paths, in order to obtain higher film conductance.

3.5 *Applications in Optoelectronic Devices*

Immediately after the successful fabrication of TCs, they were introduced to optoelectronic devices, like solar cells, OLED, touch screen, heaters, and so on. Besides, metal nanowires were also used in electromagnetic shielding, sensors, stretchable electronics. Here in this section, the application in optoelectronic devices will be discussed.

3.5.1 Flexible Solar Cells

During the past 10 years, metal nanowires (mainly Ag and Cu) basing TCs have been applied as both top and bottom electrodes in the organic solar cells or perovskite solar cells (typically, OSCs and PSCs). Compared to ITO basing devices, such devices come out with better flexibility, which is appealing for adaptive applications. In addition, when mounted on top, devices could harvest illumination from both top and bottom sides. As such, in the following part, these two kinds of devices are introduced: one is the flexible device, and the other is the semitransparent (Semi-T, for short) device. Typical progresses about the two kinds of solar cells are collected in Tables 3 and 4.

In 2019, a kind of lightweight and flexible PSCs was reported by Kang et al. [125]. As shown in Fig. 13a, they used a very thin (1.3 μm in thickness) and flexible polyethylene naphthalate (PEN) foil as substrate, orthogonally/randomly arranged AgNWs networks were deposited on it, which was used as substrate for PSCs. Power conversion efficiency (PCE) of 15.18% and power-per-weight of 29.4 W/g were achieved for orthogonal network basing TCs. This PCE was close to 16.25% of devices using ITO as TCs, but relative higher than random case (10.3%). The main

Table 3 Application of metal nanowires basing TCs in flexible solar cells

Nanowire	T[a] (%)	R_{sh} (Ω/□)	Subs. [b]	Type of solar cells	Position	PCE (%)	Year
Ag	83	15	–	OSC	Bottom	3.4	2011 [121]
Cu	88	24	Glass	OSC	Bottom	3.1	2014 [122]
Ag	92	7.7	cPI	OSC	Bottom	7.42	2015 [123]
Ag	–	–	PET	PSC	Bottom	13.36	2018 [124]
Ag	92.3	–	PEN	PSC [c]	Bottom	15.18	2019 [125]
Ag	86	11.0	Glass	PSC [d]	Bottom	15.31	2020 [126]
Cu	87.8	34.05	PES	PSC	Bottom	14.18	2020 [127]
Ag	91	22	PET	OSC	Bottom	8.94	2021 [128]
Ag	95[e]	30 [e]	PEN	PSC module	Bottom & Top	16.78	2020 [129]
Ag	–	–	Ag grid	OSC module	Bottom	16.1[e]	2021 [130]

[a]Measured at wavelength of 550 nm; [b] substrate when the nanowire network was used as bottom electrode, or bottom electrode when the nanowire network is used as top electrode; [c] lightweight device was prepared; [d] Carbon electrode was adopted as the top electrode; [e]representative data is picked from the literature; [f] OSC module was fabricated basing on a "PET/Ag grid/AgNWs/PEI-Zn" composite electrode; This efficiency was recorded using an aperture with area of 4.095 mm^2

Table 1.4 Application of metal nanowires basing TCs in semi-transparency (Semi-T) solar cells

Nanowire	Device	Position	Bottom	T% or AVT	PCE (%)	Test side	Year
Ag	OSC	Top	ITO	66% (T%)[a]	3.82 4.02	Ag ITO	2012[133]
Ag	OSC	Top	ITO	51.8% (T%)[a]	3.15	ITO[b]	2015[134]
Ag	PSC	Top	ITO	23.3% (AVT)	10.45 11.31	Ag ITO	2018[135]

[a]Measured at wavelength of 550 nm; [b]deduced from the literature, or "the top AgNWs TC was sealed before the efficiency test"

difference comes from the short circuit current density (J_{SC}). It is 20.09 mA/cm^2 (ITO), 18.63 mA/cm^2 (orthogonal) and 14.88 mA/cm^2 (random) for the three cases as shown in Fig. 13b. The authors ascribed the improved charge collection to the uniform surface of orthogonal AgNWs network. Since this could help to grow uniform perovskite (the photo-active layer, basing on organic lead halide). Also, AgNWs arrangement also affected device stability. For example, after being stored by 500 h, 85% of device efficiency was retained for orthogonal case device, while only 50% was retained for random case. Again, this was also contributed by the improved surface quality of the orthogonal network. In more, after bending at radius of 5 mm for 1000 cycles, about 90% of the start efficiency could be attained. Noting that, conducting polymer of PH1000 was used to modify the AgNW TCs, which makes the TC more conductive. However, PH1000 was observed to cause degradation on AgNWs, thus protection is needed [131].

CuNW TCs were also used in solar cells. Yang et al. (2020) prepared flexible PSCs on Al-doped ZnO (AZO) protected CuNW TCs [127]. As shown in Fig. 13b, AZO layer was coated on CuNWs by atom layer deposition (ALD) method. By tuning the Al dopant concentration, a composition-gradient AZO (g-AZO) was deposited. Such g-AZO was suggested to be favorable to device performance. It protected nanowire from being corroded by thermal treatment or chemical reactions, thus improve device stability on one hand; and provided gradient energy band alignment, and accelerate charge extraction, reduce recombination on the other hand. Subsequently, g-AZO helped flexible PSCs obtain PCE of 14.18%, close to that of ITO/PEN basing devices (16.33%), while superior to 12.34% of the device assembled from mono-AZO treated CuNWs. Again, excellent robustness was observed. 98% of the start efficiency could be attained after bending at radius of 12.5 mm for 800 cycles, compared to the sharp decrease of ITO basing devices.

More impressively, solar cell modules have also been successfully fabricated using metal nanowire basing electrodes. Gao and Meng (2020) showed that, PSC module could be fully prepared by inkjet printing, including both bottom and top electrodes, light-active layer, electron-transporting layer and hole-transporting layer [129, 132]. As shown in Fig. 14a, modules with size of 120–180 cm^2 were prepared. Such modules showed excellent power conversion efficiencies, or 16.78% for 120

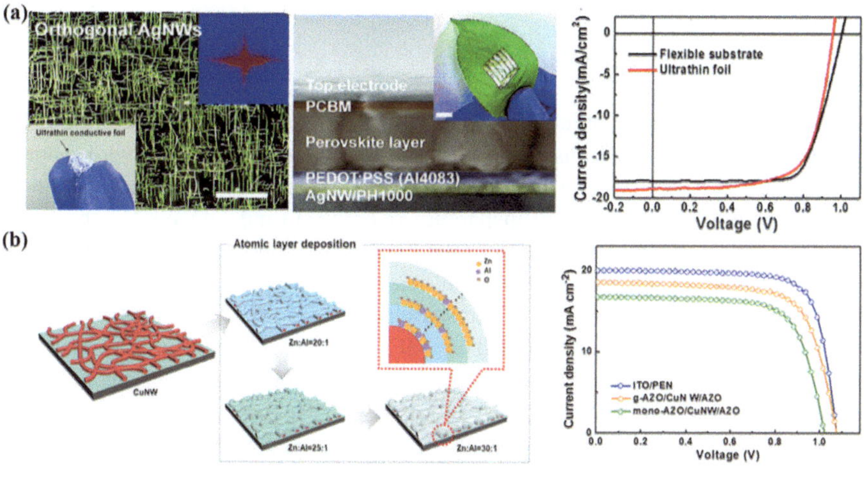

Fig. 13 Typical applications of metal nanowire basing TCs in flexible solar cells. **a** Flexible organic photovoltaic cell (OSC) assembled from AgNWs TC. The TC was made of cross-linked AgNWs on 1.3 μm-thick PEN foil; typical device structure and current density—voltage (JV) curves are shown on the right side of **a**. Reproduced with permission [125]. Copyright @ 2013, Royal Society of Chemistry. **b** Flexible perovskite solar cell (PSC) assembled from CuNWs TC which was coated by gradient-AZO (g-AZO). Typical JV curves are shown on the right side of **b**. Reproduced with permission [127]. Copyright 2020, American Chemical Society

cm^2 module, and 12.56, 10.68% for 150 and 180 cm^2 modules, respectively [129]. Moreover, less hysteresis and prolonged photo-stability were achieved [Fig. 14a]. Modules of organic photovoltaic cells (or organic solar cells, OSCs) have also been prepared using metal nanowire basing TCs. Zhou et al. (2021) fabricated large-area flexible OSC modules with size up to 54 cm^2, basing on a kind of composite flexible electrode [130]. As shown in Fig. 10b, typically the composite electrode used highly conductive Ag grid/PET as substrate (R_{sh} ~ 1.1 Ω/□), which had large void space with size up to 100 μm. Then, "AgNWs/ PEI-Zn" (noting PEI-Zn is the short form of "zinc-chelated polyethylenimine") was used to fill the void and enhance the charge collection of the solar cell. They checked the relation between R_{sh} of such "AgNWs/PEI-Zn" film (rather than the whole electrode) and the final device performance, and found that, when R_{sh} was as high as 600 Ω/□, PCE of 14.06% was achieved. Decreasing R_{sh} to 100 Ω/□ could harvest PCE of 14.74%. Such phenomenon is different to those devices basing on pure AgNWs TCs, though could be ascribed to the usage of highly conductive Ag grid. In more, the "PEI-Zn" itself is conductive. It not only flattens the surface of AgNW TCs, but also improves the thermal stability of the AgNWs basing electrode. Finally, small OSC device with efficiency of 16.1% (active area 4.095 mm^2), large area single device with efficiency of 13.1% (6 cm^2), 12.6% (10 cm^2) and module with efficiency up to 13.2% (54 cm^2) were achieved, along with excellent flexibility. Besides, flexible quantum-dots light-emitting diodes (QLEDs) with external quantum efficiency up to 13.3% were also successfully fabricated using such composite electrode.

3.5.2 Semitransparent Solar Cells

Beside flexible devices, other unique type of solar cells was also proposed. For example in 2012, visibly transparent (or Semi-T) polymer solar cells were reported by Chen et al. (2012), using AgNWs basing TC as top electrode [133]. As shown in Fig. 15a, owing to the excellent transparency of both ITO and AgNWs TC, the device actually open "windows" from both of the two sides. In more, the light-active layer is only sensitive to ultraviolet (UV) and near infrared (NIR) wavelengths. Consequently, the full device is actually "transparent" to our eyes. According the authors, transparency of the device could reach 66% at 550 nm [middle of Fig. 15a]. However, such transparent device could generate power from irradiations. Even under illumination of AM 1.5G (100 mW/cm^2, including UV, NIR, and visible wavelengths), it showed power conversion efficiency (PCE) of 4.02, and 3.82%, when it was illuminated from ITO or AgNWs TC sides. Thus, the power conversion property varies little. Following that, several works were performed. Typically, semitransparent OSC modules were reported by Guo et al. (2015), using AgNWs TCs as the top electrode [134]. As shown in Fig. 15b, the photoactive layer derived from the mixture between pDPP5T—2:$PC_{60}BM$ enables highly transparency around visible regions in the spectrum.

And series connected modules were prepared using laser pattern technique, using either rigid glass or flexible PET as substrates. Similarly, such modules could generate

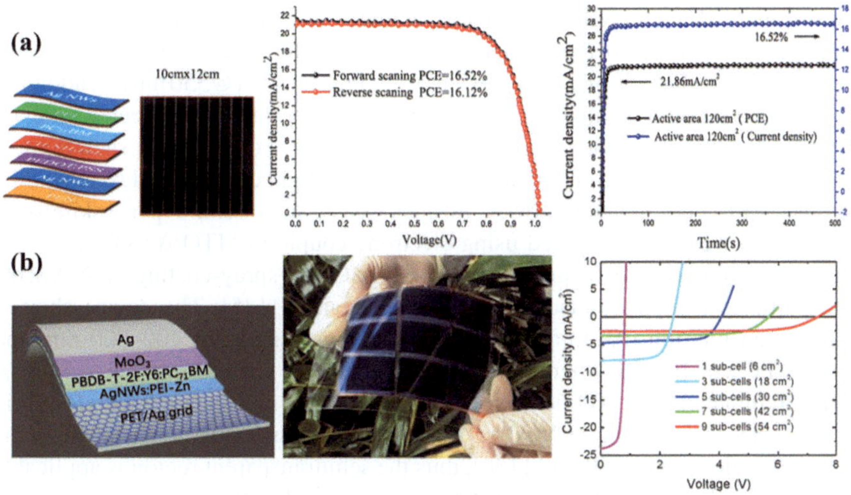

Fig. 14 Typical applications of metal nanowire basing TCs in flexible solar cells modules. **a** Flexible PSC modules assembled with AgNW TCs as both bottom and top electrodes. Typical device structure, module image, JV curves, along with photo-stability test curves is shown on the right side of **a**. Reproduced with permission [129]. Copyright @ 2021, Elsevier. **b** Flexible OSC modules assembled with basing on a "PET/Ag grid/AgNWs/PEI-Zn" composite electrode; Typical device structure, module image, JV curves are shown on the right side of **b**. Reproduced with permission [130]. Copyright @ 2021, John Wiley and Sons

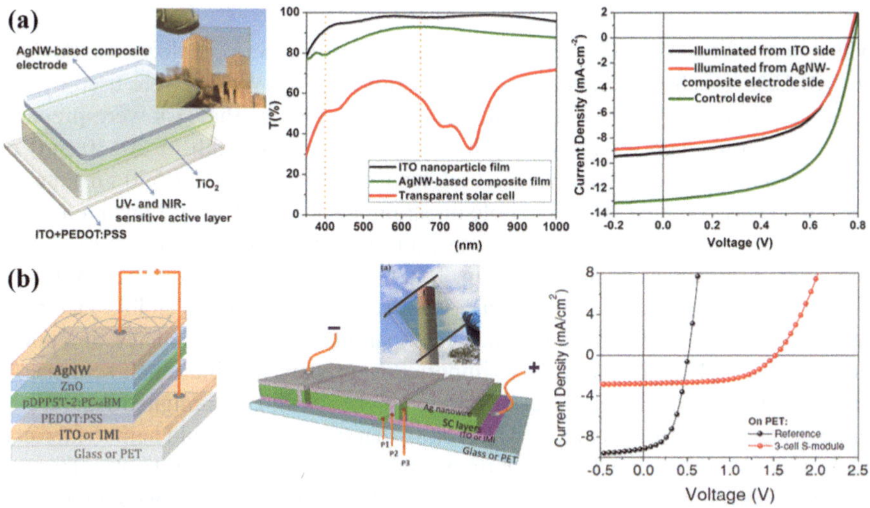

Fig. 15 Typical applications of metal nanowire basing TCs in semitransparent solar cells. **a** Semi-T OSC prepared using AgNWs TC as top electrode. Reproduced with permission [133]. Copyright @ 2012, American Chemical Society. **b** Semi-T OSC module prepared using AgNWs TC as top electrode. Reproduced with permission [134]. Copyright @ 2015, John Wiley and Sons

power from the UV and NIR wavelengths, coming out with PCE of 2.31% for single cell (with active area of 0.1 cm^2), and 2.00% for 10-cell module (with active area of 1.6 cm^2). Noting that T% of the full devices was about 51.8% (@550 nm). In addition, the efficiencies could be upgrade to 3.15 and 2.25% for devices fabricated on glass substrate. Such performance is appealing for daily use. For example, such Semi-T devices could be mounted on windows, as the visible light could pass through, while the other part could generate power. Besides OSCs, such Semi-T perovskite solar cells (PSCs) were also prepared using electrode couple of "ITO/AgNW TC". For example, Han et al. (2018) prepared Semi-T PSCs by spray-coating AgNW/ZnO nano particles composite electrode on top of devices [135]. The device showed average transmission (AVT) of 23.3% between wavelengths of 400–800 nm, and outcome PCE of 10.45% when illuminated from AgNW side, comparing to 11.31% for the illuminated from ITO side. Anyway, it is worthy to note that, lead halide hybrid perovskite layer (for example, $MAPbI_3$) usually shows high extinction behavior of the in short wavelengths (<560 nm) [138], thus the semitransparent region is applicable to longer wavelengths, which differs from those Semi-T OSCs.

In fact, using metal nanowire basing TCs becomes a hot topic in recent years. Relate progress could further refer to many publications [3, 5, 8, 136–139]. These works show that, metal nanowires basing TCs could be used as electrodes in thin film photovoltaic techniques.

3.5.3 Other Applications

Metal nanowire basing TCs have also been applied in other optoelectronic devices, like OLEDs, touch panels, heater, etc. For example, AgNW TCs were used by Lee et al. (2014) to fabricate flexible OLEDs [140]. Typically, the TCs were prepared by spin-coating AgNWs (with length of 10–20 μm and width of 20–40 nm) on either glass or PEN. Optoelectronic performance of 92.2% (T%@550 nm) and 10 Ω/□(R_{sh}) was obtained. Then the hole injection layer, hole transport layer, emission layer, hole blocking layer, electron transport layer, LiF and top Al electrode were mounted in sequence. As shown in Fig. 16a, OLED assembled from AgNW basing TC showed similar electroluminescence spectrum like that basing on ITO electrode. In addition, these devices also produced comparable current efficiencies, and power efficiencies. AgNW TCs were also imported to prepare flexible touch screens. As shown in Fig. 16b, Cho et al. (2017) prepared TCs with cross-aligned AgNWs by large-scale bar-coating method [141]. Noting that this method could help to fabricate large area (20 × 20 cm^2) TCs, with optoelectronic performance of 95.0% (T%) and 21.0 Ω/□ (R_{sh}). And then, force-sensitive touch screens were equipped, and linked to computer system. Similar works were done for application in LED [142, 143], or touch screens [65, 136, 137].

Kim et al. (2013) showed that, metal nanowire TCs could be used as transparent film heaters [144]. As shown in Fig. 17, after coating AgNWs on a large piece of PET, a heater with area of 250 × 200 mm^2 was obtained by coating two parallel electrodes on the counter sides. After bias was added between the two electrodes, current

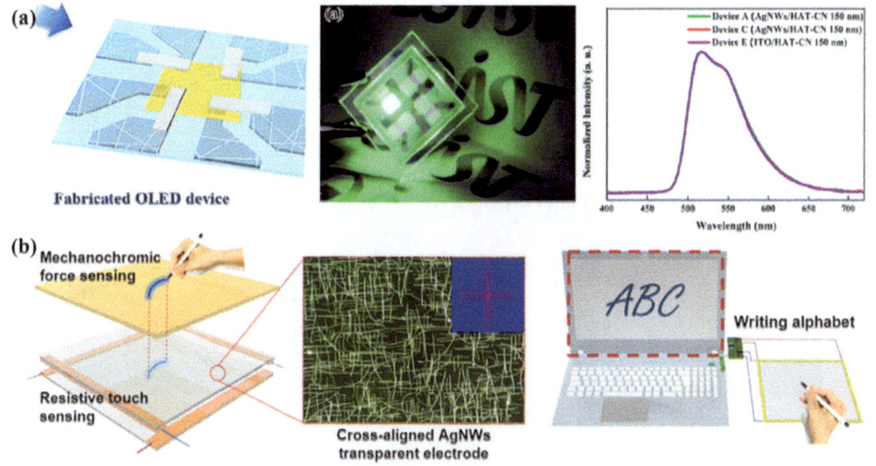

Fig. 16 **a** Application of AgNW TC in OLEDs. Reproduced with permission [140]. Copyright @ 2009, Royal Society of Chemistry. **b** Application of AgNW TC in touch screens. Reproduced with permission [141]. Copyright @ 2017, American Chemical Society

could be generated, by which joule heat was produced. They tested the temperature on the heater, and found that the temperature distribution was sensitive to nanowire distributions. Hotspots appeared at nanowire aggregations, while uniform distribution was seen at uniformly arranged networks. This heating functionality could be used to defrost. Also shown in Fig. 17b, the frost (formed by keeping film in a refrigerator for 30 min) loaded on surface of AgNW TC could be cleared in just one min, given voltage of 12 V. Moreover, relationship between the temperature, sheet resistance and voltage was studied. As could be anticipated, lower resistance and higher voltage is beneficial for achieving higher temperatures. In fact, the ability of "defrosting" is appealing for daily life, especially for windows, glasses. This has become another important application branch. More works could be found in recently published documents [9, 10, 30].

3.6 Summary

In this section, application of metal nanowire in TCs are described, including the method to prepare such TCs, the treatment to improve the optoelectronic properties of the TCs, and also the understanding towards the unique conducting mechanism of such networks. Right now, the main tasks of these TCs are lying on three cases: (i) method to protect nanowires from breakdown caused by surface diffusion, or oxidization and corrosion from outside environment; (ii) method to further decrease contact resistance between nanowires; and (iii) further understanding of the conducting behavior. Silver and copper nanowire TCs have been widely used in

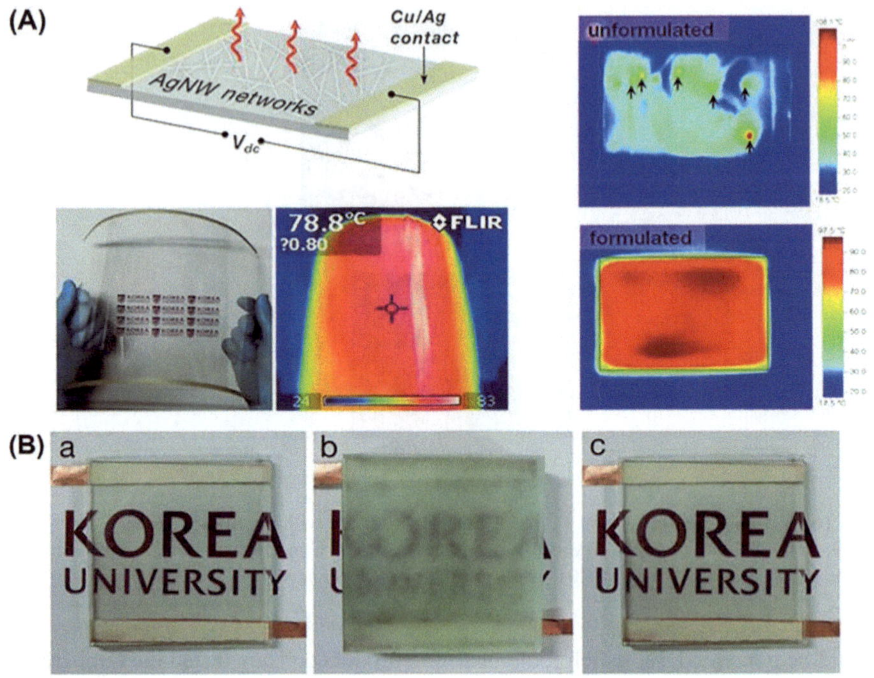

Fig. 17 Application of AgNW TC in transparent film heaters. **a** Schematic of transparent heater basing on AgNW TC; **b** defrosting test of the heater, photo images are shown for: **a** pristine TC, **b** after frost formation, and **c** after heating at 12 V for 60 s. Reproduced with permission [144]. Copyright @ 2012, John Wiley and Sons

photoelectronic devices including flexible solar cells, Semi-T solar cells, OLEDs, touch screens and heaters. High transparency, highly conductive and sound flexibility of nanowire basing TCs have opened new windows of these applications. To benefit the applications, right now, the main task might be lying on three cases: (i) In order to favor the charge transport (extraction or injection) process, surface of the nanowire basing TCs should be flattened, while keeping conductive in every position of the surface, including the void space between nanowires; (ii) Due the complex application environment, the nanowires are easily corroded, thus the protection or passivation methods are needed; (iii) Large-area production with incorporation of device modules.

4 Perspective

As long as the application in transparent conductor (TC) is considered, AgNWs and CuNWs are the mostly used candidates. These TCs have shown excellent optoelectronic performance, comparable or even superior to commonly used ITO, and better flexibility, enabling the fabrication of flexible and ultrathin photoelectric devices like solar cells, OLED, touch screens, heaters, and so on. To prompt the application of such nanowire basing TCs, more works are needed, for example, growth method of nanowires and the beneath reaction mechanism, protection towards these thin nanowires, conducting mechanism of random nanowire networks, surface roughness reduction of TC, homogenization on surface conduction, and also the large-area production of device modules. In more, new applications are welcome.

Acknowledgements C. Zhou thanks the financial support of National Scientific Foundation of China (NSFC, No. 61774170) and the Natural Science Foundation of Hunan Province (NO. 2020JJ4759). Y. Gao thanks support from National Science Foundation, United States (NSF, DMR-1903962).

References

1. Zhang R, Engholm M, Hummelgård M, Andersson H, Örtegren J, Olin H, Appl ACS (2018) Energ Mater 1:7191
2. Xia YN, Yang PD, Sun YG, Wu YY, Mayers B, Gates B, Yin YD, Kim F, Yan YQ (2003) Adv Mater 15:353
3. Scardaci V (2021) Appl Sci 11:8035
4. Hu LB, Wu H, Cui Y (2011) MRS Bull 36:760
5. Tan DC, Jiang CM, Li QK, Bi S, Song JH (2020) J Mater Sci Mater Electron 31:15669
6. Yao S, Zhu Y (2015) Adv Mater 27:1480
7. Xu X, Han G, Yu H, Jin X, Yang J, Lin J, Ma C (2019) J Phys D Appl Phys 53:05LT02
8. Angmo D, Krebs FC (2013) J Appl Polym Sci 129:1
9. Moreira IP, Sanivada UK, Bessa J, Cunha F, Fangueiro R (2021) Molecules 26:3685
10. Papanastasiou DT, Schultheiss A, Munoz-Rojas D, Celle C, Carella A, Simonato JP, Bellet D (2020) Adv Funct Mater 30:1910225
11. Li Y, Cui F, Ross MB, Kim D, Sun Y, Yang P (2017) Nano Lett 17:1312
12. Alia SM, Pivovar BS, Yan Y (2013) J Am Chem Soc 135:13473
13. Lyu Z, Xie M, Aldama E, Zhao M, Qiu J, Zhou S, Xia Y, Appl ACS (2019) Nano Mater 2:1533
14. Chen ZH, Fang R, Li W, Guan J (2019) Adv Mater 31:1900756
15. Feng Y, Zhu J (2019) Sci China-Mater 62:1679
16. Gong S, Yap LW, Zhu B, Cheng W (2020) Adv Mater 32:1902278
17. Fievet F, Lagier JP, Figlarz M (1989) MRS Bull 14:29
18. Silvert PY, Tekaia-Elhsissen K (1995) Solid State Ionics 82:53
19. Silvert P-Y, Herrera-Urbina R, Tekaia-Elhsissen K (1997) J Mater Chem 7:293
20. Fievet F, Lagier JP, Blin B, Beaudoin B, Figlarz M, Ionics SS (1989) 32–33. Part 1:198
21. Zhang QM, Li Y, Xu DS, Gu ZN (2001) J Mater Sci Lett 20:925
22. Sloan J, Wright DM, Bailey S, Brown G, York APE, Coleman KS, Green MLH, Hutchison JL, Woo H-G (1999) Chem Comm 699
23. Govindaraj A, Satishkumar BC, Nath M, Rao CNR (2000) Chem Mater 12:202
24. Braun E, Eichen Y, Sivan U, Ben-Yoseph G (1998) Nature 391:775
25. Sun Y, Yin Y, Mayers BT, Herricks T, Xia Y (2002) Chem Mater 14:4736
26. Sun Y, Xia Y (2002) Adv Mater 14:833
27. Sun Y, Gates B, Mayers B, Xia Y (2002) Nano Lett 2:165
28. Xia Y, Xiong Y, Lim B, Skrabalak SE (2009) Angew Chem Int Edit 48:60
29. Zheng Y, Zeng J, Ruditskiy A, Liu M, Xia Y (2014) Chem Mater 26:22

30. Liu Q, Tian B, Liang J, Wu W (2021) Mater Horiz 8:1634
31. Patil JJ, Chae WH, Trebach A, Carter KJ, Lee E, Sannicolo T, Grossman JC (2021) Adv Mater 33:2004356
32. Lee E-J, Chang M-H, Kim Y-S, Kim J-Y (2013) APL Mater 1:042118
33. Xu X, He S, Zhou C, Xia X, Xu L, Chen H, Yang B, Yang J (2016) RSC Adv 6:105895
34. Yang Z, Qian H, Chen H, Anker JN (2010) J Colloid Interface Sci 352:285
35. Hong BH, Bae SC, Lee CW, Jeong S, Kim KS (2001) Science 294:348
36. Wiley B, Herricks T, Sun YG, Xia YN (2004) Nano Lett 4:1733
37. Lee JH, Lee P, Lee D, Lee SS, Ko SH (2012) Cryst Growth Des 12:5598
38. Jiu J, Sugahara T, Nogi M, Suganuma K (2013) J Nanopart Res 15:1588
39. Gao Y, Jiang P, Liu DF, Yuan HJ, Yan XQ, Zhou ZP, Wang JX, Song L, Liu LF, Zhou WY, Wang G, Wang CY, Xie SS (2003) Chem Phys Lett 380:146
40. Gao Y, Song L, Jiang P, Liu LF, Yan XQ, Zhou ZP, Liu DF, Wang JX, Yuan HJ, Zhang ZX, Zhao XW, Dou XY, Zhou WY, Wang G, Xie SS, Chen HY, Li JQ (2005) J Cryst Growth 276:606
41. Sun Y, Mayers B, Herricks T, Xia Y (2003) Nano Lett 3:955
42. Niu G, Liu F, Yang Y, Fu Y, Wang W (2020) Colloid Surf A Physicochem Eng Asp 607:125490
43. Yu Y, Cui F, Sun J, Yang P (2016) Nano Lett 16:3078
44. Wiley B, Sun YG, Mayersa B, Xia YN (2005) Chem Eur J 11:454
45. Caswell KK, Bender CM, Murphy CJ (2003) Nano Lett 3:667
46. Gao P-Y, Kunath W, Gleiter H, Weiss K, Phys Z (1989) D-Atoms. Mol Clust 12:119
47. Hall BD, Flueli M, Monot R, Borel J-P (1991) Phys Rev B 43:3906
48. Bovin JO, Alfredsson V, Karlsson G, Carlsson A, Blum Z, Terasaki O (1996) Ultramicroscopy 62:277
49. Szpunar B, Erb Palumbo U, Aust KT, Lewis LJ (1996) Phys Rev B 53:5547
50. Gryaznov VG, Heydenreich J, Kaprelov AM, Nepijko SA, Romanov AE, Urban J (1999) Cryst Res Technol 34:1091
51. Zhang W, Liu Y, Cao R, Li Z, Zhang Y, Tang Y, Fan K (2008) J Am Chem Soc 130:15581
52. Im SH, Lee YT, Wiley B, Xia Y (2005) Angew Chem Int Edit 44:2154
53. Zhang WJ, Chen P, Gao QS, Zhang YH, Tang Y (2008) Chem Mater 20:1699
54. Li B, Ye S, Stewart IE, Alvarez S, Wiley BJ (2015) Nano Lett 15:6722
55. Chen DP, Qiao XL, Qiu XL, Chen JG, Jiang RZ (2010) J Colloid Interface Sci 344:286
56. Chen D, Zhu G, Zhu X, Qiao X, Chen J (2011) J Mater Sci Mater Electron 22:1788
57. Foresti ML, Innocenti M, Kobayashi H, Pezzatini G, Guidelli R (1996) J Chem Soc Faraday Trans 92:3747
58. Molares MET, Höhberger EM, Schaeflein C, Blick RH, Neumann R, Trautmann C (2003) Appl Phys Lett 82:2139
59. Shi Y, Li H, Chen L, Huang X (2005) Sci Technol Adv Mater 6:761
60. Zhang DQ, Wang RR, Wen MC, Weng D, Cui X, Sun J, Li HX, Lu YF (2012) J Am Chem Soc 134:14283
61. Cui F, Yu Y, Dou L, Sun J, Yang Q, Schildknecht C, Schierle-Arndt K, Yang P (2015) Nano Lett 15:7610
62. Jin M, He G, Zhang H, Zeng J, Xie Z, Xia Y (2011) Angew Chem Int Edit 50:10560
63. Lee J-Y, Connor ST, Cui Y, Peumans P (2008) Nano Lett 8:689
64. Garnett EC, Cai W, Cha JJ, Mahmood F, Connor ST, Greyson Christoforo M, Cui Y, McGehee MD, Brongersma ML (2012) Nat Mater 11:241
65. Lee J, Lee P, Lee H, Lee D, Lee SS, Ko SH (2012) Nanoscale 4:6408
66. Lu H, Zhang D, Ren X, Liu J, Choy WCH (2014) ACS Nano 8:10980
67. Lee SJ, Kim Y-H, Kim JK, Baik H, Park JH, Lee J, Nam J, Park JH, Lee T-W, Yi G-R, Cho JH (2014) Nanoscale 6:11828
68. Xia X, Yang B, Zhang X, Zhou C (2015) Mater Res Express 2:075009
69. Ge Y, Duan X, Zhang M, Mei L, Hu J, Hu W, Duan X (2018) J Am Chem Soc 140:193
70. Niu Z, Cui F, Kuttner E, Xie C, Chen H, Sun Y, Dehestani A, Schierle-Arndt K, Yang P (2018) Nano Lett 18:5329

71. Xu J, Wang K, Li Y, Zhuang T-T, Gao H-L, Liu Y-Y, He C-X, Yu S-H (2020) Sci China-Chem 63:1046
72. Hwang H, Kim A, Zhong Z, Kwon H-C, Jeong S, Moon J (2016) Adv Funct Mater 26:6545
73. Zhong Z, Woo K, Kim I, Hwang H, Kwon S, Choi Y-M, Lee Y, Lee T-M, Kim K, Moon J (2016) Nanoscale 8:8995
74. Huang S, Liu Y, Jafari M, Siaj M, Wang H, Xiao S, Ma D (2021) Adv Funct Mater 31:2010022
75. Niu Z, Cui F, Yu Y, Becknell N, Sun Y, Khanarian G, Kim D, Dou L, Dehestani A, Schierle-Arndt K, Yang P (2017) J Am Chem Soc 139:7348
76. Deshmukh R, Calvo M, Schreck M, Tervoort E, Sologubenko AS, Niederberger M (2018) Acs Appl Mater Interfaces 10:20748
77. Wang J, Chen H, Zhao Y, Zhong Z, Tang Y, Liu G, Feng X, Xu F, Chen X, Cai D, Kang J (2020) Acs Appl Mater Interfaces 12:35211
78. Zhang Y, Guo J, Xu D, Sun Y, Yan F (2018) Nano Res 11:3899
79. Zhang Y, Guo J, Xu D, Sun Y, Yan F (2018) Langmuir 34:3884
80. Xiang H, Guo T, Xu M, Lu H, Liu S, Yu G, Appl ACS (2018) Nano Mater 1:3754
81. Song T-B, Chen Y, Chung C-H, Yang Y, Bob B, Duan H-S, Li G, Tu K-N, Huang Y (2014) ACS Nano 8:2804
82. Selzer F, Floresca C, Kneppe D, Bormann L, Sachse C, Weiß N, Eychmüller A, Amassian A, Müller- L, Leo K (2016) Appl Phys Lett 108:163302
83. Dai S, Li Q, Liu G, Yang H, Yang Y, Zhao D, Wang W, Qiu M (2016) Appl Phys Lett 108
84. Song C-H, Ok K-H, Lee C-J, Kim Y, Kwak M-G, Han CJ, Kim N, Ju B-K, Kim J-W (2015) Org Electron 17:208
85. Jiu J, Sugahara T, Nogi M, Araki T, Suganuma K, Uchida H, Shinozaki K (2013) Nanoscale 5:11820
86. Sanders DE, DePristo AE (1992) Surf Sci 260:116
87. Khaligh HH, Goldthorpe IA (2013) Nanoscale Res Lett 8:235
88. Li Y, Tsuchiya K, Tohmyoh H, Saka M (2013) Nanoscale Res Lett 8:370
89. Khaligh HH, Xu L, Khosropour A, Madeira A, Romano M, Pradere C, Treguer M, Servant L, Pope MA, Goldthorpe IA (2017) Nanotechnology 28:425703
90. Yu H, Jin N, Wang Z, Lin J, Wei J, Luo Q, Ma CQ (2020) Nanotechnology 31:18LT01
91. Hauger TC, Al SMI, Buriak JM (2013) Acs Appl Mater Interfaces 5:12663
92. Rizza G, Attouchi F, Coulon P-E, Perruchas S, Gacoin T, Monnet I, Largeau L (2011) Nanotechnology 22:175305
93. Naik JP, Das K, Prewett PD, Raychaudhuri AK, Chen Y (2012) Appl Phys Lett 101:163108
94. Lee K, Park J, Kim H, Park HS, Song HK, Kim KH, Seo K (2018) J Mater Chem A 6:11790
95. An S, Jo HS, Kim D-Y, Lee HJ, Ju B-K, Al-Deyab SS, Ahn J-H, Qin Y, Swihart MT, Yarin AL, Yoon SS (2016) Adv Mater 28:7149
96. Kang H, Choi S-R, Kim Y-H, Kim JS, Kim S, An B-S, Yang C-W, Myoung J-M, Lee T-W, Kim J-G, Cho JH (2020) Acs Appl Mater Interfaces 12:39479
97. Wang S, Tian Y, Hang C, Wang C (2018) Sci Rep 8:5260
98. Wang S, Tian Y, Wang C, Hang C (2018) J Electrochem Soc 165:D328
99. Zhang L, Chen Y, Xu C, Liu Z, Qiu Y (2018) RSC Adv 8:14532
100. Dou L, Cui F, Yu Y, Khanarian G, Eaton SW, Yang Q, Resasco J, Schildknecht C, Schierle-Arndt K, Yang P (2016) ACS Nano 10:2600
101. Hu L, Kim HS, Lee J-Y, Peumans P, Cui Y (2010) ACS Nano 4:2955
102. De S, Higgins TM, Lyons PE, Doherty EM, Nirmalraj PN, Blau WJ, Boland JJ, Coleman JN (2009) ACS Nano 3:1767
103. Zeng Z, Wang C, Gao J (2020) J Appl Phys 127:065104
104. Behnam A, Guo J, Ural A (2007) J Appl Phys 102:044313
105. Balberg I, Binenbaum N, Anderson CH (1983) Phys Rev Lett 51:1605
106. Li J, Zhang S-L (2010) Phys Rev E 81:021120
107. Zezelj M, Stankovic I (2012) Phys Rev B 86:134202
108. Pike GE, Seager CH (1974) Phys Rev B 10:1421
109. Bauhofer W, Kovacs JZ (2009) Compos Sci Technol 69:1486

110. Kirkpatrick S (1973) Rev Mod Phys 45:574
111. O'Callaghan C, da Rocha CG, Manning HG, Boland JJ, Ferreira MS (2016) Phys Chem Chem Phys 18:27564
112. Kumar A, Vidhyadhiraja NS, Kulkarni GU (2017) J Appl Phys 122:045101
113. Li L, Holland S (2014) Nanomater Energy 3:139
114. Wu Z, López E, Buldyrev SV, Braunstein LA, Havlin S, Stanley HE (2005) Phys Rev E 71:045101
115. Rossen WR, Mamun CK (1993) Phys Rev B 47:11815
116. He S, Xu X, Qiu X, He Y, Zhou C (2018) J Appl Phys 124:054302
117. Ponzoni A (2019) Appl Phys Lett 114:153105
118. Tarasevich YY, Vodolazskaya IV, Eserkepov AV, Akhunzhanov RK (2019) J Appl Phys 125:134902
119. Balberg I (2020) J Appl Phys 128:204304
120. Zeng J, Wang Y, Zheng X, Zhou C (2022) J Phys D Appl Phys 55:414004
121. Zhu R, Chung C-H, Cha KC, Yang W, Zheng YB, Zhou H, Song T-B, Chen C-C, Weiss PS, Li G, Yang Y (2011) ACS Nano 5:9877
122. Sachse C, Weiß N, Gaponik N, Müller L, Eychmüller A, Leo K (2014) Adv Energy Mater 4:1300737
123. Kim Y, Ryu TI, Ok K-H, Kwak M-G, Park S, Park N-G, Han CJ, Kim BS, Ko MJ, Son HJ, Kim J-W (2015) Adv Funct Mater 25:4580
124. Liu X, Yang X, Liu X, Zhao Y, Chen J, Gu Y (2018) Appl Phys Lett 113:203903
125. Kang S, Jeong J, Cho S, Yoon YJ, Park S, Lim S, Kim JY, Ko H (2019) J Mater Chem A 7:1107
126. Zhou J, Li S, Lv X, Li X, Li Y, Zheng Y-Z, Tao X (2020) J Power Source 478:228764
127. Yang H, Kwon H-C, Ma S, Kim K, Yun S-C, Jang G, Park J, Lee H, Goh S, Moon J (2020) Acs Appl Mater Interfaces 12:13824
128. Liu H, Wu J, Fu Y, Wang B, Yang Q, Sharma GD, Keshtov ML, Xie Z (2021) Thin Solid Films 718:138486
129. Gao B, Meng J (2021) Sol Energy 230:598
130. Qin F, Sun LL, Chen HT, Liu Y, Lu X, Wang W, Liu TF, Dong XY, Jiang P, Jiang YY, Wang L, Zhou YH (2021) Adv Mater 33:2103017
131. Wang J, Chen X, Jiang F, Luo Q, Zhang L, Tan M, Xie M, Li Y-Q, Zhou Y, Su W, Li Y, Ma C-Q (2018) Solar RRL 2:1800118
132. Gao B, Meng J (2021) Acta Phys Sin 70:208801
133. Chen C-C, Dou L, Zhu R, Chung C-H, Song T-B, Zheng YB, Hawks S, Li G, Weiss PS, Yang Y (2012) ACS Nano 6:7185
134. Guo F, Kubis P, Przybilla T, Spiecker E, Hollmann A, Langner S, Forberich K, Brabec CJ (2015) Adv Energy Mater 5:1401779
135. Han K, Xie M, Zhang L, Yan L, Wei J, Ji G, Luo Q, Lin J, Hao Y, Ma C-Q (2018) Sol Energ Mat Sol C 185:399
136. Sun YN, Chang MJ, Meng LX, Wan XJ, Gao HH, Zhang YM, Zhao K, Sun ZH, Li CX, Liu SR, Wang HK, Liang JJ, Chen YS (2019) Nat Electron 2:513
137. Li WT, Zhang H, Shi SW, Xu JX, Qin X, He QQ, Yang KC, Dai WB, Liu G, Zhou QG, Yu HZ, Silva SRP, Fahlman M (2020) J Mater Chem C 8:4636
138. Ma C, Liu YF, Bi YG, Zhang XL, Yin D, Feng J, Sun HB (2021) Nanoscale 13:12423
139. Li D, Lai W-Y, Zhang Y-Z, Huang W (2018) Adv Mater 30:1704738
140. Lee H, Lee D, Ahn Y, Lee E-W, Park LS, Lee Y (2014) Nanoscale 6:8565
141. Cho S, Kang S, Pandya A, Shanker R, Khan Z, Lee Y, Park J, Craig SL, Ko H (2017) ACS Nano 11:4346
142. Kim BS, Won S, Seo J, Jeong SK, Kim C, Kim K-S, Kim SH, Cho SM, Kim J-H (2021) ACS Appl Mater Inter 13:26601
143. Kim T, Kang S, Heo J, Cho S, Kim JW, Choe A, Walker B, Shanker R, Ko H, Kim JY (2018) Adv Mater 30:1800659
144. Kim T, Kim YW, Lee HS, Kim H, Yang WS, Suh KS (2013) Adv Funct Mater 23:1250

Synthesis of Quantum Dots and Its Application in Heavy Metal Sensing

Atirah Tauseef and Imran Uddin

Abstract Quantum Dots (QDs) are zero-dimensional nano-particles portraying their distinguishing optical and electronic properties, they are used as nano-sensors. QDs have improved fluorescence characteristics, which comprise photostability, broad excitation spectrum, and narrow emission spectrum. QDs deal with the extensive and sensitive sensing of heavy metal ions ascribed to the presence of distinct capping agents and various functional groups lying outwardly of the QDs. These capping strata and functional moieties attune to the sensing capacity of the QDs, which influences the interactions of QDs with different analytes by various mechanisms. In this chapter, a brief overview of heavy metals as environmental contaminants, their impact on human health, and conventional techniques of detection and underlying modes are first introduced. Then, the role of QDs in sensing heavy metals such as mercury, cadmium, lead, arsenic, chromium, etc., and their progress in the multiplexed determination of heavy metal ions are explored.

Keywords Quantum dots (QDs) · Heavy metal · Sensor · Probe · Human health

1 Introduction

Quantum dots (QDs) are semiconductor nanoparticles with excitons contained in all three spatial dimensions and the ability to transform an entering light spectrum into a distinct frequency of energy output, resulting in different unique optical features. QDs contain distinct energy levels, allowing energy levels to be tuned with small variations in their size, paving way a for various applications like LEDs, photovoltaics, etc. QDs may pass through the bloodstream due to their small size, which unveils up a

A. Tauseef
Plant Pathology and Nematology Section, Botany Department, Aligarh Muslim University, Aligarh 202002, India

I. Uddin (✉)
Research Institute of Biomolecular and Chemical Engineering, Nanolab, University of Pannonia, Veszprém 8200, Hungary
e-mail: uddin.imran@mk.uni-pannon.hu

I. Uddin and I. Ahmad (eds.), *Synthesis and Applications of Nanomaterials and Nanocomposites*, Composites Science and Technology,
https://doi.org/10.1007/978-981-99-1350-3_10

whole new application facet in biomedical applications like biosensing and imaging. QDs portrayed excellent optical properties, as well as a useful surface chemistry, ligand binding capability, and the ability to encapsulate in various materials or attach to various functional moieties besides keeping their native luminescence property, due to their unique photophysical properties. The occurrence of metallic ions in the proximity of QDs has a significant impact on their optical characteristics. Optical equipment can quickly detect even minor alterations in the optical characteristics on the surface of QDs (changes in fluorescence and/or colour), making them great candidates for metal sensors. The sensitivity of a QDs to a certain metal ion causes either an increase in emission or a decrease in QD intensity. Heavy metals are one of the utmost significant pollutants in the environment, and they are rapidly becoming a major issue of concern. Because of its toxicity, non-biodegradable nature, and ability to accrue in the environment, metal contamination in the environment is a severe problem all over the world. Anthropogenic activities such as mining, smelting, industrialization, domestic and agricultural activities, as well as natural activities such as weather, metal soil erosion, and volcanic eruptions, are the primary sources of pollution heavy metal contamination is a major source of concern worldwide, and toxicity caused by a large number of heavy metal ions is far more destructive and severe than toxicity caused by a single metal species. The number of worldwide health risks related to heavy metal exposure has increased [1]. Because of their highly toxic and harmful health effects, heavy metal ions like lead, cadmium, arsenic, and mercury contamination in water has been a grave alarm around the globe. Their accumulation in the human body, agricultural, and aquaculture systems has gained widespread interest in recent years, even at extremely low concentrations. As a result of environmental contamination, the concentration of these unwanted compounds is steadily growing. The root cause is industrial effluents and garbage disposal, both of which end up in the aquatic system. Since they have non-biodegradable nature and hence persists in soil and water for prolonged periods of time, they ultimately damage humans and domestic animals, necessitating their identification and removal from biological and aquatic systems. There are a variety of instrumental methods for detecting heavy metal species, including atomic absorption spectrometry, X-Ray fluorescence spectrometry, inductively coupled plasma mass spectrometry, capillary electrophoresis, and high-sensitivity microprobes, but sample pretreatment, expensive instruments, and the need for trained people to operate these instruments are the real challenges for many of these techniques. As a result, we need a technique for detecting heavy metal ions that is simple, quick, cheap, selective, and sensitive. For metal ion detection, quantum dots have significant benefits over traditional approaches. When metal ions come into contact with the surface of quantum dots, their optical characteristics are very sensitive. Optical equipment can quickly detect slight modifications on the surface of QDs (such as fluorescence or colorimetric changes), making them a good choice for use as a metal sensor.

2 Mechanistic Insight of Human Health Effects of Heavy Metals

Human caused activities such as mining and burning of fossil fuels, heavy metals have been released into the environment by air, water and soil. Bioaccumulation of these metals in living organisms has devastating repercussions. Heavy metals interact with organelles of cell such as cell membrane, endoplasmic reticulum, mitochondria, lysosome, nuclei, and important enzymes actively participating in metabolism and detoxification reactions, leading to DNA damage and conformational changes within cells, paving way for cancer and apoptosis (programmed cell death) in biological systems. In the human body, heavy metals are compartmentalised within cells and tissues of body, here they are attached to proteins and nucleic acids alters their structure and functioning. On the other hand, heavy metal poisoning can have a number of effects on the human body. It can induce mental disease by disrupting central nervous system function, as well as harming blood components and causing damage to the lungs, liver, kidneys, and other vital organs, leading to a variety of ailments. Long-term heavy metal deposition in the body may also slow the progression of physical, muscular, and neurological healing processes, mimicking diseases like Parkinson's disease and Alzheimer's disease. Long-term exposure to many heavy metals ions and their compounds cause mutations, and mimic hormones, disordering the functioning of reproductive and endocrine systems and ultimately causing cancer. Heavy metals and metalloids are among the major environmental contaminants in both aquatic and terrestrial ecosystems. Heavy metals are particularly dangerous at low levels of exposure due to their persistence and bioaccumulation, directly alerting living species on the planet [2] (Table 1).

Table 1 Different limit values of heavy metal concentration set by WHO, EU, EPA (all expressed in mg/L)

Metal ions	EPA	WHO	EU
Silver (Ag)	–	–	0.01
Manganese (Mn)	0.05	0.05	0.05
Copper (Cu)	1.3	1	1
Zinc (Zn	5	5	0.1
Iron (Fe)	0.3	0.3	0.2
Lead (Pb)	0.005	0.015	0.05
Mercury (Hg)	0.002	0.001	0.001
Chromium (Cr)	0.1	0.05	0.05
Arsenic (As)	0.01	0.01	0.05
Cadmium (Cd)	0.005	0.005	0.005

[a] EPA—Environmental Protection Agency
[b] WHO—World Health Organization
[c] EU—European Union

3 Designing and Development of Sensors for Heavy Metals Toxicity

A range of analytical techniques are used to identify heavy metals. High-performance liquid chromatography, colourimetry, capillary electrophoresis, voltammetry, polarography, atomic emission spectroscopy, inductively coupled plasma emission spectroscopy, atomic emission spectroscopy, and inductively coupled plasma mass spectroscopy. Despite the high selectivity and specific sensitivity of all of these techniques, their use in the sensing of heavy metal ions is restricted due to high costs, time constraints, tough handling operations, the need for technical assistances for sample preparation, and the need for a variety of instruments.

Optical fluorescence-based techniques for sensing heavy metal ions in the environment have gotten a lot of attention since they offer real-time detection without the use of expensive equipment. The exceptional generation of a realistic and resilient fluorescence-based is intrinsically related to the type of the principles on which it is built especially with new breakthroughs in materials to create unique optical fluorescence fluorescence-based. Optical processes like as stroke shifts, fluorescence quenching, energy transfer, and charge transfer are exploited in fluorescence-based detection systems. These are the core concepts that provide sensitivity and selectivity to fluorescence-based sensing systems. Many different probes have been used in fluorescence-based approaches in a range of scientific disciplines, including fluorescent proteins, chemical dyes, and quantum dots [3].

3.1 Fluorescent Aptaswitch for Heavy Metal Detection

Aptamers are synthetic oligonucleotide sequences that can preferentially connect to a target molecule. A simple immobilization method, good thermal stability, and simplicity of synthesis and modification are only a few of their enticing properties for sensor design. The most noteworthy characteristic is their outstanding affinity and specificity for each of their target analytes. As a result, aptamer-based detection techniques have risen to prominence as highly selective recognition tools [4]. When thymine (T) interacts with mercury (II) ions, it generates T–Hg^{2+}–T base pairs in DNA duplexes, and when cytosine (C) interacts with Ag^{+} ions, it forms C–Ag^{+}–C mismatches [5]. Since Ono and co-workers [6] reported the first ON-based Hg^{2+} sensor, T-rich ON sequences have been widely used for the selective detection of Hg^{2+} in water samples [7, 8].

3.2 Heavy Metal Ion Sensors Based on Organic Dyes

Organic dyes are extensively employed in the construction of fluorescence-based sensors because of their attractive qualities like a high value of molar extinction coefficient, strong signal, simplicity of modification, and the existence of several potential reactive sites in their structures. The fluorophores are modified with an ion recognition unit (ionophore) that serves host for the specific metal ion in order to detect heavy metal ions. The interaction between the ionophore and the mark analyte leads to a change in the fluorophore's photophysical characteristics, which leads to a shift in its fluorescence emission, generally from "off" to "on." Mostly, crown ethers, aliphatic and aromatic amines are commonly used ionophores because they operate as electron donors, quenching fluorescence light by a photo-induced electron transfer (PET) mechanism through the fluorophore when the target metal ion is absent.

Rhodamine derivatives are the most widely used organic dyes because of their structure-dependent characteristics. Fluorescein and coumarin derivatives are two more dyes commonly used in the production of fluorescent sensors [9]. Fluorescent dyes have a narrow excitation spectrum, rapid decay, and a low quantum yield. These problems have been solved, and quantum dots can now be used in sensing devices [10].

3.3 Quantum Dots as Probe for Inorganic Metal Detection

QDs show great potential for heavy metal ion detection because of their fluorescence characteristics. QDs are proving to be superior fluorescent probes when equated to different fluorophores such as fluorescent proteins and chemical dyes. In comparison to other traditional fluorophores, QDs have greater brightness, high photostability, hefty stoke shifts, wide absorption spectra, great molar extinction coefficient, high quantum yield, long fluorescence time, photobleaching resistance, size-dependent optoelectronic properties, broad absorption spectra, and tuneable emission spectra. QDs are semiconductor nanocrystals that have lately become popular as fluorescent probes in chemical research, bio-sensing, and bio-imaging [11]. They are used as a new class of fluorescent markers as a substitute for conventional fluorescent markers (organic dyes and fluorescent proteins) due to their excellent and unique optical and electronic properties, such as broad absorption spectra, narrow and tunable emission spectra, long fluorescence lifetime, high photostability, and resistance to photodegradation [12]. The inter-band and intra-band relaxation pathways are altered by the quantum confinement effect. QDs have inspired a lot of interest as fluorescent probes for sensing processes, in vitro and in vivo bioimaging, quantum computers, light-emitting devices, photovoltaics, and, most crucially, analytics-based chemistry. QDs have remarkable electrical and optical properties which can be transformed by lowering the number of atoms present in the QDs beside preserving the chemical composition. The optical and electrical characteristics of semiconductor nanoclusters are determined by interactions amongst electrons, holes, and their immediate surroundings. When the excitation energy exceeds the bandgap of QDs, electrons move from the valence band to the conduction band, triggering photoexcitation. An electron present in an excited state has a huge amount of energy. A hole and an excited electron make form an exciton. After being recombined, electrons and holes relax to a lower energy state. QD luminescence is caused by radiation relaxation. Recombination and relaxation provide extra energy that might be radiative or non-radiative. The presence of adsorbates on the surface of QDs has been discovered to have an impact on their fluorescence efficiency. They can demonstrate a significant change in fluorescence intensity property when they approach the surrounding molecules, which is utilised in the sensing method [13]. Fluorescence intensity in QDs is caused by combination of the excitation, which in turn is a recombination of charge carriers (electron-holes), and variations in charges present on the surface would influence both the efficiency of electron–hole recombination and luminescence efficiency. Photodegradation and quenching of QD fluorescence intensity are caused by atoms with unmet valencies found on the surface of QDs. To develop QD-derived sensors for selective sensing of heavy metal ions, it's best to modify the surface of the QDs with appropriate capping layer materials and ligand molecules to make them chemically stable and photobleach-resistant [14]. The nature of capping layer and ligands affect the fluorescence response of quantum dots. A number of bifunctional ligands are used to modify the surface of QDs (TOPO, cysteine, PEG, amphiphilic polymer, glutathione, mercaptoacetic acid, thioglycolic acids, peptides, avidin, and streptavidin). These surface ligands can modify their sensitivity and choosiness for explicit analytes of interest by adjusting the fluorescence response of QDs, and this response can be further changed by changing the capping layer. Organic and inorganic materials are used as capping materials for QDs. An organic capping layer facilitates the capacity of QDs to bioconjugate. QDs which are capped by organic ligands are photo unstable, resulting in trap sites owing to weak interactions between the atoms present on the surface and the capping molecules. Inorganic capping compounds aid in the passivation of QD surfaces and the formation of the core–shell structure of QDs [15].

Quantum dots (QDs) finds explicit applications in different fields such as photovoltaics, bioimaging, and light-emitting diodes. Direct interaction between the analyte and the QDs, functionalization of the QDs, and integration of the QDs with other sensory materials are three basic methodologies for developing sensing devices [16]. Quantum dots are extensively used for detection of different heavy metals due to their aforementioned features and unique process. Metals such as mercury, lead, cadmium, arsenic, and chromium are commonly detected.

3.3.1 Detection of Mercury

Mercury (Hg^{2+}) is regarded as hugely detrimental water pollutants by the World Health Organization (WHO) because of its severe impacts on environment and human health, even at minuscule concentrations. Mercury binds to DNA, interrupting its normal function, and the toxicity is linked to mitosis across the blood–brain barrier. Clinical signs of mercury poisoning include pulmonary edema, minamata sickness, renal damage, chest pain, and chronic central nervous system damage [17]. The World Health Organization (WHO) has set a maximum acceptable amount of 1 mg/L as the upper limit based on these facts. The main factors attributed to their rising level of water resources are emissions from several enterprises and refineries. Polluted water containing mercury must be identified and rectified as soon as feasible.

Zhou and co-workers used CQDs for the detection of mercury for the foremost time. The synthesis of quantum dots was done by EDTA pyrolysis. The attachment of Hg (II) on the surface of CQD is anticipated to bring variations in surface states, paving way for recombination of non-radiative electron/hole pairs. The synthesized pristine material was capable of detecting Hg(II) to a equal of 4.2 nM, which is less than limits set by WHO. The probe was effectively used for real water samples obtained from tap water, lake water, and fountain water and against different ions [18]. Additionally, CDs synthesized from different precursors like sodium citrate, polyethene glycol (PEG), folic acid, and casein were used for the detection of Hg (II) in water with LOD values of 10 nM, 1 fM, and 6.5 nM, respectively.

GCQDs produced from vegetables (Hongcaitai and mushrooms) were used to detect Hg (II) in spiked and unspiked water samples [19–21]. Kaur and co-workers [22] synthesized a glutathione-capped CdS quantum dot with a sensing range of 0.54 nM for detection of mercury. However, the presence of copper and chromium ions obstructed mercury detection, restricting its usage in field applications [22]. Zhang and coworkers [23] used the cooperative effect of Ag_2S and ZnS quantum dots to detect Hg^{2+} electrochemically under visible light, a novel method that combines metal and semiconductor sulphides. Manna and co-workers [24] established a ratiometric sensor based upon a metal–methyl salicylaldimine complex capped on ZnS quantum dot doped with Mn^{+2} for identifying Hg^{2+} ions. The presence of the analyte in the sample is indicated by a change in luminescence color [24].

Chu and co-workers [25] designed a mercury detection system based on cerium-doped silicon sulphide quantum dots. The detection method's features include double emission, linear fluorescence dependency on mercury level, and a detection limit of 0.8 m/l [25].

Tanwar and co-workers [26] established white light emission from a combination of silicon QDs and gold nanoclusters, besides its application in mercury ion monitoring and mixture ratio management [26].

Non-metal-based quantum dots also find explicit application in heavy metal sensing. Li and co-workers [27] discussed the application of multimodal carbon dots, both blue and green carbon dots, in the detection of mercury ions with a limit of 50 nM [27]. Wang and co-workers (2019) used UV light to synthesize 1,2 dithioglycol functionalized carbon nitride quantum dots with a quantum yield of 27%, while Sahoo and co-workers [28] deployed spider silk to synthesize environmentally-friendly carbon quantum dots for recognising Hg(II) ions through the FRET mechanism [28].

Graphitic carbon nitride quantum dots (polymeric nanomaterials) for Hg(II) detection via the PL quenching process was described by Patir and Gogoi [29]. A similar technique was utilised to develop boron-nitrogen co-doped graphene quantum dots for sensing Hg^{2+} and graphene quantum dots [30]. The application of a nitrogen-doped carbon dot for sensing Hg^{2+} with a detection margin of 5.3 nM was investigated by Tadesse and co-workers [31]. Extensive research has been conducted on ratiometric fluorescence sensing of heavy metals such as Hg^{2+} and thymine-rich ssDNA for Hg^{2+} detection [32].

3.3.2 Detection of Lead (Pb)

Lead (Pb) is a heavy metal ion that poses a substantial effect on biological communities, different ecosystem, and human health, even in minute concentrations. Lead has been attributed to hypertension, developmental abnormalities, anaemia, and neurological and reproductive system malfunctions at blood concentrations more than 5 mM. As a result, the USEPA has set a Pb (II) limit of 15 mg/L in drinking water [33]. The detection of different ions, such as Pb (II) and Cu (II) ions, was examined using boron-doped CQDs (B-CQDs) with detection limits of 13.56 and 8.47 nM [34].

In another study, citric acid-based N-CQDs were coupled with Fe_2O_3 to preconcentrate and solid-phase extract little quantities of Pb (II) derived from a matrix of both vegetable and water samples. This method had a linear detection range of 0.062–62.1 mM and a good detection limit of 16.8 mM [35]. Pb (II) sensing in actual water samples was also achieved using fluorometric GCQDs made from *Ocimum sanctum* leaves (LOD-0.59 nM) and *Lantana Camara* berries (LOD-9.64 nM) [12, 36]. Sharma and Mehata [37] used MoS2 quantum dots to design a tool for lead ion detection based on quenching. This detection approach is delicate, with a detection limit of 50 M, due to dose-dependent fluorescence quenching. Mir and co-workers [38] employed fluorescence quenching to detect lead ions using ZnSe and ZnSe@ZnS core–shell quantum dots capped with thioglycolic acid [38]. Kaewprom and co-workers [39] used dithiocarbamate doped graphene quantum dots to construct a resonant light scattering sensor of complex metallic nanoparticles for the specific detection of lead ions in water samples. Carbon dots with flavonoid moieties are a very sensitive technique with a detection limit as low as 55 pM [19–21]. Another carbon dot-based method developed by Bhamore and colleagues [40] combines glutathione-capped carbon dots with an agarose gel for optical detection with the naked eye [40]. This system has a worthy linear connection with lead content and does not require a lot of heavy instrumentation for detection, making it appropriate for a field application.

3.3.3 Detection of Chromium (Cr)

The presence of high quantities of chromium in water supplies is mostly attributable to modern-day businesses (e.g., bronzing, mining, electroplating, and textile dyeing). Cr (VI) is a well-known metal that has been linked to carcinogenic and mutagenic health consequences, posing a hazard to ecological systems. According to the USEPA, total chromium in drinking water has an acceptable limit of 100 ng/ml, while Cr has a limit of 50 ng/ml. The inner filter effect and static quenching processes are also examined in relation to Cr sensing approaches employing CQDs. A detection limit of 24.6 mM for Cr (III) sensing was achieved using yellow fluorescent CQDs (y-CQDs) produced by acid carbonization of sucrose in the presence of phosphoric acid as a dehydrating agent [34].

Elmizadeh and co-workers [41] deployed a fluorescence quenching method based on synthetic ligand coated CdTe quantum dots to design a sensitive nanosensor for quick detection of Cr(III)ion [41]. Parani and Oluwafemi [42] investigated an aggregation-based quenching method for the selective detection of Cr (III) ions using AgInS2-ZnS quantum dots [42]. This sensing system used the collaborative effect of static and dynamic quenching to differentiate Cr(VI) from Cr(III) in mixed samples [7, 8]. Khan and co-workers [43] presented a ZnO quantum dot-based sensor for the selective detection of Cr^{6+}, with a detection limit of 0.18 nM [43]. Mondal and co-workers [44] make use of a white light-emitting Eu and Tb co-doped carbon fluorescent sensor for the detection of Cr(VI) ions [44]. Hu and co-workers [45] developed chlorine and nitrogen co-doped carbon nanodots for the speedy detection of Cr(VI) ions, and CQDs fluorescence was quenched using a combination of dynamic quenching and the inner filter effect [45]. Wang and co-workers [46] used a similar co-doping approach for selective detection of Cr^{3+} ions with a detection limit of 6 M by means of doping S/N in carbon dots [46]. A fluorescent probe for detection of Cr^{3+} with a limit of detection as low as 0.02 M was also designed deploying a carbon graphene allotrope.

3.3.4 Detection of Cadmium (Cd)

Cadmium (Cd) is a well-known non-essential human body component. Cd is used in sectors such as bronzing, storage batteries, plastic stabilisers, waste metal treatment, dyeing, colorants, nuclear reactor rods, and semiconductor manufacture. Cadmium uptake beyond the tolerance limit (5 mg/L) has been associated with deadly diseases such as itai-itai, prostate, kidney, and lung cancers, among others. An electrochemical sensor based on N-CQDs-graphene oxide hybrid (NCQDs-GO) was designed for numerous ion detection of lead and cadmium ions by deploying anodic stripping voltammetry [47, 48]. Because of greater electro-agile surface area and plenty of oxygen-rich functional surface moieties, the hybrid expedited the specific detection of Cd (II) and Pb (II) ions through electrostatic force, which further amplified the detection limit of the improved electrode (Cd (II): 834 mM; Pb (II): 1420 mM). CQDs also find application as a luminophore for Cd (II) detection in graphite furnace atomic absorption spectroscopy. Yin and co-workers [49] described a CdTe/CdS quantum dot inscribed with ammonium pyrrolidine dithiocarbamate, which leads to structural changes in reacting with Cd ions, causing dot quenching. The Cd ions are attached to the changed surface, restoring fluorescence and letting for effective Cd ion detection [49]. The presence of Cd ions in the material analyzed causes a variation in the ZnSe quantum dot, generating a ZnSe/CdS core–shell structure. This approach exhibited high selectivity for Cd ions, with a detection limit of 11 nM [50]. Pandey and co-workers [51] synthesized carbon dots using green technology which detect Cd (II) ions via ligand-mediated electron transfer in *Murraya koenigii* leaves. This approach may detect cadmium ions in the least value of 0.29 nM [51].

3.3.5 Detection of Arsenic (As)

Humans are prone to arsenic directly or indirectly by the absorption of contaminated water, as well as foods and crops that have been irrigated with arsenic-rich water. According to the World Health Organization, arsenic-related health problems impact up to 140 million people in around 50 nations. The inorganic form of As (III), also accepted as arsenite, is the utmost lethal of the various oxidation forms of arsenic because of its kinesis in water and is readily uptaken by cells. Skin damage (pigmentation, lesions, patches, and other symptoms), cardiovascular sickness, and skin, lung, and bladder cancer have all been associated with long-term ingestion of arsenic-contaminated water [52]. As a result of these observations, WHO has set an acceptable limit of 10 mg/L. Microwave-based pyrolysis of citric acid and sodium thiosulphate produces sulfur-doped CQDs that can be used in dual info systems, such as colorimetric and fluorescence, to detect inorganic As (III). The addition of glutathione to the SCQDs improved the sensor materials' selectivity for As (III). In fluorescence mode, these modified S-CQDs had detection limits of 32 and 48 pM, respectively, in water samples obtained from river water and tap water [53]. CQDs synthesized from decaying tomatoes and prickly pear cactus have recently been described for detecting As (III) in pond, tap, river, and industrial water samples. Pathan and colleagues [54] created a 'turn-on' sensor based on magnetic graphene oxide quantum dots for the selective detection of As^{3+} ions [54]. An analogous mechanism was studied by Wu and co-workers [55], in which CdTe/CdS core–shell quantum dots capped with cysteine make coordinating interaction with the arsenic ions, resulting in dots aggregating. This technique displayed high sensitivity and selectivity with a detection limit of 10 ng/L [55].

3.3.6 Detection of Copper (Cu)

Isarov and Chrysochoos [56] proposed a method for determining Cu^{2+} content and describing the mechanism of action using CdS QD. Cu^{2+} interacts rapidly with CdS QD surfaces, stimulating the core conduction band excitation state electrons, leading to recombinant valence band holes [56]. It would result in QD fluorescence quenching and a redshift emission peak. Xie [57] enhanced the QD modifier by coating the ZnS-coated CdSe core shell QD (CdSe/ZnS QD) surface with bovine serum albumin (BSA). So they detected Cu^{2+} and copper in Chinese herbal medicine using QD-BSA light probes [57]. Yan and co-workers [58] used cysteine and glutathione as modifiers to manufacture CdTe QDs and obtained quantitative Cu^{2+} detection with 0.15 g L^{-1} LOD. In this, Cu^{+2} binds to the surface of CdTe QDs via coordinated actions initiated by the surface's abundant carboxyl amino groups. Then it's reduced to Cu^{+}, which causes fluorescence to fade [58].

3.3.7 Detection of Silver (Ag)

CdS quantum dots were synthesized using L-cysteine-cysteine. QDs on interaction with AgC, produced florescence enhancement which could be ascribed to CdS/AgC SR complexes. Fresh L-cysteine was add on to improve imperishability. The sensitivity of this method has been improved, with a LOD of 5×10^{-9} mol/L. Xia [59] tailored water-soluble CdTe QD of four varying sizes (19, 26, 31 and 42 nm) followed by coated of mercaptopropionic acid and studied the interactions between CdTe QD of different sizes and AgC [59].

3.3.8 Miscellaneous Metal Ion Detection

Li and coworkers [60] developed CdTe QD coated with thiol and nano-rods for bivalent metal ions such as Co^{2+}, Ca^{2+}, Mg^{2+}, Zn^{2+}, Mn^{2+} and Ni^{2+}. In this study, fluorescence enhancement was caused by Zn^{2+}, whereas fluorescence quenching was caused by Mn^{2+}, Ni^{2+}, and Co^{2+} [60]. Na and Ming [61] created CdTe QD modified with thioglycolate and discovered that the mechanism could be linked to the coordination bond formed between hydroxyl and carboxyl groups and ions, besides other characteristic groups present on the surface of the QDs. This approach can be used for detection of vanadium in water samples in traces. Ali and co-workers [62] synthesized glutathione modified CdTe and CdSe/ZnSe QDs for the quantitative detection of fluorescence quenching Pb2 + with a LOD of 20 nmol L^{-1} [62].

3.3.9 Multiplexed Detection of Heavy Metals

Singh and colleagues [63] synthesized CdSe QDs which are capped with starch for collective detection of Cr along with Hg in aquatic environments. The capacity to determine the oxidation state of the analyte and the ease of synthesis are distinct features that make it a feasible field application alternative [63]. On the basis of quenching, Zhou and co-workers (2019) established a microfluidic system based on ZnSe quantum dots for the specific detection of lead and cadmium. Because of its eco-friendly nature, low cost, and high efficiency, this device is a viable alternative for field application. A comparable approach using SnO QDs for effective detection of Hg^{2+} and Pb^{2+} ions in water was discovered by Liu and co-workers [64]. Another quenching-based detecting technology is CdTe QDs which are capped with mercaptopropionic acid [65]. To detect Pb, Cr, and other metal ions, Baslak [66] employed hydroxylated CdTeS QDs coated with cetyl tri ammonium bromide. This technique had a higher sensitivity for detecting Cr (III) ions [66]. Radhakrishnan and colleagues

[67] synthesized a carbon dot–graphitic nitride-based nano-composite with fluorescence recovery in the presence of lead and chromium. This turn-on sensor depicted greater specificity and sensitivity, with detection limits of 0.2 nM, and 0.54 nM for Lead and Chromium, respectively [67]. Chini and co-workers [68] designed a FRET-based graphene quantum dot-carbon dot system for the detection of As^{5+}, Pb^{2+}, and Hg^{2+}, in which non-radiative energy transfer to carbon dots which is an acceptor causes donor-graphene dots (donor) to quench [68]. Yarur and co-workers [69] used carbon dots to construct a ratiometric detection system for the sensing of Pb^{2+} and Hg^{2+} ions, with detection limits of 37.1 and 39.5 nM, respectively [69]. Buledi and co-workers [70] studied a range of nanomaterial-based sensors for multiplexed detection of heavy metals like mercury and lead, using both metal and non-metal-based QDs [70] (Table 2).

4 Important Trends and Challenges in Present Scenario

This chapter presents big picture on the different quantum dots with special emphasis on carbon dots and their current development in the optical diagnosis of key environmental contaminants (heavy metal ions) divulges their intensified progress in topical past. Still, many tasks until now must be considered for their probable reliability and application as practically feasible sensing probes in daily life. Synthesis approaches for emerging highly stable and effective quantum dots, along with the emission from the total visible spectrum and fine bandwidth of fluorescence signal is essential for precise applications and improved sensitivity. Chemically tailored QDs have high quantum yield and FL signal intensity in comparison to green QDs. Particularly, research is in progress to find the probable inexpensively feasible methods for purification of green QDs. Numerous less explored workable precursors such as recycled waste, microbes, and residuals, are lined up for assessment of synthesis of QDs of natural doping with greater quantum yield. Above and beyond the mandatory mechanistic apprehension of green QDs synthesis, it is crucial to recognize the cause responsible for the precursor-predicated precision of the QDs for explicit heavy metal ions. On the contrary, a synchronized and upfront surface alteration might upsurge optical signal for improved pertinence. Further comprehensive studies are needed for development of a ratiometric and recyclable (by operationalizing QDs onto 2D/3D substrates) sensing probes exhibiting FL emission in the UV–visible-NIR range. We have faith in up-coming inquisition of optical detection-based systems deploying carbon dots will procure wide-ranging consideration in different fields such as agriculture, food, and textile pollutant sensing owing to its easiness, biocompatibility, profitable.

Table 2 CQD sensors for heavy metal detection in water

Metals	Carbon QDs	Source	Sensing tactics	Detection range	Detection limit	References
A. Mercury						
	CQDs	Lake water, tap water and fountain water	Fluorometric	0–3 µm	4.2 nM	Zhou et al. [18]
	CQDs	Tap and lake water	Fluorometric	1–10 µm	1 fM	Yan et al. [71]
	Ag-CQDs	Tap and lake water	Fluorometric	0.5–50 µm	85 nM	Liu et al. [72–74]
	CQDs	Aqueous solutions and real water	Fluorometric	6–79 nM	2.1 mM	Choudhary et al. [75]
B. Arsenic						
	CQDs	Aqueous solutions and real water samples	Fluorometric	–	0.2 µM	Saikia and Karak [76]
	S-CQD	Aqueous solutions	Colorimetric	–	32 pM	Gupta et al. [53]
	Thiol functionalized CQDs	Aqueous solutions and Real water samples	Fluorometric	0.1–200 mM	17.2 µM	Pooja et al. [77]
	GSH-CQDS	Aqueous solutions and real water samples	Fluorometric	2–12 nM	2.3 nM	Radhakrishnan and Panneerselvam [78]
C. Iron						
	N-CQDs	Living cells	Fluorometric	1–250 µM	0.52 µM	Song et al. [79]
	N,P-CQDs	Tap water	Fluorometric	0.05–200 µM	0.05 µM	Guo et al. [80]
	GCQDs	Aqueous solutions and real water samples	Fluorometric	0.02–40 mM	0.13 µM	Liu et al. [72–74]
D. Copper						
	B,N-CDs	Natural water samples	Fluorometric	1–25 µM	0.3 µM	Rong et al. [81]

(continued)

Table 2 (continued)

Metals	Carbon QDs	Source	Sensing tactics	Detection range	Detection limit	References
	Needle like calcium phosphate/carbon dot hybrid composites	Drinking water	Fluorometric	10–20 μM	9.82 μM	Guo et al. [82]
	CQDs	Aqueous solutions	Flourometric	0.33–66 μM	4.8 μM	Ma et al. [83]
E. Lead						
	CQDs	Aqueous solutions	Fluorometric	–	4.6 nM	Liu et al. [72–74]
	N-CQDs conjugated with Fe_3O_4	Water and vegetable samples	Solid phase extraction	0.062–62.1 mM	16.8 μM	Mashkani et al. [84]
	CQDs	Aqueous solutions and real water samples	Fluorometric	0.033–0.67 μM	12.7 μM	Liu et al. [85]
F. Cadmium						
	CQDs/AuNCsnanohybrid	Aqueous solutions and real water samples	Ratiometric Fluorometric	0–250 μM	32.5 nM	Niu et al. [86]
	NCQDs	Aqueous solutions and real water samples	Solid phase extraction	–	201 nM	Li et al. [19–21]
G. Zinc						
	Quinoline/CQDs	Aqueous solution	Fluorometric	–	6.4 nM	Zhang et al. [87]
	Quercetin/CDs	Aqueous solutions and living cells	FRET	2–100 μM	2 μM	Yang et al. [88]
	CdTe QDs/CQDs	Aqueous solutions and biological media	Ratiometric Fluorometric	0.50–40 μm	0.33 μM	Song et al. [79]
H. Chromium						
	CQDs	Environmental water samples	Fluorometric	5.1–612 mM	1.53 mM	Vaz et al. [89]

(continued)

Table 2 (continued)

Metals	Carbon QDs	Source	Sensing tactics	Detection range	Detection limit	References
	N, B-CQDs	Real water samples	Fluorometric	0–250 mM	0.5 mM	Guo et al. [90]
	N, P-CQDs	Lab prepared aqueous samples and biological samples	Fluorometric	4.2–31.7 mM	1.2 mM	Singh et al. [91]
I. Selenium						
	N-CQDs	Aqueous solutions and living cells	Fluorometric	0.2–30 mM	23 nM	Devi et al. [92]
	CQDs	Aqueous solutions and real water samples	Fluorometric	0.007–2 mM	0.78 mM	Wang et al. [93]

5 Probable Steps for Improving Sensitivity and Selectivity of Individual Metal Ions

Suitable unpretentious and handy functionalization procedures, without mislaying the biological, optical, and chemical properties of the nanoparticles, are needed to augment discrimination and sensitivity of QDs toward various heavy metal ions. Beside this, the application of maintainable genetically modified biomass with precise metabolic paths can assist in the designing of finger-printed response based on optical sensing. Like specificity, signal intensification is the unswerving process for enlightening the sensitive nature of optical sensors. Specifically easy-going and projecting ways to expand the signal strength is by passivating the QDs with various polymer-based operationalization procedures. Therefore, equilibrium amid the selectivity and sensitivity protocols can be attained. Furthermore, metal/semiconductor nanoparticles functioning as a carrier for CQDs active species might conceivably consequence in needed optical characteristics. These nanoparticles will be attaining an improved signal amplification and function as an supersensitive optical sensor for sensing analytes. There is a style for mini and simplification of detection by portable and mini devices which deliver quick and precise responses with probable importance in the point-of-care technologies. Hardware and software can be united with detection measures like colorimetric assessment and luminescent assays using portable devices adequately.

6 Concluding Remark

Escalating anthropogenic activities are responsible for releasing heavy metal pollutants in natural aquatic and soil systems, and which poses a substantial menace to human and environment. Taking into account the grave health and environmental issues of these lethal and non-biodegradable metal ions, there is immediate need of solutions like monitoring in aquatic system. Optical-chemical sensors of nano-dimensions have undeniably shown higher sensitivity and specificity for these contaminated metal ions. Yet, there is demand of environment friendly and workable sensor material in this area. Quantum dots are low-priced, ecologically friendly, biocompatible, and exhibited improved properties, and hence are considered as superior nanomaterials for sensing of heavy metals. This chapter has been systematized to brief the attempts made in the development of QD sensing systems with prime focus on the present status, challenges, and future prospects. This chapter will benefit the researchers, technologists, and engineers to scale up both biological and chemical synthesis methods along with establishment of outstanding visual color-based sensing techniques. Simplicity in designing and augmentation of specificity towards target metals needs upgradation in their performance, selectivity, and sensitivity. To conclude, challenges concerned with QD-based disposable sensors and user-friendly readout systems needs extensive investigation to inflate their use in real-world sensing applications. The field applicability of the sensors will offer feasibility in onsite metal detection and the work hassle free. It will also impart incessant monitoring of metals ions conceivable on the locations selected for observation.

Acknowledgements This research was supported by the University Grants Commission.

Declaration of Competing Interest The authors have declared no conflict of interest

References

1. Ali H, Khan E, Ilahi I (2019) Environmental chemistry and ecotoxicology of hazardous heavy metals: environmental persistence, toxicity, and bioaccumulation. J Chem
2. Jan AT, Azam M, Siddiqui K, Ali A, Choi I, Haq QMR (2015) Heavy metals and human health: mechanistic insight into toxicity and counter defense system of antioxidants. Int J Mol Sci 16(12):29592–29630
3. Wolfbeis OS (2005) Materials for fluorescence-based optical chemical sensors. J Mater Chem 15(27–28):2657–2669
4. Ruscito A, Smith M, Goudreau DN, DeRosa MC (2016) Current status and future prospects for aptamer-based mycotoxin detection. J AOAC Int 99(4):865–877
5. Ono A, Togashi H (2004) Highly selective oligonucleotide-based sensor for mercury (II) in aqueous solutions. Angew Chem 116(33):4400–4402
6. Ono A, Cao S, Togashi H, Tashiro M, Fujimoto T, Machinami T, Oda S, Miyake Y, Okamoto I, Tanaka Y (2008) Specific interactions between silver (I) ions and cytosine–cytosine pairs in DNA duplexes. Chem Commun 39:4825–4827
7. Hu J, Wang D, Dai L, Shen G, Qiu J (2020) Application of fluorescent biosensors in the detection of Hg (II) based on T-Hg (II)-T base pairs. Microchem J 159:105562
8. Hu Y, Zhang J, Li G, Xing H, Wu M (2020) Highly sensitive fluorescent determination of chromium (VI) by the encapsulation of cadmium telluride quantum dots (CdTe QDs) into Zeolitic Imidazolate Framework-8 (ZIF-8). Anal Lett 53(10):1639–1653
9. Tan JL, Zhang MX, Zhang F, Yang TT, Liu Y, Li ZB, Zuo H (2015) A novel "off–on" colorimetric and fluorescent rhodamine-based pH chemosensor for extreme acidity. Spectrochim Acta Part A Mol Biomol Spectrosc 140:489–494
10. Matea CT, Mocan T, Tabaran F, Pop T, Mosteanu O, Puia C, Iancu C, Mocan L (2017) Quantum dots in imaging, drug delivery and sensor applications. Int J Nanomed 12:5421
11. Wu P, Zhao T, Wang S, Hou X (2014) Semicondutor quantum dots-based metal ion probes. Nanoscale 6(1):43–64
12. Kumar A, Chowdhuri AR, Laha D, Mahto TK, Karmakar P, Sahu SK (2017) Green synthesis of carbon dots from *Ocimum sanctum* for effective fluorescent sensing of Pb2+ ions and live cell imaging. Sens Actuators, B Chem 242:679–686
13. Mehta VN, Desai ML, Basu H, Singhal RK, Kailasa SK (2021) Recent developments on fluorescent hybrid nanomaterials for metal ions sensing and bioimaging applications: a review. J Mol Liq 333:115950
14. Karimi H, Rajabi HR, Kavoshi L (2020) Application of decorated magnetic nano photocatalysts for efficient photodegradation of organic dye: a comparison study on photocatalytic activity of magnetic zinc sulfide and graphene quantum dots. J Photochem Photobiol, A 397:112534
15. Tomczak N, Liu R, Vancso JG (2013) Polymer-coated quantum dots. Nanoscale 5(24):12018–12032
16. Rajamanickam K (2019) Multimodal molecular imaging strategies using functionalized nano probes. J Nanotechnol Res 1:119–135

17. Hu Q, Han X, Dong G, Yan W, Wang X, Bigambo FM, Fang K, Xia Y, Chen T, Wang X (2021) Association between mercury exposure and thyroid hormones levels: a meta-analysis. Environ Res 196:110928
18. Zhou L, Lin Y, Huang Z, Ren J, Qu X (2012) Carbon nanodots as fluorescence probes for rapid, sensitive, and label-free detection of Hg^{2+} and biothiols in complex matrices. Chem Commun 48(8):1147–1149
19. Li LS, Jiao XY, Zhang Y, Cheng C, Huang K, Xu L (2018a) Green synthesis of fluorescent carbon dots from Hongcaitai for selective detection of hypochlorite and mercuric ions and cell imaging. Sens Actuators, B Chem 263:426–435
20. Li L, Liu D, Shi A, You T (2018b) Simultaneous stripping determination of cadmium and lead ions based on the N-doped carbon quantum dots-graphene oxide hybrid. Sens Actuators, B Chem 255:1762–1770
21. Li YK, Yang T, Chen ML, Wang JH (2018c) Supported carbon dots serve as high-performance adsorbent for the retention of trace cadmium. Talanta 180:18–24
22. Kaur J, Kumar V, Tikoo KB, Bansal S, Kaushik A, Singhal S (2020) Glutathione modified fluorescent CdS QDs synthesized using environmentally Benign pathway for detection of mercury ions in aqueous phase. J Fluoresc 30(4):773–785
23. Zhang L, Li P, Feng L, Chen X, Jiang J, Zhang S, Zhang C, Zhang A, Chen G, Wang H (2020) Synergetic Ag2S and ZnS quantum dots as the sensitizer and recognition probe: a visible light-driven photoelectrochemical sensor for the 'signal-on' analysis of mercury (II). J Hazard Mater 387(2019):121715
24. Manna M, Roy S, Bhandari S, Chattopadhyay A (2020) A dual-emitting quantum dot complex nanoprobe for ratiometric and visual detection of Hg 2+ and Cu 2+ ions. J Mater Chem C 8(21):6972–6976
25. Chu H, Yao D, Chen J, Yu M, Su L (2020) Double-emission ratiometric fluorescent sensors composed of rare-earth-doped ZnS quantum dots for Hg2+ detection. ACS Omega 5(16):9558–9565
26. Tanwar S, Sharma B, Kaur V, Sen T (2019) White light emission from a mixture of silicon quantum dots and gold nanoclusters and its utilities in sensing of mercury (ii) ions and thiol containing amino acid. RSC Adv 9(28):15997–16006
27. Li DY, Wang SP, Azad F, Su SC (2020) Single-step synthesis of polychromatic carbon quantum dots for macroscopic detection of Hg^{2+}. Ecotoxicol Environ Saf 190:110141
28. Sahoo NK, Das S, Jana GC, Aktara MN, Patra A, Maji A, Beg M, Jha PK, Hossain M (2019) Eco-friendly synthesis of a highly fluorescent carbon dots from spider silk and its application towards Hg (II) ions detection in real sample and living cells. Microchem J 144:479–488
29. Patir K, Gogoi SK (2018) Facile synthesis of photoluminescent graphitic carbon nitride quantum dots for Hg2+ detection and room temperature phosphorescence. ACS Sustain Chem Eng 6(2):1732–1743
30. Liu Z, Mo Z, Niu X, Yang X, Jiang Y, Zhao P, Liu N, Guo R (2020) Highly sensitive fluorescence sensor for mercury (II) based on boron-and nitrogen-co-doped graphene quantum dots. J Colloid Interface Sci 566:357–368
31. Tadesse A, Hagos M, RamaDevi D, Basavaiah K, Belachew N (2020) Fluorescent-nitrogen-doped carbon quantum dots derived from citrus lemon juice: green synthesis, mercury (II) ion sensing, and live cell imaging. ACS Omega 5(8):3889–3898
32. Aderinto SO (2020) Fluorescent, colourimetric, and ratiometric probes based on diverse fluorophore motifs for mercuric (II) ion (Hg2+) sensing: highlights from 2011 to 2019. Chem Pap 74(10):3195–3232
33. Patrick L (2006) Lead toxicity, a review of the literature. Part I: exposure, evaluation, and treatment. Altern Med Rev 11(1)
34. Devi P, Rajput P, Thakur A, Kim KH, Kumar P (2019) Recent advances in carbon quantum dot-based sensing of heavy metals in water. TrAC, Trends Anal Chem 114:171–195
35. Mehdinia A, Mashkani M, Jabbari A, Niroumand R, Ghenaatian HR, Fereidouni N, Nabid MR (2020) Extraction of trace amounts of cadmium in fish and mollusk by Fe_3O_4@ N-carbon quantum dots as adsorbent. J Food Measure Charact 14(2):725–734

36. Bandi R, Dadigala R, Gangapuram BR, Guttena V (2018) Green synthesis of highly fluorescent nitrogen–doped carbon dots from Lantana camara berries for effective detection of lead (II) and bioimaging. J Photochem Photobiol, B 178:330–338
37. Sharma P, Mehata MS (2020) Rapid sensing of lead metal ions in an aqueous medium by MoS2 quantum dots fluorescence turn-off. Mater Res Bull 131:110978
38. Mir IA, Kumar S, Bhat MA, Yuelin X, Wani AA, Zhu L (2021) Core@ shell quantum dots as a fluorescent probe for the detection of cholesterol and heavy metal ions in aqueous media. Colloids Surf, A 626:127090
39. Kaewprom C, Sricharoen P, Limchoowong N, Nuengmatcha P, Chanthai S (2019) Resonance light scattering sensor of the metal complex nanoparticles using diethyl dithiocarbamate doped graphene quantum dots for highly Pb (II)-sensitive detection in water sample. Spectrochim Acta Part A Mol Biomol Spectrosc 207:79–87
40. Bhamore JR, Park TJ, Kailasa SK (2020) Glutathione-capped Syzygium cumini carbon dot-amalgamated agarose hydrogel film for naked-eye detection of heavy metal ions. J Anal Sci Technol 11(1):1–9
41. Elmizadeh H, Soleimani M, Faridbod F, Bardajee GR (2018) A sensitive nano-sensor based on synthetic ligand-coated CdTe quantum dots for rapid detection of Cr (III) ions in water and wastewater samples. Colloid Polym Sci 296(9):1581–1590
42. Parani S, Oluwafemi OS (2020) Selective and sensitive fluorescent nanoprobe based on AgInS2-ZnS quantum dots for the rapid detection of Cr (III) ions in the midst of interfering ions. Nanotechnology 31(39):395501
43. Khan MMR, Mitra T, Sahoo D (2020) Metal oxide QD based ultrasensitive microsphere fluorescent sensor for copper, chromium and iron ions in water. RSC Adv 10(16):9512–9524
44. Mondal TK, Mondal S, Ghorai UK, Saha SK (2019) White light emitting lanthanide based carbon quantum dots as toxic Cr (VI) and pH sensor. J Colloid Interface Sci 553:177–185
45. Hu Q, Li T, Gao L, Gong X, Rao S, Fang W, Gu R, Yang Z (2018) Ultrafast and energy-saving synthesis of nitrogen and chlorine co-doped carbon nanodots via neutralization heat for selective detection of Cr (VI) in aqueous phase. Sensors 18(10):3416
46. Wang Z, Yang Y, Zou T, Xing X, Zhao R, Wang Y (2020) Novel method for the qualitative identification of chromium ions (III) using L-aspartic acid stabilized CdS quantum dots. J Phys Chem Solids 136:109160
47. Rasheed T, Bilal M, Nabeel F, Adeel M, Iqbal HM (2019a) Environmentally-related contaminants of high concern: potential sources and analytical modalities for detection, quantification, and treatment. Environ Int 122:52–66
48. Rasheed T, Nabeel F, Adeel M, Bilal M, Iqbal HM (2019b) "Turn-on" fluorescent sensor-based probing of toxic Hg (II) and Cu (II) with potential intracellular monitoring. Biocatal Agric Biotechnol 17:696–701
49. Yin Y, Yang Q, Liu G (2020) Ammonium pyrrolidine dithiocarbamate-modified CdTe/CdS quantum dots as a turn-on fluorescent sensor for detection of trace cadmium ions. Sensors 20(1):312
50. Chen L, Lu Y, Qin M, Liu F, Huang L, Wang J, Xu H, Li N, Huang G, Luo Z, Zheng B (2020) Preparation of "ion-imprinting" difunctional magnetic fluorescent nanohybrid and its application to detect cadmium ions. Sensors 20(4):995
51. Pandey SC, Kumar A, Sahu SK (2020) Single step green synthesis of carbon dots from murraya koenigii leaves; a unique turn-off fluorescent contrivance for selective sensing of Cd (II) ion. J Photochem Photobiol, A 400:112620
52. Chang Q, Chen B, Thakur C, Lu Y, Chen F (2014) Arsenic-induced sub-lethal stress reprograms human bronchial epithelial cells to CD61 cancer stem cells. Oncotarget 5(5):1290
53. Gupta A, Verma NC, Khan S, Nandi CK (2016) Carbon dots for naked eye colorimetric ultrasensitive arsenic and glutathione detection. Biosens Bioelectron 81:465–472
54. Pathan S, Jalal M, Prasad S, Bose S (2019) Aggregation-induced enhanced photoluminescence in magnetic graphene oxide quantum dots as a fluorescence probe for As (III) sensing. J Mater Chem A 7(14):8510–8520

55. Wu Y, Liu Y, Liu H, Liu B, Chen W, Xu L, Liu J (2020) Ion-mediated self-assembly of Cys-capped quantum dots for fluorescence detection of As (iii) in water. Anal Methods 12(34):4229–4234
56. Isarov AV, Chrysochoos J (1997) Optical and photochemical properties of nonstoichiometric cadmium sulfide nanoparticles: surface modification with copper (II) ions. Langmuir 13(12):3142–3149
57. Xie HY (2004) Preparation and applications of biomedical functional nanomaterials based on quantum dots. PhD diss, Wuhan University
58. Yan YX, Mou Y, Jin QH (2007) A study about new tape of CdTe quantum dots for determination of copper ions. Life Sci Instrum 5:14–18
59. Xia YS (2007) Surface modification, size selection and fluorescent sensor of CdTe quantum dots. Master thesis, Anhui Normal University
60. Li J, Bao D, Hong X, Li D, Li J, Bai Y, Li T (2005) Luminescent CdTe quantum dots and nanorods as metal ion probes. Colloids Surf, A 257:267–271
61. Na J, Ming H (2010) CdTe quantum dot fluorescence probe for the determination of trace vanadium. Anal Chem 397:3589–3593
62. Ali EM, Zheng YG, Yu HH, Ying JY (2007) Ultrasensitive Pb^{2+} detection by glutathione-capped quantum dots. Anal Chem 79:9452–9458
63. Singh A, Guleria A, Neogy S, Rath MC (2020) UV induced synthesis of starch capped CdSe quantum dots: functionalization with thiourea and application in sensing heavy metals ions in aqueous solution. Arab J Chem 13(1):3149–3158
64. Liu J, Zhang Q Xue W, Zhang H, Bai Y, Wu L … Jin G (2019) Fluorescence characteristics of aqueous synthesized tin oxide quantum dots for the detection of heavy metal ions in contaminated water. Nanomaterials 9(9):1294
65. Elfeky SA (2018) Facile sensor for heavy metals based on thiol-capped CdTe quantum dot. J Environ Anal Chem 5(01):1–5
66. Baslak C (2019) Development of fluorescence-based optical sensors for detection of Cr (III) ions in water by using quantum nanocrystals. Res Chem Intermed 45(7):3633–3640
67. Radhakrishnan K, Sivanesan S, Panneerselvam P (2020) Turn-On fluorescence sensor based detection of heavy metal ion using carbon dots@ graphitic-carbon nitride nanocomposite probe. J Photochem Photobiol, A 389:112204
68. Chini MK, Kumar V, Javed A, Satapathi S (2019) Graphene quantum dots and carbon nano dots for the FRET based detection of heavy metal ions. Nano-Struct Nano-Objects 19:100347
69. Yarur F, Macairan JR, Naccache R (2019) Ratiometric detection of heavy metal ions using fluorescent carbon dots. Environ Sci Nano 6(4):1121–1130
70. Buledi JA, Amin S, Haider SI, Bhanger MI, Solangi AR (2021) A review on detection of heavy metals from aqueous media using nanomaterial-based sensors. Environ Sci Pollut Res 28(42):58994–59002
71. Yan F, Zou Y, Wang M, Mu X, Yang N, Chen L (2014) Highly photoluminescent carbon dots-based fluorescent chemosensors for sensitive and selective detection of mercury ions and application of imaging in living cells. Sens Actuators, B Chem 192:488–495
72. Liu T, Dong JX, Liu SG, Li N, Lin SM, Fan YZ, Lei JL, Luo HQ, Li NB (2017a) Carbon quantum dots prepared with polyethyleneimine as both reducing agent and stabilizer for synthesis of Ag/CQDs composite for Hg2+ ions detection. J Hazard Mater 322:430–436
73. Liu W, Diao H, Chang H, Wang H, Li T, Wei W (2017b) Green synthesis of carbon dots from rose-heart radish and application for Fe3+ detection and cell imaging. Sens Actuators, B Chem 241:190–198
74. Liu Y, Zhou Q, Yuan Y, Wu Y (2017c) Hydrothermal synthesis of fluorescent carbon dots from sodium citrate and polyacrylamide and their highly selective detection of lead and pyrophosphate. Carbon 115:550–560
75. Choudhary R, Madhuri R, Sharma PK (2017) Detection of Hg2+ ion using fluorescent carbon dots derived from elephant foot yum via green-chemistry. In: AIP conference proceedings, vol 1832, no 1. AIP Publishing LLC, p 050010

76. Saikia A, Karak N (2018) Polyaniline nanofiber/carbon dot nanohybrid as an efficient fluorimetric sensor for As (III) in water and effective antioxidant. Mater Today Commun 14:82–89
77. Pooja D, Saini S, Thakur A, Kumar B, Tyagi S, Nayak MK (2017) A "Turn-On" thiol functionalized fluorescent carbon quantum dot based chemosensory system for arsenite detection. J Hazard Mater 328:117–126
78. Radhakrishnan K, Panneerselvam P (2018) Green synthesis of surface-passivated carbon dots from the prickly pear cactus as a fluorescent probe for the dual detection of arsenic (iii) and hypochlorite ions from drinking water. RSC Adv 8(53):30455–30467
79. Song Q, Ma Y, Wang X, Tang T, Song Y, Ma Y, Xu G, Wei F, Cen Y, Hu Q (2018) "On-off-on" fluorescent system for detection of Zn2+ in biological samples using quantum dots-carbon dots ratiometric nanosensor. J Colloid Interface Sci 516:522–528
80. Guo Y, Cao F, Li Y (2018) Solid phase synthesis of nitrogen and phosphor co-doped carbon quantum dots for sensing Fe3+ and the enhanced photocatalytic degradation of dyes. Sens Actuators, B Chem 255:1105–1111
81. Rong MC, Zhang KX, Wang YR, Chen X (2017) The synthesis of B, N-carbon dots by a combustion method and the application of fluorescence detection for Cu2+. Chin Chem Lett 28(5):1119–1124
82. Guo Y, Yang L, Li W, Wang X, Shang Y, Li B (2016) Carbon dots doped with nitrogen and sulfur and loaded with copper (II) as a "turn-on" fluorescent probe for cystein, glutathione and homocysteine. Microchim Acta 183(4):1409–1416
83. Ma X, Dong Y, Sun H, Chen N (2017) Highly fluorescent carbon dots from peanut shells as potential probes for copper ion: the optimization and analysis of the synthetic process. Mater Today Chem 5:1–10
84. Mashkani M, Mehdinia A, Jabbari A, Bide Y, Nabid MR (2018) Preconcentration and extraction of lead ions in vegetable and water samples by N-doped carbon quantum dot conjugated with Fe_3O_4 as a green and facial adsorbent. Food Chem 239:1019–1026
85. Liu Y, Zhou Q, Li J, Lei M, Yan X (2016) Selective and sensitive chemosensor for lead ions using fluorescent carbon dots prepared from chocolate by one-step hydrothermal method. Sens Actuators, B Chem 237:597–604
86. Niu WJ, Shan D, Zhu RH, Deng SY, Cosnier S, Zhang XJ (2016) Dumbbell-shaped carbon quantum dots/AuNCs nanohybrid as an efficient ratiometric fluorescent probe for sensing cadmium (II) ions and l-ascorbic acid. Carbon 96:1034–1042
87. Zhang Z, Shi Y, Pan Y, Cheng X, Zhang L, Chen J, Li MJ, Yi C (2014) Quinoline derivative-functionalized carbon dots as a fluorescent nanosensor for sensing and intracellular imaging of Zn^{2+}. J Mater Chem B 2(31):5020–5027
88. Yang M, Kong W, Li H, Liu J, Huang H, Liu Y, Kang Z (2015) Fluorescent carbon dots for sensitive determination and intracellular imaging of zinc (II) ion. Microchim Acta 182(15):2443–2450
89. Vaz R, Bettini J, Júnior JGF, Lima EDS, Botero WG, Santos JCC, Schiavon MA (2017) High luminescent carbon dots as an eco-friendly fluorescence sensor for Cr (VI) determination in water and soil samples. J Photochem Photobiol, A 346:502–511
90. Guo Y, Chen Y, Cao F, Wang L, Wang Z, Leng Y (2017) Hydrothermal synthesis of nitrogen and boron doped carbon quantum dots with yellow-green emission for sensing Cr (VI), anti-counterfeiting and cell imaging. RSC Adv 7(76):48386–48393
91. Singh VK, Singh V, Yadav PK, Chandra S, Bano D, Kumar V, Koch B, Talat, Hasan SH (2018) Bright-blue-emission nitrogen and phosphorus-doped carbon quantum dots as a promising nanoprobe for detection of Cr (VI) and ascorbic acid in pure aqueous solution and in living cells. New J Chem 42(15):12990–12997
92. Devi P, Kaur G, Thakur A, Kaur N, Grewal A, Kumar P (2017) Waste derivitized blue luminescent carbon quantum dots for selenite sensing in water. Talanta 170:49–55
93. Wang Q, Zhang S, Zhong Y, Yang XF, Li Z, Li H (2017) Preparation of yellow-green-emissive carbon dots and their application in constructing a fluorescent turn-on nanoprobe for imaging of selenol in living cells. Anal Chem 89(3):1734–1741

Synthesis of Magnetic Ferrite and TiO_2-Based Nanomaterials for Photocatalytic Water Splitting Applications

Wegdan Ramadan, Yamen AlSalka, Osama Al-Madanat, and Detlef W. Bahnemann

Abstract Clean non fossil sources of energy have an increasing urgency to support industrial and population growth to achieve this goal, the continuous development of nanostructures and nanomaterials for different applications such as photocatalytic water splitting is under intense investigation. Two of the most important materials namely, titanium dioxide, TiO_2, and ferrites having the MFe_2O_4 structure where M is transition metal are introduced in this chapter. Ferrites and titanium dioxide are two interesting nanostructures having great potential. Ferrites have many members in the family hence, offering diversity in structural and physical properties which in turn give chance for a large variety of purposes and applications. They own an energy band gap that is small enough to crop photons from the visible light region. They are also abundant on earth and have important physical properties like magnetism and multiferroicity in addition to being biocompatible which will increase their usability. On the other hand, TiO_2 has several advantages, including its stability in terms of chemical and thermal properties, in addition to its availability, photoactivity, and relatively elevated charge transfer ability. Furthermore, the nontoxicity, high oxidative strength, and cheap price are additional advantages. Despite the large band gap and its related UV-light activation, TiO_2 is one of the highly studied photocatalysts. Moreover, it has been extensively investigated in many aspects, including the kind

---

W. Ramadan (✉)
Faculty of Science, Physics Department, Alexandria University, Alexandria 21511, Egypt
e-mail: wegdan.ramadan@alexu.edu.eg

Y. AlSalka
Institut für Nanophotonik Göttingen e.V., Hans-Adolf-Krebs-Weg 1, 37077 Göttingen, Germany

O. Al-Madanat
Chemistry Department, Mutah University, Mutah, Al-Karak 61710, Jordan

D. W. Bahnemann
Institut Für Technische Chemie, Leibniz Universität Hannover, Callin Str. 3, 30167 Hannover, Germany

Laboratory "Photoactive Nanocomposite Materials", Saint-Petersburg State University, Ulyanovskaya Str. 1, 198504 Saint-Petersburg, Russia

I. Uddin and I. Ahmad (eds.), *Synthesis and Applications of Nanomaterials and Nanocomposites*, Composites Science and Technology,
https://doi.org/10.1007/978-981-99-1350-3_11

of the oxidative species (·OH radicals vs. h+), the location of the photoinduced reactions (at the surface or in the bulk), and the ways that enhance the photocatalytic performance.. Many of the synthesis techniques for both, ferrites and TiO_2 were adopted to serve definite purposes like control over phase purity, morphology, size, and dispersion which are discussed in this chapter.

Keywords Ferrites · TiO_2 · Water splitting · Photocatalysis · Green hydrogen

1 Introduction

Fossil fuels were extensively exploited for decades to drive the development of our modern society without considering the potential damages to the sustainable environmental systems. Global energy consumption is projected to rise by 28% from 2015 to 2040 [48] with the simultaneous depletion of fossil fuels. Thus, the big global challenge is to fulfill the rising energy demand while minimizing environmental pollution. The quickest developing energy sources are anticipated to be renewables, having an annual increase in consumption of 2.3% from 2015 to 2040 [48]. The sun, which falls on the earth's surface with solar energy power of 1000 W m^{-2}, is indeed the largest and cleanest renewable energy resource [79, 96]. Solar energy is considered a sustainable energy supply when its transformation into chemical energy carriers is efficiently fulfilled.

Molecular Hydrogen owns the highest energy content, i.e., 120 MJ kg^{-1}, among other fuels [4, 11]. Therefore, H_2 can be considered a sustainable fuel when it is cleanly resulted from renewable or abundant origin at reasonable prices [126, 153]. Among other methods, solar hydrogen production seems to offer the best sustainable route. There are several processes to produce H_2 through water splitting reaction, including (1) thermochemical process using solar concentrators to split water via heat from concentrated sunlight [155], (2) photobiological process, in which aerobic or anaerobic Bacteria are used [42], (3) water electrolysis process, in which an external bias generated from renewable energy is applied for the electrically driven splitting of water [59], and (4) photocatalytic process, in which a suitable photocatalyst is suspended in an aqueous solution to convert the light energy into a chemical one [100].

The photocatalytic route for water splitting offers a sustainable conversion of energy from renewables in comparison with conventional processes based on electrolyzers [62]. This method has attracted attention due to the clean transformation of energies, i.e., light into chemical energy, using a photocatalytic material. Moreover, the ambient conditions are applied to carry out its two half-reactions (Eqs. 1, 2), in which protons are reduced to H_2 and water molecule is oxidized to O_2 through a 4-electron reaction, respectively [7, 14, 62]. The splitting of a water molecule is coupled with a change in ΔG^0 equal to 237.2 kJ/mol, i.e., $\Delta E^\circ = 1.23$ V in compliance with the Nernst equation [165]. Thus, a photocatalytic system must absorb photon energy (hv) of at least 1.23 eV to thermodynamically split the water.

$$2H^+ + 2e^- \rightarrow H_2(\uparrow) \qquad E^0 = 0 \text{ V vs NHE} \tag{1}$$

$$H_2O \rightarrow {}^1\!/_2 O_2(\uparrow) + 2H^+ + 2e^- \qquad E^0 = 1.23 \text{ V vs NHE} \tag{2}$$

The overall water-splitting reaction has a $\Delta G > 0$, therefore, it absorbs energy. However, a suitable photocatalyst can indeed drive such a reaction with a positive ΔG value [128]. In contrast to a normal catalytic system, redox reactions are spatially separated using a photocatalyst, in which both ΔG_e and ΔG_h should be negative [129] as presented in Fig. 1. Accordingly, the thermodynamic demand water-splitting by photoctalysis is more cathodic energy level at the bottom of the CB compared with the standard electrode potential of (H^+/H_2) and more anodic energy level at the top of the valance band compared with the standard electrode potential of (O_2/H_2O), respectively [128, 135]. Therefore, only a few photocatalysts possess the thermodynamic driving force for the complete light induced splitting of water into H_2 and O_2.

According to the previous thermodynamic discussion, photocatalysis is a process that includes photon absorption by material to induce a reaction that has enormously slow kinetics outside this system [62]. The so-called photocatalyst absorbs certain light energy during illumination and generates e^-/h^+ as charge carriers, which are the driving forces for the next redox reactions on the surface. The thus light-absorbing material is called photocatalyst, which, in heterogeneous systems, is usually semiconducting material [62]. Consequently, heterogeneous photocatalysis essentially relies on the unique properties of the semiconducting material for an efficient harvesting of incident light and successively initiate surface reactions.

Figure 1 illustrates a typical schematic of the significant steps involved during photocatalytic water-splitting. First (1), photogeneration of electron–hole pairs via suitable irradiation of a photocatalyst to excite electrons from the valence band (VB) to the conduction band (CB). Second (2), Separation and migration of charge carriers to reach the surface. Finally (3), the electrons on the photocatalyst or the co-catalyst participate in a reduction reaction and generate hydrogen, while the holes are involved

Fig. 1 Scheme of Gibbs-energy difference during photocatalytic reactions, reprinted with permission from Elseiver [128]

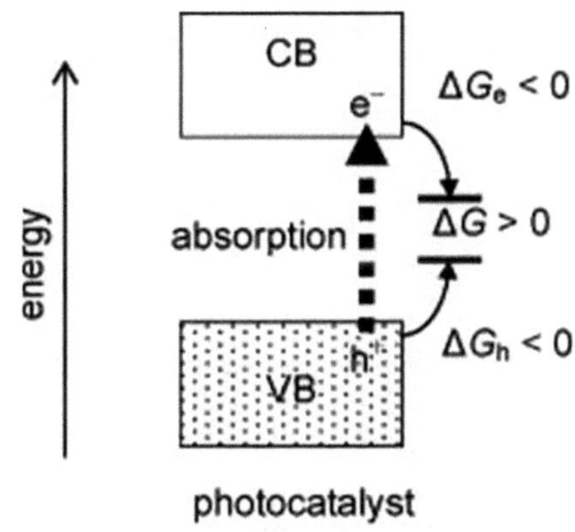

in an oxidation reaction to form oxygen or to oxidize an organic substrate. The criterion for an efficient photocatalyst is the solar–hydrogen conversion efficiency (STH) shown in Eqs. (3) [38]:

$$\eta \text{STH} = \eta \text{A} \times \eta \text{CS} \times \eta \text{CT} \times \eta \text{CR} \quad (3)$$

The STH conversion efficiency depends on the capacity of light absorption (ηA), charge separation (ηCS), charge transportation (ηCT), and reaction efficiency (ηCR) (Fig. 2).

As an alternative, the so-called photocatalytic reforming is a process that brings together two photocatalytic processes, namely the split of water and the oxidation of organics [7]. An organic substrate can be oxidized at the photocatalyst surface instead of O_2 production, while protons can be reduced to H_2 by photogenerated electrons [12]. The thus organic substrate is frequently named sacrificial reagent due to its irreversible reaction with the photogenerated holes, prohibiting, consequently, the undesired e^-/h^+ recombination [15]. Because organics are commonly greater reducing agents than water, they can be oxidized at a less positive potential. Therefore, photocatalytic reforming can more easily meet the band edge potentials of a semiconductor than the overall photocatalytic water-splitting, offering more variety of suitable materials. In such systems, O_2 is not simultaneously produced, which inhibits the back reaction forming water from O_2 and H_2, and avoids a gas separation stage [55]. A widespread range of organic substrates including alcohols and organic acids have been utilized as sacrificial reagents for H_2 evolution through photocatalysis [69].

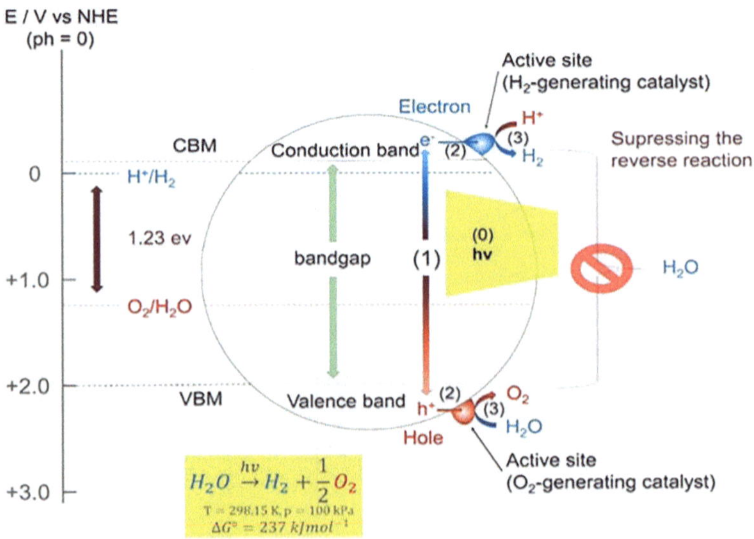

Fig. 2 The main steps of photocatalytic water-splitting on a semiconductor. Reproduced with permission Elsiever [120]

In this chapter, we concentrate primarily on the synthesis of two of the most important photocatalytic materials namely, TiO_2 and the ferrites family. They offer a lot of advantages to the field of photocatalysis in terms of their availability, stability, and response to light whether it is UV or visible. In particular, we focus on their application, performances, and the observed drawbacks in solar water-splitting. The progress witnessed in the field of materials science has been quite spectacular and this has enabled and fueled deep investigations on discovering the usage of a wide collection of the novel materials or enhancing and/or modifying the performance of some existing ones like TiO_2 and ferrites. In this chapter, different composites/heterostructures based on these photocatalysts, which are employed for water splitting, are presented.

2 Semiconductor Photocatalysis

In accordance with the molecular orbitals theory, the electronic orbitals of a semiconductor are merged and split in two bands, which are separated by an energy bandgap (E_g) as shown in Fig. 3. The valance band (VB) that is created from the highest occupied molecular orbitals (HOMO) is completely filled with electrons at 0 K. In opposition, no electrons are located in the conduction band (CB), which is built from the lowest unoccupied molecular orbitals (LUMO). Hence, such an electronic band structure offers light-induced properties for semiconductors. These involve exciting the VB electrons to partially filled CB states upon absorption of appropriate light energy [11, 89]. The family of semiconductors includes various binary compounds, but not all are used for heterogeneous photocatalysis. Several criteria should be met for these binary compounds for such applications, including the right band gap energy for effective light absorption (1–4 eV), charge carrier mobility, non-toxic properties, and the positions of their band edges, which should be suitable for the particular application [112]. Most compounds and composites that work as photocatalysts are based on d-transition-metal metal oxides such as titanium dioxide (TiO_2), zinc oxide (ZnO), tungsten oxide (WO_3), and iron oxide (Fe_2O_3) [13, 67, 135]. An electron vacancy called a positively charged hole is produced in the VB after the excitation of an electron to the CB due to the illumination of the semiconductor [9, 160]. Both electrons and holes in the CB and VB, respectively, are considered mobile entities, but due to the higher diffusion coefficients of electrons, they create an electric field that pulls them back to recombine with the slower holes [134]. However, the trapping of the electrons drastically diminishes their mobilities. The charge carriers created by the light move to reach the surface due to the near-surface electric field, then the e^-/h^+ species react with adsorbed molecules. On the surface, the holes act as oxidants (+1.0 to +3.5 V vs. NHE), while electrons are consumed in reduction reactions (+0.5 to −1.5 V vs. NHE) [61]. As illustrated in Fig. 4, one of the main limitations is the recombination of charge carriers in an indirect way via surface defects (5), or through a direct band-to-band recombination (6) [28, 113]. The crystal structure of the semiconductor is the key factor affecting this limitation. Consequently, the movement of the charge carriers to the liquid junction is essential, where the photocatalytic redox reactions with solution species on the surface take place [51]. Therefore, careful synthesis of the photocatalyst should be conducted to overcome such undesired critical issues.

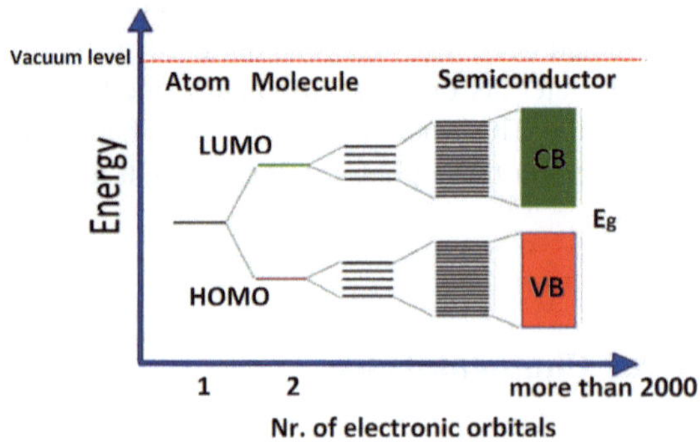

Fig. 3 Band structure of a semiconducting material based on atomic orbitals theory, adapted from the author dissertation [11]

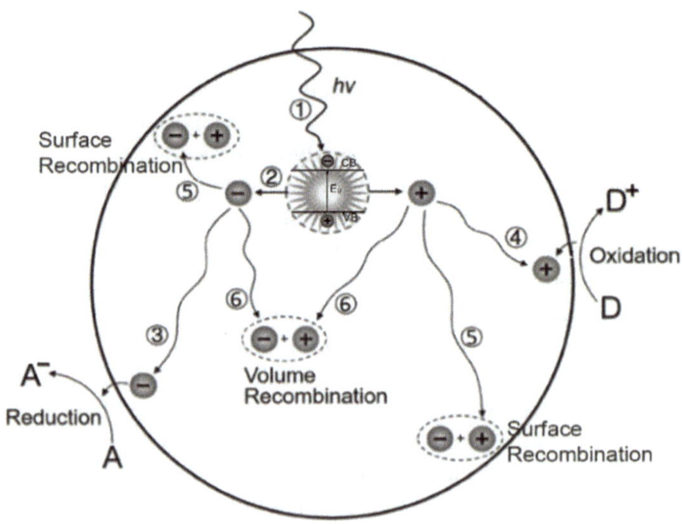

Fig. 4 Semiconducting-based photocatalyst: the possible paths of the photogenerated charge carriers, adopted with permission from ACS [102]

3 Titanium Dioxide Photocatalyst

Titanium dioxide (TiO_2) is a widely studied material for photocatalytic purposes since its reasonable reactivity, hydrophilic properties, high resistance to photo corrosion, and optimal electronic and optical capacity [15, 56, 62]. TiO_2 is considered as an intrinsic n-type semiconductor, in which one Ti^{4+} and six O^{2-} are coordinated to produce a TiO_6 octahedron [113]. TiO_2 belongs to the nonstoichiometric transition metal oxides due to the presence of oxygen vacancies as defects at ambient condition. These vacancies produce an excess of electrons, increasing the electrical conductivity [51]. Thus, the surface of n-type TiO_2 contains a lot of oxygen vacancies (Ov) defects as unpaired electrons in the CB [39, 147]. These donor-like defects lead to a downward band bending due to the creation of an accumulation layer on the surface [11, 46]. It is crucial to study the charge carrier dynamics in TiO_2 because the fast recombination of the photogenerated e^-/h^+ is a decisive factor that determines the competence of using TiO_2 in photocatalytic applications.

TiO_2 has three polymorphs: (1) brookite with an orthorhombic crystal structure and E_g of 3.3 eV, (2) anatase with a tetragonal crystal structure and E_g of 3.2 eV, and (3) rutile with a tetragonal crystal structure and E_g of 3.0 eV [28]. Therefore, the photocatalytic properties of all TiO_2 polymorphs are originated from the photogeneration of charge carriers after the excitation with UV irradiation. Table 1 presents the main properties of TiO_2 polymorphs. Various TiO_2 sources are available, but Degussa P25 is known as a standard photocatalytic material due to its well-defined structure containing anatase and rutile in a 3:1 ratio [56]. The thermodynamic stability of TiO_2 is highly related to the crystalline phase with its particle size. The highest thermodynamic stability can be achieved when the particle size is less than 11 nm for anatase, while it should be greater than 35 nm for rutile and lie between 11 and 35 nm for brookite [50]. TiO_2 in the anatase phase is considered commonly more active than rutile TiO_2 [7, 147]. Compared with rutile, anatase nanoparticles have relatively a higher surface area and more concentration of oxygen vacancies (Ovs), enhancing charge separation efficiency [11]. Anatase TiO_2 offers additionally a slightly higher redox capability than rutile due to its larger bandgap energy.

The synthesis of TiO_2 nanostructures with different morphologies has recently attracted great attention [121]. In this regard, TiO_2 can be fabricated as spheres [20, 171, 186], fibers [36, 181, 185], tubes [83, 105, 114], nanorods [40, 127, 181], sheets [19, 82, 151], and interconnected architectures [1, 63, 163] (see Fig. 5). Many factors largely affect the photocatalytic performance of TiO_2, e.g., particle size, pore volume, surface area, and surface facets. Therefore, regulating these factors is relevant for improving the photocatalytic performance of TiO_2. Structural dimension is also a crucial factor that gives TiO_2 different properties. While a sphere structure with zero dimension has a high surface area and high adsorption properties [103], one-dimensional fibers or tubes offer lower recombination rates due to the short distance for charge carrier diffusion [83]. On the other hand, two-dimensional nanosheets display smooth surfaces and high adhesion properties [82], while three-dimensional monoliths are characterized by their high carrier mobility due to the interconnecting

Table 1.1 Different TiO_2 polymorphs and their physicochemical properties. Data are adopted from Refs. [28, 113]

Property	Anatase TiO_2	Rutile TiO_2	Brookite TiO_2
Crystal structure	Tetragonal	Tetragonal	Orthorhombic
	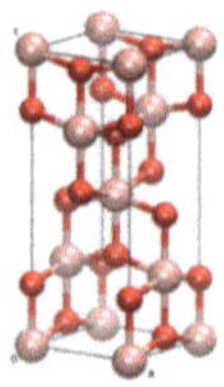	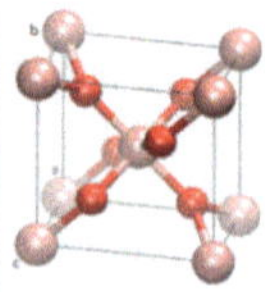	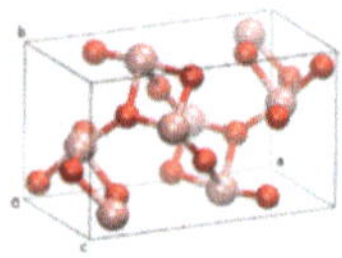
Molecules/cell	4	2	8
Crystal size (nm)	<11	>35	11–35
Lattice parameters (nm)	a = b = 0.3785, c = 0.9514	a = b = 0.4594, c = 0.2959	a = c = 0.5436, b = 0.9166
Space group	$I4_1$/amd	$P4_2$/mnm	Pbca
Volume/molecule ($A^{\circ 3}$)	34.061	31.2160	32.172
Density (g cm^{-3})	3.79	4.13	3.99
Bandgap (eV)	3.26	3.05	3.3
dielectric constant $\varepsilon(0)$	6.04	6.62	7.89

structure. Therefore, synthesizing TiO_2 with adequate dimensionality is very important as it may be reflected in better physicochemical and photocatalytic properties. The size can affect the efficiency of the photocatalytic system. Smaller nanoparticles are generally more suitable for photocatalytic water-splitting. In fact, the smaller size diminishes the distance of the charge carriers' migration to the surface leading to boosting the photocatalytic activity [81]. On the other hand, larger particles can be generally used for photoelectrochemical water-splitting due to the lower electron–hole recombination rate [137]. However, the recombination rate is related to the particle size, wherein in larger particles, the travel distance to the surface is longer, rising the likelihood of electron–hole recombination [50, 91].

TiO_2 is photoexcited and generates charge carriers within a few femtoseconds. Thereafter, these charge carriers either recombine with each other (in the bulk and/or at the surface) or move to the surface of TiO_2 for trapping or for initiating interfacial redox reactions [140]. Therefore, the photocatalytic efficiency of TiO_2 is related to the competition between charge carrier recombination, charge carrier trapping, and interfacial charge transfer [69]. It has been proven, that the quantum yield of photooxidation using TiO_2 in aqueous media is lower than 10% because more than 90% of the photogenerated charge carriers tend to recombine within 10 ns after UV irradiation [149]. However, the size of TiO_2 nanoparticles plays a crucial role, where 10–30 nm particle size is optimal for photocatalysis [24, 29] because of the

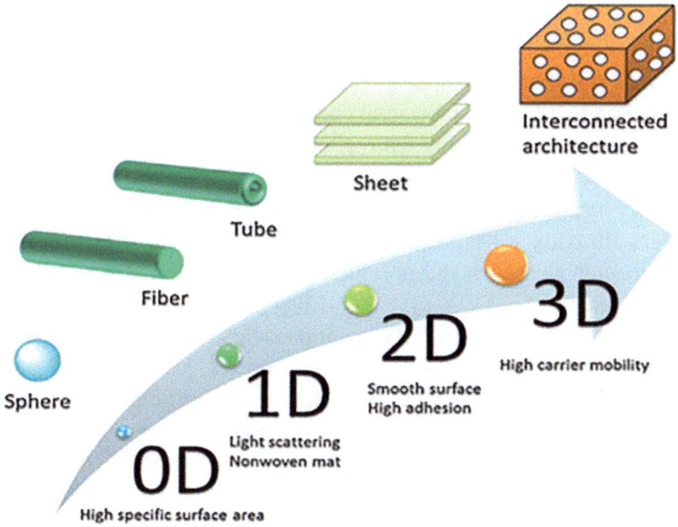

Fig. 5 Scheme of the material's structural dimensionality and their properties with permission from Elsevier [121]

faster migration and trapping of the charge carriers in the small particles. In contrast, the photogenerated e^-/h^+ can be trapped as trapped holes, trapped electrons in the bulk or at the surface or they can remain as free electrons distributed in the bulk [179]. Finally, e^-/h^+ pairs can induce a water-splitting reaction through an interfacial charge carrier transfer. Nevertheless, the incompatibility between very short lifetimes of e^-/h^+ pairs and slower kinetics of redox reactions is the main drawback in TiO_2 photocatalysis [54].

4 Synthesis of TiO_2 Photocatalytic Nanoparticles

Many synthesis procedures with diverse parameters have been successfully applied for the synthesis of TiO_2 photocatalysts. These include sol−gel, surfactant-based micelle, hydrothermal, solvothermal, chemical or physical vapor deposition, sonochemical, and microwave methods. A wide range of titanium precursors has been also applied, including titanium chloride, titanium sulfate, and titanium alkoxides. Nevertheless, the choice of the synthesis procedure in addition to the precursor type and the parameters applied has a pronounced impact on the structural, electronic, and optical properties of the synthesized TiO_2 nanoparticles [14, 176].

Nanostructured TiO_2 spheres are typically synthesized using titanium alkoxide precursors such as titanium tetra-isopropoxide or titanium tetra-butoxide. While adding an acid would accelerate the synthesis procedure, the existence of a polymer offers a porous assembly [103]. Further treatment using the hydrothermal method has been shown to produce porous or hierarchical structures. TiO_2 photocatalysts that have a one-dimensional structure in form of fibers and tubes have demonstrated a higher surface-to-volume ratio, allowing a decrease in the hole–electron recombination rate and an increase in interfacial charge carrier transfer rate [10]. One of the various methods used for making TiO_2 fibers is the electrospinning process using a high-voltage electric field (Ramakrishna). TiO_2 fibers can be tailored in terms of diameters and morphologies by regulating the synthesis parameters, including the ratio (titanium alkoxide: polymer), the concentration of the titanium precursors, the type and molecular weight of the polymer, the solvent and electric field used, and the deposition distance [121].

Mesoporous and nanoporous TiO_2 nanoparticles have been synthesized either with or without employing organic surfactant templates [15, 33]. Such porous materials have shown a larger surface area and continuous particle frameworks, offering more active sites for the adsorption of molecules and facilitating the electron transfer within the material [103]. Evaporation-Induced Self-Assembly (EISA) is one effective method of mesoporous TiO_2 synthesis. In this method, multiple titanium precursors can be used in addition to a co-polymer surfactant molecule in an alcoholic medium to form colloidal particles before the aggregation as a three-dimensional open network. The pores are then formed upon the removal of surfactant by a calcination step [11, 33].

During water-splitting, the photogenerated charge carriers can rapidly recombine unless adequately controlled. Extensive efforts have been applied including the tuning of crystal structure, defects states, dopants, and electronic structure. However, the one-dimensional photocatalytic systems still experience from a relatively low photocatalytic activity. Alternatively, 2D photocatalytic materials can be a promising strategy because the transit distance of the photogenerated e^-/h^+ is relatively shorter. Two-dimensional TiO_2 nanosheets are nanosized flake-shaped materials, having a flat surface, a high aspect ratio, and a very small thickness of 1–10 nm. Such a shape is typically prepared through an alkaline hydrothermal process starting either from TiO_2 powder [152] or from protonic titanate hydrates [151].

4.1 *Synthesis of Defective TiO_2 Photocatalytic Nanoparticles*

Black TiO_2 have lately been suggested as a potential photocatalyst for solar-driven photocatalytic hydrogen production. Lots of efforts have been made to synthesize extremely reduced TiO_2, however, even the moderate reduction degree of TiO_2, i.e., gray TiO_2, can generate peculiar defective catalytic sites, which offer the capability to use such defective materials for cocatalyst-free hydrogen production. Nevertheless, further investigations should be carried out to understand the electronic structure of defective TiO_2 and the associated mechanisms and properties, including better light absorption and enhanced charge separation. Reduced TiO_2 can be synthesized by introducing intrinsic oxygen vacancies (VOs) and generating lattice Ti^{3+} centers. Several synthesis procedures are proposed that are typically proceeded by reducing

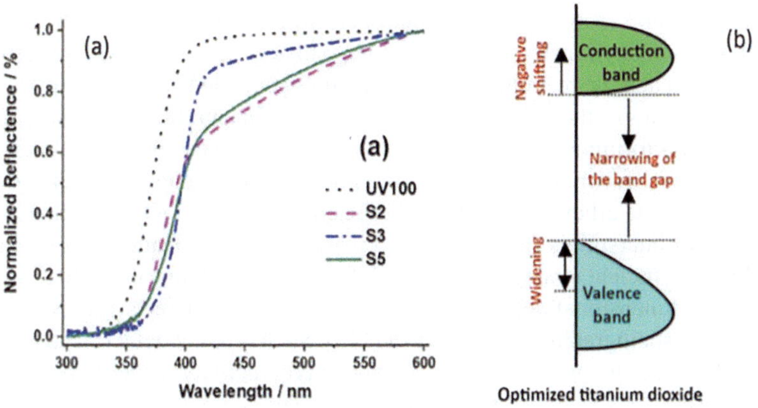

Fig. 6 UV–Visible light reflectance spectra of the prepared TiO_2 (**a**), schematic illustration of the band alignment (**b**). Adapted with permission from Elsevier [15]

TiO_2 at a high temperature using various reducing atmospheres (e.g., under vacuum, Argon, H_2/Argon mixture, and pure H_2) [31, 123]. Using such synthesis methods, gray, blue, brown, or black Ti_2O can be produced. It was found that the color obtained is related to the concentration of Ti^{3+} and VOs in the material, i.e., a strong reduction condition leads to a higher density of these defects, hence, a darker color of TiO_2 powders. The first black TiO_2 nanomaterial for photocatalysis was reported with high visible light absorption in 2011 upon thermal treatment of anatase TiO_2 at 200 °C under high pressure of H_2 (see Fig. 6a). Such an increase in the absorption of light is related to introducing electronic states and the tailing of VB and CB due to the creation of lattice disorder and H-doping (Fig. 6b) [34].

Different types of defects can be introduced that regulate many physical properties of crystalline TiO_2. Therefore, defects engineering is essential in controlling electronic, optical, and quantum properties of TiO_2, in addition to their role in the activation of TiO_2 photocatalytic processes [97]. In this regard, the synthesis method and its parameters play an essential role. As an example, the use of $TiCl_3$ as a precursor in the Evaporation-Induced Self-Assembly (EISA) method produced a defective TiO_2 for co-catalyst-free photocatalytic H_2 development [15]. $TiCl_3$ in addition to F-108 as a surfactant producing pure anatase TiO_2 that is characterized by its response to the visible region as shown in Fig. 6a. The unique photocatalytic activity of bare TiO_2 in the reforming of ethanol to generate hydrogen gas was explained by an increased carrier density alongside the extension of the edge of VB and the shift in the CB edge potential (see Fig. 6b).

The crystal defects can be generally classified regarding their dimensionalities [123, 132]. Zero-dimensional (0D) defects (points defects) are generated from a single or a few atomic positions (Fig. 7a). Such point defects can be found as Ti vacancy Ti^V (rare), Ti interstitial Ti^i (common), Oxygen vacancy VO (common in reduced TiO_2), and interstitial (e.g., hydrogen or nonmetal dopants) or substitutional impurity (e.g., metal or nonmetal dopants). Upon the introduction of crystal point defects, structural rearrangements would be initiated, leading to significant distortions in the local symmetry of Ti octahedra. This effect can, in his turn, affect the transportation and the recombination of the photogenerated charge carriers. Doping of TiO_2 by metals or the introduction of high VOs concentration contributes to additional 3d states forming below the CB [122]. However, excess electronic states below the CB can act as recombination centers for photogenerated charge carriers, lowering the photocatalytic efficiency. On the other hand, Doping of TiO_2 by nonmetal (e.g., N, C, S) and interstitials (e.g., H or Ti) creates electronic features above the VB, providing an efficient strategy to modify TiO_2 [32, 122] (Wang et al. 2017).

Another dimensional-defect type is the linear or 1D defects (dislocations), in which the crystallographic registry is lost [123]. Such types of defects lead to a lattice strain, which can be useful for TiO_2 photocatalytic activity. Otherwise, 2D crystal defects can be classified into three groups: (1) free surfaces with uncoordinated atoms, (2) interphases within a crystal such as stacking faults and antiphase boundaries, and (3) grain boundaries and two-phase boundaries (interphases). Such 2D defects are important in TiO_2 photocatalysis because they tune its surface reactivity and participate in charge transport and separation in the bulk as in the case of anatase/rutile nano-junctions [133].

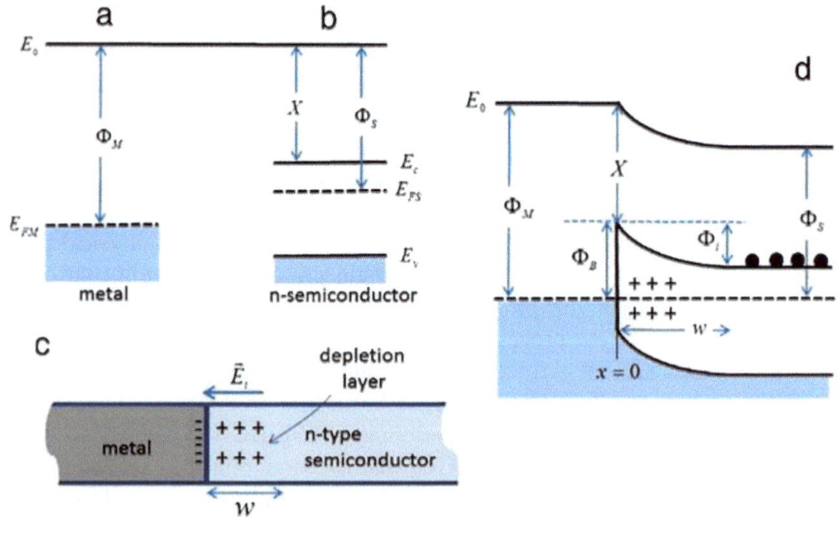

Fig. 7 A schematic represents the Schottky barrier (Sb) formation between a metal co-catalyst and n-type semiconductor. **a** Φ_M: Metal-work function and E_{FM}: Fermi level energy. **b** Φ_S: Semiconductor-work function, X: electron affinity, and E_{FS}: Fermi energy. **c** Charge distribution at the metal/semiconductor (M/S) junction. **d** M/S junction equilibrium-band-diagram. Φ_i is the energy barrier to the flow of electrons from the S to M, while Φ_B is the Sb. W is the extension of the depletion-layer. Reprinted with permission from Elsevier [45]

5 Modification of TiO_2

It is well known that the bandgap of the photocatalyst is the essential feature for its application in the reaction of water-splitting. Light harvesting governs electron and hole generation rates, recombination, and surface reaction kinetics. Regarding the TiO_2, the anatase, and rutile, the generally accepted bandgap values are 3.2 and 3.0, respectively [46]. While anatase is excited at 388 nm by UV light, the rutile extends slightly into the visible region at 415 nm, which can be related to the various positions of the CB and VB of both phases. The CB and VB of anatase are −0.5 V and 2.7 V, respectively, while for rutile they are −0.7 V and 2.33 V [2, 35, 146]. As mentioned in the previous section, however, TiO_2 satisfies the criterion for water-splitting, it suffers from (i) the rapid recombination between the electron–hole, where only ~10% of the photogenerated charge carriers successfully transfer to the interface to participate in the redox reactions which results in decreasing the reaction efficiency [147]. (ii) its large overpotential for H_2 production in water in the presence and absence of the hole scavenger. Therefore, to extend the charge carrier lifetimes, lower the overpotential of H_2 generation, and broaden the light absorption of TiO_2 towards visible light, several strategies and modifications have been employed for the TiO_2 to achieve hydrogen production from the photooxidation of water during the photocatalytic process, such as deposit metal nanoparticles like Pt, Cu, Ag, and Au on the surface of TiO_2, doping with cation or anion or sensitizing with dyes or small bandgap semiconductors.

5.1 Metal Co-Catalyst

The deposited metal co-catalyst or combination of two, such as Pd, Pt, Au, Cu, and Ag, on the surface of the semiconductor function as electrons sink that significantly reduce the recombination between the photogenerated e^-/h^+ and facilitate the reduction of water by lowering the activation energy (Yang et al. 2013). The H_2 production efficiency of different metal co-catalysts is affected by the work function of each metal. In fact, the work function of each metal plays the main role in the formation of the Schottky barrier [45]. Metal nanoparticles such as Pt (5.65 eV) and Au (5.10 eV) have higher Fermi level energy [111] compared to that TiO_2 (4.2 eV) [172]. Therefore, the photogenerated electrons move from the semiconductor to the metal nanoparticles through the interface until a thermodynamic equilibrium takes place (Fig. 7a–d) [147]. During the illumination, this equilibrium will be unsettled, thus, the generated electrons move continuously from the conduction band of the TiO_2 to the metal co-catalyst [75, 147]. The formed Schottky barrier ΦB blocks the flow of collected electrons from the metal to the TiO_2 photocatalyst which provides an efficient separation for the charge carriers. Therefore, the trapped electrons have a longer lifetime to stimulate the reduction reactions [174, 175]. In this sense, it is worth mentioning that not all the photogenerated electrons trapping by the metal

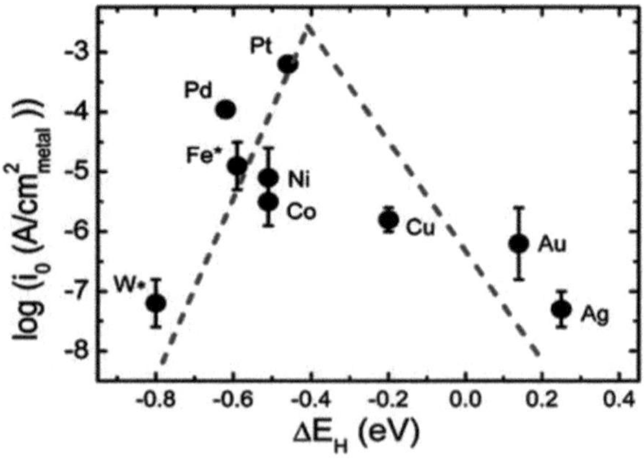

Fig. 8 Exchange current densities, log(i_0)vs the calculated hydrogen binding energy. Reprinted with permission from Ref. [150]. Copyright 2013 from the Royal Society of Chemistry

co-catalyst. By the mean of the EPR techniques, Al-Madanat et al. showed that the irradiated Pt/TiO_2 exhibited a simultaneous signal for the Ti^{3+} centers, confirming that not all the photogenerated electrons transferred to the Pt NPs [8].

As shown in Fig. 8, Pt appears as the most active noble metal for H_2 formation [150]. Despite this general conclusion, the efficiency of H_2 production is strongly dependent on its nature, preparation method, and concentration [8, 37]. Recently, it has been shown that the preparation method used for the metal co-catalyst deposition has a strong effect on the H_2 formation during the photocatalytic reforming of naphthalene and methanol [8].

There are different common techniques used for deposition the metal co-catalyst nanoparticles on the TiO_2 nanoparticles surface including a chemical reduction [124, 167], impregnation [53], deposition–precipitation [21], photodeposition [5, 6, 12, 14, 21, 53], electrode-position [80], and physical mixing [21]. The strong interaction between the metal co-catalyst and the photocatalyst is considered the key role to ensure the electron transfer from the photocatalyst to metal efficiently, otherwise increasing the electron/hole recombination rate and decreasing the H_2 formation rate. In fact, several factors are highlighted in the literature which are associated with such methods, thus, affecting their effectiveness, such as, the weak interaction between the co-catalyst and semiconductor, the large size of the metal co-catalyst nanoparticles, and the nucleation of the isolated metallic nanoparticles in the electrolyte [115, 124, 182].

Although some researchers reported the efficiency of the bare TiO_2 (anatase) toward the H_2 production, others reported the contrary. The Photocatalytic H_2 production over co-catalyst-free TiO_2 synthesized by a simple Evaporation-Induced Self Assembly (EISA) method was achieved during the photocatalytic reforming of ethanol and oxalic acid [5, 14]. While, after loading the TiO_2 with Pt and/or Au as a co-catalyst, the authors found that the activity of the H_2 production was markedly boosted. Pt/TiO_2 exhibits greater photonic efficiencies compared to Au/TiO_2, whereas Au–Pt/TiO_2 did not show any additional advantage [12]. On contrary, other research groups proposed that bimetallic co-catalyst such as Pt–Au and Pt–Ag shows a better activity towards H_2 production reaction with respect to the monometallic co-catalyst [78, 156]. On the contrary, when the bare Sachtleben Hombikat UV100 (TiO_2, anatase) and bare Aeroxide P25 (anatase/rutile mixture) were used as a photocatalyst during the photocatalytic reforming of methanol and naphthalene, the H_2 formation was not observed [5, 8]. However, loading these photocatalysts with Pt nanoparticles lead to photocatalyzed the proton to molecular H_2. The difference between both cases can be related to differences in the titanium-precursor type and the preparation method of the TiO_2 which leads to the formation high number of trapped states, that are advantageous for the photocatalytic properties and facilitated the H_2 production. Modification of the TiO_2 photocatalyst with copper (Cu), silver (Ag), and gold (Au) metal co-catalyst Nanoparticles is another important aspect from a practical point of view, because these metals extend the TiO_2 absorption to longer wavelengths (visible light) due to surface plasmon resonance (SPR). Several reports have shown that the irradiation of the Au/TiO_2 photocatalyst with visible light ($\lambda \geq 550$ nm) leads to the transfer of the electron from the Au nanoparticles to the semiconductor due to SPR [57, 157, 158]. Tanaka et al., discovered that plasmonic Au nanoparticles on TiO_2 act as a visible light driven photocatalyst for overall water-splitting in the absence of any additives [161]. Modification of the Au/TiO_2 with various types of metal co-catalyst nanoparticles boosted the H_2 and O_2 formation. Of these metals nanoparticles, Ni-modified Au/TiO_2 revealed 5-times higher rates of overall water splitting. This study showed that the authors designed and fabrication an effective novel solar energy conversion system consisting of three elemental technologies, including, (i) light harvesting with charge carrier separation and an active site for O_2 evolution (plasmonic Au particles). (ii) Electrons transfer from metallic Au nanoparticles to the active site to enable molecular hydrogen production (TiO_2), and (iii) Ni co-catalyst as an active site for hydrogen production (Fig. 9). In the other hand, the unfavorite reverse reaction between the molecular hydrogen and molecular oxygen into water could be suppressed by using NiOx.

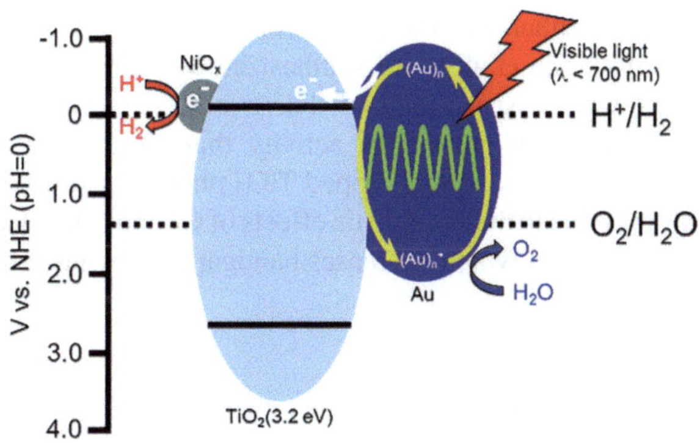

Fig. 9 Photocatalytic water-splitting over an Au/TiO_2–NiOx. Reprinted from Ref. [161] with permission from the Royal Society of Chemistry

5.2 Cation or Anion Doping

In the late 1980s, other approaches have been adopted to develop second-generation TiO_2 that extend the TiO_2 visible light response, thereby enhancing the overall efficiency, by the replacement of cations or anions in the lattice. The cations (metal ions) replacement creates impurity energy levels within the bandgap of the TiO_2, thus, facilitating the absorption in the visible range. Although a cation-doped TiO_2 can bridge both the UV and the visible radiation, it is mostly acted either as a centers for recombination of the photogenerated e^-/h^+ or ineffective in driving the surface redox reactions [148]. Moreover, the levels of impurity created by dopants in the $TiO_/$ photocatalysts are usually discrete, which would seem disadvantageous for the transfer of the photogenerated holes [94].

In the early 1990s, the first study of anion-doped TiO_2 was reported. However, it's reported in science published in 2001 by Asahi and co-workers that documented the preparation of the visible-light-active N-doped TiO_2 in heterogeneous photocatalysis. Afterward, several studies reported other visible-light-active materials such as C-doped TiO_2, and S-doped TiO_2 [148]. Asahi et al. [119] reported that the doping with Nitrogen increased the photocatalytic activity of the TiO_2 in the visible light region, which shifts the absorption edge of TiO_2-xN_x to lower energies through narrowing of the TiO_2 bandgap. Carbon- and sulfur-doped TiO_2 showed similar redshifts of the absorption edges and increased photoactivity [41, 180]. Different scenarios of the modifications that might occur to the TiO_2 bandgap (Fig. 10a) after doping with non-metals were presented in Fig. 10b–e.

By oxidizing titanium nitride under ambient conditions, a different kinds of N-doped TiO_2 nanostructured photocatalysts were prepared at different temperatures. N-TiO_2 calcined at 450 °C manifested the highest H_2 evolution activity in comparison to the other synthesized and commercial materials under UV–Vis illumination. The highest H_2 activity of the N-doped photocatalyst was attributed to the bandgap energy shifting toward the visible region [184].

Besides, loading the metal ion and the metal oxide on the surface of N-doped TiO_2 leads to enhancing the photocatalytic activity toward the production of hydrogen [49, 99]. Lin et al. [101] synthesized M–N/TiO_2 (M: Cr, Ni, Cu, Nb) by using the microwave-assisted method. These materials exhibit a higher photocatalytic molecular hydrogen production activity than the N dopedTiO_2 and the pristine TiO_2. Among all the prepared doped TiO_2 photocatalysts, Cu–N/TiO_2 showed the highest activity due to the synergistic effects of Cu and N that leading to formation an impurity energy levels which decreases bandgap energy and enhances the charge separation.

5.3 Dye Sensitizing

Another effective strategy to enhance the photoactivity of the wide bandgap semiconductor by harvesting the visible light is dye sensitizing. in this technique, the

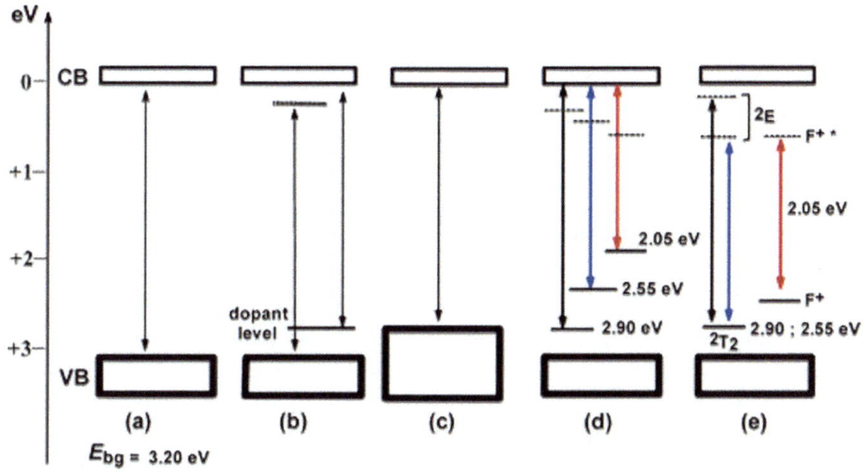

Fig. 10 Possible modification to anatase TiO_2 photocatalyst electronic structure after doping with different nonmetals: **a** pristine TiO_2; **b** localized dopant levels near the valence and conduction band of TiO_2; **c** expanding the valence band for TiO_2 narrowed the bandgap; **d** localized dopant levels and electronic transitions to the conduction band of the TiO_2; and **e** electronic transitions from localized levels near the valence band to their corresponding excited states for Ti^{3+} and F+ centers. With permission from ACS [148]

dye sensitizer acts as an energy antenna and does not change the properties of the semiconductor. The physical principle of the dye's sensitization process is based on the excitation of dye as a result to the absorption of light with suitable energy. This process is resulted to inject the electrons from the excited state of the dye into the CB of the semiconductor. In order for this process to take place, the excited state of the dye should possess a higher energy than the photocatalyst CB. In this case, the VB of the semiconductor remains unaffected in this process, whilst the CB acts as a mediator for transferring the excited electrons from the sensitizer to the electron acceptors on the surface of the photocatalyst [91]. Despite that various attempts have been done to enhance the photocatalytic activity of water-splitting, the production of molecular hydrogen and molecular oxygen in a single dye-sensitized system is still far away from the lowest accepted level (5%) of solar-to-H_2 conversion efficiency, which is represent 30% in terms of a quantum yield at 600 nm as suggested by Kudo [91]. The one-step excitation of $Ru(bpy)_3{}^{2+}$·C_{12}-sensitized $Pt/TiO_2/RuO_2$ exhibits a quantum yield of ~5% under $\lambda > 400$ nm light irradiation [183].

The photocatalytic efficiency of molecular hydrogen production from splitting the water can be improved in the dye-sensitized semiconductor suspension systems by loading the semiconductor with a metal co-catalyst and using a sacrificial electron donor in this system [35, 162]. It was reported that thiophenothiazine@Pt/TiO_2 showed a reliable performance for molecular hydrogen production from water-splitting at neutral conditions. The (E)-3-(10-hexyl-8-(methylthio)-10H-phenothiazin-3-yl)-2-cyanoacrylic acid sensitized Pt/TiO_2 photocatalyst exhibited up to 1048 μmol H_2 evolution with an apparent quantum yield around 50% [162].

6 Introduction to Ferrites

Ferrites are the most earth abundant minerals. They are promising candidate as photocatalysts due to their small band gaps and interesting magnetic properties. Most ferrites have a typical formulation of AB_2O_4, as shown in Fig. 11a. Oxygen ions are positioned in cubic closed packing, tetrahedral sites, A, and octahedral sites, B, are taken by metal cations. Such structure offers various combinations of metal cations that could show different properties. A is a transition metal and could be any of the transition metals like, Mg, Co, Mn, Ni, Cu or Zn. There is deficiency in abundance of some metals like Bi and Te making it difficult to meet large scale energy demand. On the other hand, many of the important indispensable elements used to in ferrites synthesis are abundant in earth's crust [168, 178]. Thus, earth abundance is an important factor to consider which will be reflected on the price of the metals, as presented in Fig. 11b. However, some rare earth elements frequently utilized in catalysis, like Platinum (Pt), Iridium (Ir), Ruthenium (Ru), and Rhodium (Rh), are less abundant and pose concern when large scale hydrogen generation and /or photocatalysis are intended.

7 Synthesis of Ferrites

Bulk ferrites are conventionally made by solid-state reaction. Such reaction consists of a series of thermal treatments with intermediary regrinding of stoichiometric admixtures of pulverized oxide precursors. Despite being cheap and could be easily scaled up, it has its own drawbacks, like the high synthesis temperatures that result in poly dispersed particles and less control over morphology. This is detrimental for most applications including photocatalytic water splitting. Dimension transformation of ferrites from bulk to nanometeric scale could be realized by many different approaches; physical and chemical. Most of the physical synthesis is a "top-down" approach in which bulk pure powder is transformed into nanoscale via ball milling. It is an easy, scalable, and cost-effective technique; however, contamination is one of its biggest concerns. Conversely, chemical processes are a "bottom-up" approach where nano scale particles are synthesized through reactions of reactive species.

Many advantages are offered by the wet chemical synthesis routes, they provide better control over size, dispersibility, and morphology. Ferrites surface offer great versatility for ligand functionalization because of their high reactivity which specifies the best use. These characteristics are essential in defining the magnetism of the

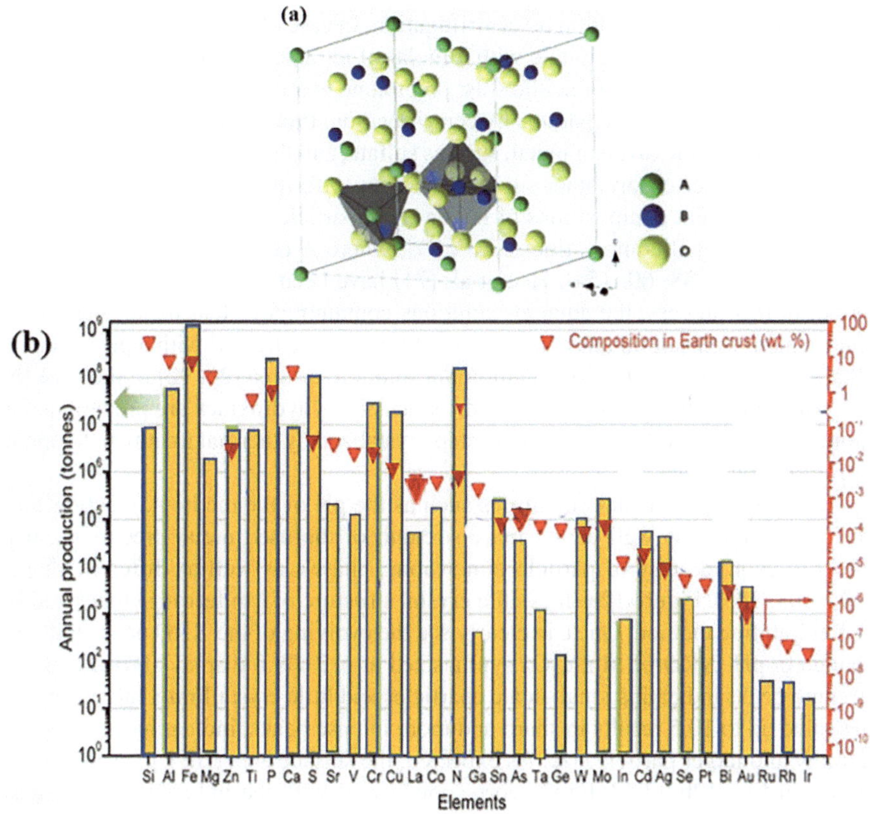

Fig. 11 **a** The crystal structure of ferrite AB_2O_4. **b** Production and inferred value data of some elements in 2010

synthesized ferrites, hence their potential applications in photocatalysis (Mendoza et al. 2016) [159], MRI [93, 131], ferrofluid [18, 141], hyperthermia and magnetic storage media [173]. Solution based chemical synthesis produce good quality ferrites with good crystallinity even though synthesized at relatively low temperature.

7.1 Co-Precipitation Method

Coprecipitation is the most popular synthetic approach used; it is straightforward, effective, and could be expanded for many of the ferrites [3, 145, 177]. It was first developed by Massart in 1981 [109]. Massart developed a method for peptizing magnetite in either alkaline or acidic media to form stable aqueous magnetic sols. It was a breakthrough such that an aqueous mixture of ferric and ferrous ions is added to the ammonia solution hence precipitates are formed. The gelatinous precipitate was then isolated from the solution by centrifugation. In case of alkaline sol: ferrofluid is done by peptizing the precipitate with 1 molar of tetramethylammonium hydroxide aqueous solution. In case of acidic sols: precipitate is mixed with 2 molars aqueous perchloric acid, which is separated by centrifugation. Precursors used in this method are mostly inorganic divalent metal, such as sulfate, and the trivalent iron salt that are dissolved in water or any other suitable medium to form a homogeneous solution. Variety of different compositions of ferrites nanoparticles like ternary metal oxides having the nominal composition of AFe^2O^4, where A could be Mn^{2+}, Fe^{2+}, Co^{2+}, Cu^{2+}, Zn^{2+}, [30, 58, 60, 154]. Hsu et al. [71] have been prepared using coprecipitation. In this process, the aqueous solutions containing divalent metal precursors, and trivalent iron salt mixed in proper stoichiometry are treated with a precipitating alkaline agent like NaOH, KOH, ammonia solution, or urea [26, 92] to increase the pH. Amorphous intermediates like hydroxides or oxyhydroxides are precipitated at room temperature and in a subsequent step crystalized by thermal treatment to obtain ferrites nanoparticles.

Changing experimental conditions such as the pH of the meduim, reaction time and temperature, as well as the ionic strength of the used metal precursors, it is likely to acquire array of particle dimensions, phases, as well as different shapes [107, 108, 143]. Figure 12a shows the crystal structure of Mn ferrite synthesized by coprecipitation method. One can clearly see the evolving of the XRD peaks with the change in pH. XRD of $MnFe_2O_4$ synthesized at pH $= 9.0$ displays broad and low strength peaks indicating small-scale cparticles and little crystallinity. Increasing the pH value up to 10.5 results in sharp and more intense peaks, indicating higher crystallinity and larger particle size. As expected, the surface area (BET measurements) shown in Fig. 12b, of $MnFe_2O_4$ corresponds well with the induced size changes. Measured surface area values were 182.82 $m^2\ g^{-1}$ and 53.99 $m^2\ g^{-1}$at pH 9.0 and 10.5, respectively. The observed decrease in surface area at pH 10.5 is related to the accumulation of small particles into larger ones. The average pore diameter, as seen from the BET measurements, increased as the pH increased and are shown in Fig. 12c. It was also reported that saturation magnetization, M_s, of the Manganese ferrite nanoparticles enhanced with the increase in the pH. At the pH value of 10.5, the recorded value of M_s is 41.48 emu g^{-1} owing to the large magnetic dipole moments of the particles. One important disadvantage of the coprecipitation method is the difficulties of complete separation of nucleation process from growth stages, that eventually contribute to the wide-ranging size distribution observed. In these cases, the resulting nano powder should be thermally annealed to enhance crystallinity and related properties like magnetization [3, 177].

7.2 Sol Gel Method

Sol Gel is a well known preparation technique that is widely used for the preparation of several types of materials like magnetic [16, 84, 87, 88, 139], ferroelectric [23, 66] and thermomechanics [52], multiferroics as well as glasses [43] and catalysts [72,

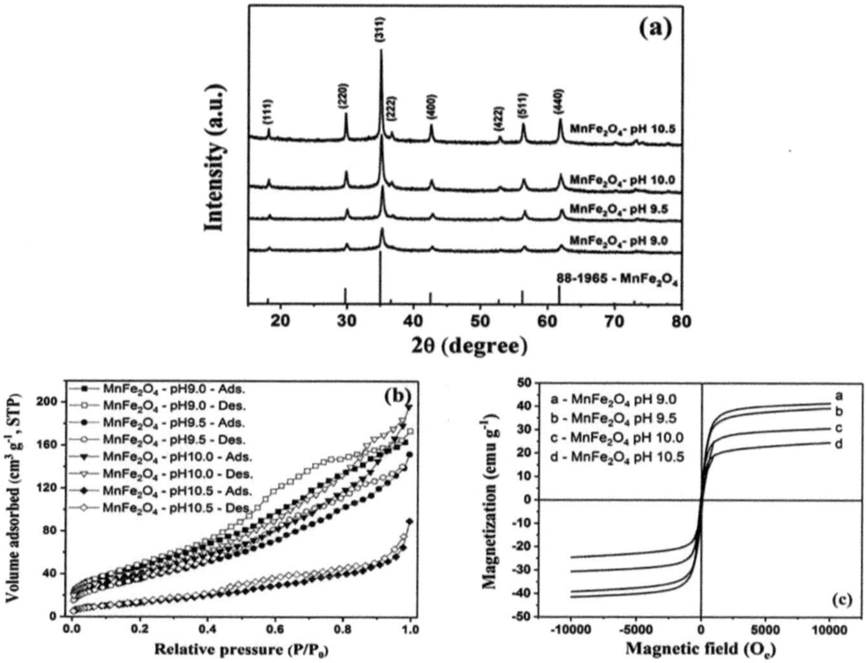

Fig. 12 **a** XRD patterns of $MnFe_2O_4$ synthesized by coprecipitation. **b** N_2 adsorption–desorption isotherms at different pH. **c** Magnetization of $MnFe_2O_4$ synthesized at various pH. With permission from RSC

164]. The emergence of sol gel process was in the year 1921, and 40 years later, in the 1960s. It was further developed and promoted due to the need for new synthesis techniques in the nuclear industry. Ever since flurry of research in materials science using sol gel was reported. Fundamentals of sol–gel technique require knowledge in the fields of physics (aspects related to fractal geometry and percolation theorem) and chemistry (hydrolysis and condensation) and ceramics (sintering). Sol gel is a multi-step synthetic technique that could involve a variety of precursors [65]. The most used precursors in sol gel are the alkoxides and the metal salts. Binary metal oxides having the formula M_2Oz, where M could be Ti, Zr, Nb, Ta, Ce or Th have been successfully synthesized using sol gel technique from the corresponding metal alkoxides [104, 164]. Metal salts precursors are usually used in the synthesis of complex structures like hexagonal ferrites (MFe_2O_4 and $MFe_{12}O_{19}$) [64, 98, 136]. Metal alkoxides are popular precursors because they react readily with water in a reaction called hydrolysis, because the hydroxyl ion becomes attached to the metal atom, as shown in the following reaction:

$$Si(OR)_4 + H_2O - HO - Si(OR)_3 + ROH \quad (4)$$

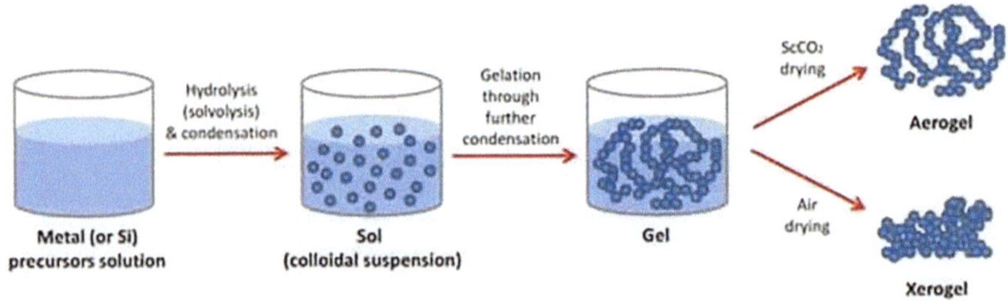

Fig. 13 Representative diagram of the Sol Gel Process. Open access from MDPI Catalysts, 2018

Here, R represents an alkyl group and ROH is alcohol. The increase in the alkyl chain length retards the hydrolysis of alkoxysilanes, while, increasing its degree of branching lowered the hydrolysis rate due to the steric effect [25]. colloid is defined as a suspension in which the dispersed phase or particles is s in the range of 1–1000 nm. In this case gravitational forces are expected to be trivial and interaction between particles is mainly governed by forces, such as van-der Waals and surface charges. TAcquired inertia by scattered phase is small enough to exhibit Brownian motion. Such motion is motivated by momentum conveyed by collisions with molecules in the medium. There are two chemical processes that are commonly participating in the sol gel method: hydrolysis of the metal alkoxide used as a precursor as well as the condensation of the intermediates. Both reactions, hydrolysis, and condensation, occur either concurrently or separately. The hydrolysis reaction takes place when a proton is eliminated from the coordination sphere of the cation hence, promoting the possibility of generation of the different types of intermediates like metal hydroxo-aqua ($M(OH)_x$ $(H_2O)_N$), metal oxyhydroxide ($MO(OH)_N$) or metal hydroxide ($M(OH)_N$) where M is the metal ion.

Condensation process liberates molecules such as water or alcohol and continues to build larger metal-containing molecules or clusters collide and link to produce giant clusters called gel. Clusters in the sol phase are mostly tangled in but not connected to the spanning cluster. However, with time, they get attached to the network causing an increase in the gel stiffness. On the other hand, the bond between clusters forming networks is responsible for the elasticity of the created solid network. Figure 13 is a graphic diagram showing the different steps of sol gel formation.

8 Water Splitting Application

Solar water splitting is one way to produce what is become known as green hydrogen. Many factors influence the performance of semiconductor materials intended to be used for hydrogen production making it a rich topic of scientific research. Finding materials with suitable properties like stability, efficient charge separation and transportation, and showing activity in the visible light region of the solar spectrum is still a challenge [169]. In this context, ferrites can offer some of the above-mentioned properties and all of them have suitable bandgap to harvest visible light ($E_g \leq 2.6$ eV) [77]. The variety of transition elements that could fit in the composition and the multiple valency states according to the metal cation used are major factors in enhancing photocatalytic water splitting [159]. This could improve the electrical conductivity of ferrites compared to the single-component iron oxide (γ/α Fe_2O_3). This improve is primarily ascribed to the existence of various metal cations which enable electrons transportation and/or establish rich redox chemistry [110]. Many inorganic compounds were used for photocatalytic water splitting [91, 106, 117] but metal oxides, in general, and ferrites, are considered materials of interest for water splitting due to stability under the different photocatalytic operation parameters and the facile synthesis techniques used with the ability to scale up for large scale application, unlike the organic materials. Basic properties like band position of the valence and conduction bands compared to water splitting redox potential were investigated and shown in Fig. 14. As shown in the figure, ferrites could be n-type, and p-type like p-$CaFe_2O_4$. They are mostly small band gaps materials which make them harvest photon in the visible range of the solar spectrum. Some of them have the suitable band positions required for full water splitting redox reactions i.e., the simultaneous generation of hydrogen and oxygen. However, and despite the many advantages, their photocatalytic performance is poor in comparison to other well-established semiconductors like TiO_2 and $BiVO_4$. The energy required for complete water splitting is 1.23 eV and this energy as well as the energy necessary to overwhelm the kinetic barrier of the reaction should be provided by a light absorber. Even though position of both CB and VB in some ferrites is suitable for full water splitting, it does not happen in practice. There are other factors that could be responsible for the inferior performance of ferrites like their indirect bandgap features, weak charge carrier dynamics, short diffusion lengths and short lived charges [130]. As far as we know, there are no reports of full water splitting achieved using ferrites alone in a photocatalytic process. And while bias free photocatalytic water splitting remains a significant challenge, a lot of advancement has been made, especially in the last decad, in electrically biased photoelectrodes. Formation of heterojunction and/or solid solution is another method used to achieve complete water splitting. Another adopted technique is the usage of a hole/electron scavenger that enables half-cycle water splitting i.e. production of either O_2 or H_2 [47, 117]. The usage of ferrites in photocatalytic water splitting, along with any other material, is commonly achieved by three different methods depending on their role in the photocatalytic process: they could be photo-cathodes, photo-anodes, or aqueous suspensions.

8.1 The Effect of Cocatalyst

There are two primary processes involved in any photocatalytic process: charge carrier separation and catalytic redox reactions on the surface. And to enhance these two processes to boost the photocatalytic water splitting performance of ferrites,

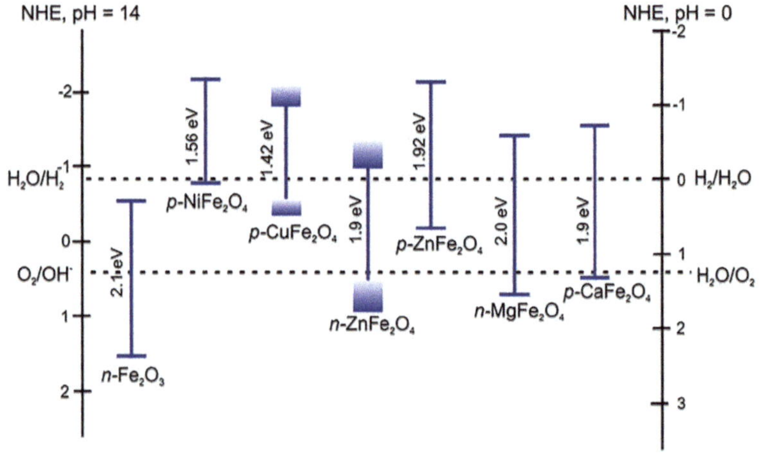

Fig. 14 Band positions of some ferrites with respect to NHE. With permission from SPIE [159]

coupling with cocatalyst seems to be a successful way to enhance charge separation [76]. Platinum (Pt) as a cocatalyst has been broadly examined and proven to be an effective cocatalyst for hydrogen evolution reaction, HER. platinum (Pt) works as a trap for electrons promoted to the surface after light irradiation thereby minimizing electrons and holes recombination, consequently enhancing the photocatalytic performance [73]. Platinum was added to magnetite, Fe_3O_4, nanoparticles from platinum salt source like chloroplatinic acid and was deposited on the ferrite surface during a photo chemical reaction. It was reported that hydrogen amount intensely increased by Pt addition a yield of 8275 μmol h^{-1} g^{-1} compared to a yield of 458 μmol h^{-1} g^{-1} without Pt addition. The optimum amount of platinum producing the largest amount of H_2 was 0.095 mg. An increased amount of metal particles on the surface of a photocatalyst could potentially reduce the absorption of photons. Moreover, cocatalyst may function as a recombination center for the photogenerated charge carriers, thus, lowering photocatalytic performance [138]. Similar behavior was observed for bismuth ferrite, $BiFeO_3$, BFO, but for oxygen evolution reaction, OER, and the usage of IrO_2 as cocatalyst [142]. OER is somehow difficult, the creation of a single molecule of oxygen involves 4holes as indicated in Eq. 7 [118]. OER occurs on a time scale approximately 5 orders of magnitude slower than H_2 evolution.

$$2H_2O_{(l)} + 4h^+ \rightarrow O_{2(g)} + 4H^+_{aq} \tag{7}$$

IrO_2 is one of the most active O_2-evolution catalysts and is located on top of the volcanic catalyst diagram shown in Fig. 15a. Other oxides like NiO, RhO_2, RuO_2, and Co_3O_4 are also at the apex of the volcanic graph [22] and some of them are more abundant and affordable in terms of their cost compared to IrO_2. However, IrO_2 has an important advantage over the other cocatalysts, it shows good stability against corrosive environment which make IrO_2 more favorable in different reaction mediums [17]. The capability of a catalyst to achieve either water oxidation or reduction depends on the CB and VB position, as mentioned earlier in this chapter, which could be constructed using Mott Schottky measurements. In the case of bismuth ferrite, BFO, the CB and the VB lie at −0.46 V and 2.69 V, respectively, against NHE (Fig. 15b). It clearly shows that the bottom of the CB is more positive than the reduction potential of the proton (H^+) to produce molecular hydrogen. But the top of the VB in BFO is also more positive than the oxidation potential of H_2O hence, it is anticipated that BFO is capable of only water oxidation (Fig. 15b). An important and facile technique to load cocatalyst oxides outside the surface of the catalyst is the impregnation method [74] in which Ir precursor in water was evaporated in a water bath, then calcined at 260 °C. In Fig. 16c, the photocatalytic oxygen evolved at different IrO_2 content is shown. Many considerations can impact the ability of OER cocatalyst activity as the type of cocatalyst and the loading amount of both the catalyst and the cocatalyst in the reaction medium. there is a bell shape type of dependance relating the amount of a cocatalyst and photo catalytic performance. Figure 15d shows the OER dependance on the loading amount of IrO_2 cocatalyst on surface of BFO, OER reaches an optimum value at 2wt % IrO_2 content. The small-scale of the cocatalyst particles makes it difficult for charge carriers to interact via recombination in the bulk of such small-scale particles. It worth mentioning here, that additional increase in the IrO_2 loading caused a reduction in the OER activity. The reduction might be related to many possibilities. The high-level of loading is likely to cover up the surface active sites on the $BiFeO_3$ surface preventing its interaction with the sacrificial agent and/or water. On the other hand, a high level of cocatalyst loading could reduce or at worst prevent light absorption, afterwards lower the photo-generated holes inside the BFO. One other key component influences the cocatalyst effectiveness is the cocatalyst dimension such that at greater IrO_2 loading content, cocatalyst accumulate forming separate islands on the surface of the catalyst, thus wasting the benefit of the nano length scale and the large surface area.

8.2 Ferrites Based Heterojunction

Constructing semiconducting heterojunctions was always an effective way to promote charge carriers separation. Ferrites show exciting photocatalytic and photo electrochemical properties towards water splitting application. Magnesuim ferrite, $MgFe_2O_4$, and zinc ferrite, $ZnFe_2O_4$, have demonstrated ability for water oxidation reactions upon formation of $CoFe_2O_3/MgFe_2O_4$ and α-$Fe_2O_3/ZnFe_2O_4$ heterojunctions [70, 85, 86]. In such heterojunctions, the photoinduced holes are moved from the VB of semiconductors like $CoFe_2O_3$, and α-Fe_2O_3 to the VB of the ferrites owing to the more negative VB location of the ferrites, this in turn will lead to efficient charge carrier separation. In the same context, the semiconductor, graphitic carbon nitride (g-C_3N_4) which has narrow band gap of the order of 2.7 eV and shows

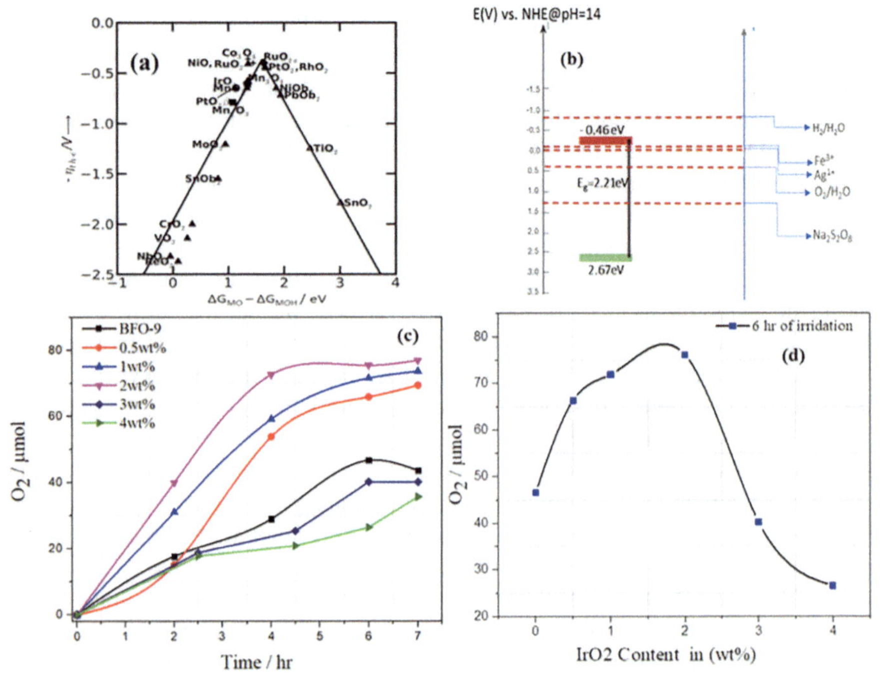

Fig. 15 **a** Volcano plots for the photocatalytic Oxygen Evolution Reaction (OER. With permission from RSC (Fabbri et al. 2014). **b** Graphic diagram showing the CB and VB in bismuth ferrite. **c** Photocatalytic OER at various IrO_2 cocatalyst on BFO under solar irradiation. **d** Relation of photocatalytic oxygen evolution of BFO/IrO_2 on the content of IrO_2. With permission from Elsevier publisher [142]

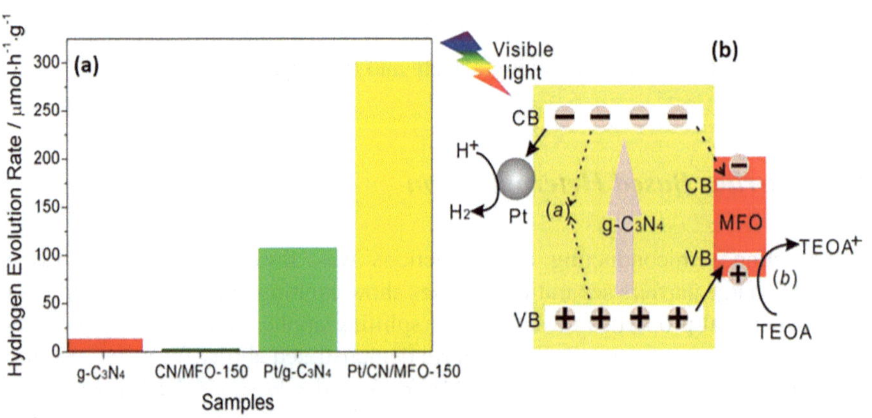

Fig. 16 **a** Histogram showing photocatalytic H_2 generation of graphitic carbon nitide loaded with 1 wt.% Pt cocatalyst and/or $MgFe_2O_4$. **b** Proposed structure of the photocatalytic H_2 generation over Pt and g-C_3N_4/$MgFe_2O_4$ Heterojunction with permission from ACS publisher [34]

suitable band positions, versus NHE. such properties making it a favorable material to produce both H_2 and O_2. However, both g-C_3-N_4 and $MgFe_2O_4$ show poor responses toward photocatalytic H_2 production (Fig. 16a). Chen and coworker [31] used a heterojunction of $MgFe_2O_4$ ferrite and g-C_3N_4 and assessed their performance for H_2 production. However, the photocatalytic activity of $gC_3N_4/MgFe_2O_4$ is quite poor, which could be explained by taking the band position of g-C_3N_4 and $MgFe_2O_4$ into consideration. The bandgaps (E_g) of g-C_3N_4 and $MgFe_2O_4$ were calculated from the UV–Vis measurements and concluded to be 2.68 and 1.78 eV, respectively.

The VB maximum location of both magnesium ferrite and graphitic carbon nitrde are 1.79 and 2.39 eV, respectively. In the same way, the CB maximum of both compounds are 0.01 eV, lower than fermi level, and 0.29 eV, over the fermi level, respectively, showing difference in the CB maxima of 0.28 eV which will assist electrons to move from the CB of g-C_3N_4 to the CB of $MgFe_2O_4$. On the other hand, the difference between the VB maxima of the two compounds is 0.60 eV which will encourage hole transport from the VB of g-C_3N_4 to the VB of $MgFe_2O_4$. Therefore, formation of heterostructure and the associated change in band structures of g-C_3N_4 and $MgFe_2O_4$ could initiate movement of photoinduced electrons and holes from g-C_3N_4 to $MgFe_2O_4$. However, such charge allocation in the heterojunction will promote charge carrier recombination at $MgFe_2O_4$, thus and again, weak photocatalytic activity of the proposed junction. The photoinduced charge carriers in the proposed heterojunction chosed to move to the magnesium ferrite, $MgFe_2O_4$, unfortunately, it was unable of photocatalytic H_2 generation. Therefore, the resolve to a good H_2 catalyst like Pt to boost H_2 production. The proposed mechanism in this approach for hydrogen production over Pt loading on g-$C_3N_4/MgFe_2O_4$ heterojunction is shown in Fig. 16b. To establish high photocatalytic activity, photogenerated electrons in g-C3N4 should be confined in their position meanwhile, holes continued in transferring from g-C_3N_4 to $MgFe_2O_4$. By using Pt, which is a good hydrogen catalyst, electrons could be entrapped in the g-C_3N_4 and could not migrate into the ferrite, hence promote charge separation and induce efficient photocatalytic activity towards H_2 production. One can conclude that the charge transport procedures in the Pt/g-$C_3N_4/MgFe_2O_4$ are quite distinct from those in g-$C_3N_4/MgFe_2O_4$. The photoinduced electrons in g-C_3N_4 prefer to move to the metal Pt for H_2 creation instead of being exhausted at the ferrite, $MgFe_2O_4$, in the Pt/g-$C_3N_4/MgFe_2O_4$. Pt as a reductive hydrogen production catalyst is essential in this case and in many similar cases.

9 Concluding Remarks

This chapter covered some aspects of the most used synthesis techniques for TiO_2 and ferrites nano-structures and their application in photocatalytic water splitting towards H_2 production. Choice of the synthesis procedures in addition to the precursor's type and the applied parameters have a great influence on the structural, electronic, and optical characteristics of the synthesized nanoparticles. Mesoporous and nanoporous nanoparticles also demonstrate a larger surface area and continuous particle frameworks, offering more active sites for the adsorption of molecules, and facilitating the electron transfer within the material. In photocatalytic water-splitting, the photogenerated charge carriers can rapidly recombine unless controlled. In this regard, synthesis techniques including the tuning of crystal structure, defects states under inert atmosphere, dopants, and electronic structure can influence the H_2 production. Synthesis and application are two synchronized processes in these two materials as well other most systems. Meanwhile, TiO_2 is a well-established photocatalyst with good photocatalytic water splitting, ferrites in its pure form, and despite the many advantages they can offer, have poor photocatalytic performance, and can perform better with the application of biased voltage, formation of heterojunction with other semiconductors to adjust bands position and/or the attachment of cocatalyst on the surface.

Acknowledgements Some of the studies presented here were performed in the laboratory "Photoactive nanocomposite materials" and supported by Saint-Petersburg State University (ID: 91696387). Wegdan Ramadan, would like to acknowledge fund received from the Alexander von Humboldt Foundation towards the purchase of laboratory equipment.

References

1. Ahn SH, Koh JH, Seo JA, Kim JH (2010) Structure control of organized mesoporous TiO_2 films templated by graft copolymers for dye -sensitized solar cells. Chem Commun 46(11):1935–1937
2. AL-Azri ZHN, Chen W-T, Chan A, Jovic V, Ina T, Idriss H, Waterhouse GIN (2015) The roles of metal co-catalysts and reaction media in photocatalytic hydrogen production: performance evaluation of M/TiO_2 photocatalysts (M=Pd, Pt, Au) in different alcohol–water mixtures. J Catal 329:355–367
3. Albuquerque AS, Tolentino MVC, Ardisson JD, Moura FCC, De Mendona R, MacEdo WAA (2012) Nanostructured ferrites: structural analysis and catalytic activity. Ceram Int 38(3):2225–2231
4. Al-Madanat O (2021) Photocatalytic transformation of water pollutants into fuels. Doctoral thesis, Gottfried Wilhelm Leibniz Universität Hannover, Germany
5. A-Madanat O, AlSalka Y, Curti M, Dillert R, Bahnemann DW (2020) Mechanistic insights into hydrogen evolution by photocatalytic reforming of naphthalene. ACS Catal 10(13):7398–7412
6. A-Madanat O, AlSalka Y, Dillert R, Bahnemann DW (2021a) Photocatalytic H_2 production from naphthalene by various TiO_2 photocatalysts: impact of Pt loading and formation of intermediates. Catalysts 11(1):107

7. A-Madanat O, AlSalka Y, Ramadan W, Bahnemann DW (2021b) TiO_2 photocatalysis for the transformation of aromatic water pollutants into fuels. Catalysts 11(3):317
8. A-Madanat O, Curti M, Günnemann C, AlSalka Y, Dillert R, Bahnemann DW (2021c) TiO_2 photocatalysis: impact of the platinum loading method on reductive and oxidative half-reactions. Catal Today 380:3–15
9. Al-Madanat O, Nunes BN, AlSalka Y, Hakki A, Curti M, Patrocinio AOT, Bahnemann DW (2021d) Application of EPR spectroscopy in TiO_2 and Nb_2O_5 photocatalysis. Catalysts 11:1514
10. Almquist CB, Biswas P (2002) Role of synthesis method and particle size of nanostructured TiO_2 on its photoactivity. J Catal 212(2):145–156
11. Alsalka Y (2020) Photocatalytic water splitting for solar hydrogen production and simultaneous decontamination of organic pollutants. Doctoral thesis, Gottfried Wilhelm Leibniz Universität
12. AlSalka Y, Al-Madanat O, Curti M, Hakki A, Bahnemann DW (2020) Photocatalytic H_2 evolution from oxalic acid: Effect of cocatalysts and carbon dioxide radical anion on the surface charge transfer mechanisms. ACS Appl Energy Mater 3(7):6678–6691
13. AlSalka Y, Granone LI, Ramadan W, Hakki A, Dillert R, Bahnemann DW (2019) Iron-based photocatalytic and photoelectrocatalytic nano-structures: facts, perspectives, and expectations. Appl Catal B 244:1065–1095
14. AlSalka Y, Hakki A, Fleisch M, Bahnemann DW (2018) Understanding the degradation pathways of oxalic acid in different photocatalytic systems: towards simultaneous photocatalytic hydrogen evolution. J Photochem Photobiol, A 366:81–90
15. AlSalka Y, Hakki A, Schneider J, Bahnemann DW (2018) Co-catalyst-free photocatalytic hydrogen evolution on TiO_2: synthesis of optimized photocatalyst through statistical material science. Appl Catal B 238:422–433
16. An SY, Shim I-B, Kim CS (2002) Mössbauer and magnetic properties of Co–Ti substituted barium hexaferrite nanoparticles. J Appl Phys 91(10):8465
17. Harriman A, Pickering I, Thomas J (1988) Metal oxides as heterogeneous catalysts for oxygen evolution under photochemical conditions. J Chem Soc Faraday Trans 1(84):2795–2806
18. Anton AIJ (1990) Measurements of turbulence suppression due to a transverse magnetic field applied on a ferrofluid motion. J Magn Magn Mater 85:137
19. Aoyama Y, Oaki Y, Ise R, Imai H (2012) Mesocrystal nanosheet of rutile TiO_2 and its reaction selectivity as a photocatalyst. Cryst Eng Comm 14(4):1405–1411
20. Bai H, Liu Z, Sun DD (2010) Hierarchically multifunctional TiO_2 nano-thorn membrane for water purification. Chem Commun 46:6542–6544
21. Bamwenda GR, Tsubota S, Nakamura T, Haruta M (1995) Photoassisted hydrogen production from a water-ethanol solution: a comparison of activities of Au-TiO_2 and Pt-TiO_2. J Photochem Photobiol, A 89(2):177–189
22. Ma B, Yang J, Han H, Wang J, Zhang X, Li C (2010) Enhancement of photocatalytic water oxidation activity on IrOx ZnO/ Zn2-x GeO4-x–3yN2y catalyst with the solid solution phase junction. J Phys Chem C 114(29):12818–12822
23. Barbooram K, Ye Z-G (2006) New soft chemical routes to ferroelectric $SrBi_2Ta_2O_9$. Chem Mater 18(2):532–540
24. Bessekhouad Y, Robert D, Weber JV (2003) Synthesis of photocatalytic TiO_2 nanoparticles: optimization of the preparation conditions. J Photochem Photobiol, A 157(1):47–53
25. Brinker CJ (1988) Hydrolysis and condensation of silicates: effects on structure. J Non-Cryst Solids 100(1):31–50
26. Bruce IJ, Taylor J, Todd M, Davies MJ, Borioni E, Sangregorio C, Sen T (2004) Synthesis, characterisation and application of silica-magnetite nanocomposites. J Magn Magn Mater 284:145–160
27. Aydın C, Al-Hartomy O, Al-Ghamdi AA, Al-Hazmi F, Yahia IS, El-Tantawy F, Yakuphanoglu F (2012) Controlling of crystal size and optical band gap of CdO nanopowder semiconductors by low and high Fe contents. J Electroceram 29(2):155–162

28. Carp O, Huisman CL, Reller A (2004) Photoinduced reactivity of titanium dioxide. Prog Solid State Chem 32(1–2):33–177
29. Chae SY, Park MK, Lee SK, Kim TY, Kim SK, Lee WI (2003) Preparation of size controlled TiO_2 nanoparticles and derivation of optically transparent photocatalytic films. Chem Mater 15(17):3326–3331
30. Chen JP, Sorensen CM, Klabunde KJ, Hadjipanayis GC, Devlin E, Kostikas A (1996) Size-dependent magnetic properties of $MnFe_2O_4$ fine particles synthesized by coprecipitation. Phys Rev B 54(13):9288
31. Chen J, Zhao D, Diao Z, Wang M, Guo L, Shen S (2015) Bifunctional modification of graphitic carbon nitride with $MgFe_2O_4$ for enhanced photocatalytic hydrogen generation. ACS Appl Mater Interfaces 7(33):18843–18848
32. Chen X, Burda C (2008) The electronic origin of the visible-light absorption properties of C-, N- and S-doped TiO_2 nanomaterials. J Am Chem Soc 130:5018–5019
33. Chen X, Mao SS (2007) Titanium dioxide nanomaterials: synthesis, properties, modifications, and applications. Chem Rev 107(7):2891–2959
34. Chen X, Liu L, Huang F (2015) Black titanium dioxide (TiO_2) nanomaterials. Chem Soc Rev 44:1861–1885
35. Chen X, Shen S, Guo L, Mao SS (2010) Semiconductor-based photocatalytic hydrogen generation. Chem Rev 110(11):6503–6570
36. Cheng Y, Huang W, Zhang Y, Zhu L, Liu Y, Fan X, Cao X (2010) Preparation of TiO_2 hollow nanofibers by electrospining combined with sol–gel process. Cryst Eng Comm 12(7):2256–2260
37. Chiarello GL, Forni L, Selli E (2009) Photocatalytic hydrogen production by liquid- and gas-phase reforming of CH_3OH over flame-made TiO_2 and Au/TiO_2. Catal Today 144(1):69–74
38. Cho S, Jang J-W, Lee K-H, Lee JS (2014) Research update: Strategies for efficient photoelectrochemical water splitting using metal oxide photoanodes. APL Mater 2:010703
39. Chretien S, Metiu H (2011) Electronic structure of partially reduced rutile TiO_2 (110) surface: where are the unpaired electrons located? J Phys Chem C 115(11):4696–4705
40. Cozzoli D, Kornowski A, Weller H (2003) Low-Temperature synthesis of soluble and processable organic-capped anatase TiO_2 nanorods. J Am Chem Soc 125:14539–14548
41. Damm C, Sakthivel S, Kisch H (2006) UV and visible light acrylate photopolymerisation initiated by nitrogen- or carbon-doped titanium dioxide. Z Phys Chem 220(4):477–486
42. Das D, Veziroglu TN (2008) Advances in biological hydrogen production processes. Int J Hydrogen Energy 33(21):6046–6057
43. Del Castillo J, Rodríguez VD, Yanes AC, Méndez-Ramos J, Torres ME (2005) Luminescent properties of transparent nanostructured Eu^{3+} doped SnO_2–SiO_2 glass-ceramics prepared by the sol–gel method. Nanotechnology 16:S300
44. Taffa D, Dillert R, Ulpe A, Bauerfeind K, Bredow T, Bahnemann DW, Wark M (2016) Photoelectrochemical and theoretical investigations of spinel type ferrites ($M_xFe_{3-x}O_4$) for water splitting: a mini-review. J Photonics Energy 7(1):012009
45. Di Bartolomeo A (2016) Graphene schottky diodes: an experimental review of the rectifying graphene/semiconductor heterojunction. Phys Rep 606:1–58
46. Diebold U (2003) The surface science of titanium dioxide. Surf Sci Rep 48(5–8):53–229
47. Dillert R, Taffa DH, Wark M, Bredow T, Bahnemann DW (2015) Research Update: Photoelectrochemical water splitting and photocatalytic hydrogen production using ferrites (MFe_2O_4) under visible light irradiation. APL Mater 3:104001
48. Doman L (2017) Today in energy [Online]. USA: U.S. Energy Information Administration. Available: https://www.eia.gov/todayinenergy/detail.php?id=32912. Accessed 08 Jan 2020
49. Du S, Lian J, Zhang F (2022) Visible light-responsive N-doped TiO_2 photocatalysis: synthesis, characterizations, and applications. Trans Tianjin Univ 28:33–52
50. Eidsvåg H, Bentouba S, Vajeeston P, Yohi S, Velauthapillai D (2021) TiO_2 as a photocatalyst for water splitting: an experimental and theoretical review. Molecules 26(6):1687
51. Enke CG (1974) Nonstoichiometry, diffusion, and electrical conductivity in binary metal oxides. Mater Corros 25(10):801–802

52. Fantozzi G, Chevalier J, Guilhot B (2001) Processing, microstructure, and thermomechanical behavior of ceramics. Adv Eng Mater 3(8):563
53. Farsinezhad S, Sharma H, Shankar K (2015) Interfacial band alignment for photocatalytic charge separation in TiO_2 nanotube arrays coated with CuPt nanoparticles. Phys Chem Chem Phys 17:29723–29733
54. Fazio G, Selli D, Ferraro L, Seifert G, Di Valentin C (2018) Curved TiO_2 nanoparticles in water: short (chemical) and long (physical) range interfacial effects. ACS Appl Mater Interfaces 10(35):29943–29953
55. Fonash SJ (2010) Chapter three: structures, materials, and scale. In: Fonash SJ (ed) Solar cell device physics, 2nd edn. Academic Press, Boston
56. Friehs E, AlSalka Y, Jonczyk R, Lavrentieva A, Jochums A, Walter J-G, Stahl F, Scheper T, Bahnemann DW (2016) Toxicity, phototoxicity and biocidal activity of nanoparticles employed in photocatalysis. J Photochem Photobiol, C 29:1–28
57. Furube A, Du L, Hara K, Katoh R, Tachiya M (2007) Ultrafast plasmon-induced electron transfer from gold nanodots into TiO_2 nanoparticles. J Am Chem Soc 129(48):14852–14853
58. Gee SH, Hong YK, Erickson DW, Park MH, Sur JC (2003) Synthesis and aging effect of spherical magnetite nanoparticles for biosensor applications. J Appl Phys 93(10):7560
59. Godula-Jopek A (2015) Hydrogen production: by electrolysis. Wiley-VCH, Weinheim, Germany
60. Grasset F, Labhsetwar N, Li D, Park DC, Saito N, Haneda H, Cador O, Roisnel T, Mornet S, Duguet E, Portier J, Etourneau JL, Grasset F, Labhsetwar N, Li D, Park DC, Saito N, Haneda H, Cador O, Roisnel T, Mornet S, Duguet E, Portier JE (2002) Synthesis and magnetic characterization of zinc ferrite nanoparticles with different environments: powder, colloidal solution, and zinc ferrite−silica core−shell nanoparticles. Langmuir 18(21):8209–8216
61. Habisreutinger SN, Schmidt-Mende L, Stolarczyk JK (2013) Photocatalytic reduction of CO_2 on TiO_2 and other semiconductors. Angew Chem Int Ed 52(29):7372–7408
62. Hakki A, AlSalka Y, Mendive C, Ubogui J, Dos Santos Claro P, Bahnemann DW (2018) Hydrogen production by heterogeneous photocatalysis. In: Wandelt K (ed) Encyclopedia of interfacial chemistry. Elsevier, Oxford
63. Hasegawa G, Morisato K, Kanamori K, Nakanishi K (2011) New hierarchically porous titania monoliths for chromatographic separation media. J Sep Sci 34(21):3004–3010
64. He X, Song G, Zhu J (2005) Non-stoichiometric NiZn ferrite by sol-gel processing. Mater Lett 59(14–15):1941–1944
65. Hench LL, West JK (1990) The sol-gel process. Chem Rev 90(1):33–72
66. Hernandez BA, Chang K-S, Fisher ER, Dorhout PK (2002) Sol-gel template synthesis and characterization of $BaTiO_3$ and $PbTiO_3$ nanotubes. Chem Mater 14(2):480–482
67. Hernández-Ramírez A, Medina-Ramírez I (2015) Semiconducting materials. In: Hernández-Ramírez A, Medina-Ramírez I (eds) Photocatalytic semiconductors: synthesis, characterization, and environmental applications. Springer International Publishing, Cham
68. Toulhoat H, Raybaud P (2020) Prediction of optimal catalysts for a given chemical reaction. Catal Sci Technol 10:2069–2081
69. Hoffmann MR, Martin ST, Choi W, Bahnemann DW (1995) Environmental applications of semiconductor photocatalysis. Chem Rev 95:69–96
70. Hou Y, Zuo F, Dagg A, Feng PA (2013) Three-dimensional branched cobalt-doped α-Fe_2O_3 nanorod/$MgFe_2O_4$ heterojunction array as a flexible photoanode for efficient photoelectrochemical water oxidation. Angew Chem Int Ed 52:1248–1252
71. Hsu WC, Chen SC, Kuo PC, Lie CT, Tsai WS (2004) Preparation of NiCuZn ferrite nanoparticles from chemical co-precipitation method and the magnetic properties after sintering. Mat Sci Eng B 111:142–149
72. Hu Q, Zhao J, Wang Y, Zhu L, Li M, Li G, Wang Y, Ge FJ (2003) Sol–gel encapsulated cobalt (III) acetylacetonate for air oxidation of penicillin derivatives. Catal A: Chem 200:271–277
73. Ida S, Yamada K, Matsuka M, Hagiwara H, Ishihara T (2012) Photoelectrochemical hydrogen production from water using p-type and n-type oxide semiconductor electrodes. Electrochim Acta 82:397–401

74. Ivanova I, Kandiel TA, Cho YJ, Choi W, Bahnemann DW (2018) Mechanisms of photocatalytic molecular hydrogen and molecular oxygen evolution over La-Doped $NaTaO_3$ particles: effect of different cocatalysts and their specific activity. ACS Catal 8:2313–2325
75. Iwata K, Takaya T, Hamaguchi H-O, Yamakata A, Ishibashi T-A, Onishi H, Kuroda H (2004) Carrier dynamics in TiO_2 and Pt/TiO_2 powders observed by femtosecond time-resolved near-infrared spectroscopy at a spectral region of 0.9− 1.5 μm with the direct absorption method. J Phys Chem B 108:20233–20239
76. Yang JH, Wang DE, Han HX, Li C (2013) Roles of cocatalysts in photocatalysis and photoelectrocatalysis. Acc Chem Res 46:1900–1909
77. Kim JH, Kim HE, Kim JH, Lee JS (2020) Ferrites: emerging light absorbers for solar water splitting. J Mater Chem A 8:9447–9482
78. Jiang Z, Zhu J, Liu D, Wei W, Xie J, Chen M (2014) In situ synthesis of bimetallic Ag/Pt loaded single-crystalline anatase TiO_2 hollow nano-hemispheres and their improved photocatalytic properties. Cryst Eng Comm 16:2384–2394
79. Kampouri S, Stylianou KC (2019) Dual-functional photocatalysis for simultaneous hydrogen production and oxidation of organic substances. ACS Catal 9(5):4247–4270. https://doi.org/10.1021/acscatal.9b00332
80. Kang J-G, Sohn Y (2011) Interfacial nature of Ag nanoparticles supported on TiO_2 photocatalysts. J Mater Sci 47:824–832
81. Kato H, Asakura K, Kudo A (2003) Highly efficient water splitting into H_2 and O_2 over lanthanum-doped $NaTaO_3$ photocatalysts with high crystallinity and surface nanostructure. J Am Chem Soc 125(10):3082–3089. Available from https://doi.org/10.1021/ja027751g
82. Katsumatak KI, Okazaki S, Cordonier CEJ, Shichi T, Sasaki T, Fujishima A (2010) Preparation and characterization of self-cleaning glass for vehicle with niobia nanosheets. ACS Appl Mater Interfaces 2:1236–1241
83. Kazuya N, Baoshun L, Yosuke I, Munetoshi S, Hidenori S, Tsuyoshi O, Hideki S, Taketoshi M, Masahiko A, Katsuhiko T, Akira F (2011) Fabrication and photocatalytic properties of TiO_2 nanotube arrays modified with phosphate. Chem Lett 40:1107–1109
84. Kim CS, Yi YS, Park K-T, Namgung H, Lee J-G (1999) Growth of ultrafine Co–Mn ferrite and magnetic properties by a sol–gel method. J Appl Phys 35(8):5223
85. Kim E, Nishimura N, Magesh G, Kim JY, Jang J-W, Jun H, Kubota J, Domen K, Lee JS (2013) Fabrication of $CaFe_2O_4$/TaON heterojunction photoanode for photoelectrochemical water oxidation. J Am Chem Soc 135:5375–5383
86. Kim HG, Borse PH, Jang JS, Jeong ED, Jung O-S, Suh YJ, Lee JS (2009) Fabrication of $CaFe_2O_4$/$MgFe_2O_4$ bulk heterojunction for enhanced visible light photocatalysis. Chem Commun 5889−5891
87. Kim WC, Kim SJ, Sur JC, Kim CS (2002) Structural and magnetic properties of $CoFe_{1.9}RE_{0.1}O_4$ (RE=Y, La) prepared by a sol–gel method. J Magn Magn Mater 242–245:197–200
88. Kim WC, Park SI, Kim SJ, Lee SW, Kim CS (2000) Magnetic and structural properties of ultrafine Ni–Zn–Cu ferrite grown by a sol–gel method. J Appl Phys 87(9):6241
89. Kisch H (2014) Molecular photochemistry. Semicond Photocatalysis 9–46
90. Kisch H (2015) Semiconductor photocatalysis principle and applications. Wiley-VCH
91. Kudo A, Miseki Y (2009) Heterogeneous photocatalyst materials for water splitting. Chem Soc Rev 38:253–278
92. Kundu A, Upadhyay C, Verma HC (2003) Magnetic properties of a partially inverted zinc ferrite synthesized by a new coprecipitation technique using urea. Phys Lett A 311:410–415
93. Lawaezeck R, Menzel M, Pietsch H (2004) Superparamagnetic iron oxide particles: contrast media for magnetic resonance imaging. Appl Organometal Chem 18:506–513
94. Lee JS (2005) Photocatalytic water splitting under visible light with particulate semiconductor catalysts. Catal Surv Asia 9:217–227
95. Lee SW, Ryu YG, Yang KJ, Jung K-D, An SY, Kim CS (2002) Magnetic properties of Zn^{2+} substituted ultrafine Co-ferrite grown by a sol-gel method. J Appl Phys 91(10):7610

96. Lewis NS, Nocera DG (2006) Powering the planet: chemical challenges in solar energy utilization. Proc Natl Acad Sci 103:15729–15735
97. Li G, Blake GR, Palstra TTM (2017) Vacancies in functional materials for clean energy storage and harvesting: the perfect imperfection. Chem Soc Rev 46:1693–1706
98. Li J-J, Xu W, Yuan H-M, Chen J-S (2004) Sol–gel synthesis and magnetization study of $Mn_{1-x}Cu_xFe_2O_4$ (x=0, 0.2) nanocrystallites. Solid State Commun 131:519–522
99. Li Y, Peng Y-K, Hu L, Zheng J, Prabhakaran D, Wu S, Puchtler TJ, Li M, Wong K-Y, Taylor RA, Tsang SCE (2019) Photocatalytic water splitting by N-TiO_2 on MgO (111) with exceptional quantum efficiencies at elevated temperatures. Nat Commun 10:4421
100. Liao C-H, Huang C-W, Wu J (2012) Hydrogen production from semiconductor-based photocatalysis via water splitting. Catalysts 2:490–516
101. Lin HY, Shih CY (2016) Efficient one-pot microwave-assisted hydrothermal synthesis of M (M=Cr, Ni, Cu, Nb) and nitrogen co-doped TiO_2 for hydrogen production by photocatalytic water splitting. J Mol Catal A: Chem 411:128–137
102. Linsebiglerl A-L, Lu G, Yates JT (1995) Photocatalysis on TiO_2 surfaces: principles, mechanisms, and selected results. Chem Rev 95:735–758
103. Liu B, Nakata K, Sakai M, Saito H, Ochiai T, Murakami T, Takagi K, Fujishima A (2011) Mesoporous TiO_2 core-shell spheres composed of nanocrystals with exposed high-energy facets: facile synthesis and formation mechanism. Langmuir 27:8500–8508
104. Livage J, Henry M, Sanchez C (1988) Sol-gel chemistry of transition metal oxides. Prog Solid St Chem 18:259–341
105. Macak JM, Zlamal M, Krysa J, Schmuki P (2007) Self-organized TiO_2 nanotube layers as highly efficient photocatalysts. Small 3:300–304
106. Marschall R (2014) Semiconductor composites: strategies for enhancing charge carrier separation to improve photocatalytic activity. Adv Funct Mater 24(17):2421–2440
107. Mascolo M, Pei Y, Ring T (2013) Room temperature co-precipitation synthesis of magnetite nanoparticles in a large pH window with different bases. Materials 6(12):5549–5567
108. Massart R, Cabuil V (1987) New trends in chemistry of magnetic colloids: polar and non polar magnetic fluids, emulsions, capsules and vesicles. J Chim Phys PCB 84:967–973
109. Massart R (1981) Preparation of aqueous magnetic liquids in alkaline and acidic media. IEEE Trans Magn 17:1247–1248
110. Matsumoto Y (1996) Energy positions of oxide semiconductors and photocatalysis with iron complex oxides. J Solid State Chem 126(2):227–234
111. Michaelson HB (1977) The work function of the elements and its periodicity. J Appl Phys 48:4729–4733
112. Mills A, Le Hunte S (1997) An overview of semiconductor photocatalysis. J Photochem Photobiol, A 108:1–35
113. Mo SD, Ching W (1995) Electronic and optical properties of three phases of titanium dioxide: rutile, anatase, and brookite. Phys Rev B 51:13023
114. Mohajerina S, Mazare A, Gongadze E, Kralj-Iglič V, Iglič A, Schmuki P (2017) Self-organized, free-standing TiO_2 nanotube membranes: effect of surface electrokinetic properties on flow-through membranes. Electrochim Acta 245:25–31
115. Mohamed HH, Bahnemann DW (2012) The role of electron transfer in photocatalysis: fact and fictions. Appl Catal B 128:91–104
116. Moniz SJA (2015) Visible-light driven heterojunction photocatalysts for water splitting—a critical review. Energy Environ Sci 8(3):731–759
117. Moniz SJA, Shevlin SA, Martin DJ, Guo Z-X, Tang J (2015) Visible-light driven heterojunction photocatalysts for water splitting: a critical review. Energy Environ Sci 8:731–759
118. Moniz SJA, Quesada-Cabrera R, Blackma, Tang CS, Southern JP, Weaver PM, Carmalt CJ (2014) A simple, low-cost CVD route to thin films of $BiFeO_3$ for efficient water photo-oxidation. J Mater Chem A 2:2922–2927
119. Morikawa T, Asahi R, Ohwaki T, Aoki K, Taga Y (2001) Band-gap narrowing of titanium dioxide by nitrogen doping. Jpn J Appl Phys 40:L561–L563

120. Moriya Y, Takata T, Domen K (2013) Recent progress in the development of (oxy)nitride photocatalysts for water splitting under visible-light irradiation. Coord Chem Rev 257:1957–1969
121. Nakata K, Fujishima A (2012) TiO_2 photocatalysis: design and applications. J Photochem Photobiol, C 13:169–189
122. Naldoni A, Allieta M, Santangelo S, Marelli M, Fabbri F, Cappelli S, Bianchi CL, Psaro R, Dal Santo V (2012) Effect of nature and location of defects on bandgap narrowing in black TiO_2 nanoparticles. J Am Chem Soc 134:7600–7603
123. Naldoni A, Altomare M, Zoppellaro G, Liu N, Kment STPN, Zboril R, Schmuki P (2019) Photocatalysis with reduced TiO_2: from black TiO_2 to cocatalyst-free hydrogen production. ACS Catal 9:345–364
124. Naldoni A, D'Arienzod M, Altomare M, Marelli M, Scottis R, Morazzoni F, Selli E, Dal Santo V (2013) Pt and Au/TiO_2 photocatalysts for methanol reforming: Role of metal nanoparticles in tuning charge trapping properties and photoefficiency. Appl Catal B: Environ 130–131, 239–248
125. Avarro erga RM, Alvarez-Galvan MC, Vaquerov F, Arenales J, Fierro JLG (2013) Chapter 3—hydrogen production from water splitting using photo-semiconductor catalysts. In: Gandía LM, Arzamendi G, Diéguez PM (eds) Renewable hydrogen technologies. Elsevier, Amsterdam
126. Ni M, Leung MK, Leung DY, Sumathy K (2007) A review and recent developments in photocatalytic water-splitting using TiO_2 for hydrogen production. Renew Sustain Energy Rev 11:401–425
127. Nian J-N, Teng H (2006) Hydrothermal synthesis of single-crystalline anatase TiO_2 nanorods with nanotubes as the precursor. J Phys Chem B 110:4193–4198
128. Ohtani B (2010) Photocatalysis A to Z—what we know and what we do not know in a scientific sense. J Photochem Photobiol C 11:157–178
129. Ohtani B (2013) Titania photocatalysis beyond recombination: a critical review. Catalysts 3:942–953
130. Osterloh FE (2013) Inorganic nanostructures for photoelectrochemical and photocatalytic water splitting. Chem Soc Rev 42:2294
131. Oswald P, Clement O, Chambon C, Schouman-Claeys E, Frija G (1997) Liver positive enhancement after injection of superparamagnetic nanoparticles: respective role of circulating and uptaken particles. Magn Reson Imaging 15:1025–1031
132. Paul A, Lauril T, Vuorinev V, Divinski SV (2014) Structure of materials. Thermodynamics, diffusion and the Kirkendall effect in solids. Springer International Publishing, Cham
133. Penn RL, Banfield JF (1999) Formation of rutile nuclei at anatase (112) twin interfaces and the phase transformation mechanism in nanocrystalline titania. Am Miner 84:871–876
134. Peter LM (2016) Photoelectrochemistry: from basic principles to photocatalysis
135. Pichat P (2007) A brief overview of photocatalytic mechanisms and pathways in water. Water Sci Technol 55:167–173
136. Plocek J, Hutlová A, Nižňansky D, Buršík J, Rehspringer J-L, Mička Z (2003) Preparation of $ZnFe_2O_4/SiO_2$ and $CdFe_2O_4/SiO_2$ nanocomposites by sol–gel method. J Non-Cryst Solids 315:70–76
137. Pokrant S, Dilgre S, Landsmann S, Trotmann M (2017) Size effects of cocatalysts in photoelectrochemical and photocatalytic water splitting. Mater Today Energy 5:158–163
138. Mangrulkar PA, Polshettiwar V, Labhsetwar NK, Varma RS, Rayalu SS (2012) Nano-ferrites for water splitting: unprecedented high photocatalytic hydrogen production under visible light. Nanoscale 4:5202–5209
139. Pullar RC, Bhattacharya AK (2002) Crystallisation of hexagonal M ferrites from a stoichiometric sol-gel precursor, without formation of the α-$BaFe_2O_4$ intermediate phase. Mat Lett 57:537–542
140. Qian R, Zong H, Schneider J, Zhou G, Zhao T, Li Y, Yang J, Bahnemann DW, Pan JH (2019) Charge carrier trapping, recombination, and transfer during TiO_2 photocatalysis: an overview. Catal Today 335:78–90

141. Raj K, Moskovitz BC (1995) Advances in ferrofluid technology. J Magn Magn Mater 149:174–180
142. Ramadan W, Feldhoff A, Bahnemann D (2021 Assessing the photocatalytic oxygen evolution reaction of $BiFeO_3$ loaded with IrO_2 nanoparticles as cocatalyst. Solar Energy Mater Solar Cells 232:111349
143. Ramadan W, Kareem M, Hannoyer B, Saha S (2011) Effect of pH on the structural and magnetic properties of magnetite nanoparticles synthesized by co-precipitation. Adv Mater Res 324:129–132. Presented at the CIMA conference, Beirut-March
144. Ramakrishna S, Kazutoshik F, Teo W-E, Lim T-C, Zuwei M (2005) An introduction to electrospinning and nanofibers
145. Roonasi P, Mater NAY (2015) A comparative study of a series of ferrite nanoparticles as heterogeneous catalysts for phenol removal at neutral pH. Chem Phys 172:143–149
146. Scanlon DO, Dunnill CW, Buckeridge J, Shevlin SA, Logsdail AJ, Woodley SM, Catlow CRA, Powell MJ, Palgrave RG, Parkin IP, Watson GW, Keal TW, Sherwood P, Walsh A, Sokol AA (2013) Band alignment of rutile and anatase TiO_2. Nat Mater 12:798–801
147. Schnider J, Matsuoka M, Takeuchi M, Zhang J, Horiuchi Y, Anpo M, Bahnemann DW (2014) Understanding TiO_2 photocatalysis: mechanisms and materials. Chem Rev 114:9919–9986
148. Serpone N (2006) Is the band gap of pristine TiO_2 narrowed by anion- and cation-doping of titanium dioxide in second-generation photocatalysts? J Phys Chem B 110:24287–24293
149. Serpone N, Lawless D, Khairutdinov R (1995) Size effects on the photophysical properties of colloidal anatase TiO_2 particles: size quantization versus direct transitions in this indirect semiconductor? J Phys Chem 99:16646–16654
150. Sheng W, Myint M, Chen JG, Yan Y (2013) Correlating the hydrogen evolution reaction activity in alkaline electrolytes with the hydrogen binding energy on monometallic surfaces. Energy Environ Sci 6:1509–1512
151. Shibata T, Sakai N, Fukuda K, Ebina Y, Sasaki T (2007) Photocatalytic properties of titania nanostructured films fabricated from titania nanosheets. Phys Chem Chem Phys 9:2413–2420
152. Shichi T, Katsumata K-I (2010) Development of photocatalytic self-cleaning glasses utilizing metal oxide nanosheets. Hyomen Gijutsu 61:30–35
153. Shimura K, Yoshida H (2011) Heterogeneous photocatalytic hydrogen production from water and biomass derivatives. Energy Environ Sci 4:2467–2481
154. Sousa MH, Tourinho FA, Depeyrot JJ, da Silva G, Lara MCFL (2001) New electric double-layered magnetic fluids based on copper, nickel, and zinc ferrite nanostructures. J Phys Chem B 105(6):1168–1175
155. Steinfeld A (2002) Solar hydrogen production via a two-step water-splitting thermochemical cycle based on Zn/ZnO redox reactions. Int J Hydrogen Energy 27:611–619
156. Su R, Forde MM, He Q, Shen Y, Wang X, Dimitratos N, Wendt S, Huang Y, Iversen BB, Kiely CJ, Besenbacher F, Hutchings GJ (2014) Well-controlled metal co-catalysts synthesised by chemical vapour impregnation for photocatalytic hydrogen production and water purification. Dalton Trans 43:14976–14982
157. Tada H (2019) Overall water splitting and hydrogen peroxide synthesis by gold nanoparticle-based plasmonic photocatalysts. Nanoscale Adv 1:4238–4245
158. Tada H (2019) Size, shape and interface control in gold nanoparticle-based plasmonic photocatalysts for solar-to-chemical transformations. Dalton Trans 48:6308–6313
159. Taffa Dereje H, Ralf D, Ulpe AC, Bauerfeind Katharina CL, Thomas B, Bahnemann DW, Michael W (2016) Photoelectrochemical and theoretical investigations of spinel type ferrites ($MxFe_{3-x}O_4$) for water splitting: a mini-review. J Photonics Energy 7(1):012009
160. Takanabe K (2017) Photocatalytic water splitting: quantitative approaches toward photocatalyst by design. ACS Catal 7:8006–8022
161. Tanka A, Teramura K, Hosokawa S, Kominai H, Tanaka T (2017) Visible light-induced water splitting in an aqueous suspension of a plasmonic Au/TiO_2 photocatalyst with metal co-catalysts. Chem Sci 8:2574–2580
162. Tiwari A, Mondal I, Pal U (2015) Visible light induced hydrogen production over thiophenothiazine-based dye sensitized TiO_2 photocatalyst in neutral water. RSC Adv 5:31415–31421

163. Tomas SA, Zelaya O, Palomino R, Lozada R, Garcia O, Yanez JM, Ferrereira Da Silva A (2008) Optical characterization of sol gel TiO_2 monoliths doped with Brilliant Green. Eur Phys J Special Topics 153:255–258
164. Tseng I-H, Wu JCS, Chou H-Y (2004) Effects of sol-gel procedures on the photocatalysis of Cu/TiO_2 in CO_2 photoreduction. J Cataly 221:432–440
165. Walter MG, Warren EL, Mckone JR, Boettcher SW, Mi Q, Santori EA, Lewis NS (2010) Solar water splitting cells. Chem Rev 110:6446–6473
166. Wanf B, Shen S, Mao SS (2017) Black TiO_2 for solar hydrogen conversion. J Materiomics 3:96–111
167. Wang C, Yin L, Zhang L, Liu N, Lun N, Qi Y (2010) Platinum-nanoparticle-modified TiO_2 nanowires with enhanced photocatalytic property. ACS Appl Mater Interfaces 2:3373–3377
168. Wang Q, Domen K (2020) Particulate photocatalysts for light-driven water splitting: mechanisms, challenges, and design strategies. Chem Rev 120:919–985
169. Ramadan W, Dillert R, Koch J, Tegenkamp C, Bahnamnn D (2019) Changes in the solid-state properties of bismuth iron oxide during the photocatalytic reformation of formic acid. Cataly Today 326:22–29
170. Wu L, Mendoza-Garcia A, Li Q, Sun S (2016) Organic phase syntheses of magnetic nanoparticles and their applications. Chem Rev 116(18):10473–10512
171. Xiang Q, Yu J, Jaroniec M (2011) Tunable photocatalytic selectivity of TiO2 films consisted of flower-like microspheres with exposed 001 facets. Chem Commun 47:4532–4534
172. Xiong G, Shao R, Droubay TC, Joly AG, Beck KM, Chambers SA, Hess WP (2007) Photoemission electron microscopy of TiO_2 anatase films embedded with rutile nanocrystals. Adv Func Mater 17:2133–2138
173. Yamaguchi K, Matsumoto K, Fujii T (1990) Magnetic anisotropy by ferromagnetic particles alignment in a magnetic field. J Appl Phys 67:4493
174. Yamakata A, Ishibashi T-A, Onishi H (2001) Water-and oxygen-induced decay kinetics of photogenerated electrons in TiO_2 and Pt/TiO_2: a time-resolved infrared absorption study. J Phys Chem B 105:7258–7262
175. Yamakata A, Ishibashi T-A, Kato H, Kudo A, Onishi H (2003) Photodynamics of $NaTaO_3$ catalysts for efficient water splitting. J Phys Chem B 107:14383–14387
176. Yan X, Chen X (2011) Titanium dioxide nanomaterials. Encycl Inorg Bioinorg Chem 1–38
177. Yang L, Zhang Y, Liu X, Jiang X, Zhang Z, Zhang T, Zhang L (2014) The investigation of synergistic and competitive interaction between dye Congo red and methyl blue on magnetic $MnFe_2O_4$. Chem Eng J 246:88–96
178. Yao T, An X, Han H, Chen JQ, Li C (2018) Photo electrocatalytic materials for solar water splitting. Adv Energy Mater 8:1800210
179. Yoshihara T, Katoh R, Furube A, Tamaki Y, Murai M, Hara K, Murata S, Arakawa H, Tachiya M (2004) Identification of reactive species in photoexcited nanocrystalline TiO_2 films by wide-wavelength-range (400–2500 nm) transient absorption spectroscopy. J Phys Chem B 108:3817–3823
180. Yu JC, Ho W, Yu J, Yip H, Wong PK, Zhao J (2005) Efficient visible-light-induced photocatalytic disinfection on Sulfur-doped nanocrystalline titania. Environ Sci Technol 39:1175–1179
181. Yun H, Lee H, Joo J-B, Wooyoung K, Yi J (2009) Influence of aspect ratio of TiO_2 nanorods on the photocatalytic decomposition of formic acid. J Phys Chem C J Phys Chem C 113:3050–3055
182. Zhang L, Mohamed HH, Dillert R, Bahnemann D (2012) Kinetics and mechanisms of charge transfer processes in photocatalytic systems: a review. J Photochem Photobiol, C 13:263–276
183. Zhang X, Peng T, Song S (2016) Recent advances in dye-sensitized semiconductor systems for photocatalytic hydrogen production. J Mater Chem A 4:2365–2402
184. Zhang X, Song P, Cui X (2013) Nitrogen-doped TiO_2 photocatalysts synthesized from titanium nitride: characterizations and photocatalytic hydrogen evolution performance. J Adv Oxid Technol 16:131–136

185. Zhao T, Liu Z, Nakata K, Nishimoto S, Murakami T, Zhao Y, Jiang L, Fujishima A (2010) Multichannel TiO_2 hollow fibers with enhanced photocatalytic activity. J Mater Chem 20:5095–5099
186. Zheng Z, Huang B, Qin X, Zhang X, Dai Y (2010) Strategic synthesis of hierarchical TiO_2 microspheres with enhanced photocatalytic activity. 16:11266–11270

Carbon and Metal Doped Polyaniline (PANI) for Energy Storage

Abdallah Ramadan and Wegdan Ramadan

Abstract With the depletion of traditional fossil fuels, rising pollution levels and fast growth of the global economy. New technology for energy conversion and storage, as well as efficient, sustainable energy sources, are all urgently needed. The development of supercapacitors (SCs) as an energy storage device has received a lot of interest in recent years. SCs are comparable to dielectric capacitors in terms of their high-power density, cyclic stability, and discharge rate. In addition, a high energy density that is comparable to batteries. In this chapter, polyaniline (PANI) based materials for electrochemical supercapacitor (ESs) electrodes are thoroughly reviewed. Pure PANI electrodes have low cycle life, low power density, and poor mechanical stability resulting from the swelling and shrinkage during the charging and discharging processes. Nevertheless, the development of nanocomposite of PANI with carbon materials or metal compounds could overcome the drawbacks of pure PANI and achieve higher electrochemical performance. Capacitance, energy, power, cycle performance, and rate capability have all been used to evaluate the performance of nanocomposites.

Graphical Abstract See Scheme 1.

Keywords Energy storage · Supercapacitors · PANI · Electrochemical double-layer capacitors · Pseudocapacitors

1 Introduction

Due to the depletion and shortage of non-renewable fossil fuels-based energy sources, a serious challenge in the current century is the energy crisis [26]. Additionally, the environment, people, and human activities are negatively impacted by the growing emissions of pollutants from nonrenewable energy (oil, gas, and coal). According to the Environmental Protection Agency (EPA), Carbon dioxide (CO_2) makes up 82%

A. Ramadan · W. Ramadan (✉)
Physics Department, Faculty of Science, Alexandria University, Alexandria 21511, Egypt
e-mail: wegdan.ramadan@alexu.edu.eg

I. Uddin and I. Ahmad (eds.), *Synthesis and Applications of Nanomaterials and Nanocomposites*, Composites Science and Technology,
https://doi.org/10.1007/978-981-99-1350-3_12

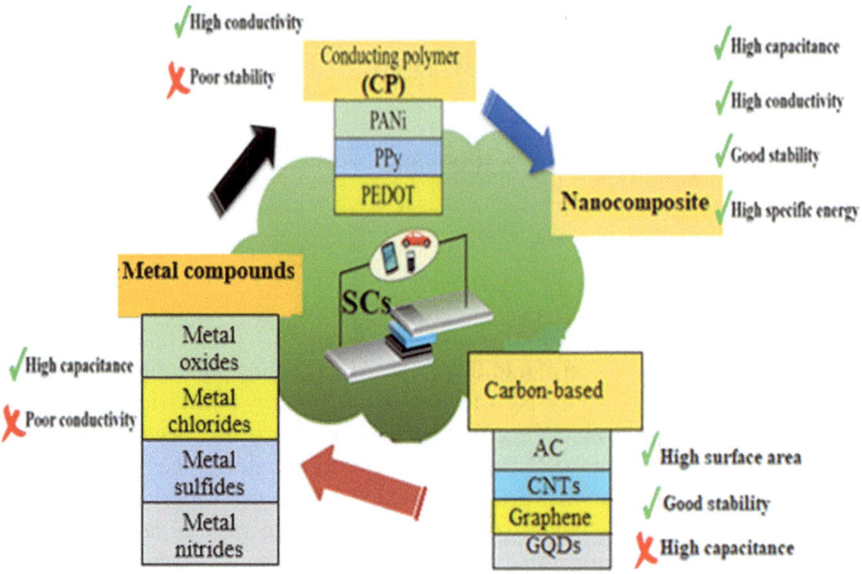

Scheme 1 Electrode materials of supercapacitor with advantages and disadvantages [35]

of emissions that are classified as greenhouse gases, followed by methane (CH_4), 9%, nitrates (NOx), 6%, and fluorinated gases (perfluorocarbons, hydrofluorocarbons, and sulfur hexafluoride), which make up 3%. Climate change consequences have been made worse by greenhouse gas emissions. These outcomes have pushed research efforts towards the creation of renewable energy sources. Energy storage devices are necessary when harvesting energy from renewable sources like photovoltaic panels or wind farms because the captured energy (such as wind or solar energy) is not always available. For example, the sun doesn't shine every day and night, and the wind doesn't always blow when we need it to. Consequently, one of the key problems that will improve the use of sustainable energies in the future is energy storage, particularly electrical energy storage. Among various energy storage technologies, lithium-ion batteries (LIBs) and supercapacitors hold great promise in broad applications such as portable electronics, smart grids and electrical vehicles [6]. Since the Sony Company first began selling LIBs in 1991, this type of secondary battery has become increasingly common in people's daily lives. Because LIBs have a high energy density of 150–200 Wh/kg, they may store electrical energy in portable devices. With a power density at least two orders of magnitude lower than that of fossil fuels, LIBs are very difficult to compare with them. Because of this, a significant number of LIBs are required to replace fuels, which sharply increases the mass of products. Different electrical energy storage devices and conversion technologies (Fig. 1) can be identified by many parameters, including energy storage mechanisms, charging and discharging processes, energy, and power densities, which define their

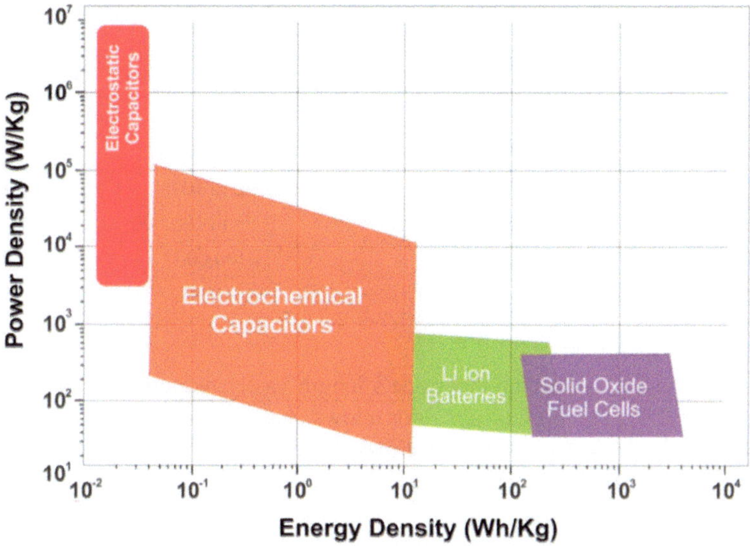

Fig. 1 Ragone plot for various electrical energy storage devices (figure used by permission of AIP Publishing) [1]

uses. Because of their slow discharge process, batteries may be employed for long-term and steady energy storage density. On the other hand, Capacitors can be used in applications that demand quick energy transmission due to their fast charging and discharging capabilities. Between traditional capacitors and batteries, supercapacitors, also known as ultracapacitors or electrochemical capacitors ESs, are a new energy storage technology. Due to its many advantages over lithium-ion batteries, such as quick charging and discharging, enormous power density, wide working temperature ranges, and extended service life cycles, supercapacitors have gained increased attention in recent years. As a result, ESs have been found to be readily applicable in a wide variety of important applications such as hybrid and electrical vehicles, portable electronics, backup power supplies, aircraft, and smart grids [9]. Batteries, capacitors, and supercapacitors are compared in Table 1 in terms of their characteristics and performance.

2 Fundamentals of Supercapacitors

A basic supercapacitor consists of two plates (current collector) coated with porous materials and separated by a porous separator soaked in an electrolyte. Each component plays an important part in determining a supercapacitor's overall performance. The supercapacitor electrode materials are deposited on conductive substrates called current collectors, it allows electrons to travel in the direction of the electrode's

Table 1 Comparison of conventional capacitor, supercapacitor and battery (table used by permission of Elsevier) [63]

Parameters	Conventional capacitor	Supercapacitor	Battery
Time of charge	10^{-6}–10^{-3} s	Sec to min	0.03–3 h
Time of discharge	10^{-6}–10^{-3} s	Sec to min	1–5 h
Energy density (Wh/kg)	<0.1	Up to 1091	Up to 1,606
Power density (W/kg)	>10,000	Up to 196,000	<1000
Cycle life	>500,000	>100,000	500–2000
Charge/discharge efficiency	~1.0	0.9–0.99	0.7–0.85

thickness instead of along the electrode's length vastly reducing the charge transport distance [86]. The porous separator allows charge transfer, and the electrolyte interacts with both electrodes. Because they determine the performance of supercapacitors, particularly the energy/power densities and life cycles of a supercapacitor, electrolytes (salt and solvent) used in supercapacitor cells are as important as the electrode materials. The electrolytes also influence the series resistance and the operating voltage of the cell. Finally, the electrode material is the most important component to achieve significant improvements in supercapacitor performance. Below is a list of the characteristics an electrode material must possess.

- To create electrochemical double-layer capacitors, a high specific surface area is required, generally between 1000 and 2000 m^2g^{-1}.
- Controlled distribution of pore size.
- Reversible redox reactions to prevent stability loss.
- Electrochemical stability beyond the limit of electrolyte decomposition.
- Surface wettability of the electrolyte.

3 Types of Supercapacitors

Based on the mechanism of charge and discharge process, there are two main types of supercapacitors as shown in Fig. 2 namely: non-Faradic, electrochemical double-layer capacitors (EDLCs) and pseudocapacitors (or redox-based electrochemical capacitors, Faradic). However, there is a hybrid type that combines the mechanism of both types mentioned above, hybrid electrochemical capacitors.

3.1 Electrochemical Double Layer Capacitors

The EDLC supercapacitor type is the first family of supercapacitors that employ porous carbon-based materials with a large specific surface area as the electrode

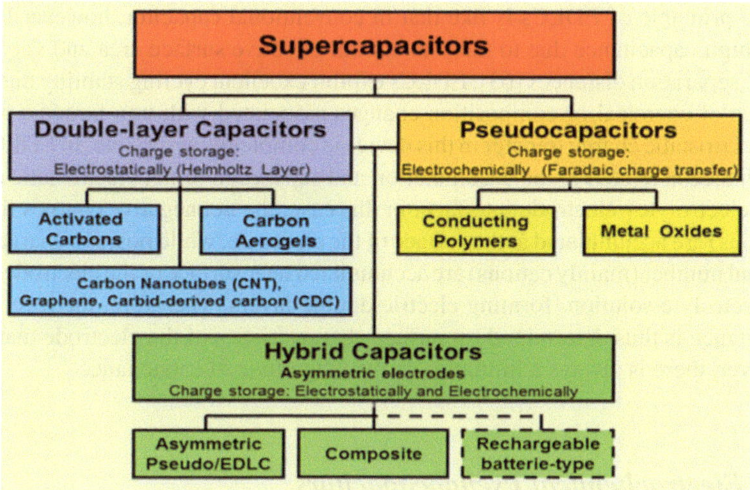

Fig. 2 Different types of supercapacitor electrodes (figure used by permission of Elsevier) [23]

material. To be more specific, the charging-discharging of an EDLC is by the accumulating of electrolyte ions onto the surface of a porous electrode by polarization. The solvated ions flow towards the two carbon electrodes when DC voltage is applied, passing across the separator to create an electric double-layer that stores the electric charges as seen in Fig. 3. The two parallel charge layers that surround the plate are referred to as the double layer.

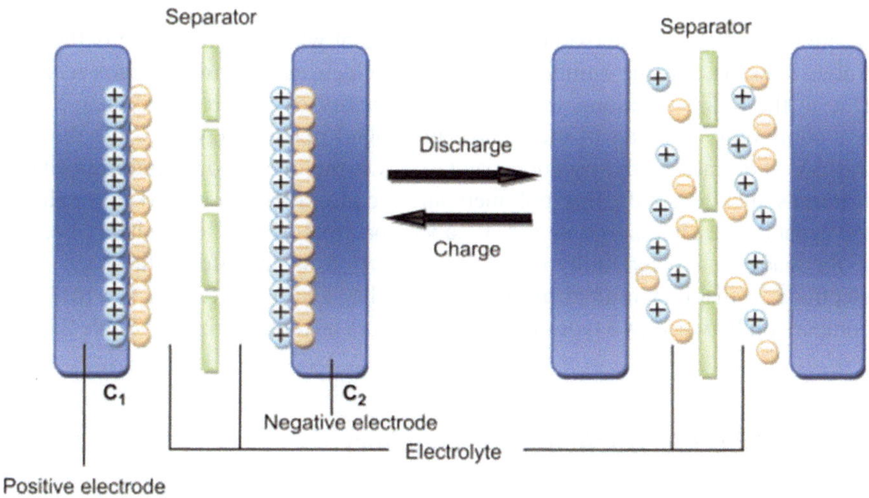

Fig. 3 Charged and discharged mechanism of an electric double layer capacitor (figure used by permission of Elsevier) [41]

The principle of EDLCs is like that in conventional capacitor, however EDLCs show high capacitance due to their maximum effective surface area and very small charge separation distances [63]. EDLCs exhibit excellent cycling stability due to the absence of chemical or composition changes associated with non-faradaic process and electrostatic charge transfer in this device is completely reversible. In EDLCs, the capacitance is related to the accumulation and separation of net electrostatic charge at the electrolyte– electrode interface. In this case, the net negative charges (mainly electrons) are accumulated at the surface of the electrode, while positive charges with an equal number (mainly cations) are accumulated near surface of the electrode facing the electrolyte solution, forming electric double-layers. Therefore, the level of the capacitance is thus determined by surface characteristics of the electrode materials. However, there is always a limitation in the magnitude of capacitance.

3.2 Electrochemical Pseudocapacitors

Unlike EDLCs, some electrochemically active materials called pseudocapacitive materials could present much higher capacitance. Pseudocapacitive materials are the second type of materials used in supercapacitors where the charging-discharging of pseudocapacitors happens at the electrode surface or close to it through fast and reversible redox reactions between electrolyte ions and electroactive species [34]. The common materials that exhibit pseudocapacitive behavior are conducting polymers and transition metal oxides/hydroxides [77]. The redox process can be described as:

$$O_X + ne^- \Longleftrightarrow R_d \tag{1}$$

where O_X and R_d are the oxidant and the reductant which are insoluble in the electrolyte solution, n is the number of transferred electrons (e^-) in the redox reaction. Due to the redox reactions, the change in quantity of the produced charge (Δq) is a continuous function of the change in applied potential (ΔV), while the ratio $\Delta q/\Delta V$ is defined as the pseudocapacitance. On the one hand, pseudocapacitance materials such as conducting polymers and metal oxides can remarkably enhance the capacitance and energy stored of supercapacitors. On the other hand, because to the inadequate reversibility of the redox reaction, which affects the shape of the electrode, it displays limited mechanical stability and a short cycle life [61]. The comparison between two types of supercapacitors are summarized in Table 2.

3.3 Hybrid Electrochemical Capacitors

The hybrid type supercapacitors are an energy storage device which exploits the advantages of EDLC and the pseudocapacitors using composites or two asymmetric electrodes. The high power density of the EDLC and higher capacitance and energy

Table 2 Comparison between electrical double layer capacitor and pseudocapacitor

	Electric double layer capacitor	Pseudocapacitor
Storage mechanism	Electrostatic adsorption (non-faradaic process)	Redox reactions on electrodes (faradaic process)
Capacitance	Capacitance remains constant	Capacitance changes with voltage
Energy density	Low	High
Power	High power due to good diffusion	Low Power due to kinetic limitations
Diffusion resistance	Low	High
Reversibility	High reversibility	Moderate reversibility
Cycle life	Good cyclability	Moderate cyclability
Potential window	Narrow (1–3 V)	Large (1–5 V)

density of the pseudocapacitors is combined to form a device with better performance [47]. A single electrode in a hybrid configuration system that combines carbon-based materials with conducting polymers or metal oxides provides the advantages of both physical and chemical charge storage mechanisms. Asymmetric supercapacitor denotes to the combination of two separate electrode materials that use two distinct storage mechanisms. Ideally, a pseudocapacitive material electrode provides high energy density while the EDLC material electrode furnishes the system with high power capability [71].

4 Electrochemical Evaluation

The electrochemical assessment is based on their specific capacitance and capacity, as well as their energy and power densities. Different approaches, including as cyclic voltammetry (CV), and galvanometric charge discharge technique (GCD) can be used to calculate the above-mentioned parameters. These parameters are measured using techniques such as three electrode and two electrode configurations.

Because it is simple to determine the cycle life, CV has emerged as a key technique for assessing the performance of SCs. Additionally, the information on the impacts of internal resistance and the ensuing dissipative losses may be discovered by analyzing the shapes of the voltammograms as a function of scan rate s. The form of the cyclic voltammogram for EDLCs should resemble a rectangle, confirming no chemical reaction as in Fig. 4a, the rectangular shape is distorted with reversible oxidation and reduction peaks for the pseudocapacitive material, indicating a Faradaic redox process, as shown in Fig. 4b. CV data can be utilized to determine the specific capacitance Cs of an electrode material by calculating the area enclosed in cyclic voltammogram $\int \boldsymbol{I}(\boldsymbol{V}) * \boldsymbol{dV}$ using the following formula [60]:

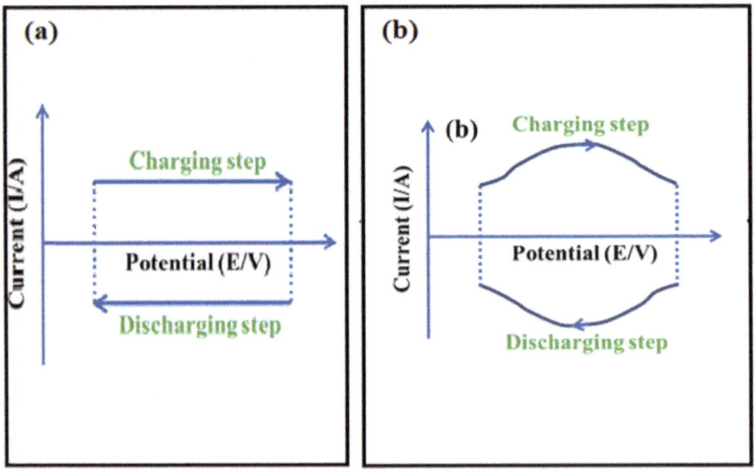

Fig. 4 Cyclic voltammograms of **a** EDL capacitor and **b** pseudo-capacitor (figure used by permission of Elsevier) [66]

$$Cs = \frac{\int I(V) * dV}{2m * s * \Delta V} \tag{2}$$

where **m** (g) is the mass loading of active material, ΔV (V) is the potential window, and **s** (mV/s) is the scan rate

The galvanostatic charge and discharge method is a more precise approach than the CV method for determining the specific capacitance of active materials. The working electrode is exposed to positive and negative continuous currents that charge and discharge the electrode within a predetermined voltage range while timing the process. Figure 5 displays a typical GCD curve that plots the voltage as a function of time. The specific capacitance Cs by F/g related to the discharge current density (I) by A/g through the relation [61, 73].

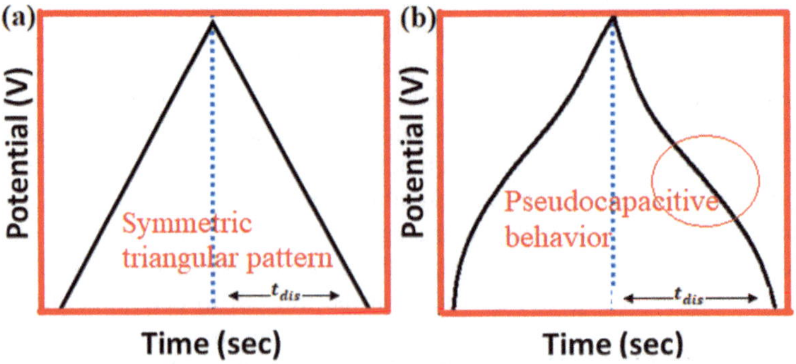

Fig. 5 GCD curves of **a** EDL capacitor and **b** pseudo-capacitor

$$Cs = \frac{I * t_{dis}}{m * \Delta V} \quad (3)$$

where t_{dis} (s) is the time of discharge and ΔV(V) is the working potential.

The characteristics of the active material are determined by the type of the GCD profiles, much like with the CV curves. Due to their consistent charge distributions during the charge–discharge process, an ideal EDLC profile has a linear GCD (Fig. 5a). The oxidation and reduction processes that occur throughout the charge–discharge process cause the pseudo-/Faradaic supercapacitors to have a non-linear GCD profile (Fig. 5b). To learn more about a supercapacitor electrode's electrochemical performance, the GCD measurement is required. It is used to do additional calculations about electrochemical characteristics including energy, power densities, and capacitance retention. The energy density (E) and power density (P) can be calculated using the following equations [91]:

$$E(Wh/Kg) = \frac{1}{2} * Cs * V^2 * \frac{1000}{3600} \quad (4)$$

$$P(W/Kg) = \frac{E}{t_{dis}} * 3600 \quad (5)$$

where Cs (F/g) is the specific capacitance, V (V) is the potential window and t_{dis} (s) is the discharge time corresponding to Cs calculated from GCD curve.

5 Polyaniline PANI

The most studied conducting polymer among the family of conducing polymers is polyaniline (PANI) due to its interesting characteristics such as ease of synthesis, doping, dedoping, low cost, mechanical flexibility, chemical properties, and environmental compatibility. There are numerous ways to synthesize polyaniline, including chemical and electrochemical polymerization. Using chemical polymerization as an example, aniline monomer is placed in an aqueous acid solution and chemical oxidants such as ammonium persulfate or ferric chloride. The polyaniline precipitates out of the chemical reaction solution. When preparing PANI nanocomposites, chemical synthesis of polyaniline provides much greater flexibility for managing the nucleation and growth mechanisms throughout the polymerization process [14]. While electrochemical polymerization is typically performed in a three-electrode electrochemical cell with an electrolyte that supports the electrochemical reaction and dissolves the aniline monomer. The monomers oxidized by repeating cycles between a potential window where the oxidation takes place and the polymer will then be electrodeposited directly onto the substrate which is a worthwhile advantage for the fabrication of binder-free supercapacitors [53].

5.1 PANI-Based Materials for Pseudocapacitors

Morphology is one of the primary factors affecting PANI's supercapacitive behavior. Varying the PANI morphology during the preparation can lead to better performance of supercapacitor [14]. Guan and coworkers [21] reported that adding a small amount of para-phenylenediamine (PPD) as additive during the polymerization of aniline can lead to the formation of longer, less entangled PANI nanofibers, which significantly enhances the electrochemical performance of PANI. Even in two electrode cells where the electroactive material is not immersed in the electrolyte to expand the electrode/electrolyte contact, the function of PANI morphology is crucially important. Also, changing the chemical polymerization temperature can produce different morphologies of PANI which could provide different cyclability. Porous PANI can also considerably enhance the performance of supercapacitor due to increasing the electrochemically accessible surface area. Sharma and coworkers [64] synthesized a nanoporous hypercrosslinked PANI with a S_{BET} of 1059 m^2/g which could deliver a specific capacitance of 410 F/g with outstanding cycling stability with 100% of the capacity retention after 1000 cycles. The electrochemical performance based on different morphologies of PANI reported in the literature are listed in Table 3.

The type of the dopant ions, electrolyte, pH of the solution, the substrate, and the deposition conditions all have an impact on the electrochemical and mechanical characteristics of the PANI supercapacitor electrode, which in turn has an impact on the capacitance of the PANI electrode [18]. The main drawbacks of PANI are the poor mechanical stability and solubility. As a result of repeated redox reactions, PANI suffer several physical/chemical changes, including swelling, shrinking, cracking or breaking, which over time reduces the material's performance. Therefore, the synthesis of composite materials may be a promising choice to improve the mechanical performance of PANI.

6 PANI/Carbon Based Material Composites

The aim of developing composite material electrodes is to incorporate the desirable characteristics of the various materials into a single electrode in order to improve stability and capacitance. Additionally, asymmetric type supercapacitors employ nanocomposite electrodes to boost operating voltage to enhance energy density. The electrodes in the PANI-carbon composites have not only efficient pseudocapacitive reactions due to synergistic effects, but also better rate capability due to the substrate's high conductivity and super cycle stability due to the excellent mechanical properties of the supporting carbon materials. Due to synergistic effects, the electrodes in PANI-carbon composites not only have effective pseudocapacitive reactions but also better rate capability and super cycle stability due to the substrate's high conductivity and the superior mechanical properties of the supporting carbon materials, respectively [45].

Table 3 Capacitance performance of different morphologies of PANI

PANI morphology	Electrolyte	Current collector	Specific capacitance	Capacitance retention	Refs.
Nanoporous PANI	1 M H_2SO_4	Carbon paper	350 F/g at 1 A/g	99% after 500 cycles at 40 A/g	[17]
PANI nanowires	1 M H_2SO_4	Stainless steel	742 F/g at 23 A/g	92% after 1500 at 3 mA/cm^2	[22]
PANI Nanofibers	1 M H_2SO_4	Stainless steel mesh	548 F/g at 0.18 A/g	75% after 1000 cycles at 1 A/g	[21]
Nanofibrous PANI	1 M H_2SO_4	Stainless steel	839 F/g at 10 mV/s	90.71% after 1000 cycles at 20 mV/s	[10]
Nanofibrous PANI	1 M H_2SO_4	Stainless steel	861 F/g at 10 mV/s	82% after 2000 cycles at 100 mV/s	[11]
Nanowire arrays	1 M $HClO_4$	Au plate	950 F/g at 1 A/g	88% after 500 cycles at 20 A/g	[72]
PANI nanowhiskers	1 M H_2SO_4	Wafer	470 F/g at 1 A/g	90.4% after 1000 cycles at 1 A/g	[81]
PANI nanobelts	1 M H_2SO_4	Ti sheet	873 F/g at 10 mV/s	96.5% after 1000 cycles at 10 mV/s	[37]
PANI nanotubes	1 M H_2SO_4	Glassy carbon	714 F/g at 0.5 mA	85% after 500 cycles at 1 mA	[65]
PANI nanospheres	1 M H_2SO_4	Graphitized carbon paper	345 F/g at 1 A/g	58% after 5000 cycles at 10 A/g	[8]
PANI nanosheets	1 M H_2SO_4	Glassy carbon	272 F/g at 1 A/g	50% after 1500 cycles at 3 A/g	[49]
PANI nanolayers	1 M H_2SO_4	ITO nanowires	738 F/g at 4 A/g	100% after 700 cycles at 4 A/g	[67]
PANI nanogranules	0.1 M H_2SO_4	Carbon paper	500 F/g at 1.5 A/g	80% after 5000 cycles at 1.5 A/g	[40]
PANI nanorods	1 M H_2SO_4	–	297 F/g at 1 A/g	65% after 1300 cycles at 100 mV/s	[74]
PANI nanofibers	1 M H_2SO_4	Stainless steel mesh	252 F/g at 0.5 A/g	76% after 1000 cycles at 2 A/g	[27]

6.1 PANI/Activated Carbon

Activated carbons ACs are one of the cheapest materials which generated by either physical or chemical activations of natural precursors (e.g. wood, coal, nutshell, etc.) [89] The specific capacitance values of supercapacitor electrode based on activated carbon are typically between 100 and 300 F/g in both aqueous and organic electrolyte operating in potential windows between 1 and 3 V. The main drawback of ACs for supercapacitor application is the wide distribution of pore size, including micropores (<2 nm), mesopores (2–50 nm) and macropores (>50 nm) [56]. Since not all these pores are efficient for supercapacitor energy storage, Micropores with a diameter around 1 nm are not accessible to most of the organic electrolyte ions. These unnecessary micropores not only increase the volume that not contributing to the charge storage, but also hinder the electrical conductivity, limit their energy density (5–8 Wh/kg) and power density [88]. To solve this issue, nanocomposite of PANI and ACs have been investigated to enhance electrochemical performance. PANI nanorods uniformly polymerized onto the cellulous-derived highly porous activated carbons (C-ACs) framework by a chemical polymerization process have been investigated as a supercapacitor electrode material by Zhang and coworkers [93]. Due to their great electron conductivity, rapid ion transit, quick and stable surface redox reactions, PANI/C-AC composites with a 3D and hierarchically porous structure were able to achieve good capacitive performance. The PANI/C-AC composites displayed a specific capacitance of 765 F/g at a current density of 1 A/g, a specific power of 14 kW/kg, and a specific energy of 22.3 Wh/kg at a scan rate of 10 mV/s. Additionally, PANI/C-ACs composites shown outstanding cycling stability with 91% capacitance retention after 5000 cycles in a symmetric two-electrode system, in contrast to PANI electrodes, which display 82% capacitance retention. Furthermore, PANI/C-ACs have a higher rate capability because the 3D C-ACs skeletons provide enough mechanical support to avoid PANI volumetric changes during the charge/discharge process. As a result, PANI/C-ACs are potential electrode materials for high-performance SCs.

To overcome the low conductivity and hydrophobicity of ACs, nitrogen doped ACs been investigated for the purpose of increasing specific capacitance. The purpose of nitrogen doping is to increase pseudo-capacitance as a result of the redox reaction that occurs in the nitrogen-contained functional groups. Furthermore, the functional groups improve the electrode's wettability, resulting in an increase in capacity.

Yu and coworkers [88] developed hierarchical nitrogen-doped porous carbon (HPC)/polyaniline (PANI) using wheat flour as carbon precursor, as shown in Fig. 6. The PANI nanowire arrays were well-ordered on both the interior and external surfaces of the hierarchical porous HPC, facilitates the electrolyte ion into the whole electrode and ensures the highly efficient charge storage, resulting in a high specific capacitance of 1080 F/g. The 3D interconnected honey-comb-like porous structure with nitrogen doping can further increase the electrode's surface wettability and provide mechanical support for PANI nanowires.

6.2 PANI/Carbon Nanotubes

Carbon nanotubes (CNTs) have gained popularity as promising materials in an energy storage device as a result of their unique properties. These properties such as large

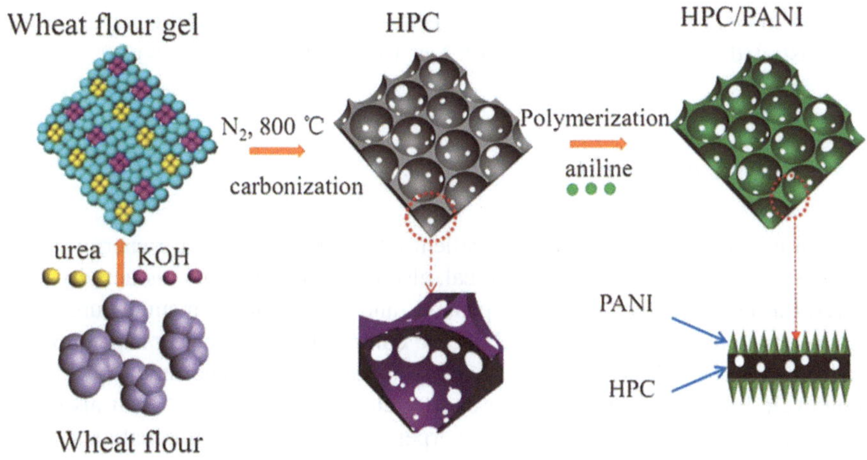

Fig. 6 Schematic illustration for the preparation of hierarchical nitrogen-doped porous carbon HPC/PANI composites (figure used by permission of John Wiley and Sons) [88]

surface area (S_{BET} = 1600 m^2 g^{-1}), different spaces for storage the electrolyte ions based on their structure, light weight, high electrical conductivity (=105 S cm^{-1}) and mechanical properties [63]. SWCNT, MWCNT, and functionalized CNT are examples of CNT layers that may be used to achieve different characteristics in supercapacitors. Surface functionalization of hydrophilic groups increased the wettability of CNTs in solvents [75]. In recent years, PANI/CNT composites have received a lot of interest in the field of energy storage based on the PANI or CNTs alone. CNT facilities the electrolyte ions to diffuse into the PANI/CNT nanocomposite electrode materials and can also improve the cycle stability of the composite during charging/discharging. Furthermore, the greater mechanical strength of PANI/CNTs electrodes allows us to construct lightweight, flexible, and foldable supercapacitors with thinner separators, eliminating the need for a binder or heavy metal foil current collectors [45]. To enhance capacitive performance, smaller size of CNT arrays with hierarchically porous structures should be aligned, for example, Zhang and coworkers [90] prepared core–shell structured PANI/CNTs arrays composite Using an electrodeposition approach. PANI/CNTs electrode material showed specific capacitances as high as 1030 F/g, excellent rate capability (95% capacity retention at 118 A/g) and great stability (5.5% capacity loss after 5000 cycles).

A plasma polymerization method was developed by Hussain and coworkers [29] to synthesize a PANI/CNT composite that is green, simple, fast, oxidizers and binder free. It produces a pinhole-free ultra-thin film with controlled thickness. Low pressure plasma accelerated chemical vapour deposition of PANI on CNT ensures a regular morphology of PANI in contrast to simple electrochemical deposition, which results in random morphology and narrows the conducting pathways. Vertically aligned CNTs and PANI/CNT electrodes have specific capacitances of 12 F/g and 1225

F/g at 5 mV/s, respectively. In a three-electrode system, the PANI/CNT electrodes demonstrated high cycling stability (65% at 15 A/g after 5800 cycles).

6.3 PANI/Carbon Fiber

Carbon nanofibers (CNFs) are an excellent alternative to CNTs for energy storage devices due to their superior mechanical, electrical, and thermal conductivity properties. Furthermore, they are less expensive, and are simpler to manufacture. There is a lot of interest in developing PANI-CNF composite with superior supercapacitor performance since PANI materials have high specific capacitance and CNFs have a long cycle life. CNFs not only offer shorter electron pathways [45] but also allow ions to enter the fiber from directions perpendicular to its longitudinal axis, which makes them ideal for CNF-based supercapacitors.

Yanilmaz and coworkers [84] used the sol–gel and electrospinning methods to fabricate binder-free flexible PANI/carbon nanofiber electrodes. The hybrid electrode outperformed an individual carbon nanofiber electrode in terms of specific capacitance (234 F/g) and cycling stability (90% after 1000 cycles). In addition, it has a high energy density of 32 Wh/kg at a power density of 500 W/kg. The remarkable pseudo-capacitive characteristics of the PANI coating on the carbon nanofiber are responsible for the high performance. The possible reason for poor rate capability of PANI/CNF nanocomposites was the high internal resistance of CNFs carbonized below 1000 °C [45], which led to a negative effect on electron transport and decrease the capacitive behavior of the nanocomposite. The electrical conductivity and hydrophilicity of CNFs should be considered in order to increase the rate capability, graphitized electrospun carbon fibers GECF which carbonized at 2200 °C were immersed in concentrated sulfuric acid for a short time [98]. Then Hou and coworkers [25] used in situ polymerization to grow long, ordered and needle-like PANI nanowires on GECF surface. 3D flexible composite of PANI/GECF electrodes without using any binders and conductive additives presented a high specific capacitance of 976.5 F/g at 0.4 A/g, an energy density approached 86.8 Wh/kg, and a capacitance retention ratio of 89.2% after 1000 cycles at 10 A/g. Furthermore, at a high current density of 50 A/g, PANI/GECF electrodes kept a specific capacitance of about 500 F/g and a coulombic efficiency of 95%.

To achieve the best possible supercapacitive performance and to push the energy density of supercapacitor electrode to a new limit; the active material, electrolyte, and substrate must all be engineered to operate together in a supercapacitor. Hashemi and coworkers [24] fabricated tubular form with rectangular pores of PANI as an active material on chemically functionalized carbon cloth FCC as a substrate as shown in Fig. 7a–b. Then, they added 1,4-napthoquinone (NQ) to the electrolyte as a redox additive, which not only offers pseudo- capacitance through direct redox processes on the electrode surface, but also forms the basis for a regeneration pathway to long-term usage of electrode active materials. After that, asymmetric configuration device used electrolyte of an acidic polymer hydrogel containing the redox additive and made up of two electrodes PANI/FCC as the positive electrode and negative electrode of AC-FCC as shown in Fig. 7a. In the presence of NQ and at current density of 1.4 A/g, this device exhibited an outstanding specific capacitance of roughly 4007 F/g, which is more than 14.5 times greater than when NQ wasn't present. (Fig. 7c). Furthermore, the device shows extremely high energy density of 1091 Wh/kg, high-power density up to 196 kW/kg, and 84% cycling stability under 35 A/g current density during 7000 cycles Fig. 7d–e. A clock, a red LED, and a rotor were all successfully operated for 1 h and 17 min, 47 min, and more than 20 s, respectively, by two asymmetric devices connected in series. (Fig. 7f).

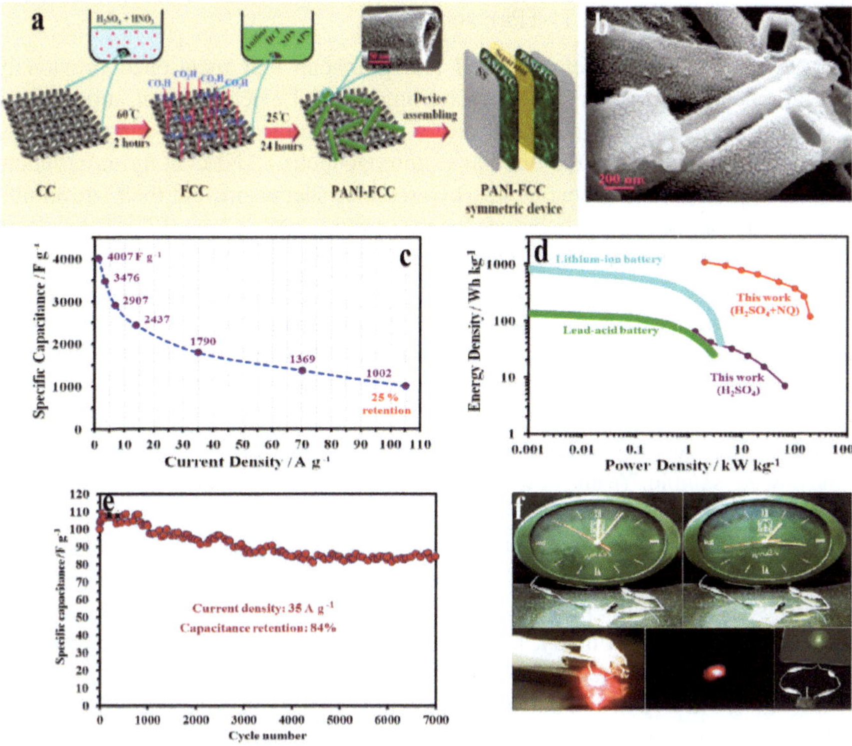

Fig. 7 **a** Schematic design of (PANI-FCC) device. **b** Field emission scanning electron microscope (FE-SEM) images of PANI. **c** estimated capacitance for an AC-FCC//PANI-FCC device from 1.4 to 105 A/g in the presence of napthoquinone NQ as a function of current density. **d** Ragone plots for an asymmetric device at different current densities compared to active mass normalized commercial devices. **e** Cyclability under 35 A/g current density during 7000 cycles; and **f** photographs of running a clock, lighting a red LED and turning a rotor with two asymmetric supercapacitors in series (figure used by permission of Elsevier) [24]

6.4 PANI/Graphene Electrode Materials

Graphene, a one-atomic thick layer of 2-D with unique morphology, was discovered in 2004 by Novoselov and Geim [55] to be a potential electrode material. Graphene has exceptional properties, such as [69]:

- High electrical conductivity (~200 Sm^{-1}),
- Great thermal conductivity (5000 W m^{-1} K^{-1}),
- Large surface area (> 2600 m^2g^{-1}),
- Great charge mobility (> 200,000 cm^2 V^{-1} s^{-1}),
- Excellent chemical and thermal stability,
- Strong Young's modulus (1 TPa); and

Based on the advantages of PANI and graphene that mentioned previously, PANI/graphene nanocomposites have become crucial electrodes for supercapacitors due to the high mechanical stability with a wide thermal operating range and the synergistic effects. On the one hand, homogeneous PANI coating on graphene sheets acts as a spacer to separate nearby graphene sheets, prevent their irreversible aggregation from occurring, and raise the usage ratio, adding more EDL capacitance to the overall capacitance [45]. On the other hand, graphene sheets offer outstanding chemical stability, high surface area, significant increase in electrical conductivity, fast electron/ion transport and a wide working potential in electrochemical devices. These advantages render it as an attractive support material to restrict the volumetric changes (expansion/contraction) and improve the cycle stability of PANI [45]. Depositing a PANI on a graphene electrode without a binder by electrochemical polymerization can overcome the drawbacks of poor stability and high resistance of the powder sample. Ye and coworkers [85] synthesized an ordered PANI nanowire array by electrochemical method on partial exfoliated graphite substrate to produce a hierarchical nanostructure. It effectively reduced the disadvantage of "dead volume" caused by the PANI stacking. Moreover, a pathway for the quick transport of electrons and ions is created by the flawless bonding of the graphene nanosheets and orderly PANI nanowire arrays. Finally, high cycle stability is ensured by the presence of graphene, which reduces the volumetric changes of PANI. A high capacitance of 3.57 F/cm^2 (607 F/g at 1 A/g) and outstanding cycling stability (80.4% after 10,000 GCD cycles) were displayed by the ordered PANI nanowire and the graphene network [85]. Adding functional groups and vacancy defects to the graphene surface provide a great beneficial in energy storage mechanism. By enriching the active sites of graphene, the specific capacitance and cycling stability of PANI/graphene composites may be greatly enhanced. Zheng and coworkers [96] developed a route for the preparation of a high electrochemical performance electrode based on three-dimensional multi-growth site graphene MSG/PANI nanocomposite. Active sites on the surface of graphene nanosheets, such as oxygen functional groups and carbon vacancy defects, were produced using chemical treatment (70% HNO_3 and 30%H_2O_2). MSG/PANI nanocomposites were found to have a specific capacitance of 912 F/g at 1 A/g, which is much greater than GO/PANI without acid treatment (432 F/g), maximum specific energy and maximum specific power of 30 Wh/kg and 3200 W/kg, respectively. The capacitance retention rate of the nanocomposite achieved 86.4% at a high current density of 20 A/g, and after 10,000 GCD cycles at 10 A/g, it attained 89.5% cycle stability. Tabrizi and coworkers [68] found that an acid-treated GO/PANI nanocomposite with a porous structure and high specific surface area revealed a maximum capacitance of 727 F/g which attributed to acid functionalization and well-defined PANI nanoarrays on GO sheets. The symmetrical device designed by combining these electrodes has a maximum energy density of 40 Wh/kg and a power density of up to 15.3 kW/kg with an outstanding stability (95.7% retention after 5000 cycles). Li and coworkers [39] used a supra-molecular in-situ assembly approach to decorate carbon nanodots on graphene-PANI (rGO@CN/PANI). The nanocomposites reveal high electrochemical performance with specific capacitance of 871.8 F/g at 0.2A/g and acceptable cycling performance (72% after 10,000 CV cycles at 30 mV/s). Mangisetti and coworkers [52] prepared nitrogen doped 3D porous carbon-graphene/PANI (3D PC-g/PANI) by a simple in-situ polymerization process. They also synthesized N-doped porous carbon/gC_3N_4 from bio- waste seeds to use as a negative electrode for ASC device. The 3D porous carbon prevents graphene nanosheets from aggregating and produces a 3D PC-g composite with a well-connected structure, that provides high conductivity, quick ion and charge transport, and high surface area (2418 m^2/g). Also, more space can be provided when using 3D PC-g as a template for PANI dispersion, which minimized the volumetric changes of PANI through charging and discharging cycles. In 0.5 M Na_2SO_4 electrolyte, symmetric SC device of 3D PC-g/PANI shows a high energy density of 117 Wh/kg and retained about 94% of initial capacitance. Also, The ASC device offered a high electrochemical performance with energy density of 97.5 Wh/kg and cyclic stability of up to 91% after 10,000 cycles in 0.5 M Na_2SO_4 as shown in Fig. 8. Table 4 lists some recent studies on graphene/PANI composites.

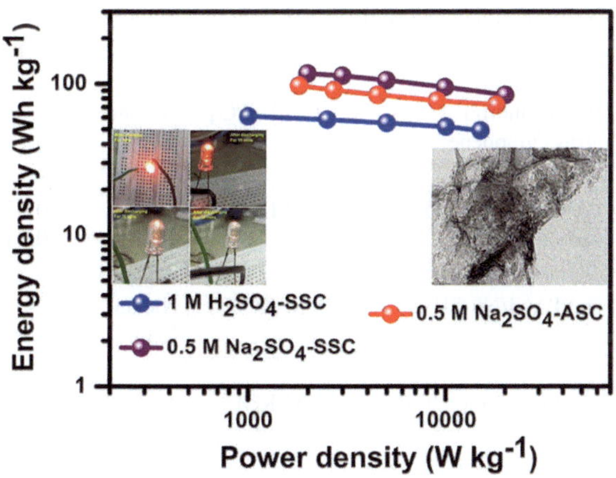

Fig. 8 Ragone plot of 3D N-doped porous carbon-graphene/Polyaniline nanohybrid as a function of current densities in different electrolytes for SSC and ASC (figure used by permission of Elsevier) [52]

Table 4 Various supercapacitor electrode materials based on graphene and PANI nanocomposites

Nanocomposite	Electrolyte	Substrate	Specific capacitance	Capacitance retention	Refs.
Graphene/PANI layers/PANI nanorods	1 M H_2SO_4	Glass carbon	578 F/g at 1 A/g	93% after 10,000 cycles at 2 A/g	[38]
PANI nanofibers/rGO	1 M H_2SO_4	Stainless steel mesh	692 F/g at 1A/g	83.3% after 1,000 cycles at 10 A/g	[32]
Flower-like PANI/graphene hybrid	1 M H_2SO_4	Glassy carbon	1510 F/g at 1 A/g	89% after 1,500 cycles at 100 mV/s	[33]
Sulfonated PANI/GO	1 M H_2SO_4	Glassy carbon	1107 F/g at 1 A/g	94% after 5,000 cycles at 3.8 A/g	[2]
Phytic acid assisted graphene/PANI	1 M H_2SO_4	Glassy carbon	865.6 F/g at 1 A/g	82% after 1,000 cycles at 5 A/g	[31]
PANI/rGO	1 M H_2SO_4	Pt foil	423 F/g at 0.8 A/g	75% after 1,000 cycles at 50 mA/cm^2	[76]
Graphene carbon sphere/PANI/rGO	1 M H_2SO_4	Stainless steel mesh	446 F/g at 5 mV/s	88.7% after 5,000 cycles at 2 A/g	[43]
Holey N-doped rGO/PANI	1 M H_2SO_4	Slice of carbon	746 F/g at 1 A/g	97% after 2000 cycles at 100 mV/s	[42]
3D rGO/self-suspended PANI	1 M H_2SO_4	glassy carbon	480 F/g at 1 A/g	94.16% after 10,000 cycles at 10 A/g	[16]
N-doped graphene/PANI hydrogels	1 M H_2SO_4	Stainless steel mesh	514.3 F/g at 1 A/g	87.1% after 1,000 cycles at 10 A/g	[99]
PANI nanorod arrays/graphene	1 M H_3PO_4	Graphene fiber	0.23 F/cm^2 at 0.1 mA/cm^2	86.9% after 8,000 cycles at 0.8 mA/cm^2	[78]
PANI/rGO/functionalized carbon cloth	1 M H_2SO_4	Carbon cloth	0.47 F/cm^2 at 0.5 mA/cm^2	75.5% after 10,000 cycles at 0.8 mA/cm^2	[13]
GO/PANI/Ni(OH)2	6 M KOH	Nickel foam	743 F/g at 1 mV/s	84.4% after 2,000 cycles at 20 mA/cm^2	[48]
Graphene/MnO_2/PANI	6 M KOH	Nickel foam	1081 F/g at 1 mV/s	99% after 1,000 cycles at 3 A/g	[70]

(continued)

Table 4 (continued)

Nanocomposite	Electrolyte	Substrate	Specific capacitance	Capacitance retention	Refs.
PGO@PANI	6 M KOH	Nickel foam	603 F/g at 1 A/g	77% after 1,500 cycles at 3 A/g	[4]

6.5 PANI/Graphene Quantum Dots Electrode Materials

The chemical interaction can introduce graphene quantum dots as a molecule dopant into PANI chains. The graphene quantum dots GQDs did not only offer the double-layer properties to the nanocomposites, but it also improved the charge transfer to the surface of PANI. Malik and coworkers [54] prepared GQDs from graphene oxide flakes, followed by synthesis GQD-PANI by the chemical oxidation method of aniline. The synthesized GQD-PANI composites show a specific capacitance of 1044 F/g at 1 A/g with 80.1% cyclic stability after 3000 cycles and this high specific capacitance is attributed to higher surface areas in nanotube morphology which provide better conductive paths for fast electron transportation. In previous studies our group showed that the Sulfur and nitrogen co-doped GQDs has a positive impact on the electrochemical properties of PANI. PANI/S,N:GQDs nanocomposite exhibited maximum specific capacitance of 2524 F/g at 2 A/g with an excellent cyclic stability of 100% after 1000 cycles [60].

7 PANI/Metal Compounds

Because of their potential for pseudocapacitance over a large range of potential and excellent stability, metal compounds are interesting candidates for supercapacitors. However, they typically have limited electrical conductivity. In recent years, the nanocomposites of PANI with high conductivity and metal compounds such as Metal oxides/hydroxide, metal sulfides, metal chlorides, and metal nitrides have all been employed as supercapacitor electrodes to improve electrochemical performance.

Metal oxides/hydroxides (MOx) electrode materials are characterized as Faradaic pseudocapacitive materials. In general, they offer higher energy density than carbon-based materials and better electrochemical stability than conducting polymers for supercapacitors [66]. The following general conditions must be satisfied before using metal oxide in supercapacitors.

(i) High specific surface area.
(ii) Electrical conductivity.
(iii) Possibility of existing in two or more oxidation states across continuous range without phase transitions; and

Among the metal oxides, RuO_2, MnO_2, V_2O_5, cobalt oxide/hydroxide, nickel oxide/hydroxide, etc. have been investigated as electrode materials for supercapacitors applications. Due to its wide working potential window, highly reversible redox processes, and three different oxidation states, RuO_2 is the most researched metal oxide. Additionally, RuO_2 has a high specific capacitance (900 F/g), metallic conductivity, good thermal stability, and extended cycle life. However, the poor performance at high current densities, toxicity, and high cost limit practical applications of RuO_2-based supercapacitors.

MnO_2, on the other hand, has a large theoretical specific capacitance (1370 F/g), is inexpensive, and is environmentally friendly, but its poor conductivity restricts its applicability [3, 79] Considering these facts, constructing a MnO_2/PANI composite is a suitable way to increase the electrochemical use of MnO_2 and PANI in supercapacitors. PANI/MnO_2 composites have been synthesized electrochemically via pulse electrodeposition by Liu and coworkers [46]. The prepared composite has rod-like structure and MnO_2 particles were uniformly distributed on PANI nanorod. MnO_2 and PANI work together synergistically to produce MnO_2/PANI composite with higher specific capacitance (810 F/g) than pure PANI (662 F/g) at 0.5 A/g. Also, after 1000 cycles, it maintained 86.3% of its original capacitance. The electrochemical properties of PANI/MnO_2 composite were studied extensively in recent years. However, it suffers from poor contact between MnO_2 and PANI membrane restricts the complete interface of electrolyte and MnO_2 resulting the loss in energy density. A super bridge between the PANI membrane and the MnO_2 nanostructure is required. The MnO_2/PANI composite with silver nanoparticle decoration has a higher specific capacitance and superior conductivity. Silver nanoparticle not only facilitate the electron transfer but also reduce the internal resistance of metal oxide pseudocapacitive materials and increase the proton diffusion throughout the electrodes [57]. Among metal oxides, double metal oxides have attracted attention for electrochemical energy storage due to their ability to create multiple oxidation states and electrical conductivity in ways that single metal oxides and carbon-based materials cannot [5]. Yu and coworkers [87] showed that the core of $NiCo_2O_4$ nanowires has strong electrical conductivity and may be used as a backbone and electron highway for charge storage, overcoming MnO_2's low electrical conductivity. In the core–shell structure, the $NiCo_2O_4$ core was also employed to improve the structural instability of PANI. Jabeen and coworkers [30] provided a new method for fabricating a core–shell $NiCo_2O_4$/PANI nanorod arrays for high stability PANI based-electrode material for SCs as illustrated in Fig. 9a. Highly porous conductive core $NiCo_2O_4$ not only acts as a strain buffer support for PANI layer but also offers rapid electron transport pathways between PANI and current collector, resulting in small electrode polarization and great power capability. The heterostructure achieves a high specific capacitance of 901 F/g at 1 A/g, exceptional cycling stability of ~91% after 3000 GCD cycles at 10 A/g and good coulombic efficiency shown in Fig. 9b.

Transition metal molybdates, such as $MnMoO_4$, $CoMoO_4$, and $NiMoO_4$, with superior pseudocapacitive characteristics have developed as promising electrode materials in recent years [50]. For example, Liu and coworkers [44] reported a facile

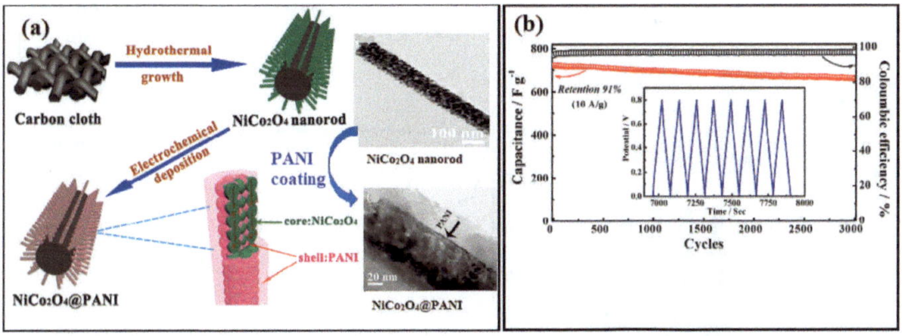

Fig. 9 **a** Schematic diagram of the fabrication process for $NiCo_2O_4$@PANI nanorod arrays. **b** Cycle performance of the $NiCo_2O_4$@PANI nanorod arrays electrode for 3000 successive charge–discharge cycles at a large current density of 10 A/g and corresponding Coulombic efficiency (inset is the typical charge–discharge curves). Reprinted with permission from [30], copyright (2016) American Chemical Society

method for fabricating $CoMoO_4$-$NiMoO_4$. xH_2O bundles with a high specific capacitance but poor cycle life (only 75.1% of the initial specific capacitance remained after 1000 GCD cycles). The structural collapse of $NiMoO_4$ during charge and discharge operations might be prevented by PANI with high mechanical stability. As a result of the synergistic effect between them, PANI/$NiMoO_4$ nanocomposite exhibited high capacitance retention of 80.7% after 2000 cycles at 5 A/g, and high specific capacitance of 1214 F/g at 1 A/g, revealing their good electrochemical stability [15].

Tungsten oxide (WO_3) has gained popularity as a viable SC electrode material in recent years due to its large specific surface area, electrochemical stability, and environmental friendliness [95]. However, low capacitance of WO_3 limit its application in high performance pseudocapacitor. WO3 modified with PANI has attracted a lot of attention for enhancing specific capacitance and cycle stability due to the synergistic effects of each component. Yang and coworkers [83] fabricated the inner/outer coating structural hexagonal WO_3/PANI through the hydrothermal-electrodeposition route. The hexagonal WO_3 nanowires were grown-up on the titanium Ti surface, and the outer PANI layer decorates the inner WO_3, resulting in WO_3-PANI hybrid nanostructures as shown in Fig. 10a. The specific capacitance (278 F/g at 1 A/g) and good cycle stability (91.9% after 1500 cycles) of WO_3/PANI electrode Fig. 10b should be attributed to its innovative nanostructure design and the interaction between the outer PANI layer with the inner WO_3 layer. On the one hand, PANI could dramatically increase the capacitance of hybrid electrodes, which is due to two factors: high specific capacitance and excellent electrical properties. On the other hand, the free spaces within the porous hexagonal-WO3 interior helps the volumetric expansion of PANI during charge and discharge cycling.

Recently, Metal tungstates such as $CoWO_4$ [7], $ZnWO_4$ [20], $MnWO_4$ [58], $FeWO_4$ [19], and $CuWO_4$ [36] are regarded the most attractive transition-metal oxides due to low toxicity, abundance, rich polymorphism, and stable multifunctional

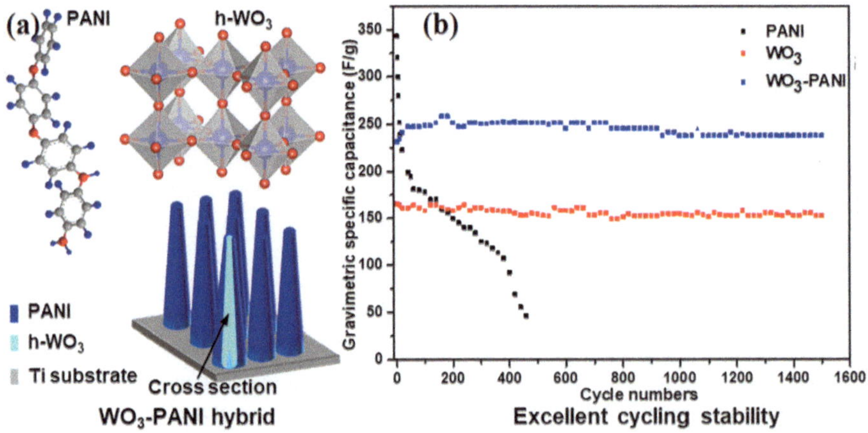

Fig. 10 **a** Schematic illustration of h-WO3/PANI hybrid nanostructures. **b** Cycling performance of the PANI, WO3, and WO3/PANI at a current density of 3 A/g, respectively (figure used by permission of Elsevier) [83]

properties. CoWO4/PANI electrode exhibited high specific capacitance of 653 F/g and outstanding cycling stability of 93.3% even after 5000 GCD cycles [59].

Supercapacitor electrode materials can also be made of other metal compounds, such as metal sulfides, metal chlorides, and metal nitrides. They are more stable in acidic electrolytes than metal oxides, but they lack the electrical conductivity and electrochemical performance of metal oxides. Majhi and coworkers [51] synthesized PANI/$CoCl_2$ composites through in situ polymerization process with different doping levels of CoCl2 (10, 15 and 20 wt%). The CV test revealed that the PANI–10% $CoCl_2$ composite had a significantly higher specific capacitance value (918 F/g) than pure PANI (382 F/g). This shows that the electrochemical performance of PANI–10% $CoCl_2$ is significantly influenced by the quantity of $CoCl_2$ present in the composite.

Because of its unique structural features and larger theoretical capacitance (than graphite 2D molybdenum disulfide MoS_2 nanosheets have increasingly attracted a lot of attention in the fields of energy storage. MoS_2 monolayers are made up of three atom layers (S-Mo-S), and their adaptability can be related to the analogous structure of graphene, which can provide a large surface area. Unfortunately, 2D MoS_2 nanosheets are easy to agglomerate due to the strong interlayer van der Waals forces [62], resulting in a decrease in active surface area and poor specific capacitance. The combination of conductive PANI with molybdenum disulfide MoS_2 not only prevents 2D MoS_2 nanosheet agglomeration, but also improves PANI cycle stability, which is beneficial to their electrochemical abilities [28]. Zhao and coworkers [94] have created a highly conductive metallic MoS_2 and PANI monolayers with unique alternating heterostructure, which will help to enhance electron/ion transfer across the electrode material while also providing great structural stability (91% capacitance retention after 2000 cycles). Zhang and coworkers [92] investigate the MoS_2/PANI core–shell structure (Fig. 11) as a supercapacitor electrode. This pompon- shaped

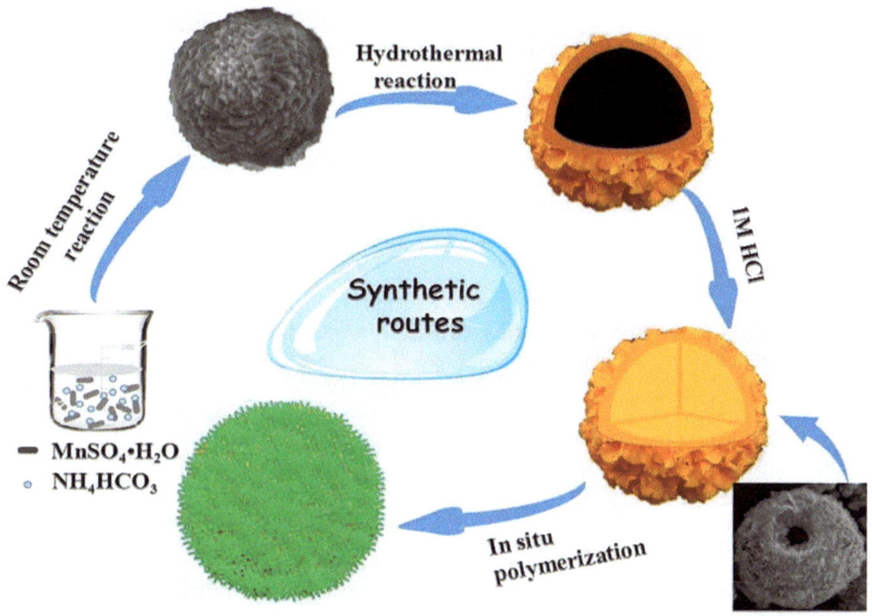

Fig. 11 Schematic illustration of the formation of MoS_2/PANI core/shell microsphere (figure used by permission of Elsevier) [92]

MoS_2/PANI composites with a high specific surface area and a more mesoporous pores presented a specific capacitance of 633 F/g with cycle stability of 86% for 1000 cycles.

Among the metal nitrides, Titanium nitrides (TiN) has an excellent candidate as an electrode material for SCs in highly corrosive electrolytes [12] due to its corrosion resistance, low-cost, thermal stability, mechanical properties and good electrical conductivity [82, 97]. Xia and coworkers [80] prepared a PANI/TiN core–shell nanowire arrays (NWAs) by electrodepositing PANI onto TiN NWAs. The TiN NWAs core was vital in improving the electrode's rate performance by providing a large surface area and rapid electron transport. The PANI shell on TiN NWAs might improve cycle stability and help to achieve excellent pseudo-capacitance performance. The PANI/TiN core–shell NWAs electrode exhibited a very high specific capacitance of 1064 F/g at 1A/g and kept 95% capacity retention after 200 cycles.

8 Conclusion

This chapter summarized current developments in PANI as a supercapacitor electrode material, including their design, and synthesis process. Pure PANI is unable to fulfil the rising demand due to its poor cycling stability and inefficient capacitance contribution. As a result, PANI must be used in conjunction with other active materials such as carbon materials and metal compounds. Due to the synergistic effect, PANI works as a conductive layer in various PANI based composite structures, and the resulting PANI based composites have showed improved electrochemical performance in supercapacitors.

References

1. Ali F, Liu X, Zhou D, Yang X, Xu J, Schenk T, Müller J, Schroeder U, Cao F, Dong X (2017) Silicon-doped hafnium oxide anti-ferroelectric thin films for energy storage. J Appl Phys 122(14):144105. https://doi.org/10.1063/1.4989908
2. Bandyopadhyay P, Kuila T, Balamurugan J, Nguyen TT, Kim NH, Lee JH (2017) Facile synthesis of novel sulfonated polyaniline functionalized graphene using m-aminobenzene sulfonic acid for asymmetric supercapacitor application. Chem Eng J 308:1174–1184. https://doi.org/10.1016/j.cej.2016.10.015
3. Bélanger D, Brousse T, Long J (2008) Manganese oxides: battery materials make the leap to electrochemical capacitors. Electrochem Soc Interface 17(1):49–52. https://doi.org/10.1149/2.f07081if
4. Bigdeli H, Moradi M, Borhani S, Jafari EA, Hajati S, Kiani MA (2018) One-pot electrochemical growth of sponge-like polyaniline-intercalated phosphorous-doped graphene oxide on nickel foam as binder-free electrode material of supercapacitor. Phys E 100:45–53. https://doi.org/10.1016/j.physe.2018.03.003
5. Cai D, Wang D, Liu B, Wang Y, Liu Y, Wang L, Li H, Huang H, Li Q, Wang T (2013) Comparison of the electrochemical performance of $NiMoO_4$ nanorods and hierarchical nanospheres for supercapacitor applications. ACS Appl Mater Interfaces 5(24):12905–12910. https://doi.org/10.1021/am403444v
6. Cao Z, Wei B (2013) A perspective: carbon nanotube macro-films for energy storage. Energy Environ Sci 6(11):3183–3201. https://doi.org/10.1039/C3EE42261E
7. Chen S, Yang G, Jia Y, Zheng H (2016) Facile synthesis of CoWO4 nanosheet arrays grown on nickel foam substrates for asymmetric supercapacitors. ChemElectroChem 3(9):1490–1496. https://doi.org/10.1002/celc.201600316
8. Chen W, Rakhi RB, Alshareef HN (2013) Morphology-dependent enhancement of the pseudocapacitance of template-guided tunable polyaniline nanostructures. J Phys Chem C 117(29):15009–15019. https://doi.org/10.1021/jp405300p
9. Conway BE (1991) Transition from "supercapacitor" to "battery" behavior in electrochemical energy storage. J Electrochem Soc 138(6):1539–1548
10. Dhawale DS, Dubal DP, Jamadade VS, Salunkhe RR, Lokhande CD (2010a) Fuzzy nanofibrous network of polyaniline electrode for supercapacitor application. Synth Met 160(5):519–522. https://doi.org/10.1016/j.synthmet.2010.01.021
11. Dhawale DS, Salunkhe RR, Jamadade VS, Dubal DP, Pawar SM, Lokhande CD (2010b) Hydrophilic polyaniline nanofibrous architecture using electrosynthesis method for supercapacitor application. Curr Appl Phys 10(3):904–909. https://doi.org/10.1016/j.cap.2009.10.020
12. Dong S, Chen X, Gu L, Zhou X, Li L, Liu Z, Han P, Xu H, Yao J, Wang H, Zhang X, Shang C, Cui G, Chen L (2011) One dimensional MnO_2/titanium nitride nanotube coaxial arrays for high performance electrochemical capacitive energy storage. Energy Environ Sci 4(9):3502–3508. https://doi.org/10.1039/C1EE01399H
13. Du P, Dong Y, Kang H, Yang X, Wang Q, Niu J, Wang S, Liu P (2018) Graphene-wrapped polyaniline nanowire array modified functionalized of carbon cloth for high-performance flexible solid-state supercapacitor. ACS Sustain Chem Eng 6(11):14723–14733. https://doi.org/10.1021/acssuschemeng.8b03278

14. Eftekhari A, Li L, Yang Y (2017) Polyaniline supercapacitors. J Power Sources 347:86–107. https://doi.org/10.1016/j.jpowsour.2017.02.054
15. Gao H, Wu F, Wang X, Hao C, Ge C (2018) Preparation of $NiMoO_4$-PANI core-shell nanocomposite for the high-performance all-solid-state asymmetric supercapacitor. Int J Hydrogen Energy 43(39):18349–18362. https://doi.org/10.1016/j.ijhydene.2018.08.018
16. Gao Z, Yang J, Huang J, Xiong C, Yang Q (2017) A three-dimensional graphene aerogel containing solvent-free polyaniline fluid for high performance supercapacitors. Nanoscale 9(45):17710–17716. https://doi.org/10.1039/C7NR06847F
17. Gawli Y, Banerjee A, Dhakras D, Deo M, Bulani D, Wadgaonkar P, Shelke M, Ogale S (2016) 3D polyaniline architecture by concurrent inorganic and organic acid doping for superior and robust high rate supercapacitor performance. Sci Rep 6:21002. https://doi.org/10.1038/srep21002. https://www.nature.com/articles/srep21002#supplementary-information
18. Ghenaatian HR, Mousavi MF, Rahmanifar MS (2012) High performance hybrid supercapacitor based on two nanostructured conducting polymers: Self-doped polyaniline and polypyrrole nanofibers. Electrochim Acta 78:212–222
19. Goubard-Bretesché N, Crosnier O, Buvat G, Favier F, Brousse T (2016) Electrochemical study of aqueous asymmetric $FeWO_4/MnO_2$ supercapacitor. J Power Sources 326:695–701. https://doi.org/10.1016/j.jpowsour.2016.04.075
20. Guan B, Hu L, Zhang G, Guo D, Fu T, Li J, Duan H, Li C, Li Q (2014) Facile synthesis of ZnWO 4 nanowall arrays on Ni foam for high performance supercapacitors. RSC Adv 4(9):4212–4217
21. Guan H, Fan L-Z, Zhang H, Qu X (2010) Polyaniline nanofibers obtained by interfacial polymerization for high-rate supercapacitors. Electrochim Acta 56(2):964–968. https://doi.org/10.1016/j.electacta.2010.09.078
22. Gupta V, Miura N (2005) Electrochemically deposited polyaniline nanowire's network a high-performance electrode material for redox supercapacitor. Electrochem Solid-State Lett 8(12):A630–A632
23. Hadjipaschalis I, Poullikkas A, Efthimiou V (2009) Overview of current and future energy storage technologies for electric power applications. Renew Sustain Energy Rev 13(6):1513–1522. https://doi.org/10.1016/j.rser.2008.09.028
24. Hashemi M, Rahmanifar MS, El-Kady MF, Noori A, Mousavi MF, Kaner RB (2018) The use of an electrocatalytic redox electrolyte for pushing the energy density boundary of a flexible polyaniline electrode to a new limit. Nano Energy 44:489–498. https://doi.org/10.1016/j.nanoen.2017.11.058
25. He S, Hu X, Chen S, Hu H, Hanif M, Hou H (2012) Needle-like polyaniline nanowires on graphite nanofibers: hierarchical micro/nano-architecture for high performance supercapacitors. J Mater Chem 22(11):5114–5120. https://doi.org/10.1039/C2JM15668G
26. Holdren JP (1991) Population and the energy problem. Popul Environ 12(3):231–255. https://doi.org/10.1007/BF01357916
27. Huang H, Zeng X, Li W, Wang H, Wang Q, Yang Y (2014) Reinforced conducting hydrogels prepared from the in situ polymerization of aniline in an aqueous solution of sodium alginate. J Mater Chem A 2(39):16516–16522. https://doi.org/10.1039/C4TA03332A
28. Huang K-J, Wang L, Liu Y-J, Wang H-B, Liu Y-M, Wang L-L (2013) Synthesis of polyaniline/2-dimensional graphene analog MoS_2 composites for high-performance supercapacitor. Electrochim Acta 109:587–594. https://doi.org/10.1016/j.electacta.2013.07.168
29. Hussain S, Kovacevic E, Amade R, Berndt J, Pattyn C, Dias A, Boulmer-Leborgne C, Ammar M-R, Bertran-Serra E (2018) Plasma synthesis of polyaniline enrobed carbon nanotubes for electrochemical applications. Electrochim Acta 268:218–225. https://doi.org/10.1016/j.electacta.2018.02.112
30. Jabeen N, Xia Q, Yang M, Xia H (2016) Unique core-shell nanorod arrays with polyaniline deposited into mesoporous $NiCo_2O_4$ support for high-performance supercapacitor electrodes. ACS Appl Mater Interfaces 8(9):6093–6100. https://doi.org/10.1021/acsami.6b00207
31. Ji J, Li R, Li H, Shu Y, Li Y, Qiu S, He C, Yang Y (2018) Phytic acid assisted fabrication of graphene/polyaniline composite hydrogels for high-capacitance supercapacitors. Compos B Eng 155:132–137. https://doi.org/10.1016/j.compositesb.2018.08.037

32. Jin K, Zhang W, Wang Y, Guo X, Chen Z, Li L, Zhang Y, Wang Z, Chen J, Sun L, Zhang T (2018) In–situ hybridization of polyaniline nanofibers on functionalized reduced graphene oxide films for high-performance supercapacitor. Electrochim Acta 285:221–229. https://doi.org/10.1016/j.electacta.2018.07.220
33. Ke F, Liu Y, Xu H, Ma Y, Guang S, Zhang F, Lin N, Ye M, Lin Y, Liu X (2017) Flower-like polyaniline/graphene hybrids for high-performance supercapacitor. Compos Sci Technol 142:286–293. https://doi.org/10.1016/j.compscitech.2017.02.026
34. Kolathodi MS, Palei M, Natarajan TS, Singh G (2020) MnO_2 encapsulated electrospun TiO_2 nanofibers as electrodes for asymmetric supercapacitors. Nanotechnology 31(12):125401. https://doi.org/10.1088/1361-6528/ab5d64
35. Kulandaivalu S, Sulaiman Y (2020) Review of the use of transition-metal-oxide and conducting polymer-based fibres for high-performance supercapacitors. Mater Des 186:108199
36. Kumar RD, Karuppuchamy S (2014) Microwave-assisted synthesis of copper tungstate nanopowder for supercapacitor applications. Ceram Int 40(8):12397–12402
37. Li G-R, Feng Z-P, Zhong J-H, Wang Z-L, Tong Y-X (2010a) Electrochemical synthesis of polyaniline nanobelts with predominant electrochemical performances. Macromolecules 43(5):2178–2183. https://doi.org/10.1021/ma902317k
38. Li J, Xiao D, Ren Y, Liu H, Chen Z, Xiao J (2019a) Bridging of adjacent graphene/polyaniline layers with polyaniline nanofibers for supercapacitor electrode materials. Electrochim Acta 300:193–201. https://doi.org/10.1016/j.electacta.2019.01.089
39. Li S, Gao A, Yi F, Shu D, Cheng H, Zhou X, He C, Zeng D, Zhang F (2019b) Preparation of carbon dots decorated graphene/polyaniline composites by supramolecular in-situ self-assembly for high-performance supercapacitors. Electrochim Acta 297:1094–1103. https://doi.org/10.1016/j.electacta.2018.12.036
40. Li X, Li X, Dai N, Wang G, Wang Z (2010b) Preparation and electrochemical capacitance performances of super-hydrophilic conducting polyaniline. J Power Sources 195(16):5417–5421. https://doi.org/10.1016/j.jpowsour.2010.03.034
41. Li X, Wei B (2013) Supercapacitors based on nanostructured carbon. Nano Energy 2(2):159–173. https://doi.org/10.1016/j.nanoen.2012.09.008
42. Liu J, Du P, Wang Q, Liu D, Liu P (2019a) Mild synthesis of holey N-doped reduced graphene oxide and its double-edged effects in polyaniline hybrids for supercapacitor application. Electrochim Acta 305:175–186. https://doi.org/10.1016/j.electacta.2019.03.049
43. Liu L, Wang Y, Meng Q, Cao B (2017) A novel hierarchical graphene/polyaniline hollow microsphere as electrode material for supercapacitor applications. J Mater Sci 52(13):7969–7983. https://doi.org/10.1007/s10853-017-1000-2
44. Liu MC, Kong LB, Lu C, Ma XJ, Li XM, Luo YC, Kang L (2013) Design and synthesis of CoMoO4-NiMoO4·xH 2O bundles with improved electrochemical properties for supercapacitors. J Mater Chem A 1(4):1380–1387. https://doi.org/10.1039/c2ta00163b
45. Liu P, Yan J, Guang Z, Huang Y, Li X, Huang W (2019b) Recent advancements of polyaniline-based nanocomposites for supercapacitors. J Power Sources 424:108–130. https://doi.org/10.1016/j.jpowsour.2019.03.094
46. Liu T, Shao G, Ji M, Wang G (2015) Polyaniline/MnO_2 composite with high performance as supercapacitor electrode via pulse electrodeposition. Polym Compos 36(1):113–120. https://doi.org/10.1002/pc.22919
47. Lokhande VC, Lokhande AC, Lokhande CD, Kim JH, Ji T (2016) Supercapacitive composite metal oxide electrodes formed with carbon, metal oxides and conducting polymers. J Alloy Compd 682:381–403. https://doi.org/10.1016/j.jallcom.2016.04.242
48. Ma L, Su L, Zhang J, Zhao D, Qin C, Jin Z, Zhao K (2016) A controllable morphology GO/PANI/metal hydroxide composite for supercapacitor. J Electroanal Chem 777:75–84. https://doi.org/10.1016/j.jelechem.2016.07.033
49. Ma Y, Hou C, Zhang H, Qiao M, Chen Y, Zhang H, Zhang Q, Guo Z (2017) Morphology-dependent electrochemical supercapacitors in multi-dimensional polyaniline nanostructures. J Mater Chem A 5(27):14041–14052. https://doi.org/10.1039/C7TA03279J

50. Mai LQ, Yang F, Zhao YL, Xu X, Xu L, Luo YZ (2011) Hierarchical MnMoO(4)/CoMoO(4) heterostructured nanowires with enhanced supercapacitor performance. Nat Commun 2:381. https://doi.org/10.1038/ncomms1387
51. Majhi M, Choudhary RB, Thakur AK, Omar FS, Duraisamy N, Ramesh K, Ramesh S (2018) CoCl2-doped polyaniline composites as electrode materials with enhanced electrochemical performance for supercapacitor application. Polym Bull 75(4):1563–1578. https://doi.org/10.1007/s00289-017-2112-1
52. Mangisetti SR, Kamaraj M, Ramaprabhu S (2019) N-doped 3D porous carbon-graphene/polyaniline hybrid and N-doped porous carbon coated gC3N4 nanosheets for excellent energy density asymmetric supercapacitors. Electrochim Acta 305:264–277. https://doi.org/10.1016/j.electacta.2019.03.043
53. Molapo KM, Ndangili PM, Ajayi RF, Mbambisa G, Mailu SM, Njomo N, Masikini M, Baker P, Iwuoha EI (2012) Electronics of conjugated polymers (I): polyaniline. Int J Electrochem Sci 7(12):11859–11875
54. Mondal S, Rana U, Malik S (2015) Graphene quantum dot-doped polyaniline nanofiber as high performance supercapacitor electrode materials. Chem Commun 51(62):12365–12368. https://doi.org/10.1039/C5CC03981A
55. Novoselov KS, Geim AK, Morozov SV, Jiang D, Zhang Y, Dubonos SV, Grigorieva IV, Firsov AA (2004) Electric field effect in atomically thin carbon films. Science 306(5696):666. https://doi.org/10.1126/science.1102896
56. Ozoemena KI, Chen S (2016) Nanomaterials in advanced batteries and supercapacitors. Springer
57. Poudel MB, Shin M, Kim HJ (2021) Polyaniline-silver-manganese dioxide nanorod ternary composite for asymmetric supercapacitor with remarkable electrochemical performance. Int J Hydrogen Energy 46(1):474–485. https://doi.org/10.1016/j.ijhydene.2020.09.213
58. Raj BGS, Acharya J, Seo M-K, Khil M-S, Kim H-Y, Kim B-S (2019) One-pot sonochemical synthesis of hierarchical MnWO4 microflowers as effective electrodes in neutral electrolyte for high performance asymmetric supercapacitors. Int J Hydrogen Energy 44(21):10838–10851. https://doi.org/10.1016/j.ijhydene.2019.03.035
59. Rajkumar S, Christy Ezhilarasi J, Saranya P, Princy Merlin J (2022) Fabrication of CoWO4/PANI composite as electrode material for energy storage applications. J Phys Chem Solids 162:110500. https://doi.org/10.1016/j.jpcs.2021.110500
60. Ramadan A, Anas M, Ebrahim S, Soliman M, Abou-Aly A (2020a) Effect of co-doped graphene quantum dots to polyaniline ratio on performance of supercapacitor. J Mater Sci Mater Electron 31(9):7247–7259. https://doi.org/10.1007/s10854-020-03297-8
61. Ramadan A, Anas M, Ebrahim S, Soliman M, Abou-Aly AI (2020b) Polyaniline/fullerene derivative nanocomposite for highly efficient supercapacitor electrode. Int J Hydrogen Energy. https://doi.org/10.1016/j.ijhydene.2020.04.093
62. Ramakrishna Matte HSS, Gomathi A, Manna AK, Late DJ, Datta R, Pati SK, Rao CNR (2010) MoS2 and WS2 analogues of graphene. Angew Chem Int Ed 49(24):4059–4062
63. Raza W, Ali F, Raza N, Luo Y, Kim K-H, Yang J, Kumar S, Mehmood A, Kwon EE (2018) Recent advancements in supercapacitor technology. Nano Energy 52:441–473. https://doi.org/10.1016/j.nanoen.2018.08.013
64. Sharma V, Sahoo A, Sharma Y, Mohanty P (2015) Synthesis of nanoporous hypercrosslinked polyaniline (HCPANI) for gas sorption and electrochemical supercapacitor applications. RSC Adv 5(57):45749–45754. https://doi.org/10.1039/C5RA03016A
65. Sk MM, Yue CY (2014) Synthesis of polyaniline nanotubes using the self-assembly behavior of vitamin C: a mechanistic study and application in electrochemical supercapacitors. J Mater Chem A 2(8):2830–2838. https://doi.org/10.1039/c3ta14309k
66. Sk MM, Yue CY, Ghosh K, Jena RK (2016) Review on advances in porous nanostructured nickel oxides and their composite electrodes for high-performance supercapacitors. J Power Sources 308:121–140. https://doi.org/10.1016/j.jpowsour.2016.01.056
67. Sumboja A, Wang X, Yan J, Lee PS (2012) Nanoarchitectured current collector for high rate capability of polyaniline based supercapacitor electrode. Electrochim Acta 65:190–195. https://doi.org/10.1016/j.electacta.2012.01.046

68. Tabrizi AG, Arsalani N, Mohammadi A, Ghadimi LS, Ahadzadeh I, Namazi H (2018) A new route for the synthesis of polyaniline nanoarrays on graphene oxide for high-performance supercapacitors. Electrochim Acta 265:379–390. https://doi.org/10.1016/j.electacta.2018.01.166
69. Tan YB, Lee J-M (2013) Graphene for supercapacitor applications. J Mater Chem A 1(47):14814–14843. https://doi.org/10.1039/C3TA12193C
70. Usman M, Pan L, Asif M, Mahmood Z (2015) Nickel foam–graphene/MnO_2/PANI nanocomposite based electrode material for efficient supercapacitors. J Mater Res 30(21):3192–3200. https://doi.org/10.1557/jmr.2015.271
71. Wang J-G, Yang Y, Huang Z-H, Kang F (2013a) A high-performance asymmetric supercapacitor based on carbon and carbon–MnO_2 nanofiber electrodes. Carbon 61:190–199. https://doi.org/10.1016/j.carbon.2013.04.084
72. Wang K, Huang J, Wei Z (2010) Conducting polyaniline nanowire arrays for high performance supercapacitors. J Phys Chem C 114(17):8062–8067. https://doi.org/10.1021/jp9113255
73. Wang L, Ye Y, Lu X, Wen Z, Li Z, Hou H, Song Y (2013b) Hierarchical nanocomposites of polyaniline nanowire arrays on reduced graphene oxide sheets for supercapacitors. Sci Rep 3:3568. https://doi.org/10.1038/srep03568
74. Wang X, Deng J, Duan X, Liu D, Guo J, Liu P (2014) Crosslinked polyaniline nanorods with improved electrochemical performance as electrode material for supercapacitors. J Mater Chem A 2(31):12323–12329. https://doi.org/10.1039/C4TA02231A
75. Wang X, Wu D, Song X, Du W, Zhao X, Zhang D (2019) Review on carbon/polyaniline hybrids: design and synthesis for supercapacitor. Molecules 24(12). https://doi.org/10.3390/molecules24122263
76. Wang Z, Qe Z, Long S, Luo Y, Yu P, Tan Z, Bai J, Qu B, Yang Y, Shi J, Zhou H, Xiao Z-Y, Hong W, Bai H (2018) Three-dimensional printing of polyaniline/reduced graphene oxide composite for high-performance planar supercapacitor. ACS Appl Mater Interfaces 10(12):10437–10444. https://doi.org/10.1021/acsami.7b19635
77. Wei W, Cui X, Chen W, Ivey DG (2011) Manganese oxide-based materials as electrochemical supercapacitor electrodes. Chem Soc Rev 40(3):1697–1721. https://doi.org/10.1039/C0CS00127A
78. Wu X, Wu G, Tan P, Cheng H, Hong R, Wang F, Chen S (2018) Construction of microfluidic-oriented polyaniline nanorod arrays/graphene composite fibers for application in wearable micro-supercapacitors. J Mater Chem A 6(19):8940–8946. https://doi.org/10.1039/C7TA11135E
79. Wu Z-S, Ren W, Wang D-W, Li F, Liu B, Cheng H-M (2010) High-energy MnO_2 nanowire/graphene and graphene asymmetric electrochemical capacitors. ACS Nano 4(10):5835–5842. https://doi.org/10.1021/nn101754k
80. Xia C, Xie Y, Wang W, Du H (2014) Fabrication and electrochemical capacitance of polyaniline/titanium nitride core–shell nanowire arrays. Synth Met 192:93–100. https://doi.org/10.1016/j.synthmet.2014.03.018
81. Yan Y, Cheng Q, Wang G, Li C (2011) Growth of polyaniline nanowhiskers on mesoporous carbon for supercapacitor application. J Power Sources 196(18):7835–7840. https://doi.org/10.1016/j.jpowsour.2011.03.088
82. Yan Y, Li B, Guo W, Pang H, Xue H (2016) Vanadium based materials as electrode materials for high performance supercapacitors. J Power Sources 329:148–169. https://doi.org/10.1016/j.jpowsour.2016.08.039
83. Yang G, Takei T, Yanagida S, Kumada N (2019) Hexagonal tungsten oxide-polyaniline hybrid electrodes for high-performance energy storage. Appl Surf Sci 498:143872. https://doi.org/10.1016/j.apsusc.2019.143872
84. Yanilmaz M, Dirican M, Asiri AM, Zhang X (2019) Flexible polyaniline-carbon nanofiber supercapacitor electrodes. J Energy Storage 24:100766. https://doi.org/10.1016/j.est.2019.100766

85. Ye Y-J, Huang Z-H, Song Y, Geng J-W, Xu X-X, Liu X-X (2017) Electrochemical growth of polyaniline nanowire arrays on graphene sheets in partially exfoliated graphite foil for high-performance supercapacitive materials. Electrochim Acta 240:72–79. https://doi.org/10.1016/j.electacta.2017.04.025
86. Yu A, Chabot V, Zhang J (2013a) Electrochemical supercapacitors for energy storage and delivery: fundamentals and applications. CRC Press
87. Yu L, Zhang G, Yuan C, Lou XW (2013b) Hierarchical $NiCo_2O_4$@MnO_2 core–shell heterostructured nanowire arrays on Ni foam as high-performance supercapacitor electrodes. Chem Commun 49(2):137–139. https://doi.org/10.1039/C2CC37117K
88. Yu P, Zhang Z, Zheng L, Teng F, Hu L, Fang X (2016) A novel sustainable flour derived hierarchical nitrogen-doped porous carbon/polyaniline electrode for advanced asymmetric supercapacitors. Adv Energy Mater 6(20):1601111. https://doi.org/10.1002/aenm.201601111
89. Zhai Y, Dou Y, Zhao D, Fulvio PF, Mayes RT, Dai S (2011) Carbon materials for chemical capacitive energy storage. Adv Mater 23(42):4828–4850. https://doi.org/10.1002/adma.201100984
90. Zhang H, Cao G, Wang Z, Yang Y, Shi Z, Gu Z (2008) Tube-covering-tube nanostructured polyaniline/carbon nanotube array composite electrode with high capacitance and superior rate performance as well as good cycling stability. Electrochem Commun 10(7):1056–1059. https://doi.org/10.1016/j.elecom.2008.05.007
91. Zhang J, Jiang J, Li H, Zhao XS (2011) A high-performance asymmetric supercapacitor fabricated with graphene-based electrodes. Energy Environ Sci 4(10):4009–4015. https://doi.org/10.1039/C1EE01354H
92. Zhang X, Ma L, Gan M, Fu G, Jin M, Zhai Y (2018) Controllable constructing of hollow MoS2/PANI core/shell microsphere for energy storage. Appl Surf Sci 460:48–57. https://doi.org/10.1016/j.apsusc.2017.10.010
93. Zhang Y, Zhang JM, Hua Q, Zhao Y, Yin H, Yuan J, Dai Z, Zheng L, Tang J (2019) Synergistically reinforced capacitive performance from a hierarchically structured composite of polyaniline and cellulose-derived highly porous carbons. Mater Lett 244:62–65. https://doi.org/10.1016/j.matlet.2019.02.045
94. Zhao C, Ang JM, Liu Z, Lu X (2017) Alternately stacked metallic 1T-MoS2/polyaniline heterostructure for high-performance supercapacitors. Chem Eng J 330:462–469. https://doi.org/10.1016/j.cej.2017.07.129
95. Zheng H, Ou JZ, Strano MS, Kaner RB, Mitchell A, Kalantar-zadeh K (2011) Nanostructured tungsten oxide—properties, synthesis, and applications. Adv Func Mater 21(12):2175–2196. https://doi.org/10.1002/adfm.201002477
96. Zheng X, Yu H, Xing R, Ge X, Sun H, Li R, Zhang Q (2018) Multi-growth site graphene/polyaniline composites with highly enhanced specific capacitance and rate capability for supercapacitor application. Electrochim Acta 260:504–513. https://doi.org/10.1016/j.electacta.2017.12.100
97. Zhou Y, Guo W, Li T (2019) A review on transition metal nitrides as electrode materials for supercapacitors. Ceram Int 45(17, Part A):21062–21076. https://doi.org/10.1016/j.ceramint.2019.07.151
98. Zhou Z, Liu K, Lai C, Zhang L, Li J, Hou H, Reneker DH, Fong H (2010) Graphitic carbon nanofibers developed from bundles of aligned electrospun polyacrylonitrile nanofibers containing phosphoric acid. Polymer 51(11):2360–2367. https://doi.org/10.1016/j.polymer.2010.03.044
99. Zou Y, Zhang Z, Zhong W, Yang W (2018) Hydrothermal direct synthesis of polyaniline, graphene/polyaniline and N-doped graphene/polyaniline hydrogels for high performance flexible supercapacitors. J Mater Chem A 6(19):9245–9256. https://doi.org/10.1039/C8TA01366G

Ceria-Based Nano-composites: A Comparative Study on Their Contributions to Important Catalytic Processes

Kumer Saurav Keshri and Biswajit Chowdhury

Abstract CeO_2 has been an important functional material due to its unique oxygen storage capacity and ability to form Ce^{3+}/Ce^{4+} redox system. The abilities of CeO_2 can be further modified by forming composites with noble metals and metal oxides, which can lead to the design of unique catalysts for different catalytic processes. Hereby, the use of CeO_2 based composites in the important gas phase, liquid phase and photocatalytic reactions, namely—water–gas shifting, oxidation of alcohol, water splitting, photocatalytic degradation of organic pollutants, Suzuki–Miyaura coupling etc. have been discussed.

Keywords Catalysis · Ceria · Photocatalytic degradation · Water splitting · Organic pollutant

1 Introduction

1.1 Catalysis: General Overview

Catalysis is a highly demanding field of chemistry due to its direct relation with chemical productivity [1]. The term "Catalysis" was first coined by Berzelius in 1835, after investigation of various related chemical phenomena [2]. Catalysis involves materials known as catalysts, that can increase or decrease the rates of certain reactions, without facing considerable alterations in its own structure. A society without sufficient abundance of food, medicines, fuel or other necessary chemicals is hard to progress. Catalysis revolutionizes the productive force of such materials by providing sustainable, cheaper and greener routes of their production. From economic point of view, catalysts have generated products with 200–300 times of worth than their own cost [1].

K. S. Keshri · B. Chowdhury (✉)
Department of Chemistry and Chemical Biology, Indian Institute of Technology (Indian School of Mines), Dhanbad, Jharkhand 826004, India
e-mail: biswajit72@iitism.ac.in

I. Uddin and I. Ahmad (eds.), *Synthesis and Applications of Nanomaterials and Nanocomposites*, Composites Science and Technology,
https://doi.org/10.1007/978-981-99-1350-3_13

The first use of catalysis is almost as old as the human civilization, alcohol fermentation being one of such early examples. The first documentation and proper research on inorganic catalysts was done by Valerius Cordus in 1575, by utilizing sulphuric acid to convert alcohol to ether [2]. In 1817, catalytic activity was reported by Humphry Davy, who noticed that combustion of fuel gasses by oxygen is accelerated in the presence of platinum wire. The heavy industrial usage of catalysts started from the end of the nineteenth century, influenced by the pioneering contributions of eminent scientists like Ostwald, Nernst, Haber, Bosch and Mittasch [2].

The catalysts influence chemical reactions kinetically by reducing their kinetic energy. An efficient catalyst can-not only provide high conversion of the substrate material at ambient conditions, but can also improve the yield of the desired products by increasing their selectivity. Although in ideal cases the catalysts cannot get consumed or get structurally altered after the ending of a reaction, in reality they have certain limitations in such capabilities. Thus, immunity against deactivation is also a deciding factor against the efficiency of a catalyst.

Catalysts are basically divided into two groups:

(i) Homogenous Catalysts: These are the catalysts that remain in the same phase as the reactants in a reaction. Some advantages of this type of catalysts are: easy accessibility of the catalytic active sites, controllable enantio-, regio- and chemo-selectivity and easier mechanistic approach for optimization of the catalyst. However, one of the major disadvantages of the homogenous catalysis is the difficulty in separating the catalyst from the reaction medium [3].

(ii) Heterogenous Catalysts. The heterogenous catalysts remain in separate phase from that of the reactants. Heterogeneous catalysts are highly utilized in chemical industries due to easier separation and optimization techniques. The heterogenous catalysts depend on high surface area and tunable adsorptive properties, as the catalyst surface acts as the hub for chemical reactions. Thus, the performance of heterogenous catalysts vary as per size, shape, exposed crystal planes, surface functionality, lattice defects etc. The activity of a heterogenous catalyst is determined from the Turn-Over Frequency (TOF), which is defined as the moles of the reactant converted per mole of active sites in the catalyst. Usually, the reactants follow three kinds of pathways to form products over the heterogenous catalysts, as described below (Fig. 1) [4]:

    (i) **Eley–Rideal Mechanism**: In this kind of mechanism, only one of the reactants is chemisorbed on the catalyst surface while the other reactant is not chemisorbed. The products get desorbed after their formation.

    (ii) **Langmuir–Hinshelwood Mechanism**: In this kind of mechanism, all the reactant molecules are adsorbed on the catalyst surface. The product is desorbed after formation.

    (iii) **Mars–van Krevelen Mechanism**: In this kind of mechanism, the catalyst itself actively takes part into the reaction by providing reactive species to the reactants and aid in the product formation.

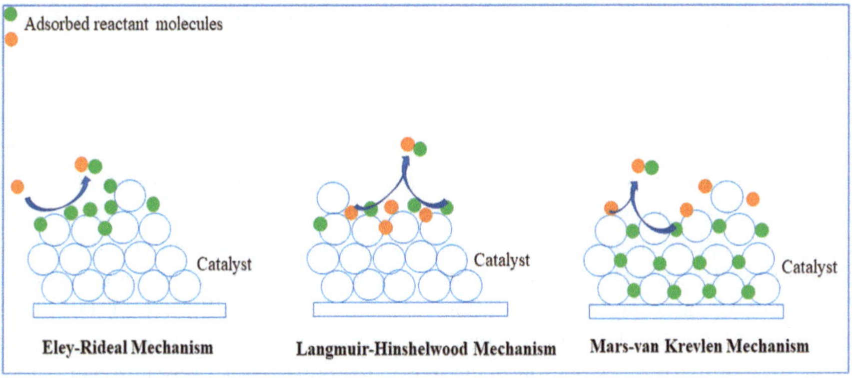

Fig. 1 Different reaction pathways in heterogenous catalysis

1.2 Ceria Based Catalysts

Ceria based materials have high demands in both industrial and academic research due to their versatility in heterogenous catalysis. The wide range of use and studies on ceria materials for over the recent twenty years confirm the irreplaceable dignity of ceria as unique "functional material" [5–11]. This dignity can be attributable to both geological and chemical reasons. Geologically, the abundance of cerium is highest among the apparent "rare earth elements", even surpassing the abundance of copper and nickel [10]. Chemically, ceria is unique in many prospects: besides its cubic fluorite structure and F_m3_m space group, it can change its oxygen states while unaltering structural integrity, can activate and modify supported metals and can hinder their sintering. One of the effects of the structural integrity of ceria, which is unaffected by its changing oxidation states, is that it can smoothly store and release oxygen species through the creation of Ce^{3+}/Ce^{4+} redox cycle. This property is quantized as Oxygen Storage Capacity (OSC) [5]. Thus, in suitable conditions, ceria can easily donate its weakly bound oxygen to a reactant species, creating oxygen vacancies in its own surface; on the other hand, it can easily refill those vacancies if oxygen is supplied externally. This aspect promotes lucrativeness of ceria specially as an oxidation catalyst, as ceria can directly involve in the reactions via its own active oxygen species through Mars–van Krevlen mechanism, other than Langmuir–Hinshelwood or Eley–Rideal mechanisms, where the catalyst surface acts as spectator [12].

Despite these interesting features, most of the oxygen vacancies of bulk ceria are inaccessible due to low surface area. Thus, to further exploit the oxygen vacancies to enhance oxygen storage capacity of ceria, a further increase in surface area and uncoordinated bonds in necessary. This goal can be easily achieved by stabilizing ceria materials in the nano regime [13]. Ceria is known to contain three stable crystal planes, namely (111), (110) and (100) with stability order of (111) > (110) > (100) [14]. These planes provide hugely different surroundings to the Ce^{4+} species. The

(111) plane exhibits Tasker Type 2 oxide surface, where O–Ce–O–O–Ce–O–O–Ce–O repetitive layers are found [15]. Therefore, although these planes are charged, the surface dipole is nullified due to O–Ce–O three layer units. On the other hand, the (110) planes are charge neutral as they contain both anions and cations in each plane; this structure is denoted as Tasker Type 1. The (100) planes follow Tasker Type 3 pattern with charged surface and Ce–O–Ce–O layer sequence. Therefore, the (100) planes are polar in nature and possess very high surface energy [15]. As a result, the Ce (111) planes contain single co-ordination vacancies on both O and Ce, often denoted as (1CUS). The Ce (110) planes possess double co-ordination vacancies (2CUS) on Ce, and single co-ordination vacancies on O (1CUS). In Ce (100) planes both O and Ce have 2CUS [15]. Thus, the unique natures of these stable planes of ceria are worthy of investigation in catalytic viewpoint. Novel research works have revealed that different planes of ceria can be predominantly exposed by synthesizing different morphologies of ceria nanocrystals [16]. Especially, the nanorods, nanocubes and nano-polyhedra are of paramount interest due to competitive and unique contributions in different catalytic reactions [5].

Although ceria materials are well known for their versatile catalytic applications, they often get sintered at high temperatures, which reduces their surface area and oxygen storage capacity and reduce their catalytic effectivity [17]. Due to this reason, CeO_2 is often doped or composited with other metal oxides [18] like TiO_2, Al_2O_3 and ZrO_2, and other materials like SiO_2 zeolites and MOFs. Deposition and impregnation of noble metal nanoparticles on ceria materials form another class of useful nanocomposites, with special abilities like the activation of molecular O_2. Thus, the summarization of the key roles of ceria-based nanocomposites for industrially and environmentally relevant reactions like Water–Gas shifting, Alcohol oxidation. reduction of organic pollutants, photocatalytic water splitting, Suzuki coupling etc. are expected to contribute in the better understanding over this important field.

2 Ceria-Based Composites

2.1 Ceria-Metal / Metal Oxide Composites

The composites of CeO_2 with other metals can be designed in different ways, e.g., (i) metal on top of CeO_2 surfaces, (ii) CeO_2 on top of metal surfaces and (iii) CeO_2 and metal particles on top of other metal oxides [7]. The first kind of composites mentioned here are the most common ones, which usually comprises the noble metal particles deposited on CeO_2 support. For the second kind, CeO_2 films are deposited on metal surfaces as overlayers [19]. This type of CeO_2 based inverse oxide metal composites show unique features of CeO_2 thin films and ultra-thin films compared to the bulk CeO_2 materials. The catalytically active sites of these inverse model catalysts can be situated on the CeO_2-metal interphases [20] or on the isolated ultrathin CeO_2 islands themselves [21].

2.2 Metal Support Interaction

An amazing property of ceria is its capacity to support a variety of metallic species, ranging from nanoparticles to clusters and single atoms [10]. A major challenge in utilizing metal (specially transition metal) nanoparticles as catalysts is their rapid deactivation due to sintering and Ostwald ripening [22]. Therefore, Strong Metal Support interaction (SMSI), a term explained by Tauster in 1978, is important to achieve maximum catalytic performance from supported metal particles [23] SMSI can be majorly attributed to electronic redox interaction between metal nanoparticles and the support [24]. Ceria is already acknowledged for its affinity to redox cycles due to non-stoichiometric nature [25]. Therefore, it can easily facilitate complete utility of the supported metal particles by enhancing their dispersion and letting them escape being shadowed by bigger particles [26].

2.3 CeO_2-Coinage Metal Composites

Investigation for interaction between ceria and coinage metals (i.e. metals from Group 10 and 11 in the periodic table) have been done extensively [27]. As it was discussed earlier, stabilization and high dispersion of noble metal nanoparticles is important for high catalytic applicability. It was found that, morphology of ceria plays a major role in determining its interaction with coinage and noble metals [27]. In general, the coinage metals also help in the activation of molecular O_2 which is essential for many reactions [28]. Moreover, to further understand the chemical and electronic interaction between coinage metals and ceria polymorphs, it is important to compare interactions with other metals that share the same group as those noble metals in the periodic table. Therefore, interactions between ceria and two major groups consisting of coinage metals, i.e., Groups 10 and 11 are discussed below.

2.3.1 Cu/CeO_2 Systems

The contacts between copper and ceria is noteworthy because of formation of Cu^{2+}/Cu^{+} and Ce^{3+}/Ce^{4+} redox couples, increase in reducibility and metal-support interfacial reactivities [29]. Moreover, Copper can help in the activation of molecular oxygen by forming adduct with O_2 via different ways, namely: superoxo (η^1 and η^2), hydroperoxo (η^1), μ-1,2-peroxo, μ-η^2: η^2-peroxo, bis-μ-oxo, bis-μ_3-oxo, μ_4-peroxo etc. [30]. copper-based enzymes can be attributed to the ability of copper to donate electrons to the π^* HOMO of O_2, which activates O_2 to O_2^{2-}. As O_2^{2-} also contain singlet ground state as most organic substrates, spin allowed bonding becomes easier [31]. The inadequacy of Strong Metal Support Interactions (SMSI) theory in explaining Cu/CeO_2 interactions, resulted in an alternate explanation in terms of Electronic Metal Support Interactions (EMSI) [29]. The EMSI theory

Table 1 Interactions between CeO_2 polymorphs and Group 11 metals [33–35]

Group 11 metal	Crystal plane of CeO_2		
	111	110	100
	Most stable sites for adsorption of Group 11 metal		
Cu	O_d	O_u–O_u 2NN	O_u–O_u 1NN
Ag	O_d	Ce–Ce 3NN	O_u–O_u 1NN
Au	O_u–Ce	O_u–O_u 2NN	O_u–O_u 1NN

describes that the catalytic activity of metal sites are result of electronic perturbations arising from bonding interactions between metal and ceria support. James et al., the group responsible for coining the term EMSI, have shown that the affinity of copper towards stoichiometric sites of ceria results in rapid increase in Ce^{3+} species in the expense of Ce^{4+} [32]. The most stable sites of Cu, Ag and Au on different CeO_2 planes have been summarized in Table 1.

Branda et al. have done theoretical calculations to investigate the nature of interactions between Cu, Ag, Au and the (111) surface of CeO_2 [33]. As per that study, Cu is most stably adsorbed with threefold co-ordination at O_d sites, i.e. with O atoms at 3rd atomic layer. The second most stable site is the O_u–O_u sites where Cu bridges between to O atoms of the uppermost layer [33].

As for of $CeO_2(110)$ surfaces, theoretical investigations by Cui et al. suggest that Cu mainly gets adsorbed at O_u–O_u and Ce–Ce sites. Moreover, the electronic charge transfer from Cu is not only limited to its first nearest neighboring Ce atom (1NN site), but can be transferred to 2nd and 3rd nearest neighbor (2NN and 3NN) Ce atoms too. In these cases, the charge transfer results in the oxidation of Cu to Cu^+. However, in some cases, Cu donates two electrons to two 3NN Ce atoms, which results in oxidation of Cu to Cu^{2+}. The most energetically favorable site for the adsorption of Cu is the O_u–O_u site [34]. In the case of $CeO_2(100)$ planes, Cu is adsorbed at O_u–O_u and O_u and Ce-top sites with adsorption energy order of O_u–O_u > Ce top > O_u [35].

2.3.2 Ag/CeO_2 Systems

The interactions between Ag and the (111) planes of CeO_2 have been extensively studied by many research groups [33, 36–39]. Some works have suggested the presence of Ag_2O layers between Ag metallic and CeO_2 phases [39]. DFT calculations have shown that O_2molecules at the interface between Ag and $CeO_2(111)$ surface have lower O_2dissociation barrier due to formation of triple phase boundaries (TPB), which can be successfully utilized in Solid Oxide Fuel Cells (SOFC) [38]. The strong bonding between Ag atoms and the (111) terraces and defects of CeO_2, resulted in the formation of sinter resistant Ag nanoparticles [40]. Synchrotron Radiation Photoemission Studies (SRPS) suggest that Ag nanoparticles grow three dimensionally on properly ordered (111) surfaces of CeO_2. However, in defect enriched $CeO_2(111)$

surfaces Ag nanoparticles show two dimensional island growth around the defect sites and three dimensional island growth around the other parts of the surface [36].

It is evident that, similar to Cu, Ag also partially reduces the CeO_2 surface [37]. There are, however, different opinions about the process of reduction of the CeO_2(111) surface in the presence of Ag nanoparticles. SPRS studies by Kong et al. suggest that CeO_2 is reduced due to reverse spillover by O atoms from Ag–CeO_2 interface to Ag nanoparticles [36]. On the other hand, Luches et al. has ascribed such reduction to electron transfer from Ag to CeO_2 [37]. In general, Ag has average lower work function compared to Cu and Au [41, 42]. As a result, electron transfer from Ag to $CeO_2$is much more facile [43].

The ideas revealed from the theoretical and experimental studies on Ag/CeO_2 interactions has been further applied and explored on different ceria polymorphs. CeO_2 nanorods, which predominantly expose (110) plane are known to support majorly cationic Ag^+ species [44]. On the other hand, ceria nanocubes, that predominantly expose (100) planes, mainly support metallic Ag species [44, 45].

2.3.3 Au/CeO_2 Systems

Au/CeO_2systems appear as one of the most elite groups of catalysts that are much explored due to academic enthusiasm as well as wide-spread industrial efficiency. Similar to Cu and Ag, the interactions between Au and the CeO_2(111), (110) and (100) planes have been theoretically studied by Branda et al. [33], Cui et al. [34] and Chen et al. [35] respectively. However, the nature of the active Au species and its interaction with CeO_2 supports have been matter of continuous debate. While some groups claim that the active Au species on CeO_2(111) is predominantly cationic [46], some other groups have questioned the stability $Au^{\delta+}$ species on CeO_2(111) and have suggested the presence of neutral Au as active species [47]. Investigations and comparison by Branda et al. suggest that, the theoretical results obtained about the oxidation states of Au species on CeO_2(111) is completely method dependent. Therefore, there might be statistical distribution of metallic and cationic Au species dependent on temperature [48]. Castellani et al. inferred that Au on most of the CeO_2(111) sites is neutral, however, some Au species situated in the junction between CeO_2(111) terraces are oxidized to $Au^{\delta+}$ due to greater interaction [49]. Zhang et al. have investigated the diverse nature of interaction between Au and CeO_2(111) in stoichiometric and non-stoichiometric forms. In the stoichiometric CeO_2, the Au species are most stabilized in the bridge like section between two O atoms, due to greater hybridization between O 2p and Au 5d. The oxidation of Au species is aided by the empty 4f orbitals of Ce atoms near Fermi level. On the other hand, in non-stoichiometric CeO_{2-x}, Au gets reduced to $Au^{\delta-}$ due to strong ionic bonding between Ce and Au [50]. However, in contrast to the stable sites of Ag and Cu mentioned above, Au is predominantly stabilized at O_u–Ce site of CeO_2(111), which is a bridging site between Ce and O_u sites [33]. The fact that, some theoretical methods predict the O_u–O_u and O_u sites as the most stable site for Au, is largely due to use of a larger lattice parameter than experimental value and the energetic similarity between

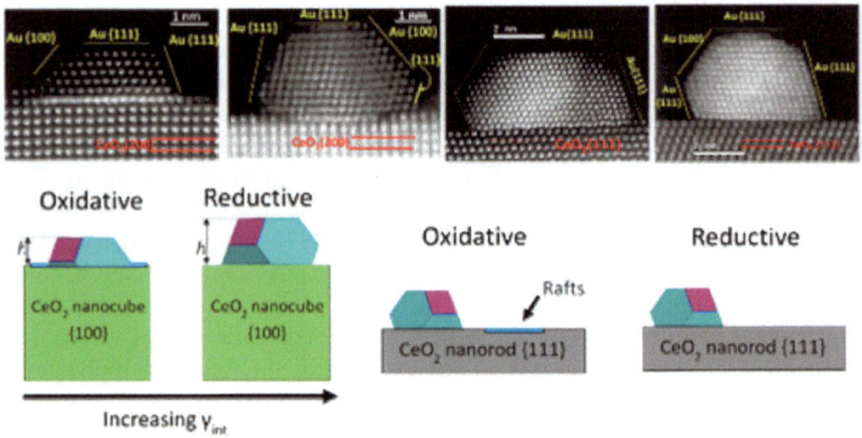

Fig. 2 Interaction between different planes of CeO_2 and Au (Reproduced with permission from Ref. [52], Copyright © 2015, American Chemical Society)

neutral and cationic Au species [48]. Engel et al.'s recent study on the interactions between Au clusters and non-stoichiometric $CeO_{2-x}(111)$ surface demonstrates that Au clusters only interact with oxygen vacancies in their vicinity [51]. Study by Lin et al. have revealed that the adhesion as well as atomic structure of adsorbed gold nanoparticles rely on the exposed plane and reaction environment (Fig. 2). In oxidative condition, the Au–CeO_2 nanorods with (111) exposed plane containing Au nanoparticles and rafts. However, in reducing condition the rafts get vanished due to increase in interface free energy [52]. As for $CeO_2(110)$ and (100) surfaces, Au is predominantly stabilized at O_u–O_u site [34, 35] and mostly remains in the oxidized state [34].

Hu et al. have performed detailed study on the interaction of lower and higher valent dopants with the (111) plane of CeO_2 [53, 54]. It was observed that lower valent dopants can create delocalized or localized holes. During generation of oxygen vacancies, some of the formed unpaired electrons find these holes to stabilize themselves. Therefore, the energy of oxygen vacancy formation is reduced. These dopant-induced electronic movements can work in both short and long ranges. However, the dopants with similar or higher valences than Ce usually show short-range effects, mostly limited to the oxygen atoms in their adjacent neighborhood [54].

2.3.4 Ni/CeO_2 Systems

The Ni/CeO_2 composites have been efficient in a variety of reactions, like reverse water gas shifting, CO_2 methanation etc. [55–57]. Chafi et al. have calculated the interactions between Ni and CeO_2 using VASP calculations [58]. The study has shown that Ni is adsorbed at top position of the (1 1 1) plane of CeO_2, but it resides at the bridging position for (1 1 0) surface of CeO_2. The study also revealed that Ni is incorporated in the tetrahedral site of CeO_2 when the amount of Ni is ¼. When the amount of Ni is increased from ¼ to 1, the incorporation of Ni becomes exothermic and a trigonal environment comprising of 3 O atoms is preferred for Ni insertion. Both insertion and adsorption of Ni in CeO_2 occurs slightly in greater amount on the (1 1 1) surface, compared to the (1 1 0) surface of CeO_2. Also, Ni_2Ce alloys might form in the system, which might cause decrease in the Ni–Ce bond length. Wang et al. have postulated that Ni atoms present at the interstitial and substitutional point defects of CeO_2 helps in promoting its oxygen buffering ability [59]. The Ni atoms present at the interstitial positions form Ni^{+}–Ce^{3+} defect complexes, which causes the reduction of Ce^{4+} to Ce^{3+} and elevated oxygen vacancy formation in these regions. On the other hand, the substituted Ni^{2+} species create oxygen vacancies in a different manner, without the need for the reduction of Ce^{4+}. In case of substituted Ni^{2+} species, unsaturated O2p valence bands and unoccupied gap states are formed in the O atoms in the vicinity of the Ni^{2+} species. Oxygen vacancies in these cases are formed by quenching of these gap states.

2.3.5 Pd/CeO_2 and Pt/CeO_2 Systems

The industrial efficiency of Pd/CeO_2 systems is well acknowledged in scientific society. The core activity of Pd is due to formation of PdO, which has good oxygen donor capacity at limited temperature range [60]. It was proposed by Shyu et al. that the interactions between Pd and CeO_2 can be attributed to the presence of Pd–O–O–Ce–O species [61]. He et al. have analyzed the interaction between Pd atoms with different crystal planes of CeO_2 [62]. It was revealed that the Pd species most strongly interacts with the (100) plane of CeO_2, followed by the (110) and (111) planes. Also, the stabilization of different oxidation state of the Pd species were found to be dependent on those crystal planes. It was found that $Pd^{\delta+}$ with ($0 \leq \delta \leq 1$) can be stabilized on the CeO_2 (111) planes, where-as the CeO_2 (110) planes can support Pd^{2+} species under certain conditions. The CeO_2 (100) can sustain the Pd^{1+} and $Pd^{\delta+}$ ($\delta < 1$) species with almost equal probability.

Pt/CeO_2 materials are another novel class of catalytic systems versatile usage [63]. Pt/CeO_2 systems have been widely used for electro-oxidation of ethanol [64–68], an important reaction in Direct Alcohol Fuel Cells (DAFC). One of the most studied combinations in this respect, are the Pt–CeO_2/C systems [68]. The review paper by Mori et al. gives detailed information about various Pt–CeO_2/C based systems in Direct Alcohol Fuel Cells [68]. Derevyannikova et al. have done an extensive study on the interactions between Pt and CeO_2over a wide range of Pt loading (1–30%) and wide range of Pt species size (single atoms to clusters) [69]. The Pt species is found to exist in two major oxidation states, Pt^{2+} and Pt^{4+}.

3 Application of CeO_2 Based Nanocomposites in Different Reactions

3.1 *Water Gas Shift Reaction*

Water gas shift (WGS) is a very important CO utilization reaction where CO and Water is converted to CO_2 and H_2. The significance of WGS lies on the fact that, it can simultaneously reduce the amount of harmful CO from the environment and provide H_2 as a useful alternative fuel [47]. The traditional WGS approach involves Fe–Cr catalysts for high temperature [70] and Cu–Zn catalysts for low temperature [71]. However, the practical application of most of the traditional catalysts in fuel cells is limited due to their low stability and activity [72]. There have been multiple efforts to use nanocomposites and doped metal oxides catalysts containing metals from group 10 and 11 and oxides of Ce, Ti, Zr, Al etc. [73] to facilitate WGS reaction. Yildirim et al. [74] have done an extensive statistical study and predictive modelling on different water gas shift catalysts reported in scientific publications in the years from 2002 to 2012 by the means of decision tree analysis and artificial neural networks. The study shows that CeO_2 has been the most reported catalyst support and an efficient promoter for Pt, for at-least the time frame defined by the study. The decision tree analysis (Fig. 3) postulated that the reaction temperature lower than 165 °C is insignificant for any water gas shifting to occur with supported gold catalysts. The study also shows that the method of catalyst preparation also highly affects the reactivity of the catalysts, e.g., deposition–precipitation method is quite unfavorable to achieve high WGS reactivity for Au catalysts, as there is lower concentration of active sites that restricts strong metal support interactions with CeO_2 catalysts. On the other hand, methods like gelation and co-precipitation facilitates high active site concentration and Au–CeO_2 interaction. The Pt catalysts usually required higher temperature than the gold ones, usually higher than 260–300 °C for moderate CO conversion and greater than 300 °C for high conversion.

Therefore, it can be clearly noticed that, there are multiple contributing factors behind the activity of noble metal-CeO_2 nanocomposites. The studies on Au–CeO_2 catalysts for WGS were started by Andreeva et al. [75]. The first reported work on Au–CeO_2 catalysts for WGS by the research group [75] consists of detailed analysis on the effects of gold amount, vapor/CO ratio, space velocity etc. on WGS activity. Pt/CeO_2 catalysts are also quite efficient in water–gas shifting reactions. Theoretical calculations reveal that the interactions between Pt and CeO_2 aid in the dissociation of water molecules and also prevent the deposition of carbon on Pt and its deactivation by Boudouard reaction [7]. The inverse CeO_2 catalysts like CeO_x/Au [76] and CeO_x/Cu [77] have also acted as good water gas shift catalysts. The CeO_x/Cu systems are particularly useful as CeO_x aids in the dissociation of water molecules, Cu adsorbs the CO and the reaction proceeds through the HOCO intermediates [78].

There have been extensive studies on the effect of CeO_2 morphology on water gas shift reaction. Flytzani-Stephanopoulos et al. have explored the water gas shifting on Cu/CeO_2 [80], Au/CeO_2 [81] composites with different CeO_2 morphologies like

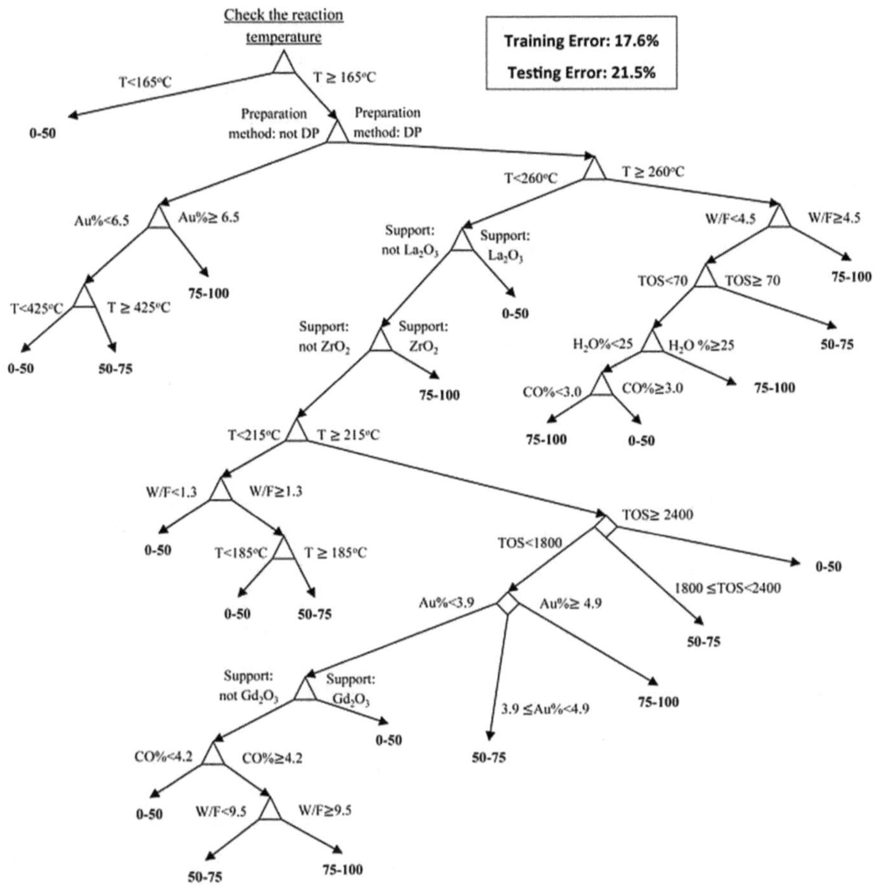

Fig. 3 Decision tree analysis for water gas shift over noble metal catalysts. Reprinted with permission from [74] Copyright © 2014 Hydrogen Energy Publications, LLC. Published by Elsevier Ltd. All rights reserved

nanorods and nanocubes. Their studies indicate the existence of different types of CuO_x species on the Cu/CeO_2 nanocrystals [80]. The weakly bonded CuO_x clusters and CuO nanoparticles don't participate in the low temperature water gas shifting. On the other hand, the regions where strong interactions between Cu and CeO_2 lead to the formation of Cu–[O_x]–Ce species, act as active sites for the low temperature water gas shift. The Cu^{2+} species that are doped in the CeO_2 support in scattered and secluded manner, can also indirectly contribute to the reaction process, by providing oxygen vacancies for the reaction and by supplying Cu when the surface Cu species is removed by leaching or sintering. The morphology of the CeO_2 support also plays a key role in the process. The abundance of oxygen vacancies on the {1 0 0} and {1 1 0} facets of the CeO_2 nanorods provide better catalytic performance compared to the {1 0 0} facets of the nanocubes. The works of Flytzani-Stephanopoulos et al.

also suggest that Au nanoparticles stabilized on the {1 1 0} facets of the CeO_2 nanorods provided the optimum activity for water gas shifting [81]. Apart from, the nanorod, nanocube morphologies of CeO_2, some other morphologies have also been explored recently. Chen et al. have prepared CeO_2–TiO_2 hybrid nanotubes to support Pt nanoparticles for effective water gas shifting [82]. The CeO_2–TiO_2 hybrid nanotubes caused the generation of PtO_x species with good redox capability and excellent interaction between the Pt^0 species and oxygen vacancies, which lead to its superior ability in water gas shift reaction.

3.2 *Preferential Oxidation of CO (CO-PROX)*

The excess CO remaining in the water–gas shift reaction can be oxidized to CO_2 via Preferential Oxidation of CO or CO-PROX process. CeO_2 based composites have also been effective as CO-PROX catalysts. The nanoparticles of group 10 and group 11 metals like Pd, Pt, Ni, Au, Ag and Cu supported on CeO_2 have been quite effective in CO oxidation [5, 7]. Among those, the Pt/CeO_2 composites have been very useful due to strong electronic interactions between Pt and CeO_2. The Pt(111) can convert CO to CO_2 via disproportionation reaction ($2CO \rightarrow C + CO_2$) and also increase the sticking coefficient of O_2 [83]. The nature of CO oxidation also depends on the amount of CO coverage on the material. As per Polster et al.'s study [79], the CO oxidation on Pt/CeO_2 catalyst follows Langmuir–Hinshelwood mechanism at lower CO coverage while the reaction proceeds through Mars–van Krevlen mechanism at high CO coverage (Fig. 4). Apart from Pt and other expensive metals, Cu is recently being used as a cheaper alternative for CO oxidation. Some recent studies suggest that reduction treatment on Cu/CeO_{2-x} catalysts prior to the CO oxidation change the shape of Cu nanoparticles from 3-dimensional to 2-dimensional, which increases the interactions between the Cu and CeO_{2-x} species, oxidizes Cu from 0 to +1 oxidation state, promotes the transport of O atoms to the border of the Cu species and increases the CO oxidation [84]. The effect of different morphologies of CeO_2 on its catalytic activities towards CO oxidation have also been explored. CO oxidation reactions have been studied on Au/CeO_2 catalysts with different CeO_2 morphologies, like nanorods, nanocubes and nanopolyhedrea. In general, the CeO_2 nanorod based systems had shown the highest catalytic activities, due to the stabilization of cationic Au species on the {1 1 0}, {1 0 0} and {1 1 1} facets of the nanorods [85, 86]. Recently, Li et al. have used Au clusters (Au_{25}) stabilized on CeO_2 nanorods, nanocubes and nano-polyhedra for catalyzing the oxidation of CO [85]. Their work suggests that Au_{25} clusters are stabilized on the {2 0 0} facets of the nanocubes and on the {0 0 2} and {1 1 1} facets of the nanorods and nanopolyhedra. The higher adsorption of [O] species on the {0 0 2} and {1 1 1} facets of the CeO_2 nanorods and nanopolyhedra compared to the {2 0 0} facets of the nanocubes, and partial charge of Au on the Au_{25} clusters lead to the highest catalytic activity of Au_{25}/CeO_2 nanorods.

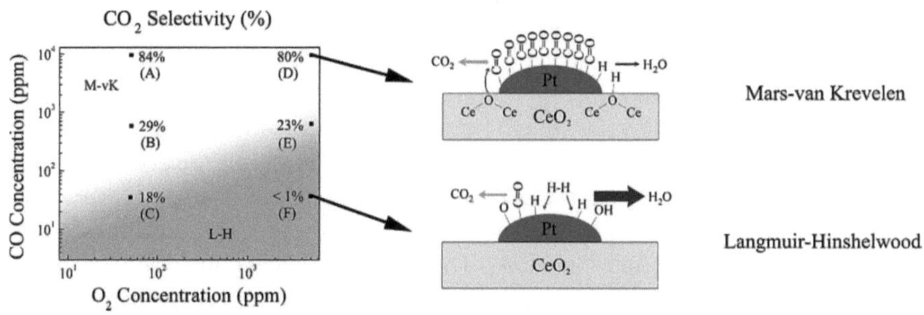

Fig. 4 The dual reaction pathways (Langmuir–Hinshelwood and Mars–van Krevlen) for the oxidation of CO on Pt/CeO_2 composites. Adapted with permission from [79] Copyright © 2010 Elsevier Inc. Published by Elsevier Inc. All rights reserved

3.3 *CO_2 Mitigation Reactions*

The enhancement in concentration of CO_2 in the atmosphere due to uncontrolled usage of carbon sources in various anthropogenic activities is a major reason for global warming, elevated ocean water level and sea-water acidification [87, 88]. Therefore, mitigation of CO_2 and its conversion to useful chemicals is a useful approach for the greener future. One of the foremost routes for CO_2 utilization have been the reduction of CO_2 to CO [89, 90], methanol [7, 91, 92], hydrocarbons [55, 93] etc. The reduction of CO_2 to CO is usually performed by reverse water gas shifting (RWGS) reaction, where mixture of CO_2 and H_2 is transformed to CO and H_2O [89]. Ni based catalysts have been found to be efficient for RWGS reactions, and Ni/CeO_2 systems have been widely explored for this purpose [89, 90]. Winter et al. have done isotopic studies on the RWGS reaction catalyzed by Ni/CeO_2 catalysts [90]. The study suggests that CeO_2 is initially reduced partially by the H_2 provided during the reaction. The partly reduced CeO_2 (which contains both Ce^{3+} and Ce^{4+} species) exchanges O atoms with CO_2, although the reduced CeO_2 system doesn't get completely re-oxidized. The exchange of O atoms leads to the formation of CO from CO_2. The Ni in the system maintains a metallic state and aids in the hetero-exchange of atoms and the dissociation of CO_2 to CO via hydrogenation route. Shen et al. also studied the RWGS reaction on Ni/CeO_2 catalysts with lower Ni doping [56]. Their investigations by density functional theory (DFT) and diffuse reflectance infrared Fourier transform spectroscopy (DRIFTS) revealed an alternate mechanism, where H_2 gets dissociated heterolytically on CeO_2 surface. Due to this type of dissociative chemisorption of H_2 on the surface of CeO_2, some unique species like Ce^{4+}–OH and Ce–H are formed. The study suggests that Ce^{4+}–OH species can activate the CO_2 molecules and bicarbonate (*HCO_3) intermediates are formed. The active H^- in the Ce–H species helps in the reduction of the bicarbonate intermediate to formate intermediate (*HCOO) and subsequent reduction to CO. Along-with Ni, the effects of some other metals have also been studied. Yang et al. have incorporated oxides of Cr and Fe into Ni/CeO_2–Al_2O_3 composites [57]. The incorporation of Fe in oxide form developed electronic enrichment of the Ni surface via good FeO_x–Ni interaction. This resulted in elevated CO_2 conversion in RWGS and excellent catalytic stability. On the other hand, incorporation of CrO_x in $Ni/CeO_2–Al_2O_3$ composites lead to increase in selectivity towards CO_2 methanation instead of RWGS. Apart from Ni/CeO_2 systems, some other systems like Cu/CeO_2 composites are also efficient in RWGS reactions. Studies by Yang et al. suggest that strong metal support interactions (SMSI) between CeO_2 and Cu result in high dispersion of Cu, which causes high H-spillover on Cu–CeO_2 interface [94]. This in turn, causes reduction of oxidized Cu and CeO_x systems to Cu^0 and Ce^{3+}. These reduced species can easily dissociate CO_2 to CO via formation of Cu^+–CO intermediates.

In the fossil fuel crisis of the present world, methanol is gradually taking the role of one of the most important alternate fuels. Therefore, the generation of methanol from CO_2 is getting special attention from researchers around the world. The CO_2 to methanol reactions are usually dominated by Cu based catalysts, although Cu shows no significant catalytic activity when used alone. CeO_2 has been composited with Cu to achieve better production of methanol from CO_2 [7]. Specially CeO_x/Cu(1 1 1) inverse oxide/metal catalyst have shown lower activation energy for methanol production compared to Cu(1 1 1) alone and ZnO/Cu(1 1 1) composites as per the studies of Senenayake et al. [91]. Their study also investigated the surface chemistry of CeO_x/Cu(1 1 1) catalysts by AP-XPS and AP-IRRAS techniques. This revealed that Ce^{3+} can stabilize the important $CO_2^{\delta-}$ intermediate during the reaction, which leads to the facile formation of methanol. The study also suggests that the methanol production rate is maximum when CeO_x covers 30–40% area of the Cu(111) surface. However, the CeO_x/Cu inverse catalysts often show lower selectivity to methanol despite having high activity. The reason for the low selectivity is the formation of CO by reverse water gas shifting, which competes with the formation of methanol [92]. Therefore, instead of inverse catalysts, Cu/CeO_2 composites have also been used. Zhu et al. have investigated the effect of Cu–CeO_2 interactions in the conversion of CO/CO_2 mixture to methanol in detail [92]. The study indicates that Cu–CeO_2 catalysts can provide greater yield of methanol from CO/CO_2 mixture due to greater dispersion of Cu on CeO_2 as well as inhibition of reverse water gas shifting (RWGS) during the reaction. The inhibition of RWGS has been attributed to the formation and high adsorption of formate species on the Cu surface, which decreases the abundance of active sites for the RWGS reaction. The study also indicates that CO_2 can act as poison to inhibit the formation of methanol from CO, via formation of carbonate-type intermediates at Cu–CeO_2 interface. In that case, methanol is mainly generated from CO_2 directly, instead of CO intermediate.

Another way for the mitigation of CO_2 is the generation of hydrocarbons from it. Ni/CeO_2 systems have been quite efficient in the generation of methane by the reduction of CO_2. Zhang et al. have done theoretical calculations to predict the mechanistic pathways for CO_2 methanation [55]. The study suggests that CO_2 can get transformed to CH_4 via three different pathways, namely (i) the formate pathway, (ii) reverse water gas shifting (RWGS) + CO hydrogenation pathway and (iii) direct C–O bond cleavage pathway (Fig. 5). CO_2 methanation on Ni/CeO_2 is usually dictated by the RWGS + CO hydrogenation pathway, where the adsorbed CO_2 is first converted to CO* and then is hydrogenated to CH_4. The breakage of the C–O bond in H_2CO^* intermediate with the aid of H* species acts as the rate determining step in the process. Wang et al. used CeO_2/TiO_2 composites to prepare photocatalysts for the production of methane from CO_2 [93]. Their study suggests that the adsorption of CO_2 along-with H_2O on the CeO_2 surface leads to the formation of bidentate CO_3^- and HCO_3-species, which can easily get transformed to surface CO_2^- when simulated sunlight is irradiated. The surface CO_2^- can easily be converted to CH_4 in the presence of H_2O.

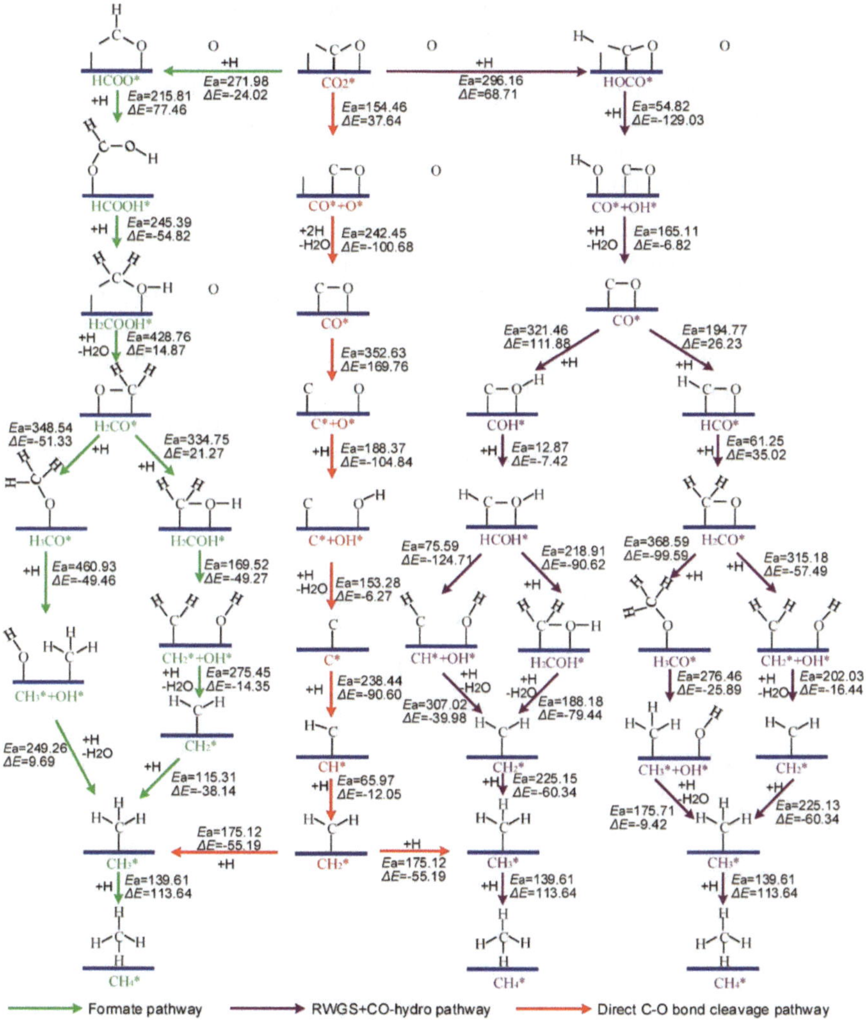

Fig. 5 Different reaction pathways for CO_2 methanation on Ni/CeO_2 catalysts. Adapted with permission from [55] Copyright © 2021 Elsevier Inc. Published by Elsevier Inc. All rights reserved

3.4 Alcohol Oxidation Reaction

Alcohol oxidation is one of the most studied reactions in chemistry with both academic and industrial applications. The selective oxidation (also named as "selox") products of alcohols, e.g. aldehyde, ketone, acid, ester etc. have a wide scale of usage; ranging from agrochemical, pharmaceutical industry to plastic, clothing, cosmetics manufacturing processes and even bio-fuel production [95].

Metal oxide supported noble metal catalysts have been widely used to oxidize a wide range of alcohols, Au [96, 97] and Pd [98, 99] based catalysts being the most prominent among them. Typically, alcohol oxidation on Au/Metal Oxide catalysts follow the following steps [96] (shown in Fig. 6):

Step 1. The deprotonation of the alcohol on the metal oxide surface
Step 2. The β-hydride elimination of the alcohol by the Au nanoparticles and formation of aldehyde product
Step 3. The re-oxidation of Au nanoparticles by supplied oxygen.

Au/CeO_2 system is widely explored in alcohol oxidation reactions. One of the first instances of utilizing Au/CeO_2 systems in alcohol oxidation belongs to the work of Abad et al. on different types of alcohols [97]. The study reveals that Au nanoparticles deposited on CeO_2 nanoparticles can perform better than conventional CeO_2, as the requirement of base and solvents become obsolete. The study by Abad et al. also demonstrate that the secondary alcohols provide selective and high conversion of corresponding aldehydes in a shorter time compared to the allylic or benzylic alcohols. Also, it shows that the aliphatic primary alcohols have lesser chance to

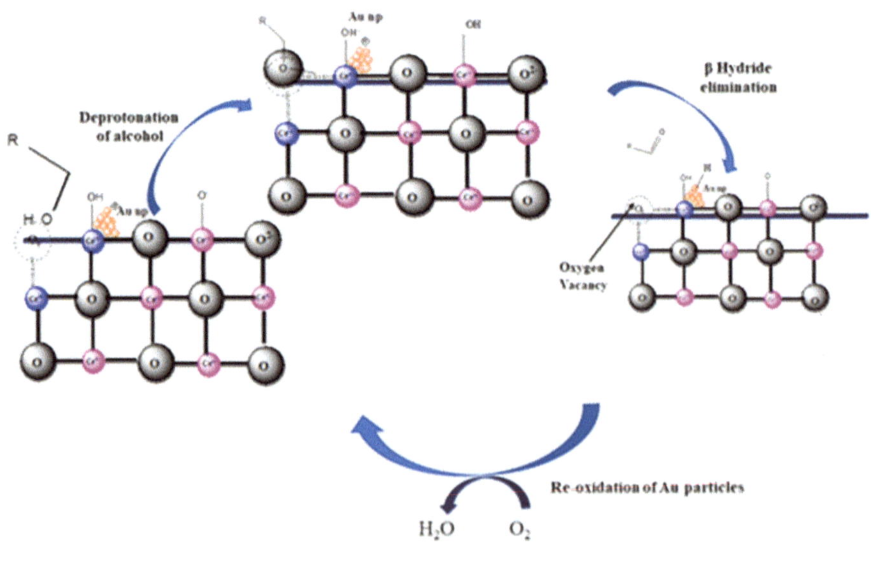

Fig. 6 Reaction mechanism of alcohol oxidation [96]

undergo reaction in solvent-free condition and have higher chance to be converted to esters via hemiacetal intermediate. While comparing the performance of Au/CeO_2 catalysts with that of Pd/CeO_2, it was observed that Au nanoparticles are more active and selective towards "–OH" groups than the Pd particles. This chemo-selectivity is especially important while oxidizing allylic alcohols, as the allylic double bonds must be protected from oxidation [96]. However, the reactions were usually carried out in basic aqueous solution or in solvent-free conditions, in the latter case the catalysts needed to be washed in basic medium before reuse, to prevent poisoning from carboxylic acid groups. A major drawback of basic washing was, the leaching of gold, which affected the catalytic activity. Investigation by Abad et al. has revealed many important aspects of alcohol oxidation reaction using Au/CeO_2 catalysts [100]. As a result of this study, toluene was granted as the most efficient solvent, partly because it doesn't promote the leaching of gold. Moreover, the catalysts were treated with different reduction procedures, however the catalytic activity was observed to be more dependent on the gold content and the nature of supports, rather than the reduction methods.

Simple alcohols like methanol and ethanol are currently being globally produced from different biological sources due to their huge demand as potential biofuels and as precursors to industrially important chemicals like acetaldehyde, acetic acid, ethyl acetate, diethyl ether etc. Gazsi et al. have studied the decomposition of methanol on Au/CeO_2 catalysts, along-with other gold deposited supports like TiO_2, MgO, Norrit, Al_2O_3, SiO_2 etc. [101]. It was observed that the efficiency of these different catalysts in MeOH decomposition follow the order of $Au/CeO_2 > Au/Al_2O_3 > Au/TiO_2 > Au/MgO > Au/Norit > Au/SiO_2$. Au/CeO_2 was also found to be more effective than other supported Au catalysts in the decomposition of ethanol [102]. Idriss et al. have studied the reaction and interactive nature of ethanol with different metal nanoparticles like Au [103], Pd [104], Pt [105] etc. supported by CeO_2, by means of FT-IR (Fourier Transform Infra-Red Spectroscopy) and TPD (Temperature Programmed Desorption). The presence of acetaldehyde was detected all these catalysts. However, unlike Pd/CeO_2 and Pt/CeO_2, the Au/CeO_2 catalysts didn't show any peak of adsorbed acetaldehyde in FT-IR. This indicates that the energy requirement of acetaldehyde formation is higher than the adsorption energy of acetaldehyde on the Au/CeO_2 catalyst, which facilitates the desorption of acetaldehyde. Overall, the Au/CeO_2 catalysts were found to provide acetaldehyde as the major oxidized product of ethanol at 573 K, where-as at higher temperature like 773 K, acetone was the major product [103].

For a long time, benzyl alcohol has been used to evaluate the catalytic activities of several CeO_2 based catalysts in alcohol oxidation reactions. Even from the early stages of investigation, Au/CeO_2 systems have shown favorable activities towards benzyl alcohol oxidation [100, 106]. However, to enrich the catalytic output Au/CeO_2 was doped with different types of metal dopants. Investigations by Guan et al. with Au/CeO_2 nanorods and nanocubes revealed that, the rate limiting steps in liquid phase benzyl alcohol oxidation is different in base-free and base-assisted reaction conditions [106]. The base-free condition follows O–H bond cleavage as rate limiting step, but this switches to the C–H bond cleavage in the basic condition.

The activity of the gold nanoparticles are also altered in these two cases. While the metallic gold nanoparticles are more active on the CeO_2 nanorods in base-free reaction condition, the nanocubes contain more active gold species in the base-assisted reaction. Haider et al. found that medium sized gold nanoparticles (approx. 6.9 nm) can provide optimum activity in benzyl alcohol conversion [107]. The same study also showed that supercritical CO_2 solvent can provide better selectivity of benzaldehyde compared to organic solvents like toluene, mesitylene etc. The nature of gold particles was further modified by using different synthesis procedures. Two such methods, i.e. 'homogenous deposition precipitation' and 'direct anionic exchange', were used by Sudarshanam et al. to synthesize Au/CeO_2 catalysts [108]. The homogenous deposition precipitation method was proved to provide more catalytically efficient material due to creation of more oxygen vacancies and smaller gold nanoparticles compared to the direct anionic exchange method. The gold particle variation has been extended to the synthesis of single Au atoms and Au clusters over CeO_2 nanorod supports in some recent works by Lei et al. [109]. The CeO_2 nanorod supported Au single atoms provided more than twice turnover frequency of benzyl alcohol conversion compared to CeO_2 nanorod supported Au clusters or nanoparticles. This efficiency was achieved due to the presence of smaller Au species, high concentration of oxygen vacancies and the existence of [O–O_v–Ce–O–Au] sites in the interfaces between Au and CeO_2. The catalyst pre-treatment conditions also have great impact on benzyl alcohol oxidation.

Some early comparative studies by Abad et al. demonstrated the lesser efficiency of Pd/CeO_2 catalysts compared to Au/CeO_2 in catalyzing alcohol oxidation [96]. However, Xin et al. carried out investigations on the nature of active species in Pd/CeO_2 for alcohol oxidation reactions [99]. In that work, single atomic Pd and Pd cluster containing 6 atoms were deposited on CeO_2 nanorods. The catalysts with atomically dispersed Pd showed very high conversion of benzyl alcohol (TOF = 6739 h^{-1}) with very high selectivity to benzaldehyde (>99%) and high recyclability. On the other hand, the samples with Pd clusters didn't show any activity. This result indicates that single Pd atoms are the true active species in alcohol oxidation. Zheng et al. have studied the effect of CeO_2 morphologies on the activity of Pd/CeO_2 systems [98]. That investigation revealed that Pd/CeO_2polyhedra are the most active catalysts due to the presence of high amount of $Pd^{2+}$and Ce^{3+} species. On the other hand, combining Pd/CeO_2 with N-doped graphene resulted in solvent free oxidation of benzyl alcohol with very high turnover frequency (110664 h^{-1} or 30.74 s^{-1}). Moreover, CeO_2 helped to prevent the oxidation of Pd on N-doped graphene, which increased the stability of the catalyst.

Combining Pd with Au in a single system is also a very good approach in this regard. $AuPd/CeO_2$ systems have shown higher activities than their Au or Pd based counterparts. The high activity of $AuPd/CeO_2$ system has been thought to be the accumulation of electron density towards Au from Pd, which increases the interaction of Pd to the alcohol substrate [110]. The synergy between Au and Pd systems have been explained by the migration of Pd onto Au nanoparticles during the reaction, thus forming Au@Pd and AuPd@Pd nanoparticles [111].

3.5 *Suzuki–Miyaura Coupling*

Suzuki–Miyaura coupling is a very efficient and powerful method to produce C–C bonds where aryl halides react with aryl boronic acids in the presence of Pd based catalysts [112]. Although the initial experiments on Suzuki Miyaura coupling were done in the homogeneous catalytic route, this route has some major disadvantages like: formation of undesirable palladium black and the the difficulty in the separation of catalysts [113]. Therefore, heterogenous catalysts are synthesized by depositing Pd nanoparticles on different support materials. CeO_2 has been very useful support for Suzuki coupling. Carrettin et al. have studied the catalytic effect of Au(III) and Pd(II) species stabilized on CeO_2 on Suzuki homo-coupling and cross-coupling reactions. It was observed that Au(III) facilitates homo-coupling while Pd(II) species promotes cross-coupling [114]. Borkowski et al. have used CeO_2–Al_2O_3 supported $PdCl_2$ and imidazolium type ionic liquids as catalysts to get high conversion in Suzuki reaction [115].

Apart from the thermocatalytic route, Suzuki coupling has also been done in the photocatalytic route. Zhang et al. reported the Pd and Au deposited porous CeO_2 nanorods as effective photocatalysts for Suzuki–Miyaura coupling under visible light (>400 nm) and room temperature [112] In the presence of visible light, hot electrons and holes are generated in the porous CeO_2 nanorods (Fig. 7). The hot electrons are transferred to the nearby Au species, which further transfers it to the neighboring Pd species. This process increases the concentration of hot electrons in the Pd particles and those electrons are transferred to the chlorobenzene—which results in the activation of the C–Cl bond. On the other hand, the holes present in the valence band of the porous CeO_2 nanorods are transferred to the phenyl boronic acid and aid in breaking the C–B bonds. In this way, the combined role of Pd, Au and CeO_2 helps in C–C bond formation with high yield.

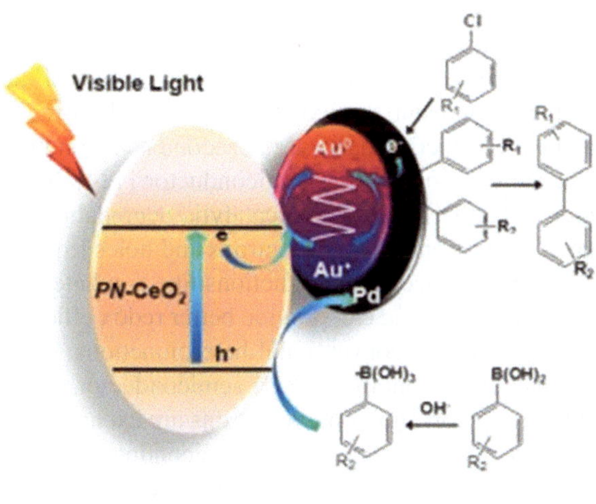

Fig. 7 Suzuki–Miyaura coupling over Au–Pd/porous CeO_2 nanorod catalysts. Adapted with permission from [112] Copyright © 2015, American Chemical Society

3.6 Photocatalytic Reactions with CeO_2 Based Nanocomposites

CeO_2, having its unique redox property, wide band gap (~3.2 eV) and n-type semiconducting capacity, act as excellent photocatalyst for different photocatalytic processes like degradation of organic pollutants and water splitting reactions [5]. The photocatalytic oxidation and reduction of organic pollutants are viable ways of removal of these hazardous molecules and their conversion to useful chemicals [8]. Thus CeO_2 morphologies like nanorods, nanocubes, nanodiscs, hollow-spheres etc. have been used in the chemical conversion of phenol, nitrophenol, organic dyes like methyl orange, methylene blue, Rhodamin-B etc. [116]. However, low visible light absorption capacity and high electron–hole recombination rate of CeO_2 are major drawbacks to use it for such reactions [117]. To overcome this limitation, CeO_2 has been composited with other non-metal, transition metal and noble metal based systems [8]. These modifications can bring the light absorption capacity of CeO_2 materials to visible light range. For example, N and S doped ZnO–CeO_2 nanocomposites can aid in the degradation of methyl orange in the visible range, by decreasing the electron hole recombination [118]. CeO_2 has been coupled with various carbon based [119, 120] silica based [121], and zeolite [122, 123] materials. These composites possess high surface area, thermal stability, facile electron transport and lower band gaps, which makes them effective catalysts for the degradation of organic dyes [8]. heterojunctions are really effective in this type of photocatalytic degradations. The CeO_2 based heterojunction composites can be classified into type I, type II and type III heterojunctions. In the type I heterojunctions (Fig. 8), two semiconductors are coupled in such a way that their bands are aligned in straddling position, i.e. the valence band and conduction band of the first semiconductor is lower and higher than the valence band and conduction band of the second semiconductor respectively. The type II heterojunctions (Fig. 8) show staggered alignment of the energy bands of the semiconductors, i.e. both the valence and conduction bands of the first semiconductor possess higher energy than the valence and conduction band of the of the second semiconductor respectively. In type III heterojuctions (Fig. 8) the bands of the component semiconductors are in broken alignment, i.e. both the valence and the conduction band of the second semiconductor are lower in energy than the valence band of the first semiconductor [124]. Type II heterojunctions are usually more effective for the photocatalytic degradation of organic pollutants, because of good separation between electrons and holes [8].

Some advanced heterojunctions like p–n heterojunctions, Schottky junctions, Z-Scheme, S-Scheme etc. provide better redox efficiency and electron–hole migration compared to the conventional heterojunctions. The p–n heterojunctions consist of one p-type and another n-type semiconductor. In the interface of the p and n-type semiconductors, holes and electrons are exchanged-which leads to the formation of negative charge on the p-type semiconductor and positive charge on p-type semiconductor. Thus, an internal electric field is generated in the material. Under light irradiation, the electrons and holes can be excited simultaneously, which leads to the accumulation of electrons and holes in the conduction band of n-type semiconductor and the valence band of the p-type semiconductor. This leads to greater electron–hole separation and result in better photocatalytic efficiency. BiOI and TiO_2 are effective p-type semiconductors that can be composited with CeO_2 which is n-type semiconductor in nature [125]. On the other hand, CeO_2–CdS heterojunctions are effective as n–n heterojunctions [126].

The Z-scheme heterojunctions (Fig. 9a) have similar structure compared to the type-II semiconductors but different electron and hole migration pathway. Here the electrons and holes of weaker redox capability from the first and second semiconductor respectively are combined but the holes and electrons with stronger redox power remain undiminished. The S-scheme heterojunctions (Fig. 8b) contain one oxidation semiconductor and another reduction semiconductor. The oxidation semiconductor has lower level of conduction band and Fermi level compared to the reduction semiconductor. In the presence of light, the S-scheme heterojunctions show electron–hole recombination of weaker electrons and the stability of stronger electrons and holes similar to the Z-scheme heterojunctions [127]. Various CeO_2 based Z-scheme and S-scheme photocatalysts have been reported. TiO_2 and g-C_3N_4 have been composited with $CeO_2$to form both Z-type [128, 131] and S-type [132] heterojunctions and have been used for the degradation of methylene blue p-nitrophenol, methyl orange and other organic pollutants. The Schottky junctions (Fig. 9c) are another type of heterojunctions, a co-catalyst is combined with the catalyst material such that electrons are migrated from the catalyst to the co-catalyst, thus generating free electrons and holes in the catalyst/cocatalyst interfaces [8]. For CeO_2 usually noble metals like Au, Ag, Pt, Pd etc. are used to generate Schottky junctions. However, transition metal carbides and carbonitrides are currently being explored as cheaper alternatives of these noble metals for the design of Schottky junctions. Shen et al. have used Ti_3C_2-MXene as to be coupled with CeO_2 and generate Schottky junctions for the effective photocatalytic degradation of tetracycline [129]. The surface plasmon resonance (Fig. 9d) of noble metals are also utilized by combining them with CeO_2. As a result of the surface plasmon resonance, the noble metals generate hot electrons in the presence of light irradiation. These hot electrons can be transferred to CeO_2 to get greater photocatalytic ability. Wen et al. synthesized $Ag/Ag_2O/CeO_2$ heterojunctions for the effective degradation of enrofloxacin [133]. Here the metallic Ag particles acted as the provider of hot electrons by surface plasmon resonance whereas the Ag_2O and CeO_2 acted as p and n-type semiconductors respectively and aided in the electron transportation for the pollutant degradation procedure.

Different morphologies of CeO_2 have also been explored in this regard. Magdalane et al. have prepared CeO_2/Y_2O_3 hierarchical nanorods for the degradation of rhodamine-B dye [134]. The one-dimensional nature and high surface area of the CeO_2/Y_2O_3 composite nanorods provided higher catalytic activity in this regard compared to other CeO_2 based composite catalysts. Kohantorabi et al. synthesized Ni nanoparticles supported on CeO_2 nanorods along with Ag, Co and Cu nanoparticles for photocatalytic degradation of rhodamine-B and p-nitrophenol [135, 136]. Langmuir–Hinshelwood model was used to explain the reaction mechanism. It was proposed that BH_4^- ions from the $NaBH_4$ reducing agent got absorbed on the metal

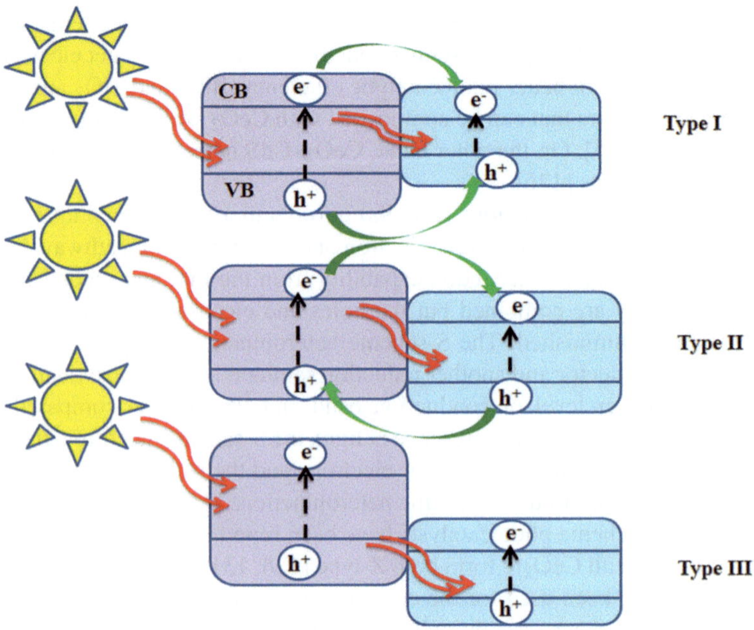

Fig. 8 Type I Type II and Type III semiconductors [8]

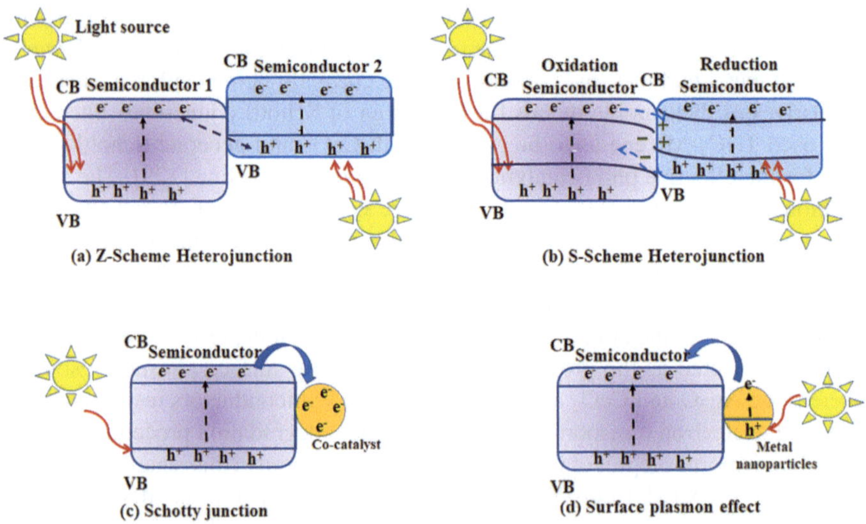

Fig. 9 **a** Z-scheme, **b** S-scheme, **c** Schottky junction and **d** Surface plasmon effect [127–130]

nanoparticles at the catalyst surface. The adsorbed BH_4^- species interact with the solvent water molecules and lead to the formation of surface –H species. The p-nitrophenol molecules on the other get transformed to nitrophenolate ions in the presence of BH_4^- species and get adsorbed on the catalyst surface. Thereafter p-nitrophenol gets reduced to p-aminophenol by the surface –H groups generated on the catalyst. On the other hand, Karuthapandian et al. used CeO_2 nanocubes immobilized on zeolites for the degradation of Congo red and methyl orange dyes [137].

Photocatalytic water splitting is an important reaction for hydrogen production, and thus has good prospect in alternative fuel techniques. CeO_2 based photocatalysts have efficiently helped in H_2 production from water due to their easy reducibility CeO_2 has been composited with materials like CdS, Fe_2O_3, MoS_2, TiO_2 etc. to form effective Z-scheme and p–n heterojunction photocatalysts for the effective production of H_2 by water splitting [11]. Graphitic carbon nitride (g-C_3N_4) is also being coupled with CeO_2, despite the fact that C–N bond has quick electron–hole recombination rate and low quantum yield. The g-C_3N_4/CeO_2 based 0D/2D heterojunctions and ternary composites have shown improved ability compared to many previous water-splitting catalysts [138, 139]. The photocatalytic water splitting with nano-composites of different CeO_2 morphologies have also been explored. Hao et al. used CeO_2–ZrO_2 composites with dendritic crosslink for photocatalytic water splitting [140]. The unique morphology of the catalyst provided high surface area, lower grain size and lower band gap, which led to higher electron–hole separation for efficient reaction.

3.7 Electrocatalytic Reactions with CeO_2 Based Nanocomposites

The ability of CeO_2 to propagate Ce^{3+}/Ce^{4+} redox cycles and the possession of high amount of oxygen vacancies facilitate its electronic and ionic conductivity. Moreover, alkaline solution CeO_2 can exchange oxygen ion reversibly in the following manner [9].

$$1/2O_2(\text{gas}) + 2e^-(\text{from } CeO_2) \Leftrightarrow O^{2-}(\text{in } CeO_2)$$

Therefore, CeO_2 gathered attention as efficient electrocatalyst for different reactions like oxygen reduction reaction (ORR), hydrogen reduction reaction (HOR) oxygen evolution reaction (OER) and hydrogen evolution reaction (HER). The ORR, HOR, OER and HER reactions are used in lithium ion batteries, hydrogen fuel cells and proton exchange membrane fuel cells [141, 142]. In the ORR and HOR reactions O_2 is reduced and H_2 is oxidized to water or H_2O_2. In the OER and HER reactions O_2 and H_2 are generated in electrochemical routes, e.g. electrocatalytic water splitting [9]. However, CeO_2 has several limitations in the electro-catalysis of these reactions too. The high charge–discharge potential and low-rate capability of CeO_2 limits its usage in the ORR reaction. The HOR reaction in CeO_2 is limited by the difficulty in managing the adsorption and desorption of H atoms on the CeO_2 surface. high anodic overpotential of CeO_2 also limit its usage in OER reaction. The HER reaction also proceeds in slower reaction rate on the CeO_2 material [9]. Therefore, to overcome these challenges, CeO_2 is composited with different materials.

Platinum nanoparticles have been effective companions for CeO_2 to increase its ability to catalyze ORR reaction. Chauhan et al. synthesized Pt–CeOx nanowire/C composite catalysts for the effective electrocatalytic reduction of O_2 to H_2O (Fig. 10) [143]. The reaction utilized the Frenkel and Schottky defects of the Pt–CeOx nanowire interface for superior catalytic activity. The Frenkel defects formed in the surface of the Pt–CeOx nanowire resulted in increased oxygen affinity of the system. On the other hand, the Schottky defect clusters formed in the bulk region of the Pt–CeOx nanowire helped in propagating the Ce^{3+}/Ce^{4+} redox cycle and aided in the reaction process. Apart from the composites with noble metals, CeO_2 has also been coupled with other important electrocatalysts like $La_{1-x}Sr_xMnO_3$ (LSM), NOSCF (N, O, S doped Carbon foam) and reduced graphene oxide (rGO) for effective oxygen reduction [9, 11].

The Pd–CeO_2/C systems have been quite effective for the HOR reactions too. Apart from Pd; Ni and Ir have also been used for the effective oxidation of H_2 [25]. The OER activity of CeO_2 has been improved by compositing it with RuO_2, which increased its electron conductivity [144]. Apart from RuO_2, Ni based materials are also being used as the cheaper alternatives [11, 145]. The HER reaction have been improved by the composites of CeO_2 with group 10 metals like Pd and Ni [9]. Also, transition metal phosphides like NiP and CoP have been effective materials to be

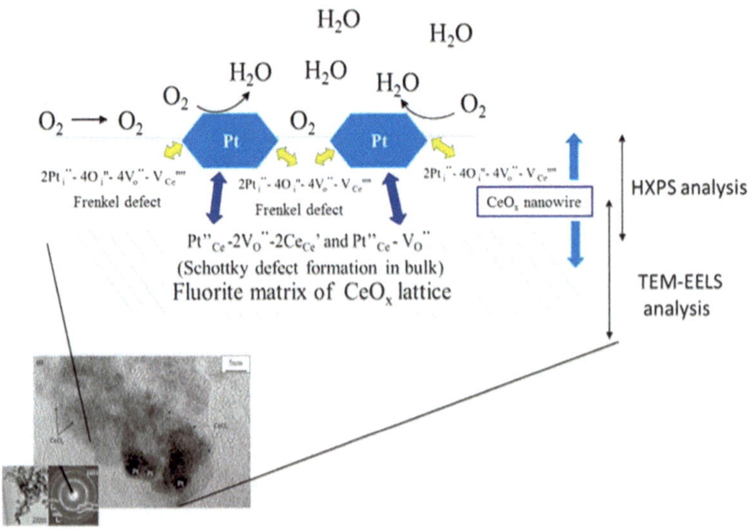

Fig. 10 Electrocatalytic Reduction of Oxygen on Pt–CeOx/C composite catalyst. Adapted with permission from [143] Copyright © 2016, American Chemical Society

combined with CeO_2 for effective HER and OER for electrocatalytic water splitting [146, 147]. The density functional theory calculation on CoP–CeO_2 composites suggests that these types of materials have lower water splitting free energy and favorable hydrogen adsorption energy which leads to their optimal catalytic activity [146].

4 Conclusion

This chapter elaborates the design and usage of several ceria-based composites in versatile reactions. The high activity of CeO_2 based composites can be utilized to catalyze multiple types of reactions like alcohol oxidation, water–gas shifting, photocatalytic degradation of pollutants etc. CeO_2 is unique because of its non-stoichiometric nature, oxygen storage capacity and capability of propagating redox cycle in its lattice. CeO_2 acts as efficient support for stabilizing the noble metal nanoparticles and is capable to act as an n-type semiconductor in photocatalytic reactions. Thus, a huge number of research works on CeO_2 based nanocomposite catalysts are getting published by different research groups. Our goal was to summarize some key aspects of these research works.

Acknowledgements KSK would like to acknowledge IIT (ISM) for research fellowship. BC would like to acknowledge Indo-German center for Science and Technology for funding in the project IGSTC/Call 2018/CO2 BioFeed/15/2019-20.

References

1. The Intelligence Report: Business Shifts in the Global Catalytic Process Industries, 2017–2023. In: Catal. Group. https://www.catalystgrp.com/multiclient_studies/intelligence-report-business-shifts-global-catalytic-process-industries-2017-2023/. Accessed 8 Sep 2020
2. Wisniak J (2010) The history of catalysis. From the beginning to nobel prizes. Educ Quím 21:60–69. https://doi.org/10.1016/S0187-893X(18)30074-0
3. Wang D, Astruc D (2017) The recent development of efficient Earth-abundant transition-metal nanocatalysts. Chem Soc Rev 46:816–854. https://doi.org/10.1039/C6CS00629A
4. Dumesic JA, Huber GW, Boudart M (2008) Principles of heterogeneous catalysis. In: Handbook of heterogeneous catalysis. American Cancer Society
5. Montini T, Melchionna M, Monai M, Fornasiero P (2016) Fundamentals and catalytic applications of CeO2-based materials. Chem Rev 116:5987–6041. https://doi.org/10.1021/acs.chemrev.5b00603
6. Huang X, Zhang K, Peng B, Wang G, Muhler M, Wang F (2021) Ceria-based materials for thermocatalytic and photocatalytic organic synthesis. ACS Catal 11:9618–9678. https://doi.org/10.1021/acscatal.1c02443
7. Rodriguez JA, Grinter DC, Liu Z, Palomino RM, Senanayake SD (2017) Ceria-based model catalysts: fundamental studies on the importance of the metal–ceria interface in CO oxidation, the water–gas shift, CO2 hydrogenation, and methane and alcohol reforming. Chem Soc Rev 46:1824–1841. https://doi.org/10.1039/C6CS00863A

8. Fauzi AA, Jalil AA, Hassan NS, Aziz FFA, Azami MS, Hussain I, Saravanan R, Vo D-VN (2022) A critical review on relationship of CeO2-based photocatalyst towards mechanistic degradation of organic pollutant. Chemosphere 286:131651. https://doi.org/10.1016/j.chemosphere.2021.131651
9. Wang J, Xiao X, Liu Y, Pan K, Pang H, Wei S (2019) The application of CeO2-based materials in electrocatalysis. J Mater Chem A 7:17675–17702. https://doi.org/10.1039/C9TA04804A
10. Trovarelli A, Fornasiero P (2013) Catalysis by ceria and related materials. World Scientific
11. Li Q, Song L, Liang Z, Sun M, Wu T, Huang B, Luo F, Du Y, Yan C-H (2021) A review on Ceo2-based electrocatalyst and photocatalyst in energy conversion. Adv Energy Sustain Res 2:2000063. https://doi.org/10.1002/aesr.202000063
12. Ruiz Puigdollers A, Schlexer P, Tosoni S, Pacchioni G (2017) Increasing oxide reducibility: the role of metal/oxide interfaces in the formation of oxygen vacancies. ACS Catal 7:6493–6513. https://doi.org/10.1021/acscatal.7b01913
13. Guozhong C (2004) Nanostructures and nanomaterials: synthesis, properties and applications. World Scientific
14. Jiang Y, Adams JB, van Schilfgaarde M (2005) Density-functional calculation of CeO2 surfaces and prediction of effects of oxygen partial pressure and temperature on stabilities. J Chem Phys 123:064701. https://doi.org/10.1063/1.1949189
15. Mullins DR, Albrecht PM, Calaza F (2013) Variations in reactivity on different crystallographic orientations of cerium oxide. Top Catal 56:1345–1362. https://doi.org/10.1007/s11244-013-0146-7
16. Mai H-X, Sun L-D, Zhang Y-W, Si R, Feng W, Zhang H-P, Liu H-C, Yan C-H (2005) Shape-selective synthesis and oxygen storage behavior of ceria nanopolyhedra, nanorods, and nanocubes. J Phys Chem B 109:24380–24385. https://doi.org/10.1021/jp055584b
17. Alcala R, DeLaRiva A, Peterson EJ, Benavidez A, Garcia-Vargas CE, Jiang D, Pereira-Hernández XI, Brongersma HH, ter Veen R, Staněk J, Miller JT, Wang Y, Datye A (2021) Atomically dispersed dopants for stabilizing ceria surface area. Appl Catal B Environ 284:119722. https://doi.org/10.1016/j.apcatb.2020.119722
18. Khan MA, Enam-Ul-Haq JMS, Xu C, Ahmad Shah SS, Nazir MA, Imran M, Assiri MA, Ahmad A, Hussain S (2021) Facile synthesis of ceria-based composite oxide materials by combustion for high-performance solid oxide fuel cells. Ceram Int 47:22035–22041. https://doi.org/10.1016/j.ceramint.2021.04.223
19. Tovt A, Stetsovych V, Dvořák F, Johánek V, Mysliveček J (2019) Ordered phases of reduced ceria as inverse model catalysts. Appl Surf Sci 465:557–563. https://doi.org/10.1016/j.apsusc.2018.09.068
20. Rodriguez JA, Graciani J, Evans J, Park JB, Yang F, Stacchiola D, Senanayake SD, Ma S, Pérez M, Liu P, Sanz JFdez, Hrbek J (2009) Water-gas shift reaction on a highly active inverse CeOx/Cu(111) catalyst: unique role of ceria nanoparticles. Angew Chem Int Ed 48:8047–8050. https://doi.org/10.1002/anie.200903918
21. Castellarin-Cudia C, Surnev S, Schneider G, Podlucky R, Ramsey MG, Netzer FP (2004) Strain-induced formation of arrays of catalytically active sites at the metal–oxide interface. Surf Sci 554:L120–L126. https://doi.org/10.1016/j.susc.2004.01.059
22. Hansen TW, DeLaRiva AT, Challa SR, Datye AK (2013) Sintering of Catalytic nanoparticles: particle migration or Ostwald ripening? Acc Chem Res 46:1720–1730. https://doi.org/10.1021/ar3002427
23. Tauster SJ, Fung SC (1978) Strong metal-support interactions: occurrence among the binary oxides of groups IIA–VB. J Catal 55:29–35. https://doi.org/10.1016/0021-9517(78)90182-3
24. Grabchenko MV, Mamontov GV, Zaikovskii VI, La Parola V, Liotta LF, Vodyankina OV (2020) The role of metal–support interaction in Ag/CeO2 catalysts for CO and soot oxidation. Appl Catal B Environ 260:118148. https://doi.org/10.1016/j.apcatb.2019.118148
25. Li P, Chen X, Li Y, Schwank JW (2019) A review on oxygen storage capacity of CeO2-based materials: influence factors, measurement techniques, and applications in reactions related to catalytic automotive emissions control. New Insight Environ Catal 18th NCC 327:90–115. https://doi.org/10.1016/j.cattod.2018.05.059

26. Cargnello M, Doan-Nguyen VVT, Gordon TR, Diaz RE, Stach EA, Gorte RJ, Fornasiero P, Murray CB (2013) Control of metal nanocrystal size reveals metal-support interface role for ceria catalysts. Science 341:771–773. https://doi.org/10.1126/science.1240148
27. Spezzati G, Benavidez AD, DeLaRiva AT, Su Y, Hofmann JP, Asahina S, Olivier EJ, Neethling JH, Miller JT, Datye AK, Hensen EJM (2019) CO oxidation by Pd supported on CeO2(100) and CeO2(111) facets. Appl Catal B Environ 243:36–46. https://doi.org/10.1016/j.apcatb.2018.10.015
28. Boronat M, Corma A (2010) Oxygen activation on gold nanoparticles: separating the influence of particle size, particle shape and support interaction. Dalton Trans 39:8538–8546. https://doi.org/10.1039/C002280B
29. Konsolakis M (2016) The role of Copper-Ceria interactions in catalysis science: recent theoretical and experimental advances. Appl Catal B Environ 198:49–66. https://doi.org/10.1016/j.apcatb.2016.05.037
30. ndi LIS, Simándi LI (2003) Advances in catalytic activation of dioxygen by metal complexes. Springer Science & Business Media
31. Solomon EI, Ginsbach JW, Heppner DE, Kieber-Emmons MT, Kjaergaard CH, Smeets PJ, Tian L, Woertink JS (2010) Copper dioxygen (bio)inorganic chemistry. Faraday Discuss 148:11–39. https://doi.org/10.1039/C005500J
32. James TE, Hemmingson SL, Ito T, Campbell CT (2015) Energetics of Cu adsorption and adhesion onto reduced CeO2(111) surfaces by calorimetry. J Phys Chem C 119:17209–17217. https://doi.org/10.1021/acs.jpcc.5b04621
33. Branda MM, Hernández NC, Sanz JFdez, Illas F (2010) Density functional theory study of the interaction of Cu, Ag, and Au atoms with the regular CeO2(111) Surface. J Phys Chem C 114:1934–1941. https://doi.org/10.1021/jp910782r
34. Cui L, Tang Y, Zhang H, Hector LG, Ouyang C, Shi S, Li H, Chen L (2012) First-principles investigation of transition metal atom M (M = Cu, Ag, Au) adsorption on CeO2(110). Phys Chem Chem Phys 14:1923–1933. https://doi.org/10.1039/C2CP22720G
35. Chen L-J, Tang Y, Cui L, Ouyang C, Shi S (2013) Charge transfer and formation of Ce3+ upon adsorption of metal atom M (M = Cu, Ag, Au) on CeO2 (100) surface. J Power Sources 234:69–81. https://doi.org/10.1016/j.jpowsour.2013.01.121
36. Growth, structure, and stability of Ag on CeO2(111): synchrotron radiation photoemission studies. J Phys Chem C. https://pubs.acs.org/doi/abs/10.1021/jp112392y. Accessed 25 Mar 2020
37. Luches P, Pagliuca F, Valeri S, Illas F, Preda G, Pacchioni G (2012) Nature of Ag islands and nanoparticles on the CeO2(111) surface. J Phys Chem C 116:1122–1132. https://doi.org/10.1021/jp210241c
38. (2006) Oxygen reduction reactions in the SOFC cathode of Ag/CeO2. Solid State Ion 177:939–947. https://doi.org/10.1016/j.ssi.2006.02.029
39. (2009) Soot combustion over silver-supported catalysts. Appl Catal B Environ 91:489–498. https://doi.org/10.1016/j.apcatb.2009.06.019
40. Farmer JA, Campbell CT (2010) Ceria maintains smaller metal catalyst particles by strong metal-support bonding. Science 329:933–936. https://doi.org/10.1126/science.1191778
41. Green M (1969) Work function: measurements and results. In: Solid state surface science. M. Dekker
42. Hölzl J, Schulte FK, Wagner H (1979) Work functions of metals. Solid Surf Phys
43. Saravanan R, Agarwal S, Gupta VK, Khan MM, Gracia F, Mosquera E, Narayanan V, Stephen A (2018) Line defect Ce3+ induced Ag/CeO2/ZnO nanostructure for visible-light photocatalytic activity. J Photochem Photobiol Chem 353:499–506. https://doi.org/10.1016/j.jphotochem.2017.12.011
44. Chang S, Li M, Hua Q, Zhang L, Ma Y, Ye B, Huang W (2012) Shape-dependent interplay between oxygen vacancies and Ag–CeO2 interaction in Ag/CeO2 catalysts and their influence on the catalytic activity. J Catal 293:195–204. https://doi.org/10.1016/j.jcat.2012.06.025
45. (2016) Soot oxidation over CeO2 and Ag/CeO2: factors determining the catalyst activity and stability during reaction. J Catal 337:188–198. https://doi.org/10.1016/j.jcat.2016.01.019

46. Guzman J, Carrettin S, Fierro-Gonzalez JC, Hao Y, Gates BC, Corma A (2005) CO oxidation catalyzed by supported gold: cooperation between gold and nanocrystalline rare-earth supports forms reactive surface superoxide and peroxide species. Angew Chem Int Ed 44:4778–4781. https://doi.org/10.1002/anie.200500659
47. Rodriguez JA, Liu P, Hrbek J, Evans J, Pérez M (2007) Water gas shift reaction on Cu and Au nanoparticles supported on CeO2(111) and ZnO(000$\bar 1$): intrinsic activity and importance of support interactions. Angew Chem Int Ed 46:1329–1332. https://doi.org/10.1002/anie.200603931
48. Brandą MM, Castellani NJ, Grau-Crespo R, de Leeuw NH, Hernandez NC, Sanz JF, Neyman KM, Illas F (2009) On the difficulties of present theoretical models to predict the oxidation state of atomic Au adsorbed on regular sites of CeO2(111). J Chem Phys 131:094702. https://doi.org/10.1063/1.3216102
49. Castellani NJ, Branda MM, Neyman KM, Illas F (2009) Density functional theory study of the adsorption of au atom on cerium oxide: effect of low-coordinated surface sites. J Phys Chem C 113:4948–4954. https://doi.org/10.1021/jp8094352
50. Zhang C, Michaelides A, King DA, Jenkins SJ (2008) Structure of gold atoms on stoichiometric and defective ceria surfaces. J Chem Phys 129:194708. https://doi.org/10.1063/1.3009629
51. Engel J, Schwartz E, Catlow CRA, Roldan A (2020) The influence of oxygen vacancy and Ce 3+ ion positions on the properties of small gold clusters supported on CeO 2-x (111). J Mater Chem A
52. Lin Y, Wu Z, Wen J, Ding K, Yang X, Poeppelmeier KR, Marks LD (2015) Adhesion and atomic structures of gold on ceria nanostructures: the role of surface structure and oxidation state of ceria supports. Nano Lett 15:5375–5381. https://doi.org/10.1021/acs.nanolett.5b02694
53. Hu Z, Li B, Sun X, Metiu H (2011) Chemistry of doped oxides: the activation of surface oxygen and the chemical compensation effect. J Phys Chem C 115:3065–3074. https://doi.org/10.1021/jp110333z
54. Hu Z, Metiu H (2011) Effect of dopants on the energy of oxygen-vacancy formation at the surface of ceria: local or global? J Phys Chem C 115:17898–17909. https://doi.org/10.1021/jp205432r
55. Zhang J, Yang Y, Liu J, Xiong B (2021) Mechanistic understanding of CO2 hydrogenation to methane over Ni/CeO2 catalyst. Appl Surf Sci 558:149866. https://doi.org/10.1016/j.apsusc.2021.149866
56. Shen H, Dong Y, Yang S, He Y, Wang Q, Cao Y, Wang W, Wang T, Zhang Q, Zhang H (2022) Identifying the roles of Ce3+−OH and Ce−H in the reverse water-gas shift reaction over highly active Ni-doped CeO2 catalyst. Nano Res. https://doi.org/10.1007/s12274-022-4207-8
57. Yang L, Pastor-Pérez L, Gu S, Sepúlveda-Escribano A, Reina TR (2018) Highly efficient Ni/CeO2-Al2O3 catalysts for CO2 upgrading via reverse water-gas shift: effect of selected transition metal promoters. Appl Catal B Environ 232:464–471. https://doi.org/10.1016/j.apcatb.2018.03.091
58. Chafi Z, Keghouche N, Minot C (2007) DFT study of Ni–CeO2 interaction: adsorption and insertion. Surf Sci 601:2323–2329. https://doi.org/10.1016/j.susc.2007.03.041
59. Wang X, Shen M, Wang J, Fabris S (2010) Enhanced oxygen buffering by substitutional and interstitial Ni point defects in ceria: a first-principles DFT+U study. J Phys Chem C 114:10221–10228. https://doi.org/10.1021/jp101100f
60. Trovarelli A (2002) Fundamentals and applications of ceria in combustion reactions. In: catalysis by ceria and related materials. World Scientific
61. Shyu JZ, Otto K (1989) Characterization of Pt/γ-alumina catalysts containing ceria. J Catal 115:16–23. https://doi.org/10.1016/0021-9517(89)90003-1
62. He B, Wang J, Ma D, Tian Z, Jiang L, Xu Y, Cheng S (2018) Interaction of Pd single atoms with different CeO2 crystal planes: a first-principles study. Appl Surf Sci 433:1036–1048. https://doi.org/10.1016/j.apsusc.2017.10.134

63. Xin Y, Zhang N, Lv Y, Wang J, Li Q, Zhang Z (2020) From nanoparticles to single atoms for Pt/CeO2: synthetic strategies, characterizations and applications. J Rare Earths. https://doi.org/10.1016/j.jre.2020.03.007
64. Sedighi M, Rostami AA, Alizadeh E (2017) Enhanced electro-oxidation of ethanol using Pt–CeO2 electrocatalyst prepared by electrodeposition technique. Int J Hydrog Energy 42:4998–5005. https://doi.org/10.1016/j.ijhydene.2016.12.014
65. Xu C, Shen PK (2004) Novel Pt/CeO2/C catalysts for electrooxidation of alcohols in alkaline media. Chem Commun 2238–2239. https://doi.org/10.1039/B408589B
66. Kakaei K, Rahimi A, Husseindoost S, Hamidi M, Javan H, Balavandi A (2016) Fabrication of Pt–CeO2 nanoparticles supported sulfonated reduced graphene oxide as an efficient electrocatalyst for ethanol oxidation. Int J Hydrog Energy 41:3861–3869. https://doi.org/10.1016/j.ijhydene.2016.01.013
67. Xu C, Shen PK (2005) Electrochamical oxidation of ethanol on Pt-CeO2/C catalysts. J Power Sources 142:27–29. https://doi.org/10.1016/j.jpowsour.2004.10.017
68. Mori T, Ou DR, Zou J, Drennan J (2012) Present status and future prospect of design of Pt–cerium oxide electrodes for fuel cell applications. Prog Nat Sci Mater Int 22:561–571. https://doi.org/10.1016/j.pnsc.2012.11.010
69. Derevyannikova EA, Kardash TY, Stadnichenko AI, Stonkus OA, Slavinskaya EM, Svetlichnyi VA, Boronin AI (2019) Structural insight into strong Pt–CeO2 interaction: from single Pt atoms to PtOx clusters. J Phys Chem C 123:1320–1334. https://doi.org/10.1021/acs.jpcc.8b11009
70. Devaiah D, Smirniotis PG (2017) Effects of the Ce and Cr contents in Fe–Ce–Cr ferrite spinels on the high-temperature water-gas shift reaction. Ind Eng Chem Res 56:1772–1781. https://doi.org/10.1021/acs.iecr.6b04707
71. Pal DB, Chand R, Upadhyay SN, Mishra PK (2018) Performance of water gas shift reaction catalysts: a review. Renew Sustain Energy Rev 93:549–565. https://doi.org/10.1016/j.rser.2018.05.003
72. Ebrahimi P, Kumar A, Khraisheh M (2020) A review of recent advances in water-gas shift catalysis for hydrogen production. Emergent Mater 3:881–917. https://doi.org/10.1007/s42247-020-00116-y
73. Andreeva D, Tabakova T, Ilieva L (2012) Ceria-based gold catalysts: synthesis, properties, and catalytic performance for the wgs and prox processes. In: Catalysis by ceria and related materials. Imperial College Press, pp 497–564
74. Odabaşı Ç, Günay ME, Yıldırım R (2014) Knowledge extraction for water gas shift reaction over noble metal catalysts from publications in the literature between 2002 and 2012. Int J Hydrog Energy 39:5733–5746. https://doi.org/10.1016/j.ijhydene.2014.01.160
75. Andreeva D, Idakiev V, Tabakova T, Ilieva L, Falaras P, Bourlinos A, Travlos A (2002) Low-temperature water-gas shift reaction over Au/CeO2 catalysts. Catal Today 72:51–57. https://doi.org/10.1016/S0920-5861(01)00477-1
76. Rodriguez JA, Ma S, Liu P, Hrbek J, Evans J, Pérez M (2007) Activity of CeOx and TiOx nanoparticles grown on Au(111) in the water-gas shift reaction. Science. https://doi.org/10.1126/science.1150038
77. López Cámara A, Cortés Corberán V, Martínez-Arias A, Barrio L, Si R, Hanson JC, Rodriguez JA (2020) Novel manganese-promoted inverse CeO2/CuO catalyst: In situ characterization and activity for the water-gas shift reaction. Catal Today 339:24–31. https://doi.org/10.1016/j.cattod.2019.01.014
78. Mudiyanselage K, Senanayake SD, Feria L, Kundu S, Baber AE, Graciani J, Vidal AB, Agnoli S, Evans J, Chang R, Axnanda S, Liu Z, Sanz JF, Liu P, Rodriguez JA, Stacchiola DJ (2013) Importance of the metal-oxide interface in catalysis. In situ studies of the water-gas shift reaction by ambient-pressure X-ray photoelectron spectroscopy. Angew Chem Int Ed 52:5101–5105. https://doi.org/10.1002/anie.201210077
79. Polster CS, Zhang R, Cyb MT, Miller JT, Baertsch CD (2010) Selectivity loss of Pt/CeO2 PROX catalysts at low CO concentrations: mechanism and active site study. J Catal 273:50–58. https://doi.org/10.1016/j.jcat.2010.04.017

80. Si R, Raitano J, Yi N, Zhang L, Chan S-W, Flytzani-Stephanopoulos M (2012) Structure sensitivity of the low-temperature water-gas shift reaction on Cu–CeO2 catalysts. Catal Today 180:68–80. https://doi.org/10.1016/j.cattod.2011.09.008
81. Si R, Flytzani-Stephanopoulos M (2008) Shape and crystal-plane effects of nanoscale ceria on the activity of Au-CeO2 catalysts for the water-gas shift reaction. Angew Chem Int Ed 47:2884–2887. https://doi.org/10.1002/anie.200705828
82. Chen Y, Li X, Li J, Du Y, Peng Q, Wu L, Xinjun L (2021) CeO2-TiO2 hybid-nanotubes with tunable oxygen vacancies as the support to confine Pt nanoparticles for the low-temperature water-gas shift reaction. ChemistrySelect 6:11900–11907. https://doi.org/10.1002/slct.202102823
83. Suchorski Y, Wrobel R, Becker S, Weiss H (2008) CO oxidation on a CeOx/Pt(111) inverse model catalyst surface: catalytic promotion and tuning of kinetic phase diagrams. J Phys Chem C 112:20012–20017. https://doi.org/10.1021/jp806033v
84. Matte LP, Thill AS, Lobato FO, Novôa MT, Muniz AR, Poletto F, Bernardi F (2022) Reduction-driven 3D to 2D transformation of Cu nanoparticles. Small 2106583. https://doi.org/10.1002/smll.202106583
85. Li Z, Zhang X, Shi Q, Gong X, Xu H, Li G (2021) Morphology effect of ceria supports on gold nanocluster catalyzed CO oxidation. Nanoscale Adv 3:7002–7006. https://doi.org/10.1039/D1NA00680K
86. Yi G, Xu Z, Guo G, Tanaka K, Yuan Y (2009) Morphology effects of nanocrystalline CeO2 on the preferential CO oxidation in H2-rich gas over Au/CeO2 catalyst. Chem Phys Lett 479:128–132. https://doi.org/10.1016/j.cplett.2009.08.011
87. Martens JA, Bogaerts A, De Kimpe N, Jacobs PA, Marin GB, Rabaey K, Saeys M, Verhelst S (2017) The chemical route to a carbon dioxide neutral world. Chemsuschem 10:1039–1055. https://doi.org/10.1002/cssc.201601051
88. Keshri KS, Bhattacharjee S, Singha A, Bhaumik A, Chowdhury B (2022) Synthesis of cyclic carbonates of different epoxides using CO2 as a C1 building block over Ag/TUD-1 mesoporous silica catalyst: a solvent free approach. Mol Catal 522:112234. https://doi.org/10.1016/j.mcat.2022.112234
89. Chang K, Zhang H, Cheng M, Lu Q (2020) Application of ceria in CO2 conversion catalysis. ACS Catal 10:613–631. https://doi.org/10.1021/acscatal.9b03935
90. Winter LR, Chen R, Chen X, Chang K, Liu Z, Senanayake SD, Ebrahim AM, Chen JG (2019) Elucidating the roles of metallic Ni and oxygen vacancies in CO2 hydrogenation over Ni/CeO2 using isotope exchange and in situ measurements. Appl Catal B Environ 245:360–366. https://doi.org/10.1016/j.apcatb.2018.12.069
91. Senanayake SD, Ramírez PJ, Waluyo I, Kundu S, Mudiyanselage K, Liu Z, Liu Z, Axnanda S, Stacchiola DJ, Evans J, Rodriguez JA (2016) Hydrogenation of CO2 to methanol on CeOx/Cu(111) and ZnO/Cu(111) catalysts: role of the metal-oxide interface and importance of Ce3+ sites. J Phys Chem C 120:1778–1784. https://doi.org/10.1021/acs.jpcc.5b12012
92. Zhu J, Su Y, Chai J, Muravev V, Kosinov N, Hensen EJM (2020) Mechanism and nature of active sites for methanol synthesis from CO/CO2 on Cu/CeO2. ACS Catal 10:11532–11544. https://doi.org/10.1021/acscatal.0c02909
93. Wang Y, Zhao J, Wang T, Li Y, Li X, Yin J, Wang C (2016) CO2 photoreduction with H2O vapor on highly dispersed CeO2/TiO2 catalysts: surface species and their reactivity. J Catal 337:293–302. https://doi.org/10.1016/j.jcat.2015.12.030
94. Yang S-C, Pang SH, Sulmonetti TP, Su W-N, Lee J-F, Hwang B-J, Jones CW (2018) Synergy between ceria oxygen vacancies and Cu nanoparticles facilitates the catalytic conversion of CO2 to CO under mild conditions. ACS Catal 8:12056–12066. https://doi.org/10.1021/acscatal.8b04219
95. Wilson K, Lee AF (2013) Chapter 2: Mechanistic studies of alcohol selective oxidation. In: Heterogeneous catalysts for clean technology: spectroscopy, design, and monitoring. Wiley
96. Abad A, Almela C, Corma A, García H (2006) Efficient chemoselective alcohol oxidation using oxygen as oxidant. Superior performance of gold over palladium catalysts. Tetrahedron 62:6666–6672. https://doi.org/10.1016/j.tet.2006.01.118

97. Abad A, Concepción P, Corma A, García H (2005) A collaborative effect between gold and a support induces the selective oxidation of alcohols. Angew Chem Int Ed 44:4066–4069. https://doi.org/10.1002/anie.200500382
98. Zheng H, Wei Z-H, Hu X-Q, Xu J, Xue B (2019) Atmospheric selective oxidation of benzyl alcohol catalyzed by Pd nanoparticles supported on CeO2 with various morphologies. ChemistrySelect 4:5470–5475
99. Xin P, Li J, Xiong Y, Wu X, Dong J, Chen W, Wang Y, Gu L, Luo J, Rong H, Chen C, Peng Q, Wang D, Li Y (2018) Revealing the active species for aerobic alcohol oxidation by using uniform supported palladium catalysts. Angew Chem 130:4732–4736. https://doi.org/10.1002/ange.201801103
100. Abad A, Corma A, García H (2008) Catalyst parameters determining activity and selectivity of supported gold nanoparticles for the aerobic oxidation of alcohols: the molecular reaction mechanism. Chem Eur J 14:212–222. https://doi.org/10.1002/chem.200701263
101. Gazsi A, Bánsági T, Solymosi F (2009) Hydrogen formation in the reactions of methanol on supported Au catalysts. Catal Lett 131:33–41. https://doi.org/10.1007/s10562-009-0052-6
102. Gazsi A, Koós A, Bánsági T, Solymosi F (2011) Adsorption and decomposition of ethanol on supported Au catalysts. Catal Today 160:70–78. https://doi.org/10.1016/j.cattod.2010.05.007
103. Sheng P-Y, Bowmaker GA, Idriss H (2004) The reactions of ethanol over Au/CeO2. Appl Catal Gen 261:171–181. https://doi.org/10.1016/j.apcata.2003.10.046
104. Yee A, Morrison SJ, Idriss H (1999) A study of the reactions of ethanol on CeO2 and Pd/CeO2 by steady state reactions, temperature programmed desorption, and in situ FT-IR. J Catal 186:279–295. https://doi.org/10.1006/jcat.1999.2563
105. Yee A, Morrison SJ, Idriss H (2000) A study of ethanol reactions over Pt/CeO2 by temperature-programmed desorption and in Situ FT-IR spectroscopy: evidence of benzene formation. J Catal 191:30–45. https://doi.org/10.1006/jcat.1999.2765
106. Guan Y, Ligthart DAJM, Pirgon-Galin Ö, Pieterse JAZ, van Santen RA, Hensen EJM (2011) Gold stabilized by nanostructured ceria supports: nature of the active sites and catalytic performance. Top Catal 54:424–438. https://doi.org/10.1007/s11244-011-9673-2
107. Haider P, Kimmerle B, Krumeich F, Kleist W, Grunwaldt J-D, Baiker A (2008) Gold-catalyzed aerobic oxidation of benzyl alcohol: effect of gold particle size on activity and selectivity in different solvents. Catal Lett 125:169–176. https://doi.org/10.1007/s10562-008-9567-5
108. Sudarsanam P, Mallesham B, Durgasri DN, Reddy BM (2014) Physicochemical and catalytic properties of nanosized Au/CeO2 catalysts for eco-friendly oxidation of benzyl alcohol. J Ind Eng Chem 20:3115–3121. https://doi.org/10.1016/j.jiec.2013.11.053
109. Lei L, Liu H, Wu Z, Qin Z, Wang G, Ma J, Luo L, Fan W, Wang J (2019) Aerobic oxidation of alcohols over isolated single Au atoms supported on CeO2 nanorods: catalysis of interfacial [O–Ov–Ce–O–Au] sites. ACS Appl Nano Mater 2:5214–5223. https://doi.org/10.1021/acsanm.9b01091
110. Nahm T-U, Jung R, Kim J-Y, Park W-G, Oh S-J, Park J-H, Allen JW, Chung S-M, Lee YS, Whang CN (1998) Electronic structure of disordered Au-Pd alloys studied by electron spectroscopies. Phys Rev B 58:9817–9825. https://doi.org/10.1103/PhysRevB.58.9817
111. Olmos CM, Chinchilla LE, Villa A, Delgado JJ, Hungría AB, Blanco G, Prati L, Calvino JJ, Chen X (2019) Size, nanostructure, and composition dependence of bimetallic Au–Pd supported on ceria–zirconia mixed oxide catalysts for selective oxidation of benzyl alcohol. J Catal 375:44–55. https://doi.org/10.1016/j.jcat.2019.05.002
112. Zhang S, Chang C, Huang Z, Ma Y, Gao W, Li J, Qu Y (2015) Visible-light-activated Suzuki-Miyaura coupling reactions of aryl chlorides over the multifunctional Pd/Au/porous nanorods of CeO2 catalysts. ACS Catal 5:6481–6488. https://doi.org/10.1021/acscatal.5b01173
113. Chen Z, Vorobyeva E, Mitchell S, Fako E, Ortuño MA, López N, Collins SM, Midgley PA, Richard S, Vilé G, Pérez-Ramírez J (2018) A heterogeneous single-atom palladium catalyst surpassing homogeneous systems for Suzuki coupling. Nat Nanotechnol 13:702–707. https://doi.org/10.1038/s41565-018-0167-2
114. Carrettin S, Corma A, Iglesias M, Sánchez F (2005) Stabilization of Au(III) on heterogeneous catalysts and their catalytic similarities with homogeneous Au(III) metal organic complexes. Appl Catal Gen 291:247–252. https://doi.org/10.1016/j.apcata.2005.01.047

115. Borkowski T, Dobosz J, Tylus W, Trzeciak AM (2014) Palladium supported on Al2O3–CeO2 modified with ionic liquids as a highly active catalyst of the Suzuki-Miyaura cross-coupling. J Catal 319:87–94. https://doi.org/10.1016/j.jcat.2014.08.007
116. Ali S, Basak S, Sikdar S, Roy M (2021) Synergetic effects of green synthesized CeO2 nanorod-like catalyst for degradation of organic pollutants to reduce water pollution. Environ Nanotechnol Monit Manag 16:100539. https://doi.org/10.1016/j.enmm.2021.100539
117. Zhao W, Dong Q, Sun C, Xia D, Huang H, Yang G, Wang G, Leung DYC (2021) A novel Au/g-C3N4 nanosheets/CeO2 hollow nanospheres plasmonic heterojunction photocatalysts for the photocatalytic reduction of hexavalentchromium and oxidation of oxytetracycline hydrochloride. Chem Eng J 409:128185. https://doi.org/10.1016/j.cej.2020.128185
118. Rahemi Ardekani S, Sabour Rouh Aghdam A, Nazari M, Bayat A, Saievar-Iranizad E, Liavali MN (2019) Synthesis and characterization of photocatalytically active crumpled-shape nanocomposites of nitrogen and sulfur co-doped ZnO–CeO2. Sol Energy Mater Sol Cells 203:110195. https://doi.org/10.1016/j.solmat.2019.110195
119. Kesarla MK, Fuentez-Torres MO, Alcudia-Ramos MA, Ortiz-Chi F, Espinosa-González CG, Aleman M, Torres-Torres JG, Godavarthi S (2019) Synthesis of g-C3N4/N-doped CeO2 composite for photocatalytic degradation of an herbicide. J Mater Res Technol 8:1628–1635. https://doi.org/10.1016/j.jmrt.2018.11.008
120. Madkour M, Ali AA, Abdel Nazeer A, Al Sagheer F, Belver C (2020) A novel natural sunlight active photocatalyst of CdS/SWCNT/CeO2 heterostructure: in depth mechanistic insights for the catalyst reactivity and dye mineralization. Appl Surf Sci 499:143988. https://doi.org/10.1016/j.apsusc.2019.143988
121. Taratayko A, Larichev Y, Zaikovskii V, Mikheeva N, Mamontov G (2021) Ag–CeO2/SBA-15 composite prepared from Pluronic P123@SBA-15 hybrid as catalyst for room-temperature reduction of 4-nitrophenol. Catal Today 375:576–584. https://doi.org/10.1016/j.cattod.2020.05.001
122. Zuo X, Ma S, Wu Q, Xiong J, He J, Ma C, Chen Z (2021) Nanometer CeO2 doped high silica ZSM-5 heterogeneous catalytic ozonation of sulfamethoxazole in water. J Hazard Mater 411:125072. https://doi.org/10.1016/j.jhazmat.2021.125072
123. Mittal H, Babu R, Dabbawala AA, Stephen S, Alhassan SM (2020) Zeolite-Y incorporated karaya gum hydrogel composites for highly effective removal of cationic dyes. Colloids Surf Physicochem Eng Asp 586:124161. https://doi.org/10.1016/j.colsurfa.2019.124161
124. Hu W, Yang J (2017) Two-dimensional van der Waals heterojunctions for functional materials and devices. J Mater Chem C 5:12289–12297. https://doi.org/10.1039/C7TC04697A
125. Cano-Franco JC, Álvarez-Láinez M (2019) Effect of CeO2 content in morphology and opto-electronic properties of TiO2-CeO2 nanoparticles in visible light organic degradation. Mater Sci Semicond Process 90:190–197. https://doi.org/10.1016/j.mssp.2018.10.017
126. Taddesse AM, Bekele T, Diaz I, Adgo A (2021) Polyaniline supported CdS/CeO2/Ag3PO4 nanocomposite: An "A-B" type tandem n-n heterojunctions with enhanced photocatalytic activity. J Photochem Photobiol Chem 406:113005. https://doi.org/10.1016/j.jphotochem.2020.113005
127. Xu Q, Zhang L, Cheng B, Fan J, Yu J (2020) S-scheme heterojunction photocatalyst. Chem 6:1543–1559. https://doi.org/10.1016/j.chempr.2020.06.010
128. Munawar T, Mukhtar F, Nadeem MS, Manzoor S, Ashiq MN, Mahmood K, Batool S, Hasan M, Iqbal F (2022) Fabrication of dual Z-scheme TiO2-WO3-CeO2 heterostructured nanocomposite with enhanced photocatalysis, antibacterial, and electrochemical performance. J Alloys Compd 898:162779. https://doi.org/10.1016/j.jallcom.2021.162779
129. Shen J, Shen J, Zhang W, Yu X, Tang H, Zhang M, Zulfiqar LQ (2019) Built-in electric field induced CeO2/Ti3C2-MXene Schottky-junction for coupled photocatalytic tetracycline degradation and CO2 reduction. Ceram Int 45:24146–24153. https://doi.org/10.1016/j.ceramint.2019.08.123
130. Tanaka A, Hashimoto K, Kominami H (2012) Preparation of Au/CeO2 exhibiting strong surface plasmon resonance effective for selective or chemoselective oxidation of alcohols to aldehydes or ketones in aqueous suspensions under irradiation by green light. J Am Chem Soc 134:14526–14533. https://doi.org/10.1021/ja305225s

131. Wei X, Wang X, Pu Y, Liu A, Chen C, Zou W, Zheng Y, Huang J, Zhang Y, Yang Y, Naushad M, Gao B, Dong L (2021) Facile ball-milling synthesis of CeO2/g-C3N4 Z-scheme heterojunction for synergistic adsorption and photodegradation of methylene blue: characteristics, kinetics, models, and mechanisms. Chem Eng J 420:127719. https://doi.org/10.1016/j.cej.2020.127719
132. Vignesh S, Chandrasekaran S, Srinivasan M, Anbarasan R, Perumalsamy R, Arumugam E, Shkir M, Algarni H, AlFaify S (2022) TiO2-CeO2/g-C3N4 S-scheme heterostructure composite for enhanced photo-degradation and hydrogen evolution performance with combined experimental and DFT study. Chemosphere 288:132611. https://doi.org/10.1016/j.chemosphere.2021.132611
133. Wen X-J, Niu C-G, Zhang L, Liang C, Zeng G-M (2018) A novel Ag2O/CeO2 heterojunction photocatalysts for photocatalytic degradation of enrofloxacin: possible degradation pathways, mineralization activity and an in depth mechanism insight. Appl Catal B Environ 221:701–714. https://doi.org/10.1016/j.apcatb.2017.09.060
134. Magdalane CM, Kaviyarasu K, Vijaya JJ, Siddhardha B, Jeyaraj B, Kennedy J, Maaza M (2017) Evaluation on the heterostructured CeO2/Y2O3 binary metal oxide nanocomposites for UV/Vis light induced photocatalytic degradation of Rhodamine - B dye for textile engineering application. J Alloys Compd 727:1324–1337. https://doi.org/10.1016/j.jallcom.2017.08.209
135. Kohantorabi M, Gholami MR (2017) MxNi100−x (M = Ag, and Co) nanoparticles supported on CeO2 nanorods derived from Ce–metal organic frameworks as an effective catalyst for reduction of organic pollutants: Langmuir-Hinshelwood kinetics and mechanism. New J Chem 41:10948–10958. https://doi.org/10.1039/C7NJ03009F
136. Kohantorabi M, Gholami MR (2017) Kinetic analysis of the reduction of 4-nitrophenol catalyzed by CeO2 nanorods-supported CuNi nanoparticles. Ind Eng Chem Res 56:1159–1167. https://doi.org/10.1021/acs.iecr.6b04208
137. Latha P, Karuthapandian S (2017) Novel, facile and swift technique for synthesis of CeO2 nanocubes immobilized on zeolite for removal of CR and MO dye. J Clust Sci 28:3265–3280. https://doi.org/10.1007/s10876-017-1292-z
138. Zhang S, Guo J, Zhang W, Gao H, Huang J, Chen G, Xu X (2021) Dopant and defect doubly modified CeO2/g-C3N4 nanosheets as 0D/2D Z-scheme heterojunctions for photocatalytic hydrogen evolution: experimental and density functional theory studies. ACS Sustain Chem Eng 9:11479–11492. https://doi.org/10.1021/acssuschemeng.1c03683
139. Zhu C, Wang Y, Jiang Z, Xu F, Xian Q, Sun C, Tong Q, Zou W, Duan X, Wang S (2019) CeO2 nanocrystal-modified layered MoS2/g-C3N4 as 0D/2D ternary composite for visible-light photocatalytic hydrogen evolution: Interfacial consecutive multi-step electron transfer and enhanced H2O reactant adsorption. Appl Catal B Environ 259:118072. https://doi.org/10.1016/j.apcatb.2019.118072
140. Hao Y, Li L, Zhang J, Luo H, Zhang X, Chen E (2017) Multilayer and open structure of dendritic crosslinked CeO2-ZrO2 composite: enhanced photocatalytic degradation and water splitting performance. Int J Hydrog Energy 42:5916–5929. https://doi.org/10.1016/j.ijhydene.2017.01.093
141. Li X, Li Z, Yang X, Jia L, Qing FuY, Chi B, Pu J, Li J (2017) First-principles study of the initial oxygen reduction reaction on stoichiometric and reduced CeO2(111) surfaces as a cathode catalyst for lithium–oxygen batteries. J Mater Chem A 5:3320–3329. https://doi.org/10.1039/C6TA10233F
142. Tang T, Ding L, Yao Z-C, Pan H-R, Hu J-S, Wan L-J (2022) Synergistic electrocatalysts for alkaline hydrogen oxidation and evolution reactions. Adv Funct Mater 32:2107479. https://doi.org/10.1002/adfm.202107479
143. Chauhan S, Mori T, Masuda T, Ueda S, Richards GJ, Hill JP, Ariga K, Isaka N, Auchterlonie G, Drennan J (2016) Design of low Pt concentration electrocatalyst surfaces with high oxygen reduction reaction activity promoted by formation of a heterogeneous interface between Pt and CeOx nanowire. ACS Appl Mater Interfaces 8:9059–9070. https://doi.org/10.1021/acsami.5b12469

144. Demir E, Akbayrak S, Önal AM, Özkar S (2019) Ceria supported ruthenium(0) nanoparticles: highly efficient catalysts in oxygen evolution reaction. J Colloid Interface Sci 534:704–710. https://doi.org/10.1016/j.jcis.2018.09.075
145. Fang G, Cai J, Huang Z, Zhang C (2019) One-step electrodeposition of cerium-doped nickel hydroxide nanosheets for effective oxygen generation. RSC Adv 9:17891–17896. https://doi.org/10.1039/C9RA02682G
146. Zhang R, Ren X, Hao S, Ge R, Liu Z, Asiri AM, Chen L, Zhang Q, Sun X (2018) Selective phosphidation: an effective strategy toward CoP/CeO2 interface engineering for superior alkaline hydrogen evolution electrocatalysis. J Mater Chem A 6:1985–1990. https://doi.org/10.1039/C7TA10237B
147. Ma G, Du X, Zhang X (2021) Selective sulfuration, phosphorization and selenylation: a universal strategy toward Co-Ni-M@CeO2/NF (M = O, S, P and Se) interface engineering for efficient water splitting electrocatalysis. J Alloys Compd 864:158486. https://doi.org/10.1016/j.jallcom.2020.158486

MIX
Papier aus verantwortungsvollen Quellen
Paper from responsible sources
FSC® C105338

If you have any concerns about our products,
you can contact us on
ProductSafety@springernature.com

In case Publisher is established outside the EU,
the EU authorized representative is:
Springer Nature Customer Service Center GmbH
Europaplatz 3, 69115 Heidelberg, Germany

Printed by Libri Plureos GmbH
in Hamburg, Germany